Sustainable Civil Infrastructures

Sustainable Civil Infrastructures (SUCI) is a series of peer-reviewed books and proceedings based on the best studies on emerging research from all fields related to sustainable infrastructures and aiming at improving our well-being and day-to-day lives. The infrastructures we are building today will shape our lives tomorrow. The complex and diverse nature of the impacts due to weather extremes on transportation and civil infrastructures can be seen in our roadways, bridges, and buildings. Extreme summer temperatures, droughts, flash floods, and rising numbers of freeze-thaw cycles pose challenges for civil infrastructure and can endanger public safety. We constantly hear how civil infrastructures need constant attention, preservation, and upgrading. Such improvements and developments would obviously benefit from our desired book series that provide sustainable engineering materials and designs. The economic impact is huge and much research has been conducted worldwide. The future holds many opportunities, not only for researchers in a given country, but also for the worldwide field engineers who apply and implement these technologies. We believe that no approach can succeed if it does not unite the efforts of various engineering disciplines from all over the world under one umbrella to offer a beacon of modern solutions to the global infrastructure. Experts from the various engineering disciplines around the globe will participate in this series, including: Geotechnical, Geological, Geoscience, Petroleum, Structural, Transportation, Bridge, Infrastructure, Energy, Architectural, Chemical and Materials, and other related Engineering disciplines.

SUCI series is now indexed in SCOPUS and EI Compendex.

Bhalchandra V. Khode · Sudarshan Kurwadkar ·
P. Y. Pawade · Sujesh D. Ghodmare ·
Hasim Ali Khan
Editors

Innovations in Technologies: Pioneering Sustainable Infrastructure for a Resilient Future

Infrastructural Development and Sustainable Materials

Springer

Editors
Bhalchandra V. Khode
Department of Civil Engineering
G H Raisoni College of Engineering
Nagpur, Maharashtra, India

P. Y. Pawade
Department of Civil Engineering
G H Raisoni College of Engineering
Nagpur, Maharashtra, India

Hasim Ali Khan
Department of Civil Engineering
G H Raisoni College of Engineering
Nagpur, Maharashtra, India

Sudarshan Kurwadkar
Department of Civil and Environmental Engineering
California State University
Fullerton, CA, USA

Sujesh D. Ghodmare
Department of Civil Engineering
G H Raisoni College of Engineering
Nagpur, Maharashtra, India

ISSN 2366-3405 ISSN 2366-3413 (electronic)
Sustainable Civil Infrastructures
ISBN 978-3-031-82132-5 ISBN 978-3-031-82133-2 (eBook)
https://doi.org/10.1007/978-3-031-82133-2

This Springer imprint is published by the registered company Springer Nature Switzerland AG
The registered company address is: Gewerbestrasse 11, 6330 Cham, Switzerland

Foreword

It is a great pleasure to present the Sustainable Civil Infrastructures (SUCI) Springer Book Series, featuring proceedings from the 5th International Conference on "Recent Trends in Infrastructural Development and Sustainable Materials (IC-RTIDSM-2024)" held on 9 and 10 February 2024 at G. H. Raisoni College of Engineering, Nagpur. This conference brings together leading scholars, industry experts, and emerging researchers from around the world to share their latest findings, innovative ideas, and cutting-edge developments in the fields of infrastructural development and sustainable materials.

In an era of unprecedented change and rapid technological advancements, fostering a collaborative environment for knowledge exchange is more important than ever. This conference serves as a crucial platform for intellectual dialogue, promoting the cross-pollination of ideas that drive progress and innovation. The diversity of perspectives and the richness of discussions at this event are a testament to the vibrant and dynamic nature of our field.

The excellent caliber of research presented at the conference is reflected in the articles included in this book. They span a wide range of subjects, from theoretical investigations to real-world applications, addressing some of the most pressing challenges and opportunities we face today. Each contribution has been rigorously peer-reviewed to ensure it meets the highest standards of academic excellence.

I sincerely appreciate the diligence and commitment of the authors, as well as the insightful and constructive criticism of the reviewers.

It is my hope that the insights and discoveries contained within these pages will inspire further research, spark new ideas, and foster continued dialogue within our community. As you read through these proceedings, we encourage you to engage with the content, reflect on the findings, and consider how they might influence your own work.

I would like to thank everyone for the participation and for contributing to the enrichment of the proceeding of the Recent Trends in Infrastructural Development and Sustainable Materials (IC-RTIDSM-2024). Expecting the continued advancement of our field and the exciting possibilities that lie ahead.

Dr. Sachin Untawale
Director
G H Raisoni College of Engineering
Nagpur, India

Preface

We are immensely honored to assume the role of editors for the esteemed proceedings of the "International Conference on Recent Trends in Infrastructural Development and Sustainable Materials (IC-RTIDSM-2024), hosted by the Civil Engineering Department of G H Raisoni College of Engineering, Nagpur, conducted on 9 and 10 February, 2024. This monumental event, facilitated by Springer Nature, marks a pivotal moment in the discourse surrounding contemporary Civil Engineering. IC-RTIDSM-2024 emerges as a beacon of collaborative endeavor, drawing together luminaries from academia, industry stalwarts, research scholars, and budding engineering graduates. At its core, the conference endeavors to dissect and deliberate upon scientific inquiries germane to infrastructure technologies, environmental sustainability, smart transportation and the deployment of sustainable materials. As guardians of this scholarly forum, we find ourselves humbled by the prospect of fostering dialogue on the myriad facets of sustainable infrastructure development.

The genesis of IC-RTIDSM-2024 is rooted in a profound commitment to amalgamate diverse research and innovation endeavors, thereby nurturing the seeds of sustainable infrastructural growth. By providing a fertile ground for deliberation, the conference aspired to engender holistic solutions to the environmental exigencies precipitated by rapid urbanization. Within the confines of this intellectually charged arena, academia, industry luminaries, governmental policymakers, and community stakeholders contributing their unique perspective towards the shared goal of sustainable development. It is incumbent upon us to recognize the imperative for an integrated approach to development, one that transcends disciplinary boundaries and embraces the ethos of sustainability.

IC-RTIDSM-2024 serves as an apotheosis of this collective endeavor, offering a nexus for the exchange of innovative ideas, groundbreaking research, and pragmatic solutions. With a keen eye towards recent trends and advancements, the conference champions the cause of smart technologies tailored to the exigencies of modern urban landscapes. As stewards of this seminal event, we extend our heartfelt gratitude to the authors, plenary speakers, participants, and the multifaceted cadre of individuals whose unwavering dedication culminated in the success of

IC-RTIDSM-2024. Special acknowledgment is reserved for the International Advisory Committee, national and international reviewers, technical committee members, and organizing committee of G. H. Raisoni College of Engineering, Nagpur, whose collective efforts paved the way for this momentous occasion.

This book will serve as a valuable reference for students, researchers, and professionals interested in Smart Materials and Sustainability and Sustainable Infrastructure Development IC-RTIDSM-2024.

Nagpur, India

Dr. Sudarshan Kurwadkar
Dr. Bhalchandra V. Khode
Dr. P. Y. Pawade
Dr. Sujesh D. Ghodmare
Dr. Hasim Ali Khan

Acknowledgements

We would like to express our sincere gratitude to all those who contributed to the success of the *5th International Conference on Recent Trends in Infrastructural Development and Sustainable Materials (IC-RTIDSM-2024)*.

First and foremost, we would like to extent our heartfelt thanks to *Springer Publication* for accepting our proposal and providing us an opportunity for publication of conference proceeding in the *Scopus Indexed Springer Book Series; Sustainable Civil Infrastructures (SUCI)*.

We extend our sincere thanks to Dr. Sudarshan Kurwadkar, Professor, Department of Civil and Environmental Engineering California State University—Fullerton, CA, whose invaluable guidance was instrumental for the conduction of this conference and delivering insightful keynote address. The support and encouragement of all the international and national advisory committee members was remarkable.

We are also deeply grateful to Patron Shri. Sunil Raisoni, Chairman, Raisoni Group Institutes, Smt. Shobha Raisoni, Trustee, Raisoni Group of Institutions, Mr. Shreyas Raisoni, Executive Director, Raisoni Group of Institutions and Dr. Sachin Untawale, honorary chair and Director, G. H. Raisoni College Engineering, Nagpur, for providing the necessary resources and a conducive environment for this conference.

We are highly indebted to Dr. S. B. Jaju, Dy. Director and Dean R&D, Dr. P. V. Walke, Dean Academics, G. H. Raisoni College Engineering, Nagpur, for his valuable suggestions throughout the conference.

Special thanks to organizing committee members Dr. P. J. Wadhai, Dr. K. R. Dabhekar, Dr. Isha. P. Khedikar, Dr. S. S. Solanke, Prof. P. B. Patil, Dr. S. S. Sanghai, Dr. M. P. Bhorkar, Prof. N. H. Pitale, Dr. A. A. Gulghane, and Dr. S. Gaur for their untiring efforts and contribution towards successful of conference.

Further, we extend thanks to the all authors for their collaboration and participation, which significantly enriched the quality of this conference.

Furthermore, we appreciate the efforts of the conference reviewers, whose dedication and hard work made this event possible. Their meticulous review and valuable comments helped improve the quality of book.

Nagpur, India

Dr. Sudarshan Kurwadkar
Dr. Bhalchandra V. Khode
Dr. P. Y. Pawade
Dr. Sujesh D. Ghodmare
Dr. Hasim Ali Khan

Contents

Chapter 1
Finite Element Analysis of Composite Slab with Profiled Steel Deck and Comparison with Conventional Slab

Tejaswi Patode and Prashant Hiwase

Abstract Composite deck slab floors with profiled steel decking are becoming common place in many nations due to their ability to facilitate quicker, lighter, and more affordable construction. In order to enhance its shear-bond properties, rolled embossments are incorporated into the concrete as well as steel deck to form the composite interaction that creates the slab. The strength of the contact between the concrete layer and profiled steel sheeting determines the composite deck slabs' strength. A composite deck slab with a profiled steel sheet is made by pouring concrete over the steel sheet. Two primary purposes of a profiled steel sheet are that it serves as formwork during the concrete-setting process and as tension reinforcement once the concrete has solidified. Before moving further with the design of the entire composite slab, it is imperative to verify the sheet's suitability for use during the building phase. This study is focused on examining the effects of various parameters on the behavior of composite slabs. ANSYS software has been used to create 3-D Finite Element Models of Conventional Slab and Composite slabs. ANSYS was used to analyze these models in order to examine their structural behavior and functionality and it demonstrates their ability as sustainable alternatives to the standard slab. The use of Profile steel sheet resulted in a decrease of Volume by about 29% when compared to Conventional slab which further resulted in 3% lesser Deformation. Moreover, the stress experienced by the Composite slab decreased by about 32%. These results demonstrate the structural and financial advantages of using these cutting-edge, environmentally friendly building techniques.

Keywords Composite slab · Trapezoidal steel deck · Conventional slab · ANSYS · Finite element analysis

T. Patode (✉) · P. Hiwase
Shri Ramdeobaba College of Engineering and Management, Nagpur, Maharashtra, India
e-mail: patodetk@rknec.edu

P. Hiwase
e-mail: hiwasep1@rknec.edu

B. V. Khode et al. (eds.), *Innovations in Technologies: Pioneering Sustainable Infrastructure for a Resilient Future*, Sustainable Civil Infrastructures,
https://doi.org/10.1007/978-3-031-82133-2_1

1.1 Introduction

1.1.1 General

These days, composite slabs—a mixture of profiled steel sheets and concrete—are frequently utilized in building construction. A composite slab can be described as follows: "a slab in which profiled steel sheets are used as permanent shuttering initially and then combine structurally with the hardened concrete and act as tensile reinforcement in the finished floor." The elements of a composite slab with profiled steel decking are shown in Fig.1.1.

Concrete in a composite steel floor deck acts compositely with the metal sheet after the concrete has solidified. For the steel slab to function as tensile reinforcement, there must be a sufficient transfer of horizontal shear forces between it and the concrete slab for the composite action to occur. The bending movement causes the concrete and the reinforcement to separate vertically in addition to horizontally through shearing forces. As such, the profiled sheet needs to be engineered not only to transfer the horizontal shears but also to withstand vertical separation. Embossments or stiffeners are used to achieve resistance to vertical separation.

Advantages of Composite Slab Construction

1. Once in place and before concrete is poured, the steel deck instantly becomes a platform to support construction loads, providing a stable and safe surface for construction workers.
2. By acting as a formwork for in-situ concrete, steel decking reduces the need for costly, time-consuming removable formwork construction.
3. Once in service, the steel deck acts as a tensile steel, eliminating the need for time-consuming rebar placement and attachment to the slab. The only things that need to be increased are shrinkage, temperature, and negative moments.
4. The form of the bridge deck profile, which can be done very efficiently, can help reduce by about 13% the amount of backfilling concrete required for the deck. The result is a significant reduction in dead weight leading to a lighter superstructure and reduced load on the foundation.

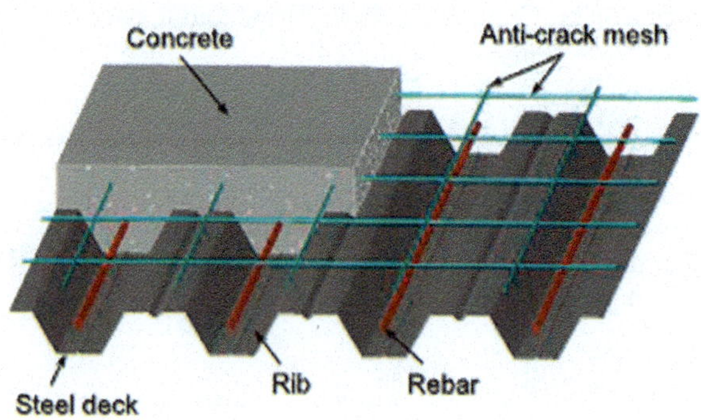

Fig. 1.1 Components of composite slab

5. The cellular shape of the bridge allows the formation of conduit cells in the flooring so that services (telephone, plumbing, communications wiring, etc.) can be integrated and distributed within the depth of the floor. This increases the space or reduces the height of the building.
6. Steel bridge is made of light gauge steel plate, lightweight, and easy to handle during transportation and installation.
7. The aesthetics of the bottom plate surface have been significantly improved by reducing the plate finishing work.
8. This decreases development time since the pouring of adjoining floors can proceed without having to hold up for already poured floors to pick up quality.
9. Significantly increases the construction economy. With the above advantages, composite panels have significantly improved the competitiveness of steel frame projects.

1.2 Literature Review

1.2.1 Introduction

The literature shows the considerable work and study on a composite slab with a profiled steel sheet and its behavior under flexural loads. Several experimental numerical analyses were performed on the composite slab by various authors. This chapter aims to review of literature related to flexure tests on composite slabs with profiled steel decking. These works are reviewed keeping in view the methodology, principles, and various aspects related to composite slabs with steel profiled deck. Some of the related works are discussed below.

Review of Numerical and Analytical Work

The literature that has been referred from various research papers is illustrated and the brief review of work related to the performance of flat slab structures studied by the researcher is presented below.

In a 2013 publication, Abas and Gilbert et al. discussed the strength as well as usability of continuous composite panels that combine steel fiber reinforced concrete and deep trapezoidal steel decks. By adjusting the steel fibers with and without steel mesh in the panel's negative moment area, a test program on two-span continuous composite panels was carried out to measure the strength and serviceability of SFRC composite panels. A comprehensive evaluation is carried out on the impact on deflection, crack width, and moment redistribution under service load.

An article examining the longitudinal shear behavior of steel and concrete composite panels was presented by Johnson and Shepherd (2013). The suggested approach assesses how the stiffness of the concrete layer and the slab in its entirety is affected by the plastic deformation of the concrete layer and concrete cracks. It is feasible to investigate the behavior of the contact from the beginning of loading to rupture by examining the deformation and resistance to contact between the steel plate and concrete. Experimental research was done on the resistance of the cross-section of six composite panel pieces made of profiled steel panels of the "Holorib" type. This investigation included theoretical estimations of the test slab's cross-sectional resistance.

An investigation report on the shear bond characteristics of composite panels in accordance with Eurocode 4 was provided by Cifuentes and Medina (2013). In order to examine the shear bond characteristics of composite panels and assess the Eurocode's requirements for the panels' longitudinal shear resistance, this study involves testing thirty composite panels. Analysis was done on the partial shear and m–k shear connection techniques. Up to three distinct thicknesses of two distinct types of profiled steel plates were taken into consideration. The study evaluated the influence of previous cyclic loading and the location of cracking agents on the manufactured part. resistance to vertical cutting. Bending tests with different thicknesses of steel plates were performed.

In 2014, Gholamhoseini and Gilbert et al. submitted a paper on adhesion-slip correlations and longitudinal shear stress in composite concrete slabs. The findings of short-term damage testing on four popular profiled flooring types in Australia are presented in the current investigation. Large, simply supported specimens with span/4 or span/6 shear spans were subjected to four-point bending tests. Testing establishes the bond-slip relationship. The finite element model, which models the bond qualities between a steel slab and concrete, is put forth and validated by experimental data through the use of interface elements.

A paper on the experimental investigation of concrete slabs with curved steel decks was presented by Manjunath and Sureshchandra (2014). This paper's primary goals are to examine the behavior of composite panels in bending tests (both with and without embossed terrace panels); to ascertain the impact of bond strength between chemical and mechanical bonds through end-slip testing, and to examine the behavior of non-fine-grained concrete and design mixtures under composite action. AralditeGY257 IN and Aradur 140 adhesive chemicals were employed in a 2:1 ratio in the experiment. Ten simply supported composite deck panels were tested in the test program until two points at L/4 of the span were reached when linear loading was applied. The composite panels' bending characteristics, ultimate strength, and bond strength are determined.

Merool et al. (2014) presented an experimental research paper on composite decks with different aspect ratios. In this paper, two synthetic spherical samples with different aspect ratios of one wavelength width and three wavelength widths were tested. With bolts to increase adhesion and load-bearing capability, the composite interaction of the steel deck and concrete forms the panels. Experiments were performed to estimate the load-bearing capacity of composite decks with bolts as shear connectors. A comparison of the load-carrying capacity is made for one-wavelength and three-wavelength plates with bolts as shear connectors.

A cold-formed steel composite metal bridge plate was the subject of an experimental investigation by Patil and Shahezad (2014). Composite deck panels are a complex form of metal decking utilized in tensile regions that are cast with concrete panels. The concrete supports the weight of use, allowing the deck to function as a composite material. Composite profiles have a highly intricate behavior that is directly influenced by their contact resistance and deformation capacity. Because of slippage at the interface between the profile and the concrete member, the precise nature of the bond between concrete and steel is still not well understood. Examining the structural characteristics of composite panels with stainless steel decks is the primary goal of this paper.

The paper on composite concrete panels with profiled steel decks: A comparison between experimental study and simulation research was presented by Hedaoo and Raut in 2015. Through simulation and experimental research, the investigation of composite concrete slabs with profiled steel decks of the CRIL DECKSPAN type is part of the research study. The steel deck with multiple-layer relief panels and concrete combine synthetically to form the slab. However, because of the intricate shear phenomenon, it does not enhance the adhesion qualities. Consequently, in accordance with Eurocode 4: Part 1.1, full-scale experimental tests were conducted to examine the shear and flexural strength. The eighteen specimens were split up into six sets of three, and each set was tested on simply supported slabs under both static and cyclic loading conditions with varying shear span lengths. This paper presents full-scale finite element (FE) modeling and analysis of composite panels. Contact elements display is used to simulate the shear bond interaction between the concrete and steel deck. Comparing the experimental data validates and verifies the FE analysis.

1.2.2 Concluding Remark

The literature is reviewed and composite panels with profiled steel decks are studied. Various parameters are studied. Research composite panel design methods, mk method, and PSC method. Many experimental and analytical works have been done by many investigators in the field of nonlinear analysis of composite panels with profiled steel decks. The concept of composite panels with different profile shapes is currently being rapidly developed. The design methods are based on full-scale composite panel tests; Therefore, many researchers have conducted experimental tests to determine the longitudinal shear force. However, limited work has been done on software that determines the state of composite panels with different parameters. In this thesis, we try to study the behavior of composite panels with trapezoidal steel deck using ANSYS.

1.3 Methodology

1.3.1 Introduction

In India, in recent years, the application of cold-formed steel decking as structural components in deck construction has increased. Profiled steel deck panels are cold-formed from low-thickness, high-strength panels. These sections are subjected to more complex bending and nonlinear behavior than hot-rolled sections; therefore, their use requires further research and development. The lightest, most efficient, and economical deck system is created with a mixture that combines the structural properties of concrete and profiled steel decks.

- This system aims to offer economic advantages by reducing concrete consumption while ensuring comparable structural integrity and performance to the traditional slab.
- The study will analyze and compare this alternative to determine its viability and effectiveness as a cost-effective and sustainable solution for modern construction projects.
- This research project's primary goal is to identify a more affordable substitute for the traditional slab that would nonetheless have equivalent or superior structural performance.
- A substitute, a composite slab with a profile steel deck, has been selected for this purpose.
- With a lesser use of concrete and equivalent or superior structural integrity and performance to the conventional slab, this technology seeks to provide financial benefits.
- This alternative will be analyzed and contrasted in the study to ascertain its efficacy and viability as an affordable and environmentally friendly option for contemporary building projects.

The Following Tasks Will Be Carried Out Sequentially in Order to Meet the Project's Goals:

1. Compiling the required data for analytical work and carrying out earlier studies on the design of composite slabs using profile steel sheets.
2. Outlining the characteristics of the substance.
3. Determining the slab's final measurements.
4. Using ANSYS Workbench 2022 R2, designing and modelling the conventional slab and composite slab.
5. Comparing the outcomes of the ANSYS program, such as volume change, stress intensities, and deflection behavior.
6. Selecting the best slab from the conventional slab and composite slab in order to draw a conclusion.

1.4 Material Properties

Tables 1.1 and 1.2, respectively, contain a list of the features of the Profile steel deck and Concrete that were considered throughout the model's development.

1.4.1 Geometry of Composite Slab and Conventional Slab

Conventional Slab. A 3D slab model was created in ANSYS with dimensions of 4000 × 3000 × 125 mm. The volume of the slab is 1.5 m^3

Composite Slab with Profiled Steel Sheet. The size of the slab was kept the same as that of conventional slab i.e.,4000 × 3000 × 125 mm.

Common Parameters

Loading—The loading condition was kept the same for both the slabs which is shown in Figs. 1.6 and 1.8.

- Dead Load- 1 KN/m^2 (Including Gravity)
- Live Load—5 KN/m^2 (As pe IS 875 Part 2: 1987)
- Considering Factor of Safety 1.5.
- Total Load = 1.5 x (D.L. + L.L.) i.e., 9 KN/m^2

Reinforcement–As the slab is a two-way slab the distribution bars also act as Main reinforcement.8 mm ϕ bars @ 300 mm c/c were used while modelling of composite slab.

Support Condition–Fixed support was provided on all four sides of the slab panel, i.e. the slab was considered to be resting on beams on all four sides. This is shown in Fig. 1.5.

Table 1.1 Properties of concrete

Concrete properties	Values
Grade	M20
Density	25 KN/m^3
Characteristics compressive stress	20 N/mm^2
Elastic modulus	21,572 N/mm^2
Poisson's ratio	0.3

Table 1.2 Properties of profiled steel deck

Steel decking properties	Values
Yield stress	250 N/mm^2
Density	78.50 KN/m^3
Elastic modulus	210,000 N/mm^2
Poisson's ration	0.3

Meshing-The concrete in the composite slab was modelled in ANSYS using Tetrahedral elements—Tet10. The reinforcement as well as the profile sheet was modelled using the Hexahedral Elements-Hex8. The meshing used in the composite slab is shown in Fig. 1.7 whereas that used in conventional slab is shown in Fig. 1.9.

1.4.2 Modelling

The modeling of the Composite slab and the Conventional slab is depicted in the following figures, which also demonstrate how the components are arranged and configured within the slab.

Composite Slab. The cross-section of the profile steel sheet used for the research purpose is displayed in Fig. 1.2. Figure 1.3 shows the profile steel sheet modelled in ANSYS whereas Fig. 1.4 shows the arrangement of steel reinforcement and profile sheet.

Conventional Slab. Figure 1.8 shows the loading condition of conventional slab and meshing is shown in Fig. 1.9.

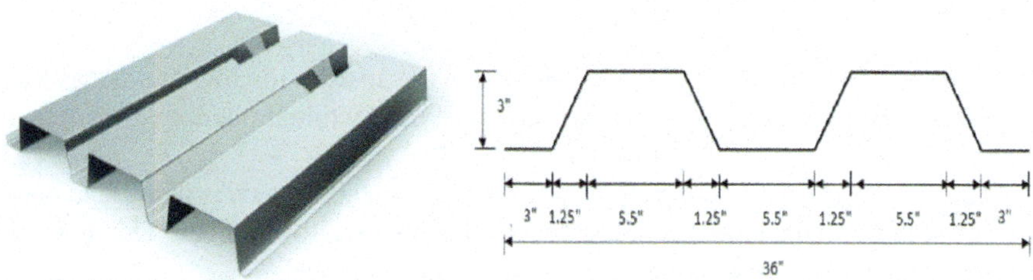

Fig. 1.2 Profile sheet *Source* S.M. Zahurul Islam et al. Structural strength and behavior of profiled steel sheet-concrete composite slab

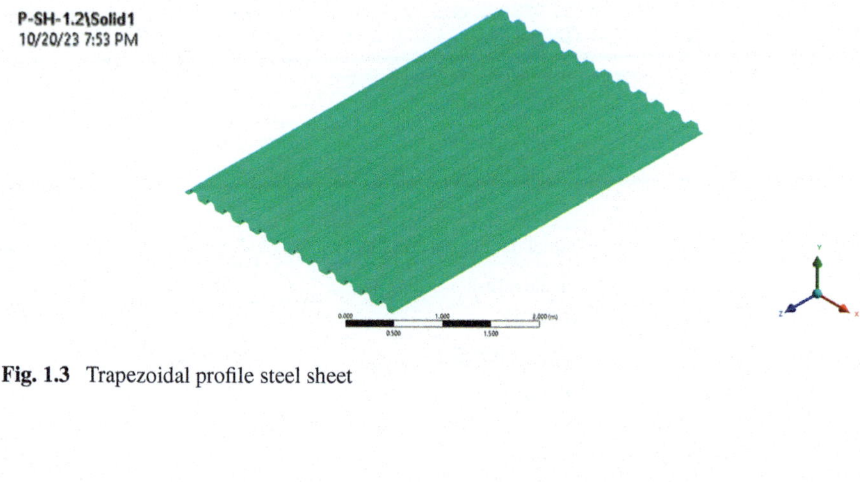

Fig. 1.3 Trapezoidal profile steel sheet

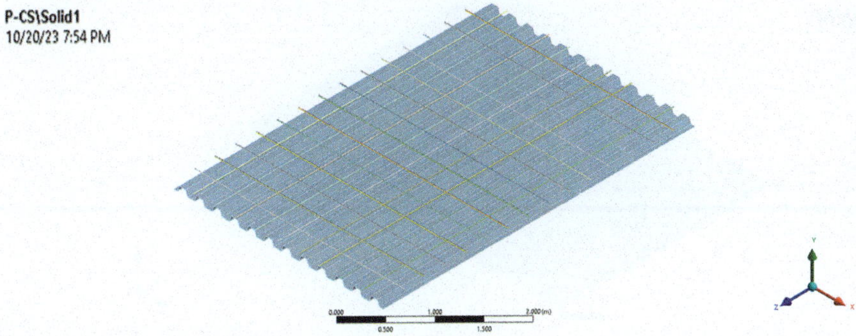

Fig. 1.4 Reinforcement placed on profile steel sheet

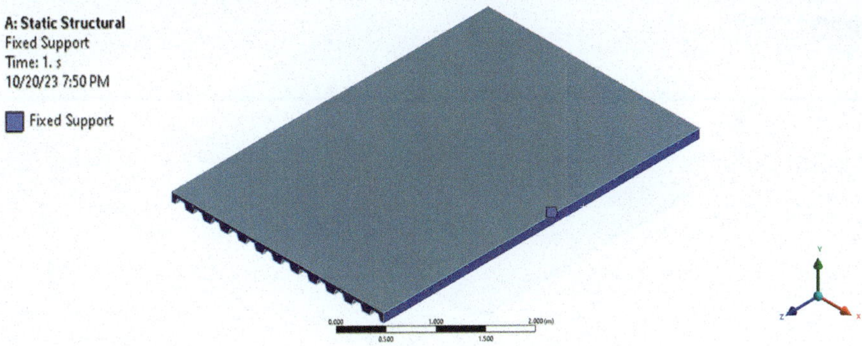

Fig. 1.5 Fixed support

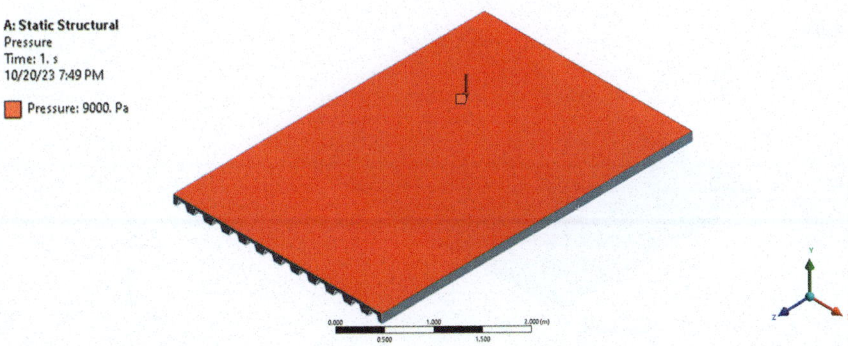

Fig. 1.6 Loading condition

Fig. 1.7 Meshing of composite slab

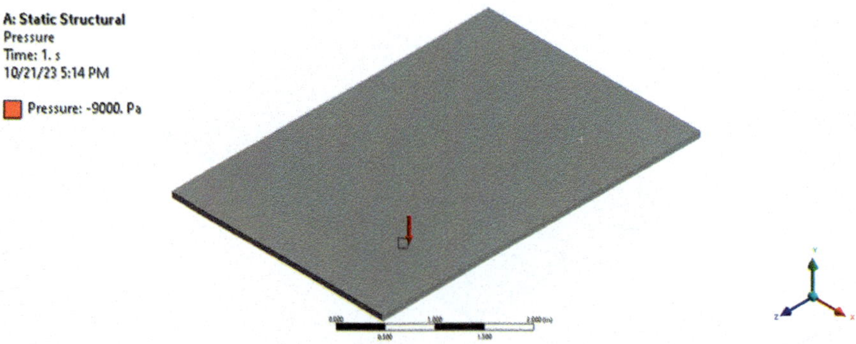

Fig. 1.8 Conventional slab

Fig. 1.9 Meshing of conventional slab

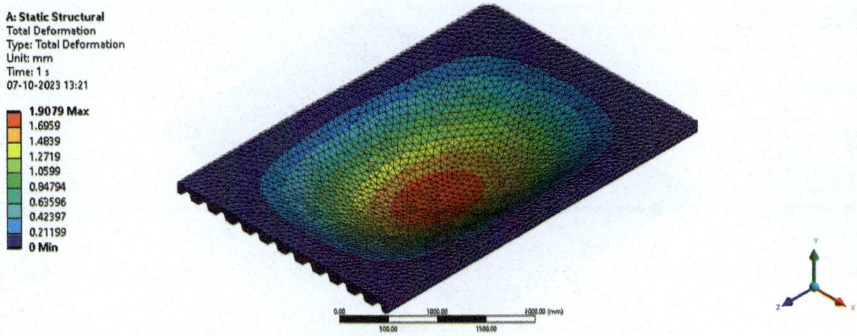

Fig. 1.10 Total deformation of composite slab

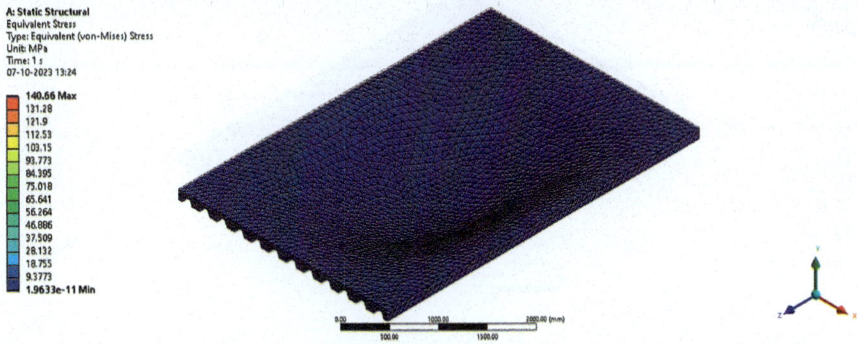

Fig. 1.11 Equivalent elastic stress of composite slab

1.5 Performance Analysis

Here, the composite slab with profiled steel deck and conventional slab is modelled and analyzed in ANSYS 2022 R2. The deformed shape, stress–strain values, and volume of concrete are determined. The results obtained from ANSYS for the total deformation of the composite slab and conventional slab are displayed in Figs. 1.10 and 1.12 correspondingly. Figures 1.11 and 1.14 show the Equivalent elastic stress of composite slab and conventional slab respectively (Figs. 1.13 and 1.15).

1.6 Results

- The Deformation of the composite slab with 0.8 mm thick profiled steel sheet was found to be 1.9079 mm whereas that of the conventional slab was 1.965 mm.

Fig. 1.12 Equivalent elastic strain of composite slab

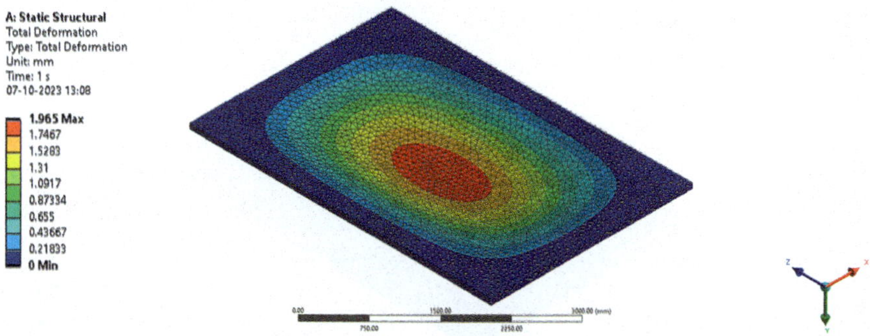

Fig. 1.13 Total deformation of conventional slab

Fig. 1.14 Equivalent elastic stress of conventional slab

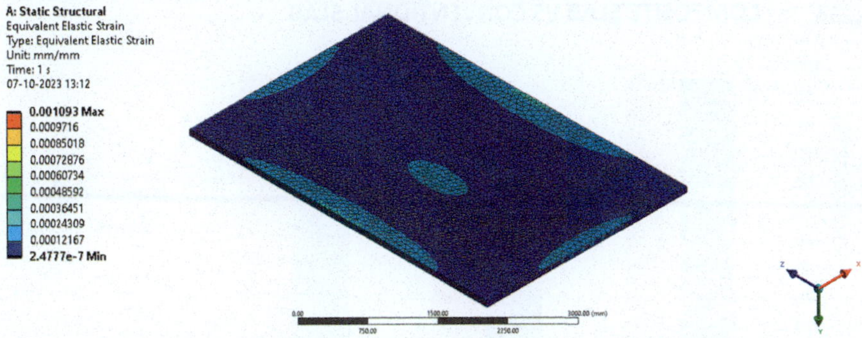

Fig. 1.15 Equivalent elastic strain of conventional slab

Table 1.3 Comparison between composite slab and conventional slab

Parameters	Composite slab	Conventional slab
Total deformation (mm)	1.9079	1.965
Equivalent stress (MPa)	140.66	206.33
Volume (m^3)	1.066	1.5

- The Equivalent Stress of the composite slab with 0.8 mm thick profiled steel sheet was found to be 140.66 MPa whereas that of conventional slab was 206.33 MPa.

1.7 Conclusion

Table 1.3 shows the final results for the conventional slab and the composite slab. It can be seen that both slabs' deflections are within acceptable bounds. We can determine which slab is best suited for use in construction using the study's findings. Finding the best option based on considerations like cost-effectiveness, structural integrity, and sustainability is made easier by the research's detailed examination of each slab type's performance (Fig. 1.16).

(1) Smaller deflection values indicate better structural performance. The Composite slab deflects 3% less than the conventional slab.
(2) It is preferred to use slabs with lower equivalent stress values. There is 32% less stress on the composite slab than there is on the conventional slab.
(3) The Composite Slab results in overall cost savings since it uses 29% less concrete than Conventional slabs when comparing the amounts of concrete used.
(4) The benefits of the composite slab over the conventional slab regarding deflection performance and lower concrete volume make it a better option overall, according to the findings.

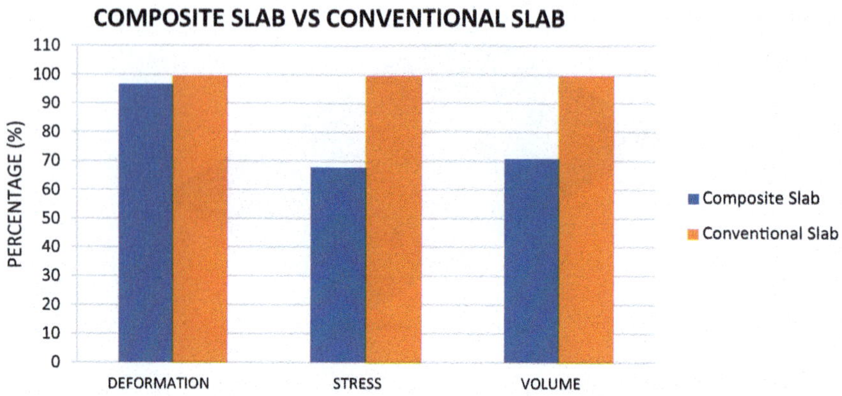

Fig. 1.16 Comparison between composite slab and conventional slab

References

Abas FM, Gilbert RI, Bradford MA (2013) Strength and serviceability of continuous composite slabs with deep trapezoidal steel decking and steel fibre reinforced concrete. Eng Struct 49:866–875

Cifuentes H, Medina F (2013) Experimental study on shear bond behaviour of composite slabs. J Constr Steel Res

Eurocode 4 (2003) Design of composite steel and concrete structures–Part 1.1: general rules and rules for buildings. EN 1994-1.1: 2003, European Committee for Standardization, Brussels

Ghalomhoseini A, Gilbert RI, Bradford MA, Chang ZT (2014) Longitudinal shear stress and bond slip relationship in composite concrete slabs. Eng Struct

Namdeo A. Hedaoo, Laxmikant M. Gupta, Girish N. Ronghe (2012) Design of composite slabs with profiled steel decking: a comparison between experimental and analytical studies. Int J Adv Struct Eng

Jhonson RP, Shepford AJ (2013) Resistance to longitudinal shear of composite slabs with longitudinal reinforcement. J Constr Steel Res

Manjunath TN, Sureshchanadra BS (2014) Experimental study on concrete slab with profiled steel decking. Int J Eng Res Technol (IJRET) 3(7). ISSN: 2278-0181

Merool V, Harshvadan KMP (2014) Experimental study on composite deck with different aspect ratio 3(2)

Patil S, Shahezad M (2014) Experimental study on cold formed steel composite metal deck slab. Int J Eng Sci Res Technol. ISSN: 2277-9655

Chapter 2
Effect of Sugarcane Juice and Its Products on the Setting Time of Cement When Used as Admixtures

Sudhir Sharma and Pushpendra Kumar Sharma

Abstract Concrete is mostly used in building construction and large-scale projects all around the world. Sand, aggregate, cement, and water are the ingredients of concrete. The admixture is used to enhance the qualities of concrete in addition to these components. The main factor in hardening or setting is cement. Hydration, which happens as a result of cement and water reacting chemically, leads to hardening. It is crucial to modify the setting process's pace in order to accommodate the unique requirements and weather circumstances. Chemical admixtures are used to achieve this goal. Therefore, Cement's initial setting time can be increased or decreased by using chemical admixtures such as retarders or accelerators. In construction projects, chemical admixtures are often used to improve the properties of concrete. Contrarily, chemical admixtures pose a risk to human health as well as the environment. Sugarcane is completely natural, has no negative environmental impact, and is completely safe for humans. In this study, the effect of using sugarcane juice and its products as organic admixtures on cement setting time is investigated.

Keywords Sugarcane juice · Chemical admixtures · Initial setting time of cement

2.1 Introduction

Concrete is formed when water is combined with a mixture of cement and aggregate. The cement undergoes a reaction with the water, and, under typical weather conditions, this mixture gradually solidifies and becomes hard over time. Once the mixture is placed in formwork and undergoes the curing process, it solidifies into a rigid structure similar to stone. The mixture hardens because of the way that cement and water

S. Sharma (✉) · P. K. Sharma
School of Civil Engineering, Lovely Professional University, Phagwara, Punjab, India
e-mail: nl.sudhir@gmail.com

P. K. Sharma
e-mail: pushpendra.23437@lpu.co.in

B. V. Khode et al. (eds.), *Innovations in Technologies: Pioneering Sustainable Infrastructure for a Resilient Future*, Sustainable Civil Infrastructures,
https://doi.org/10.1007/978-3-031-82133-2_2

react chemically (Neville and Brooks 2004). In almost all construction projects, concrete is used as a building material (Memon and Memon 2022). If certain standards and procedures are followed, such as taking temperature into account, construction tasks can be completed. The desired quality can be attained by carefully following detailed procedures. Engaging in physical activities involving concrete speeds up the cement hydration process at temperatures above 100°F. In hot weather, concrete gains initial strength quickly, but over time, it loses strength noticeably. Plastic shrinkage from the rapid evaporation of water causes the development of tensile stresses and subsequent cracking (Neville and Brooks 2006; Usman et al. 2012). Similar problems can arise when pouring concrete in a cold environment. As the water hardens into the concrete's plastic state, the concrete's volume rises. The mechanisms by which concrete sets and hardens are slowed down by a lack of water due to the effect it has on chemical reactions. In turn, this causes a large number of pores to form, which reduces the concrete's strength (Neville and Brooks 2006; Usman et al. 2013). Consequently, the admixtures are used to maintain the established standards (Lea 1988). Admixture is the other component used to make concrete besides cement, aggregates, and water (Maheswari et al. 2021). Admixture, referring to substances besides cement, aggregates, and water, is mixed with the concrete either during or just prior to its preparation. Admixture is utilized in order to regulate the properties of the concrete as stated by the requirements of the construction. An admixture is used to enhance the characteristics of the concrete, such as its ability to be shaped and handled, etc., and to address the challenges presented by extremely high or low temperatures (Shah et al. 2014). Despite its many favourable qualities, such as binding, strength, and durability, concrete's usability is limited in varying weather conditions and locations. The time it takes for the initial setting may differ depending on the weather conditions. As a result, retarders and accelerators are utilised in hot and cold weather conditions to prolong or speed up the initial setting time (Kawade et al. 2017a). Accelerator is an admixture that, if mixed with mortar, concrete, or grout, accelerates the pace at which hydraulic cement hydrates, reduces the amount of time it takes to set, or quickens the process of hardening or developing strength (IS 1999). Retarder is an admixture that slows the cement paste's setting and, by extension, the setting of cement-containing materials like mortar or concrete (IS 1999). It is obvious that chemical admixtures (such as retarders and accelerators) are required to enhance concrete's properties; nevertheless, these admixtures are harmful to the environment and cause leaching. When chemical admixtures are exposed, when a structure is demolished and its rubble is discarded, or when its granulates are used as gravel, the ecosystem suffers. Because the chemical admixtures are water-soluble, they cause environmental problems. Chemical admixtures used in the production, storage, shipment, and usage of concrete all contribute to environmental contamination. Because of these constraints, research into bio-admixtures is expanding in order to reduce the cost of synthetic admixtures. It is critical to use a better alternative to chemical admixtures as an admixture that may increase the properties of concrete while posing no environmental risk (Faqe et al. 2020; Patel and Deo 2016; Dandotiya et al. 2022; Togero 2006; Abdulrahman and Mohammad 2011; Chandra et al. 1998; Dwivedi et al. 2008). In high-performance concrete, the admixture's cost is occasionally greater than the cement's cost. Therefore, the advancement of organic admixtures is more important than the advancement of cement for the improvement of concrete (Hanehara and Yamada 1999).

As a result, using natural admixtures to enhance the properties of both hardened and fresh concrete is desirable. Organic admixtures are preferable to artificial admixtures since they are less expensive and better for the environment (Mahmood et al. 2021). Finding substitute admixtures will therefore help create concrete that is ecologically sustainable. Researchers and scientists working on concrete all around the world are trying to create non-chemical admixtures that are thought of as ecologically conscious and environmentally acceptable building materials. Because they are so plentiful and their manufacture is not particularly sophisticated, natural or green admixtures can be a practical choice and enable green building. These combinations are created with fruit, herpes, or animal products. Typically, biopolymers and natural admixtures created by biotechnological procedures are referred to as "bio-admixtures." Although there isn't much study on the subject, using natural admixtures based on fruit or plant extracts looks like a decent option. According to studies on historic structures, lime and plant extract were used in construction (Thirumalini and Sekar 2013). Admixture use has a history that is very similar to that of concrete and mortar. Around 3000 BC, molasses, bananas, fruit juices, dairy products, olive oil, and other natural admixtures were often utilised in India. For a variety of purposes, glycerol, pine resin, protein, reducing agents (bee wax), and Plasticizers (sugar, mineral oils, and olive oils) were all naturally occurring additives that were used to lime mortar in India (Chandra 2003; Kurdowski 2014). Plants were used as admixture in lime mortars in ancient times (Thirumalini and Sekar 2013). Using jaggery, powdered shells, and different types of nine herbs, the age-old Vadakkunnathan temple in Thirrsur, Kerala, India, was retrofitted (Thirumalini and Sekar 2013). The numerous plaster mixture compositions employed in the restoration work at Fort Vettimurichakotta, East Fort, West Fort, Pazhavangadi, Puthen Street, Virakupurakkotta, and Sreevaraham in India were identified in a manuscript of palm leaf recovered in the Padmanabhapuram Palace. A mixture of ingredients, including different fruits and herbs as well as a specific kind of cactus, was combined with palm jaggery and allowed to ferment for 15 days. The plaster was prepared by combining this mixture with lime (Thirumalini and Sekar 2013). Jaggery was a common admixture in ancient times (Michelle 2009). The building of the England-France Channel has long employed molasses (sugar) as a retarder (Neville 1995). Molasses can be added to building materials as a water reducer, retarder, and grinding aid. Molasses has been widely used in China as a cement and concrete admixture for water reduction and retardation, as well as a grinding aid (Olbrich 2006).

2.2 Materials

2.2.1 *Cement*

The term cement is used to define a substance that undergoes a reaction with water, resulting in the formation of a solid, durable mass. In the general sense, cement is typically described as the ability of any substance to form bonds with other substances through a range of chemical reactions (Paul et al. 2018; Bahhou et al. 2021; Neville and Brooks 2010).

The clinker nodules form as a result of partial fusion when the raw material mixture of calcareous (such as lime stone) and argillaceous (such as clay) is heated at a high temperature, i.e., 1400–1450 °C. The clinker is combined with a portion of calcium sulphate, and this mixture is then finely processed to create cement. Controlling the setting process is regulated by calcium sulphate. And it is frequently referred to as gypsum (Neville and Brooks 2010; Singh 2003; Taylor 1997).

The main raw materials used in the manufacture of portland cement are silica, iron oxide, lime, and alumina (Table 2.1).

As these substances are heated in the kiln, they interact with one another, producing a range of intricate effects. The four main components of Portland cement are included in Table 2.2, together with their shortened symbols. Each oxide is denoted by a letter in the abbreviated notations used by cement chemists, such as $CaO \rightarrow C$, $SiO_2 \rightarrow S$, $Al_2O_3 \rightarrow A$, and $Fe_2O_3 \rightarrow F$. In hydrated cement, the letter H also stands for H_2O.

Table 2.1 Approximate limits of Portland cement's composition

Oxide	Content percentage
CaO(Lime)	60–67
SiO_2(Silica)	17–25
Al_2O_3 (Alumina)	3–8
Fe_2O_3(Iron oxide)	0.5–6
MgO (Magnesia)	0.1–4
Alkalis	0.2–1.3
SO_3 (Sulphur trioxide)	1–3

Table 2.2 Principal ingredients in Portland cement

Name of Compound	Oxide composition	Abbreviation
Tricalsium silicate	$3CaOSiO_2$	C_3S
Dicalsium silicate	$2CaOSiO_2$	C_2S
Tricalsium aluminate	$3CaOAl_2O_3$	C_3A
Teracalcium aluminoferrite	$4CaOAl_2O_3$ Fe_2O_3	C_4AF

The silicates, specifically C3S and C2S, are crucial substances because they give Portland cement its hydrated state strength. Portland cement contains the undesirable chemical C3A. The strength of the cement is largely or entirely unaffected by the C3A (Neville and Brooks 2010). The most significant component of Portland cement, which makes up between 50 and 70 percent of the cement's composition, is tricalcium silicate (C3S, also referred as alite). 15%–30% dicalcium silicate (C2S, also referred as belite), 5%–10% tricalcium aluminate (C3A, also referred as aluminate), and 5%–15% tetracalcium aluminoferrite (C4AF, also referred as ferrite) are the other main minerals found in Portland cement (Taylor 1997).

2.2.1.1 Hydration of Cement

Hydration refers to the changes that occur when anhydrous cement or its constituent elements are exposed to water (Taylor 1997). A chemical interaction between anhydrous substances and water molecules is involved in hydration, leading to the creation of superfluid, hydrated compounds. Mechanical, physical, and chemical transformations emerge as a consequence of the chemical reactions occurring between non-hydrated substances and water molecules (Whittaker et al. 2010). Several chemical reactions occur when cement and water are mixed directly. As a result, the cement paste transforms from a manageable paste to a firm solid. The process of hydration, which results in the production of the various components, requires a number of simultaneous chemical interactions between regular Portland cement and water molecules. Its complexity is dictated by the clinker's original composition, which is directly influenced by the transformation that occurs in the cement matrix (Lavagna and Nistico 2023). The way that regular Portland cement hydrates has been well investigated. Ordinary Portland cement's clinker is divided into four phases: two types of calcium silicates (C2S and C3S), one type of calcium aluminate (C3A), and one type of ferrite (C4AF). To control how quickly Portland cement used in conventional construction sets, calcium sulphate is added to these phases. The main factors in the early hydration are C3S and C3A, although C2S hydration is more significant for the later development of cement properties (Taylor 1990).

To properly comprehend the hydration process, it is necessary to monitor the four key clinker phases, namely alite, belite, aluminate, and ferrite. The hydration of cement is frequently summarised as follows (Otto 1983):

Clinker phases + water → Hydrated phase + energy
(High in energy with no water) **(low energy contains water)** **(heat of hydration)**

Silicates and aluminates react with water at first, causing Portland cement to hydrate. Calcium hydroxide, commonly known as portlandite, and calcium silicate hydrate are formed when silicates react with water.

$$\text{Silicates} + \text{water} \rightarrow C - S - H + \text{portlandite}$$

When aluminates and gypsum (calcium sulphate) combine with water, ettringites are created (Taylor 1997).

$$\text{Aluminate}(C_3A) + \text{calcium Sulphate} + \text{water} \rightarrow \text{ettringite}$$

Cement undergoes a complicated chemical process called hydration. The hydration of cement occurs in four phases, This might be utilized to illustrate how hydration works: During the hydration process, the silicates and aluminates in Portland cement react with water to produce a solid mass. Silicates and aluminates react with water at first, causing Portland cement to hydrate. When aluminates and gypsum (calcium sulphate) combine in the presence of water, ettringites occur.

Stage 1

$$C_3A + 3CaSO_4 + 32H_2O \rightarrow 3CaO.Al_2O_3.3CaSO_4.32H_2O$$

(Trisulphate or ettringite)

In addition, Calcium hydroxide, also known as portlandite Ca(OH), and calcium silicate hydrate are created when silicates and water are combined. Portland cement gains its strength mainly from C_3S, which stands for alite, the major ingredient in cement. Upon contact with water, Alite quickly undergoes a reaction. The surface ions of alite begin to dissolve, releasing silicate and hydroxyl ions. As a result, the positive charge of Ca + ions is balanced by the negative charge of ions such as hydroxyl and silicate anions. Because alite dissolves faster than surface ions can permeate through the liquid phase, the liquid phase becomes oversaturated. As a result of this hydration, the CSH, or tobermorite, is formed. Dicalcium silicates (C_2S or belite) hydrate slowly in the same way that alite does. Tricalcium silicate (alite), which is affected by the w/ c ratio and particle size distribution, sets and hardens similarly to Portland cement. Calcium hydroxide (CH) and C-S–H (gel-like) are therefore created. The hydration of cement results in the formation of ettringite, also known as crystalline calcium sulfoaluminate hydrates ($3CaO.Al_2O_3.3CaSO_4.32H_2O$, an AFt-phase column structure), and monosulfate, also known as $3CaO.Al_2O_3.CaSO_4.12H_2O$, an AFm-phase lamellar structure. When Portland cement is still in the early stages of hydration, ettringite, the most important phase of the AFt, is produced.

Stage 2

$$2C_3S + 6H_2O \rightarrow 3CaO.2SiO_2.3H_2O \quad + \quad 3Ca(OH)_2$$

(Calsium silicate hydrate or C – S – H) (portlandite)

$$2C_2S + 4H_2O \rightarrow 3CaO.2SiO_2.3H_2O \quad + \quad Ca(OH)_2$$

(Calsium silicate hydrate (portlandite)

or C − S − H)

Stage 3

$$C_3A + Ca(OH)_2 + 12H_2O \rightarrow 4CaO.Al_2O_3.13H_2O$$

$$C_3AF + 4Ca(OH)_2 + 22H_2O \rightarrow 4CaO.Al_2O_3.13H_2O + 4CaO.Fe_2O_3.13H_2O$$

Stage 4

$$\underset{\text{(ettringite)}}{2C_3A + 3CaO.Al_2O_3.3CaSO_4.32H_2O} + 4H_2O \rightarrow \underset{\text{(monosulfate)}}{3CaO.Al_2O_3.CaSO_4.12H_2O}$$

Ettringite is a critical stage of AFt and occurs when cement first begins to hydrate. Typically, when cement is hydrated, ettringite is the first material to precipitate. Ettringite is created when gypsum is introduced and interacts with calcium aluminates when there is water present. Cement setting is slowed down by ettringite production. In the subsequent phase, calcium trisulfate and calcium disulfate react with water, calcium hydroxide (portandlite) and calcium silicate hydrate are produced. Calcium silicate hydrate, which makes about 60–65% of the solid mass produced during cement hydration, is the main by-product (Neville and Brooks 2010; Taylor 1997; Cocke et al. 1993; Collepardi 2003; Batic et al. 2000; Wang et al. 2018).

2.2.1.2 Setting of Cement

A paste is created when cement and water are mixed; this paste undergoes a gradual stiffening process before it ultimately starts to solidify (Baradan 1998). Setting occurs when cement stiffens without significantly increasing its compressive strength, typically within a few hours (Taylor 1997). Anhydrous cement is initially found in a suspension of scattered particles, and the process of setting occurs when these dispersed particles gradually bond together, resulting in a strengthened system. This process of alteration is usually accomplished by means of a chemical response initiated by the introduction of water to cement, leading to hydration. Essentially, it entails a reaction occurring between the cement particles and water (Jiang et al. 1995).

2.2.1.3 Initial Setting Time of Cement

When applying cement-containing substances, the initial setting time is an important consideration. The transportation schedule, concrete placement, concrete finishing process, and lastly, the consolidation of the concrete are all controlled by the initial setting time (Lee et al. 2004; Zheng et al. 2012; Aydin and Gül 2007). The cement setting time is crucial for the placement of concrete and carrying out quality checks while placing concrete. Every stage of the concrete process, including mixing, transportation, placement of concrete, and finishing, is governed by the setting time (Pessiki and Carino 1988). It reacts immediately and strongly when pure C_3A is mixed with water. Flash setting, which happens as a result, causes cement paste to instantly harden. Gypsum is mixed with cement to prevent this flash-setting. As shown, when gypsum is lacking in the cement, tricalcium aluminate (C_3A) reacts very fast.

$$3CaO.\ Al_2O_3 + 6H_2O \rightarrow 3CaO.\ Al_2O_3.6H_2O \quad (2.1)$$

It reacts similarly swiftly in the presence of calcium hydroxide, a result of the hydration of calcium silicate.

$$3CaO.\ Al_2O_3 + Ca(OH)_2 + 12H_2O \rightarrow 4CaO.Al_2O_3.13H_2O \quad (2.2)$$

Both of the preceding processes show the extremely rapid setting of Portland cement paste. In the grinding mill, sulphate in the form of gypsum is added to the clinker. In cement paste, gypsum functions as a retarder, while calcium aluminate combines with gypsum in the presence of water to form ettringite;

$$\underset{\text{Calsium}}{3CaO.\ Al_2O_3} + \underset{\text{Gypsum}}{3(CaSO_4.2H_2O)} + \underset{\text{Water}}{26H_2O} \rightarrow \underset{\text{Trisulphate or ettringite Aluminate}}{3CaO.\ Al_2O_3.\ 3CaSO_4.32H_2O} \quad (2.3)$$

The calcium aluminate hydrate is produced in the aforementioned processes 1 and 2 quite quickly, and the cement paste has some strength analogous to its initial set.

Ettringites are created initially when calcium aluminate comes into contact with calcium sulphate, or gypsum. These ettringites encapsulate the cement particles and create a thin layer on them during the first few hours of hydration. This coating does not prevent the particles from slipping around. and as a result, the cement paste continues to be plastic (Otto 1983).

2.2.2 *Use of Admixtures to Regulate the Initial Setting Time*

Admixture is a component that is added to a concrete mixture to change or improve the properties of the concrete in its fresh or hardened state (Li 2011). Chemical admixtures are largely used to control concrete fluidity, setting time, and hardening time (Sakai et al. 2006). The right use of admixtures can be used to regulate the setting time in terms of acceleration or retardation (Shah et al. 2014). Chemical admixtures regulate how fresh concrete behaves rheologically. Retarders and accelerators can change how quickly cement hydrates (Nkinamubanzi et al. 2016). The rheological characteristics of concrete are crucial. Depending on the admixture and cement combination, the interaction between the two can induce abrupt variations in the fluidity of concrete as well as stiffness (Hanehara and Yamada 1999; Nocun –Wczelik W, Trybalska B. 2007; Du and Folliard 2005; Yang et al. 2000; Mohamed et al. 2022). For the formation of concrete, admixture plays a significant role. Admixtures can be used for a variety of reasons, depending on their properties. The many forms of chemical admixtures and the properties of their particles are shown in Table 2.1 (Patel and Shah 2013).

Chemical admixtures	Function
Superplasticizer	To produce dense, high-strength concrete by reducing the water required by 15% to 20% without compromising workability
Accelerator	When concreting in cold weather, it helps to shorten the concrete's setting time, which allows for the early removal of forms
Retarder	Recommended in locations where high temperatures are involved in the concreting process, it prolongs the setting time by slowing down the hydration of cement
Admixture for water reduction	To reduce cement usage by achieving a particular workability (slump) at a low water-to-cement ratio for a given strength
Air entraining admixture	Small air bubbles that function as rollers are introduced into concrete to improve workability. Because they act as a cushion for the expanding water in the concrete during freeze–thaw cycles, these bubbles are especially helpful in cold areas

2.2.2.1 Use of Sugarcane Juice and Its Products as Admixtures

Sugarcane is one of the most significant crops. It is grown in tropical and subtropical climates worldwide. After Brazil, India is the second-biggest producer of the crop. Water makes up 75–85% of sugarcane juice, with reducing sugar making up 0.3–3% and nonreducing sugar making up 10–21% (Phanikumar 2011; Swaminathan 1995).The microstructure of C-S–H and the formation process both alter with the mixing of sucrose (Maria et al. 2002). When sucrose is added to cement paste, the process of acceleration or retardation, or the beginning of cement setting, is triggered. This acceleration or retardation is dependent on the production of Aft (Wu and Wang 1998). Admixture dosages that produce retardation are critical; in simple terms, retardation and acceleration are dose-dependent. Sucrose, glucose, and other additives induce retardation. Admixtures such as organic chemicals, inorganic salts, and so on accelerate cement hydration. When used in low dosages, additives such as triethanolamine and triisopropanolamine work as accelerators (Gartner and Myers 1993). The amount of admixture used affects the rate of cement hydration. Some admixtures employed as water reducers might delay cement hydration, subject to the mixing ratio (Marchon and Flatt 2016). When the solution of sucrose was mixed with Portland cement in various amounts, a significant quantity of AFt was produced. The cement's hydration process accelerated when a large amount of sucrose solution was added, accelerating the cement's setting (Wang 1994). A larger quantity of molasses causes the induction phase to last longer. A small quantity of sugarcane molasses may accelerate the aluminate phase's dissolution, while an excessive amount could enhance the nucleation and development of AFt in the early stages. 1.0% molasses speeds up the generation of AFt from the hydration of C3A and Ferrite. However, the hydration of cement paste with 1.0% molasses considerably prevents C3A and Ferrite's second hydration, and the hydration of C3S is delayed at the same time (Weifeng et al. 2014). The consistencies of Ca2 + and OH- are reduced, and Aft (Ettringite) is produced when sucrose, which makes up 0.045% of the weight of cement, is added. Because the AFT covers the surface of the clinker particles and manifests itself as a gel-like product, sucrose serves as a retarding admixture. The hydration processes of C2S and C3S are noticeably poisoned by sucrose when the sucrose content is increased to 0.24%, and the consistency of Ca2 + and OH- accelerates the setting process because the Aft now behaves like a needle (Zhenghao et al. 2014).

Ettringite formation and cement paste flow behaviour were found to be significantly correlated. Ettringite formation has been found to be directly correlated with cement paste's loss of workability. The impact of the added ettringite on the paste's workability increased as the cement paste's ettringite content rose. Low-density ettrigite development causes a shift in the cement paste's water-to-solid ratio (Jakob et al. 2019).

The addition of sugar as an additive has a major impact on the cement's initial and final setting times. If a small amount of sugar was added to the cement, the setting times rose sharply. The cement's initial and final setting times increased to their maximum values when about 0.05% sugar was added. However, when more than 0.05% of sugar was added, the setting times began to shorten. It was discovered that when sugar content exceeded 0.1%, the cement's initial setting time dropped less than when sugar content was zero (Ahmad et al. 2020). In the studies, it was found that adding any sort of sugar—granular, caster, or brown—in ratios of 0.5%, 1.5%, or 0.5% had the effect of retarding the setting of cement. However, when added in ratios of 2.5% and 5%, all of these sugars acted as a cement accelerator. The final setting time of cement was shown to rise when sugar was added up to a specific ratio, or the ideal 1.5%, but to decrease when sugar was added in larger quantities. With the addition of any kind of sugar, cement's initial setting time was also shortened and increased up (Al Khafaji and Ruddock 2018). It was discovered that adding up to 0.2% sugar to the cement paste increased the initial setting time compared to adding 0% sugar and that adding up to 0.15% sugar increased the final setting time. At 0% sugar addition, the initial and final settings occurred at 120 and 349 min, respectively. The cement initially set in 174 min when 0.2% sugar was added, while it took 112 min at 0.15% sugar. The final settings occurred at 491, 316, and 294 min when 0.15%, 0.2%, and 0.25% sugar were added (Kawade et al. 2017b).

2.3 Conclusion

Based on the aforementioned studies, it can be inferred that the presence of sugarcane juice and its product in cement affects the hydration process and the setting time. When a small quantity of sugarcane juice and its products are added, it prolongs the cement's setting time by slowing down the hydration process. Conversely, larger amounts of sugarcane juice and its products expedite the hydration process, thereby reducing the setting time of the cement.

The same pattern was also found in earlier investigations.

A small quantity of sugarcane molasses slows the cement paste's setting and hardening process while increasing the fluidity of cement mortar; however, excess molasses accelerates the process of setting and hardening (Weifeng et al. 2014). The addition of sugarcane juice as an admixture in the cement paste reduced the setting time significantly. Experiments revealed that higher sugarcane juice proportions in the cement paste resulted in even shorter setting times than the controlled cement paste without any sugarcane juice addition. When compared to the controlled cement paste, the initial setting time of the cement was reduced by 33, 57, 70, and 90% after adding 2, 5, 10, and 20% sugarcane juice by weight of cement, respectively. These findings clearly show that sugarcane juice acts as an accelerator, hastening the cement-setting process (Paul et al. 2022). Sugar was added to cement paste in varying quantities, ranging from 0.00%, 0.04%, 0.05%, 0.06%, 0.07%, 0.10%, 0.20%, 0.30%, 0.40%, and 0.05% by cement weight. When the setting durations of cement paste were tested under normal conditions, it was discovered that adding up to 0.25% sugar enhanced the initial and final setting times of cement when compared to adding 0% sugar. At 0% sugar, the initial and final setup times were 143 and 204 min, respectively. At 0.25% sugar, the initial setting and final setting times were extended to 260 and 649 min, respectively. When the sugar levels exceeded these percentages, both setting times began to fall enormously. At 0.30% sugar, the initial and final settings occurred at 42 and 72 min, respectively. Both setting times were 20 and 40 min at 0.40 to 0.50% sugar (Khan and Baradan 2002).

References

Abdulrahman AS, Mohammad I (2011) Green plant extract as a passivation promoting inhibitor for reinforced concrete. Int J Eng Sci Technol 3(8):6484–6490

Ahmad S, Lawan A, Al-Osta M (2020) Effect of sugar dosage on setting time, microstructure and strength of Type I and Type V Portland cements. Case Stud Constr Mater 13:e00364

Al Khafaji ZS, Ruddock F (2018) Study the retardant effect of using different sugar's types on setting time and temperature of cement paste. Int J Civ Eng Technol 9:519–530

Aydin AC, Gül R (2007) Influence of volcanic originated natural materials as additives on the setting time and some mechanical properties of concrete. Constr Build Mater 21:1277–1281

Bahhou A, Taha Y, Khessaimi YE, Hakkou R, Tagnit-Hamou A, Benzaazoua M (2021) Using calcined marls as non-common supplementary cementitious materials—a critical review. Minerals 11(5):517

Baradan B (1998) Construction materials–II. 5th.ED. Dokuz Eylul University, Technical faculty publication section, Izmir Turkey
Batic OR, Milanesi CA, Maiza PJ, Marfil SA (2000) Secondary ettringite formation in concrete subjected to different curing conditions. Cem Concr Res 30:1407–1412
Chandra S (2003) Architecture and ancient building materials in India. Tech. Books International, New Delhi
Chandra S, Eklund L, Villarreal RR (1998) Use of cactus in mortars and concrete. Cem Concr Res 28(1):41–51
Cocke DL, Mollah MYA (1993) The chemistry and leaching mechanisms of hazardous substances in cementitious solidification/stabilization systems. In: Spence RD (ed) Chemistry and microstructure of solidified waste forms. Lewis Publishers, Boca Raton, FL, pp 187–241
Collepardi M (2003) A state-of-the-art review on delayed Ettringite attack on concrete. Concrete cement and concrete composites 25(4–5):401–407
Dandotiya A, Parihar RS, Jha AK (2022) An investigational properties of concrete using jaggery. Int J Trend Sci Res Dev (IJTSRD) 6(5). Available Online: www.ijtsrd.com e-ISSN: 2456–6470
Du L, Folliard KJ (2005) Mechanisms of air entrainment in concrete. Cem Concr Res 35:1463–1471
Dwivedi VN, Das SS, Singh NB, Rai S, Gajbhiye NS (2008) Portland cement hydration in the presence of admixtures black gram pulse and superplasticizer. Mater Res 11(4):427–431
Faqe H, Dabbagh H, Mohammed A (2020) Natural admixture as alternative for chemical admixture in concrete technology: a review. J Duhok Univ 23(2):301–308
Gartner E, Myers D (1993) Influence of tertiary alkanolamines on portland cement hydration. J Am Ceram Soc 76:1521–1530
Hanehara S, Yamada K (1999) Interaction between cement and chemical admixture from the point of cement hydration, absorption behaviour of admixture, and paste rheology. Cem Concr Res 29:1159–1165
IS 9103 (1999) Specification for concrete admixtures -[CED 2: Cement and Concrete]
Jakob C, Jansen D, Ukrainczyk N, Koenders E, Pott U, Stephan D, Neubauer J (2019) Relating ettringite formation and rheological changes during the initial cement hydration: A comparative study applying XRD analysis. Rheol Meas Modeling Mater 12:2957. https://doi.org/10.3390/ma12182957
Jiang SP, Mutin JC, Nonat A (1995) Studies on mechanism and phisco-chemical parameters at the origin of cement setting. Cem Concr Res 25(4):779–789
Kawade A, Kamthe I, Khemalapure AB (2017) Experimental study of effect of sugar on properties of concrete. J Struct & Transp Stud 2(1)
Kawade A, Kamthe I, Khemalapure AB (2017) Experimental study of effect of sugar on properties of concrete. J Struct & Transp Stud 2(1)
Khan B, Baradan B (2002) The effect of sugar on setting-time of various types of cements. Q Sci vis 8:71–78
Kurdowski W (2014) Cement and concrete chemistry. Springer Dordrecht Heidelberg New York London. ISBN 978–94–007–7944–0
Lavagna L, Nistico R (2023) An Insight into the Chemistry of Cement—A Review. Appl Sci 13:203
Lea F (1988) The chemistry of cement and concrete, 3rd ed. Edward Arnold, London, UK
Lee HK, Lee KM, Kim YH, Yim H, Bae DB (2004) Ultrasonic in-situ monitoring of setting process of high-performance concrete. Cem Concr Res 34:631–640
Li Z (2011) Advanced concrete technology
Maheswari AU, Harshavardhan R, Tejas JS,Chandradeep K, Monisha SK, Padal S, Basheer SK (2021) Utilization of sugarcane bagasse as retarder in concrete. IJIRT 8(3). ISSN: 2349–6002
Mahmood HF, Dabbagh H, Mohammed AA (2021) Comparative study on using chemical and natural admixtures (grape and mulberry extracts) for concrete. Case Stud Constr Mater 15:e00699
Marchon D, Flatt RJ (2016) Mechanisms of cement hydration. In: Aïtcin P-C, Flatt RJ (eds) Science and technology of concrete admixtures. Elsevier (Chapter 8), pp 129e146

Maria C, Garci J, Jennings H (2002) New insights into the effects of sucrose on the hydration and microstructure of cement pastes. Cem Concr Res 32(3):393–399

Memon RA, Memon FA (2022) Experimental study on the effect of sugarcane juice as an admixture on fresh and hardened properties of concrete. Int Res J Mod Eng Technol Sci. e-ISSN: 2582–5208, 04(05)

Michelle S (2009) Did our church builders used egg white as mortar ingredients? The bimolecular reason behind collecting mortar samples from historical sites. (Ref: https://www.academia.edu)

Mohamed AK, Weckwerth SA, Mishra RK, Heinz H, Flatt RJ (2022) Molecular modeling of chemical admixtures; opportunities and challenges. Cem Concr Res 156:106783

Neville AM (1995) Properties of concrete. Vol 4, Longman, London

Neville AM, Brooks JJ (2006) Concrete technology. Dorling Kindersleg, New Delhi, India

Neville AM, Brooks JJ (2010) Concrete technology. Prentice Hall. ISBN 978–0–273–73219–8, Malaysia

Neville AM, Brooks JJ (2004) ConcreteTechnology. Prentice Hall, Malaysia

Nkinamubanzi PC, Mantellato S, Flatt RJ (2016) Superplasticizers in practice. In: Aïtcin P-C, Flatt RJ (eds) Science and technology of concrete admixtures. Elsevier (Chapter 16), pp 353e378

Nocun–Wczelik W, Trybalska B (2007) Effect of admixtures on the rate of hydration and microstructure of cement paste. Cement Wapno Beton 6:284–9

Olbrich H (2006) The molasses. Biotechnologie-Kempe GmbH, Berlin

Originated by Otto L,Transl. by Amerongen C (1983) V from the 6. German ed.-4. Engl. ed. by Kohlhaas, Cement engineers handbook, Wiesbaden, Berlin, Bauverlag

Otunyo AW, Koate I (2015) Sugar cane juice as a retarding admixture in concrete production. Glob J Eng Res 14:17–23. ISSN 1596-292x

Patel GK, Deo SV (2016) Effect of natural organic materials as admixture on properties of concrete. Indian J Sci Technol 9(37)

Patel V, Shah N (2013) A survey of high performance concrete developments in civil engineering field. Open J Civ Eng 3:69–79

Paul SC, Van Rooyen AS, van Zijl GP, Petrik LF (2018) Properties of cementbased composites using nanoparticles: A comprehensive review. Constr Build Mater 189(2018):1019–1034

Paul SC, Babafemi AJ, Md Miah J, Md Basit A, Md. Sadiqul Hasan N, Ying Kong S (2022) Role of sugarcane juice as a natural admixture on setting time and hardened properties of cementitious materials. Infrastructures 7:145

Pessiki SP, Carino NJ (1988) Setting time and strength of concrete using the impact-echo method. ACI Mater J 85:389–399

Phanikumar HK (2011) Sugarcane juice powder by spray drying technique, Science tech enterpreneur, national research development corporation (An Enterprise of DSIR, Min. of S&T, Government of India), Malleswaram, Bengaluru

Sakai E, Ishida A, Ohta A (2006) New trends in the development of chemical admixtures in Japan. J Adv Concr Technol 4(2):211–223

Shah DS, Shah PM, Pitroda JK (2014) Chemical admixtures: A major role in modern concrete materials and technologies. National Conference on: trends and Challenges of Civil Engineering in Today's Transforming World. ISBN: 978–81–929339–0–0

Singh NB, Shivani C, Sarita R (2003) Chemistry of portland cement. J Indian Chern Soc 80:319–325

Swaminathan V (1995) Food science chemistry & experimental foods, the Bangalore Printing & Publishing Co. Ltd., Bagtalore

Taylor HFW (1990) Cement chemistry. Academic Press, Harcourt Brace Jovanovich Publishers, London

Taylor HFW (1997) The chemistry of cements, 2nd edn. Thomas Telford Services Ltd., New York

Thirumalini P, Sekar SK (2013) Review on herb used as admixture in lime mortar used in ancient structures. Indian J Appl Sci 3(8):295–298

Togero A (2006) Leaching of hazardous substances from additives and admixtures in concrete. Environ Eng Sci 23(1):102–116

Usman ND, Idusuyi FO, Ojo EB, Simon BM (2012) The use of sawdust and palm kernel shell as substitute for fine and coarse aggregates in concrete construction in developing countries. J Chem, Mech Eng Pract 2(3):51–62

Usman ND, Adebitan EO, Kwari JK, Gyang JB (2013) Determination of engineering properties of locust bean waste and wild vine use in construction in Nigeria. In: Architecture, environment, technology and soceity in the global south, sabs conference (pp 10th–12th October) 2013, Nairobi, Kenya: Jomo Kenyatta University of Agriculture and Technology

Wang P (1994) Formation of hydration on C3A mineral and Portland clinker with and without sucrose. J Shangh Inst Build Mat 7(2):107–117

Wang L, Yang HQ, Zhou SH, Chen E, Tang SW (2018) Hydration, mechanical property and C-S-H structure of early-strength low-heat cement-based materials. Mater Lett 217:151–154

Weifeng LI, Suhua MA, Shengbiao Z, Xiaodong S (2014) Physical and chemical studies on cement containing sugarcane molasses. J Therm Anal Calorim 118:83–91

Whittaker M, Dubina E, Plank J, Black L (2010) The effects of prehydration on cement performance. In: Proceedings of the 30th cement and concrete science conference 2010, Birmingham, UK, 13–15 September

Wu J, Wang P (1998) Mechanism of regulation set of sucrose in Portland cement. J Chin Ceram Soc 26(2):164–170

Yang Q, Zhu P, Wu X et al (2000) Properties of concrete with a new type of saponin air-entraining agent. Cem Concr Res 30:1313–1317

Zheng JJ, Zhang J, Scherer GW (2012) Prediction of the degree of hydration at initial setting time of cement paste with particle agglomeration. Cem Concr Res 42:1280–1285

Zhenghao LI, Weiguo S, Qing L, Zili LU, Lianghong CAO, Zhiyuan XU (2014) Influence of sucrose on hydration process of Portland cement. Key Eng Mater 599:46–51

Chapter 3
Application of Drone Photogrammetry for Mapping of a Tunnels and Excavation Site

P. A. Kokate, G. L. Sarode, A. K. Raina, and Sagar N. Nimsadkar

Abstract Unmanned aerial vehicle (UAV) technology asdvancements in recent years have drastically lowered the price of aerial photography. Drone-based mapping applications are therefore becoming more and more commonplace. Compared to satellite imaging, drone photos may be used to quickly and accurately create digital maps of specific terrain formations, down to a few centimeters.The purpose of this article is to highlight the benefits of drone-based photogrammetry over satellite image technology. Then prospects for aerial 3D mapping are mentioned stepwise using image processing software. A case study of a tunnel excavation site is highlighted in this research article. The result shows that drone-based photogrammetry is a very good option for developing 3D maps of excavation sites within a short time. Drone aided mapping is mentioned stepwise using image processing software. A case study of a tunnel excavation site is highlighted in this research article. The result shows that drone-based photogrammetry is a very good option for developing 3D maps of excavation sites within a short time.

Keywords UAV · Drone mapping · 3D model · 3D reconstruction · Drone photogrammetry

---

P. A. Kokate (✉) · G. L. Sarode · S. N. Nimsadkar
CSIR- National Environmental Engineering Research Institute, Nagpur 440020, India
e-mail: pa_kokate@neeri.res.in

A. K. Raina
CSIR-Central Institute of Mining and Fuel Research, Nagpur Research Center (Mining Technology), Nagpur 440001, India

Academy of Scientific and Innovative Research, Ghaziabad 201002, India

B. V. Khode et al. (eds.), *Innovations in Technologies: Pioneering Sustainable Infrastructure for a Resilient Future*, Sustainable Civil Infrastructures,
https://doi.org/10.1007/978-3-031-82133-2_3

3.1 Introduction

The collection of aerial images of a geographical area is always a challenging task in the field of engineering (Esposito et al. 2017; Budiharto et al. 2021). The traditional way of capturing images was the balloon-based method. The helicopter-based image capture was a time-consuming and costly activity. Many of them are complicated, area specific, and manpower dependent. Satellite images are readily available from various agencies. There is a rapid growth in remote sensing techniques; mapping can be done for parameters like water bodies, their characteristics, etc. (Colomina and Molina 2014). The satellite images are low resolution data in the public domain. It is very difficult to collect a relatime, highresoultion data at low cost.

Currently, the UAV based image processing systems have become an alternative for different engineering applications. One of these applications (Cruzan et al. 2016; Kokate et al. 2023a; Urbanová et al. 2017; Singh et al. 2019) is the estimation of the area of barren land, the identification of real-time activities on the ground, and the extraction of multiple topographic parameters from the processed data in the geotagged images. This technology is promising in environmental engineering, ecological management, and forest management. It shows better results as well as the accuracy of the identified location compared with satellite images.

3.2 Literature Review

Many researchers work in the field of multi-rotor and fixed-wing UAVs to explore photogrammetric and remote sensing applications in environmental engineering (Singh et al. 2019; Kokate et al. 2023b; Kokate et al. 2023b; Sarode et al. 2023) reported that the UAV is a prominent device for the various purposes of aerial image processing and data collection, etc. The study by Goussanou et al. 2016a reported that the tropical forest contributions reduce carbon dioxide and provide carbon stock. The algometric equation is used for mitigation of climate change. The study was done for eighteen tree species, whose volume, biomass and density were determined. The study done by Arango and Morales (2015), reported that UAVs are becoming the most important tool that can be used in different engineering applications, mostly in surveying. In the paper, the author calculated the volume of stockpile material, and found its accuracy and efficiency. The author compared the volume measured by the total station and the multi-copter UAV. One study by Arango and Morales (2015) conveys that the volume estimation method uses a UAV. They found the volumes of stockpiles by involving themselves in many surveys, by using Topcon Imaging Station (IS) for three-dimensional modeling (3D) & compare it with the amount calculated by the conventional method of the same stockpile.

The study by Azari et al. (2018) reported the limitations of photogrammetric technologies. It can be overcome by using LIDAR technology like multiple target capability or does not dependent on the environmental light conditions (Amon et al.

2014). discussed the open-source software that made the remote sensing so inexpensive. In this paper, two landslides were taken for UAV investigations the first one was the supersized and the second is (Goussanou et al. 2016a) reported the findings of canopy area, tree height, and crown volume by using the quadcopter. The visible-light camera and multispectral sensors were used as sensors (Niethammer et al. 2012). reported UAVs are helpful in finding volume and the area of the sand stock material. So, the study area in this paper is undisclosed sand mine located in western Wisconsin and the imagery is processed afterward by using pix4D software. In (Goussanou et al. 2016b), there is no such comparison between the areas taken by UAVs and the satellite image to strengthen the quality as well as the accuracy of the reviewed that UAVs are becoming very popular in the field of photogrammetric and remote sensing.

Guisado-Pintado et al. 2019) discussed the accuracy of the satellite scale and the field scale by generating data and reducing the data gap in soil erosion monitoring. In this researcher used a fixed wing aircraft which is of type Siri- us I (MAV inci, Germany) which is also supplied with the digital system camera (Panasonic) employed with it. In paper (Rocca 2016), it has been found that moving vehicles with the help of unmanned aerial vehicles integrated with video processing techniques. Using GPS technology researchers investigated the usefulness of Unmanned Aerial Vehicles (Cruzan et al. 2016). investigated the application of too small aerial drones or micro-unmanned aerial vehicles can be used for low-elevation surveys and they have a wide range of applications in plant ecology which also includes mapping vegetation over small to medium-sized regions. Small drones were used for taking aerial images of the interesting area.This study (Bendig et al. 2014) estimated the bio- mass of barley using the crop surface models (CSMs) which is taken from the UAV- Based RGB imaging. Predicting the yield of the agricultural field is the most important as well as a great relief for the farmers, and estimating aboveground biomass as well as monitoring the crop gives the bonus to the farmers. In this study (Guido et al. 2016) Li Wei (China) about the unmanned airship photogrammetry system used for the low altitude. In this paper (Liu et al. 2020), the author explains that satellite images are time-consuming as well as they did not provide the picture with higher accuracy (Fig. 3.1).

In this work, the first part discusses the importance of aerial images and a few drones aided photogrammetric studies in the field of ecology and environmental engineering. The second part discusses the case study of railway tunnel site where the low-cost drone was deployed for the collection of aerial images and development of 3D map. The final part discussed the results of the work.

3.3 Methodology

A commercial photogrammetric camera drone was used over a proposed railway line near Mumbai. The flight planning and navigation were performed with drone mapping software along with the path of the railway line. GNSS positions and height

Fig. 3.1 Images of drone and flight (before and after)

for the location were also cross verified through the GPS survey data. Multiple flights for three days were performed, with main east–west flights covering the entire area. A total of 133 images were captured on the site locations with 85% image overlapping. The RGB images were captured at an angle of 45 Degree and 90 Degree in mesh structure. Al the images were processed using a images processing software Pix4D (Edu Version). The QGIS (Open Source) software was used to generate the digital map of the site.

3.4 Result and Discussion

The geotagged images were collected using RGB camera drone. Then the images were processed using the images processing software. Table 3.1 shows the stitching of the images was performed with minimum level of errors.

Minimum Error and Maximum Error Indicate the range of error intervals for geolocation. Columns X, Y, and Z display the percentage of photos that contain geographical location mistakes within the predetermined intervals of error. The geo location error refers to the discrepancy between the initial and calculated positions of a picture depicted in Table 3.1. It should be noted that the mistakes in the geographical position of the image do not align with the precision of the observed 3D points. The relationship of the camera and internal parameters are mentioned in Fig. 3.2a. The color white indicates a complete connection between the parameters, meaning that any alteration in one parameter can be entirely offset by the other. The color black signifies that the parameter is completely independent. The nadir and 45^0 images were processed by using image processing software as depicted in Fig. 3.3.

Weak links are represented by bright colors and can be established through manual tie points or by adding additional images. The bundle block adjustment result is shown by dark green ellipses, which indicate the level of uncertainty in the camera's relative position (Fig. 3.4).

Table 3.1 Most minimum errors while stitching the aerial images

Min error [m]	Max error [m]	Geolocation error X [%]	Geolocation error Y [%]	Geolocation error Z[%]
–	–15.000	0.000	0.000	0.000
–15.000	–12.000	0.000	0.000	0.000
–12.000	–9.000	0.000	0.000	0.000
–9.000	–6.000	0.000	0.000	0.000
–6.000	–3.000	8.270	6.020	0.000
–3.000	0.000	33.080	42.860	45.110
0.000	3.000	46.620	46.620	54.890
3.000	6.000	12.030	4.510	0.000
6.000	9.000	0.000	0.000	0.000
9.000	12.000	0.000	0.000	0.000
12.000	15.000	0.000	0.000	0.000
15.000		0.00	0.000	0.000
Mean [m]		–0.025	–0.001	–0.002
Sigma [m]		2.063	1.586	0.751
RMS Error		2.063	1.586	0.751

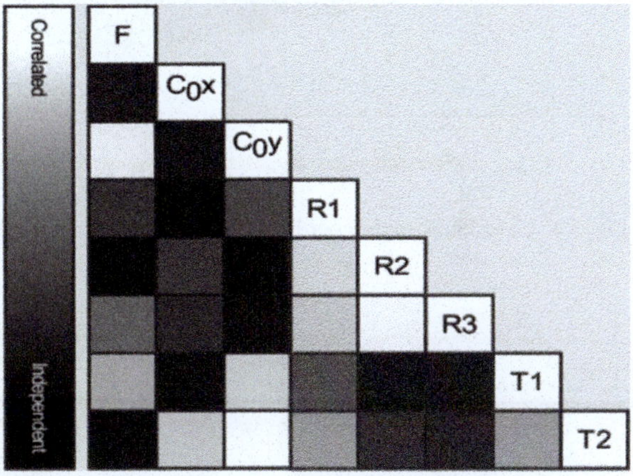

Fig. 3.2a Internal camera parameters (a grayscale matrix diagram)

The Ortho mosaic (true Orthophoto) was generated to develop an entire 2-D map of the site. A maximum number of key points (more than 1000) used to create the model. This method handles all types of terrain, as well as large datasets. The 2-D map, as per Fig. 3.5 generated using the image processing software. The DTM shows elevation data of a terrain structure in digital format. The digitalization was performed after adding 7 ground control points. Then the data was collected in the mapping software (QGIS) (Fig. 3.6).

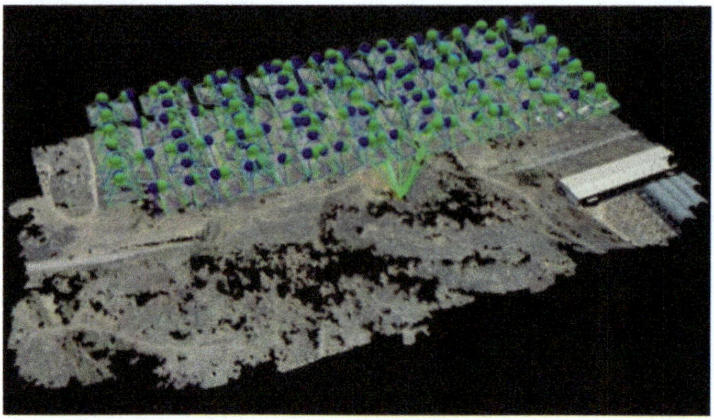

Fig. 3.3 Mosaic view during the process of image stitching

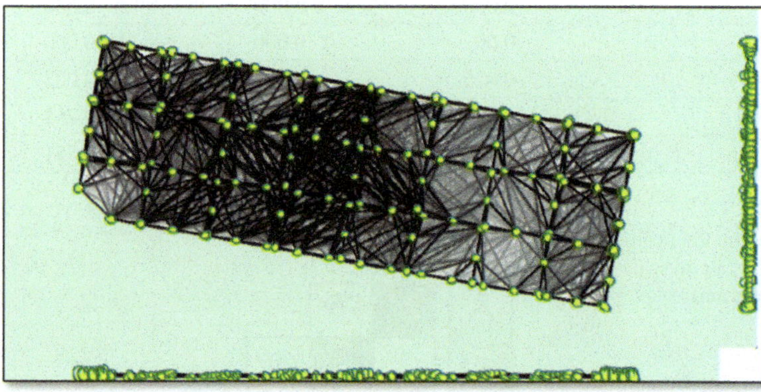

Fig. 3.4 Calculated image coordinates with associated image links

Fig. 3.5 DTM view of the tunnel

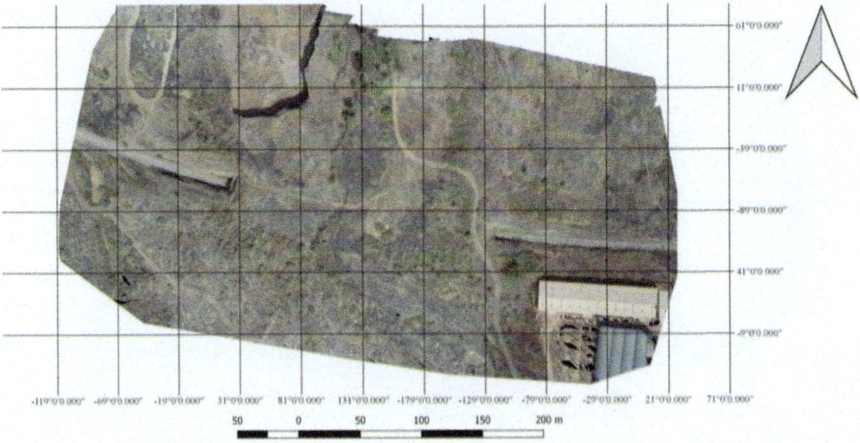

Fig. 3.6 Detailed view of the site after image processing

This data presents novel perspectives of the chosen topographical location. The current state of the sites can be readily observed through the utilization of the 3D map. This map has been created using aerial pictures captured by drones at 45 and 90-degree angles from the vertical (NADIR) (Fig. 3.7).

In order to correct for artifacts that arise from depending just on the original data, a Digital Terrain Model (DTM) is simply a Digital Elevation Model (DEM) that has been supplemented with extra elements, such as break lines and supplemental

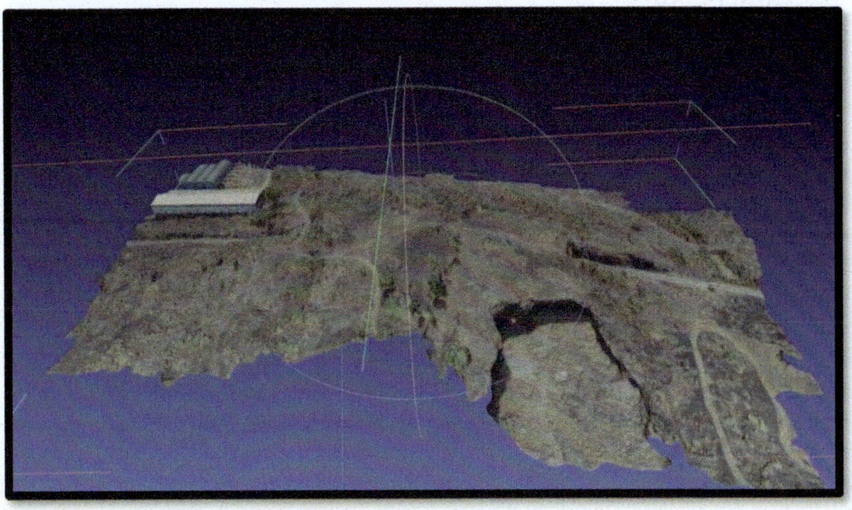

Fig. 3.7 3D aerial map of the tunnel

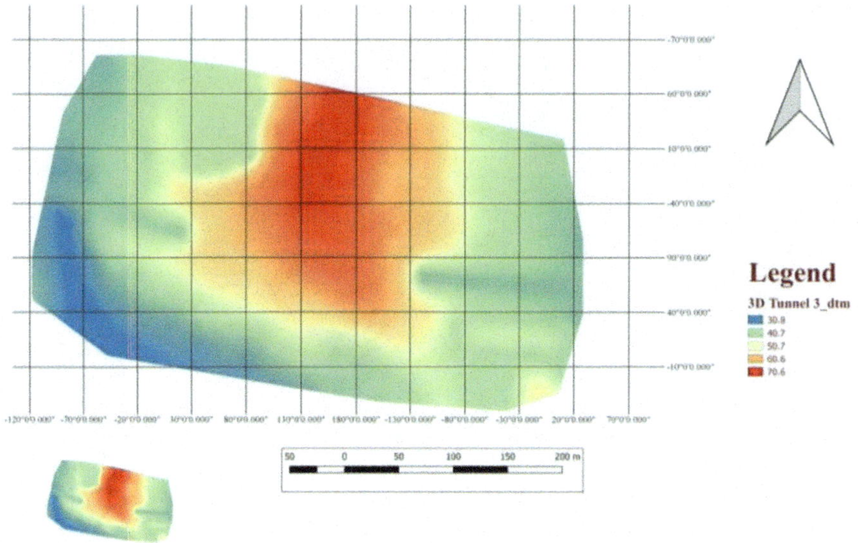

Fig. 3.8 DTM of the tunnel

observations. The Digital Terrain Model (DTM) of the three tunnels is shown in Fig. 3.5. This shows the average ground height for the three tunnels (Fig. 3.8).

3.5 Conclusion

The case study was successfully completed at the railway line tunnel site. The horizontal accuracy of 2 cm (+-10%) can be obtained with this methodology. This technology can be useful for large area mapping within a short time. The dependency of the satellite can be minimized, and centimeter-level resolution data can be collected within a limited time span.

In the future, this method can be coupled with AI algorithms for perimeter, work progress, civil construction site surveillance. They could also be used for continuous monitoring of mining sites, stockpile management, equipment tracking, and safety inspections. Drones will very certainly be combined with other new technologies like robots, virtual reality, and augmented reality. This partnership will allow mining experts to visualize and interact with data in real-time, boosting decision-making processes and the overall efficiency of mineral exploration efforts.

Acknowledgements Authors are thankful to the Mumbai Railway Vikas Corporation (MRVC) officials for their support during this study at Karjat- Panvel Railway Route. Authors would like express their gratitude to the Director CSIR-NEERI and CSIR-CIMFR for their support and encouragement to conduct novel research work.

References

Amon P, Riegl U, Rieger P, Pfennigbauer M (2014) Introducing a new class of survey-grade laser scanning by use of unmanned aerial systems (UAS)

Arango C, Morales CA (2015) Comparison between multicopter uav and total station for estimating stockpile volumes. Int Arch Photogramm Remote Sens Spat Inf Sci XL-1-W4, 131–135. https://doi.org/10.5194/isprsarchives-XL-1-W4-131-2015

Azari MM, Sallouha H, Chiumento A, Rajendran S, Vinogradov E, Pollin S (2018) Key technologies and system trade-offs for detection and localization of amateur drones. IEEE Commun Mag 56:51–57. https://doi.org/10.1109/MCOM.2017.1700442

Bendig J, Bolten A, Bennertz S, Broscheit J, Eichfuss S, Bareth G (2014) Estimating biomass of barley using crop surface models (CSMs) Derived from UAV-based RGB imaging. Remote Sens 6:10395–10412. https://doi.org/10.3390/rs61110395

Budiharto W, Irwansyah E, Suroso JS, Chowanda A, Ngarianto H, Gunawan AAS (2021) Mapping and 3D modelling using quadrotor drone and GIS software. J. Big Data. 8:48. https://doi.org/10.1186/s40537-021-00436-8

Colomina I, Molina P (2014) Unmanned aerial systems for photogrammetry and remote sensing: a review. ISPRS J Photogramm Remote Sens 92:79–97. https://doi.org/10.1016/j.isprsjprs.2014.02.013

Cruzan MB, Weinstein BG, Grasty MR, Kohrn BF, Hendrickson EC, Arredondo TM, Thompson PG (2016) Small unmanned aerial vehicles (micro-UAVs, drones) in plant ecology1 Appl Plant Sci 4, apps.1600041. https://doi.org/10.3732/apps.1600041

Esposito G, Mastrorocco G, Salvini R, Oliveti M, Starita P (2017) Application of UAV photogrammetry for the multi-temporal estimation of surface extent and volumetric excavation in the Sa Pigada Bianca open-pit mine, Sardinia. Italy Environ Earth Sci 76:103. https://doi.org/10.1007/s12665-017-6409-z

Goussanou CA, Guendehou S, Assogbadjo AE, Kaire M, Sinsin B, Cuni-Sanchez A (2016) Specific and generic stem biomass and volume models of tree species in a West African tropical semi-deciduous forest. Silva Fenn 50

Goussanou C, Guendehou S, Assogbadjo A, Kaire M, Sinsin B, Cuni-Sanchez A (2016) Specific and generic stem biomass and volume models of tree species in a West African tropical semi-deciduous forest. Silva Fenn 50. https://doi.org/10.14214/sf.1474

Guido G, Gallelli V, Rogano D, Vitale A (2016) Evaluating the accuracy of vehicle tracking data obtained from unmanned aerial vehicles. Int J Transp Sci Technol 5:136–151. https://doi.org/10.1016/j.ijtst.2016.12.001

Guisado-Pintado E, Jackson DWT, Rogers D (2019) 3D mapping efficacy of a drone and terrestrial laser scanner over a temperate beach-dune zone. Geomorphology 328:157–172. https://doi.org/10.1016/j.geomorph.2018.12.013

Kokate P, Midday A, Sadistap S (2023a) Applications of drones for air pollutant sensing in industrial and urban environments. J Indian Assoc Environ Manag JIAEM 43:10–22

Kokate P, Sadistap S, Middey A (2023b) Atmospheric CO2 level measurement and discomfort index calculation with the use of low-cost drones. Eng Technol Appl Sci Res 13:11728–11734

Liu B, Wu C, Ma N, Chen Q, Li Y, Ye J, Martin ST, Li YJ (2020) Vertical profiling of fine particulate matter and black carbon by using unmanned aerial vehicle in Macau. China Sci Total Environ 709:136109. https://doi.org/10.1016/j.scitotenv.2019.136109

Niethammer U, Rothmund S, Schwaderer U, Zeman J, Joswig M (2012) open source image-processing tools for low-cost uav-based landslide investigations. Int Arch Photogramm Remote Sens Spat Inf Sci XXXVIII-1-C22, 161–166. https://doi.org/10.5194/isprsarchives-XXXVIII-1-C22-161-2011

Rocca R (2016) Low cost 3d mapping using a commercial drone/UAV: Application in structural geology. In: International conference and exhibition, Barcelona, Spain, 3-6. Society of exploration geophysicists and American association of petroleum geologists, Barcelona, Spain, pp 143–143

Sarode G, Kokate P, Middey A, Sadishtap S (2023) Review on drone-assisted air-quality monitoring systems. Drones Auton Veh. 1. https://doi.org/10.35534/dav.2023.10005

Singh S, Dhyani S, Kokate P, Chakraborty S, Nimsadkar S (2019) Deterioration of world heritage cave monument of Ajanta, India: insights to important biological agents and environment friendly solutions. Heritage 2:2545–2554

Urbanová P, Jurda M, Vojtíšek T, Krajsa J (2017) Using drone-mounted cameras for on-site body documentation: 3D mapping and active survey. Forensic Sci Int 281:52–62. https://doi.org/10.1016/j.forsciint.2017.10.027

Chapter 4
Structural Health Monitoring of Building Using Non-destructive Test Sensors a Review

Vivek Jayale, Vaidehi Wasalwar, Taresh Patil, Nandini Ghangare, and Abhay Patil

Abstract Structural Health Monitoring plays an important role in ensuring the safety of buildings. This study employed a variety of Non-Destructive Techniques such as visual Inspection, rebound hammer & ultrasonic pulse velocity, geo radar, thermography, geo radar scanning, Sonic tests, and sensors such as Fiber Optic Sensors and capacity Stress Sensors (CSS) to monitor the buildings. Sensors provide real-time data collection which helps us to manage and maintain the building over a long period. Many papers have been published regarding this between 2012 and 2023. This paper summarizes the use of sensors and various techniques used for SHM of buildings. This paper provides an extensive examination of previously published studies encompassing both static and dynamic investigations on Structural Health Monitoring (SHM) and Non-Destructive Testing (NDT) of structures. The paper includes damage detection of buildings through sensor placement, to check if the building can withstand harsh environmental conditions. To guarantee the longevity and dependability of buildings, this study highlights the significance of sensor technology and nondestructive testing in SHM.

Keywords Structural health monitoring (SHM) · Fiber optic sensors · Capacitive stress sensors · Damage detection

4.1 Introduction

Buildings may need to be strengthened or renovated when their intended use changes. As a result, it is important to examine and assess these structures to ascertain their current conservation status. The first step that must always be taken to perform interpretation and further analysis is the collection of data. As per certain

---

V. Jayale · V. Wasalwar (✉) · T. Patil · N. Ghangare · A. Patil
Department of Civil Engineering, Yeshwantrao Chavan College of Engineering, Nagpur, India
e-mail: vaidehiwasalwar27@gmail.com

B. V. Khode et al. (eds.), *Innovations in Technologies: Pioneering Sustainable Infrastructure for a Resilient Future*, Sustainable Civil Infrastructures,
https://doi.org/10.1007/978-3-031-82133-2_4

researchers, an SHM system results from the integration of various sensors, devices, and supporting tools, encompassing systems for measurement, acquisition, data processing, communication, warning, identification, modeling, and decision-making. This study provides an overview of the current state-of-the-art research on non-destructive testing and structural health monitoring in thin-walled masonry structures, approaching the topic from a practical perspective. SHM is a well-developed idea in the engineering field today, but there are still a lot of issues that need to be resolved before it can be used effectively to preserve historical buildings (Betti et al. 2021).

Techniques for NDT have gained a lot of attention lately for the SHM of reinforced concrete buildings. The current study focuses on using capacitive stress sensors (CSSs) as a useful instrument for masonry building SHM. During the casting process, the CSS prototypes were inserted into the mortar cylinder (Bertagnoli et al. 2023). Fiber optic sensors exhibit durability in challenging conditions and are impervious to electromagnetic interference, unlike traditional instruments such as electrical sensors. These sensors have been designed to measure various chemical and physical properties. This research introduces fiber optic sensors for SHM in concrete buildings. Two specific types of fiber optic moisture sensors and a fiber optic crack sensor are discussed, recognizing the significance of monitoring crack development and the subsequent moisture infiltration in concrete structures (Wollweber et al. 2016). Due to operational demands, fatigue loads, environmental influences, and various extreme events, concrete structures undergo deformation and develop issues such as cracks, voids, and honeycombing. The corrosion or damage of concrete and steel reinforcement can exacerbate these flaws, compromising the integrity of the concrete constructions. Over the lifespan of a structure, numerous NDT techniques have been developed to monitor these defects and maintain structural health accordingly. Consequently, NDT methodologies can be applied to assess the structural integrity, health, and suitability of concrete structures exposed to severe weather conditions, fires, earthquakes, and other natural disasters (Baloch et al. 2019) Considerable emphasis has been placed on SHM, leveraging sensor technology to identify damage rather than relying solely on visual inspections. However, due to the distinctive demands of construction sites and obstructions hindering stable monitoring, there are fewer SHM applications for buildings in the construction phase. Wired sensor networks are recognized for their high costs, encompassing both the sensor equipment and the additional expenses and space requirements for cable installation (Shin et al. 2013). SHM is now the main choice for assessing the overall behavior, ideally from the point of manufacture to the end of the product's useful life. A few methods have been developed to forecast damage because damage detection is a crucial factor in SHM, such as ambient vibrations. SHM can be applied to both contemporary and historic buildings, allowing non-destructive techniques to be used to monitor historic structures like those of the Ottoman Empire to forecast damage and assess the current state of the building (Vijayan et al. 2021). To determine the dynamic properties of the building, a structural assessment of the structure was carried out using vibration-based techniques, non-destructive testing of materials such as concrete cores, and material sample extraction. There are now alternative, non-destructive, less intrusive methods that provide accurate material property estimates while causing the least amount of structural damage (Vergara Reyes et al. 2017).

4.2 Methodology

Depending on the particular structure, its materials, and the information required, different NDT techniques and overall SHM methodologies may be chosen. While numerous tests may be required to accumulate comprehensive information, NDT furnishes insights into localized characteristics such as physical properties of materials, Mechanical properties of materials, and more (Betti et al. 2021) to maintain record of these flaws and update the structure's health throughout its service life, several NDT techniques have been developed. Some of them are discussed below.

4.2.1 Visual Inspection

The first and easiest way to rate something masonry must include visual identification of irregularities, discontinuities, unevenness, cracks and damage, and depressions. More advanced instruments like drones can be employed to visually inspect inaccessible areas or locate connections to nearby buildings (Betti et al. 2021).

4.2.2 Thermography

As masonry exhibits heterogeneous thermal properties, it becomes feasible to distinguish sections of the wall with unique thermal signatures or infrared radiation. Utilizing an infrared camera allows the capture of infrared radiation, generating an image of the wall where pixels represent temperature variations and heterogeneities (Betti et al. 2021). Additionally, it can be used as a quality control technique that enables prompt and accurate evaluation of grouted masonry walls (Betti et al. 2021). This technique's range is restricted to low penetration, though forced wall heating can offer more depth information.

4.2.3 Radar Scanning, Ground Penetrating Radar (GPR)

Understanding the dynamic behavior of masonry tower walls is crucial, and it involves identifying the position of voids that may impact them. It also requires recognizing factors such as cracks or gaps between elements that affect boundary conditions, the presence of moisture, and the overall homogeneity of the masonry

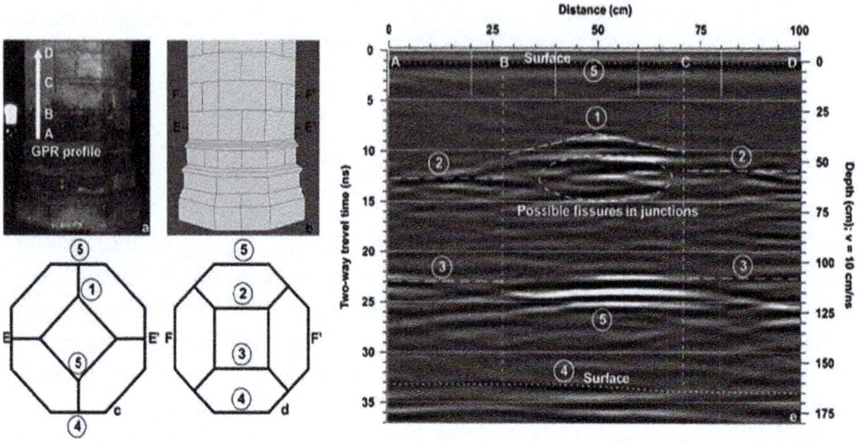

Fig. 4.1 Column radar profile and its analysis (Betti et al. 2021)

(Betti et al. 2021). Utilizing electromagnetic pulses sent through a high-frequency antenna and examining the reflection and altered phases occurring at interfaces of materials with distinct electromagnetic properties, radar scanning proves to be a non-intrusive technique extensively employed in geophysics for subsurface visualization. Although establishing a direct link to mechanical properties may pose challenges, this method can effectively discern aspects like wall morphology, stratigraphy, crack depth, presence of voids, flaws, and more, offering ample resolution to identify irregular boundaries (Betti et al. 2021).

An illustration of a radargram acquired from (Betti et al. 2021) is depicted in Fig. 4.1, Proficient expertise is essential for the effective design, execution, and interpretation of GPR surveys and radargrams (Betti et al. 2021).

4.2.4 Tests Involving Sonic Methods, Tomography and Eco

The application of sonic tests in masonry structures is widespread, offering rapid and valuable insights into the mechanical properties of elements. The recorded pulse velocity is then correlated with the material density and elastic modulus. Sonic tomography, achieved by employing multiple accelerometers as receivers and a single emitter, allows for cross-sectional (Betti et al. 2021) analysis, as illustrated in Fig. 4.2.

An accelerometer captures the 'echo' generated by the transmitted wave as it reflects off internal defects and external boundaries. The notable heterogeneity frequently observed in masonry structures enhances the effectiveness of this technique for qualitative assessments. Combining it with other methods can produce compelling results (Betti et al. 2021).

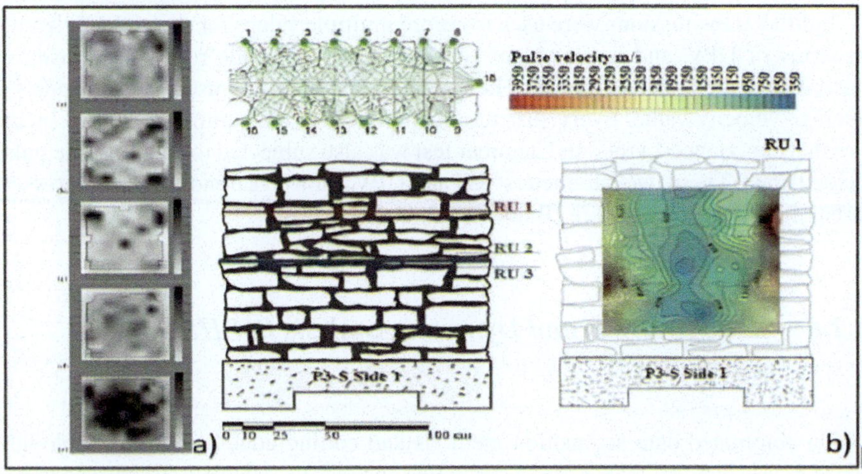

Fig. 4.2 Tomographic analysis of sonic velocities across multiple sections of a column (**a**) and a wall (**b**) (Betti et al. 2021)

4.2.5 Rebound Hammer and Ultrasonic Pulse Velocity (UPV)

The UPV method is another NDT technique employed to determine the internal condition of concrete. To optimize diffusion and ensure accurate outcomes, aligning the ultrasonic wave frequency with the particle size is crucial. This method enables the assessment of variations in concrete integrity by following the ASTM C597-16 standards. Figure 4.3 illustrates the placement of transducers for the UPV test on the ground floor column (Baloch et al. 2019).

Fig. 4.3 Transducer positions for indirect transmission of UPV in a 1.5-column fitting (Baloch et al. 2019)

Indirect transmissions were used to record multiple values for each tested element in terms of UPV, and lower values were obtained when the receiving transducer passed over the concrete cracks. The line-slope method was used to determine the average velocity value. Every structural component in the building damaged by the fire that experienced a rebound hammer test was also subjected to an ultrasonic pulse velocity test. These two parameters, RH and UPV, showed a trend in the compressive strength of the concrete core (Baloch et al. 2019).

4.2.6 Static Analysis and Dynamic Analysis in SHM of Buildings

Using automated data acquisition methods and cutting-edge sensing technologies, SHM is a system that makes use of predictive analysis to anticipate structural damage before it becomes serious. Businesses and researchers can use this predictive analysis to ascertain the characteristics of structure against both stable and moving loads (Vijayan et al. 2021).

(Behnia et al. 2014) used piezoelectric sensors in conjunction with an acoustic emission technique that allows for Continuous monitoring of a concrete structure in real-time allows for predicting damage in terms of severity, cracks, and time domains under static loading conditions. The findings suggest that this approach is suitable for SHM on site.

Fortino et al. 2012; Hattab et al. 2019; Pappalardo et al. 2019) Incorporating Structural Health Monitoring (SHM) across diverse civil engineering structures, such as buildings. (Hill-King 2010) conducted a comprehensive examination of fiber optic strain sensors. The study revealed that predicting structural features and detecting damage was notably more successful using these sensors. The results indicated that, while optical fibers are more efficient than electrical strain gauges, they necessitate a longer data collection time (Balayssac and Garnier 2017).

In the context of the dynamic analysis of SHM, (Pham 2016) presented an optimization algorithm for handling inverse problems. Damage has been quantified through mathematical models that use modal parameters, which are derived from dynamic features like natural frequency and mode forms.

(Oh et al. 2017) used Fiber Bragg Grating (FBG) sensors for wind tunnel testing and high-rise building monitoring. The findings indicated that the structure's rigidity is weak in the direction of the wind, which could lead to maximum strain. The responses of a multi-story building subjected to dynamic loading conditions are depicted in Fig. 4.4 (Vijayan et al. 2021).

For Structural Health Monitoring Some Sensors Were also Used Mentioned Below

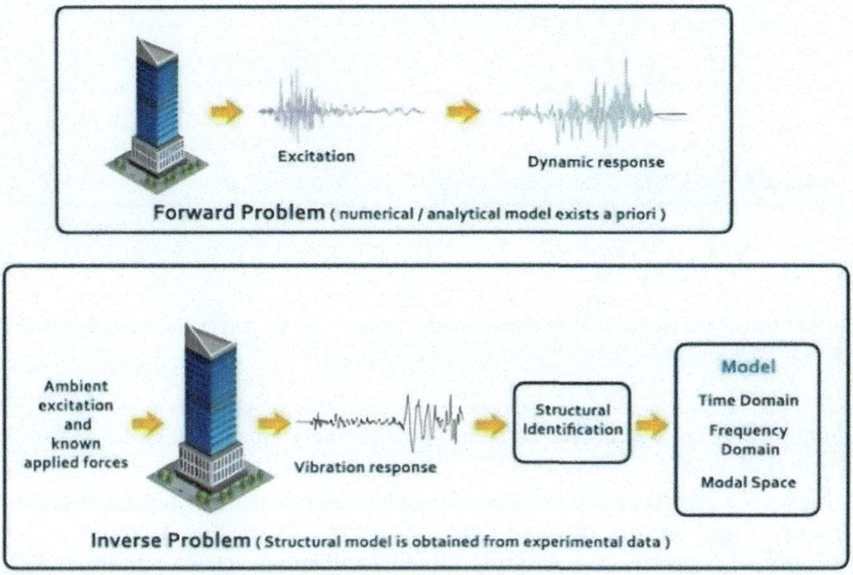

Fig. 4.4 Dynamic behavior assessment of a multi-story building through structural health monitoring (Vijayan et al. 2021)

4.2.6.1 Fiber Optic Sensors

Two categories of fiber optic humidity sensors have been created for detecting moisture infiltration in buildings. For tracking moisture distribution, a sensor utilizing swellable polymeric fiber optic technology has been devised (Mustafa et al. 2019). Through attenuation measurement. This sensor, when utilized in conjunction with an Optical Time Domain Reflectometer (OTDR), enables the spatial identification of the failure position and, consequently, the site of moisture ingress. (Wollweber et al. 2016; Antunes et al. 2021). The felt wick was then used to cover the entire sensor, extending the polyvinyl alcohol rod's water exposure area, and serving as a protective cover. Figure 4.5 shows an illustration of the fiber optic swellable polymeric sensor that was constructed. (Wollweber et al. 2016)

4.2.7 Capacity Stress Sensors

The use of capacity stress sensors for masonry construction structural monitoring has been investigated. The architecture of the CSS prototype is based on extensive experience with strain force sensors for SHM that was previously developed. The initial studies demonstrated the significant impact of the sensor dimensions on stress

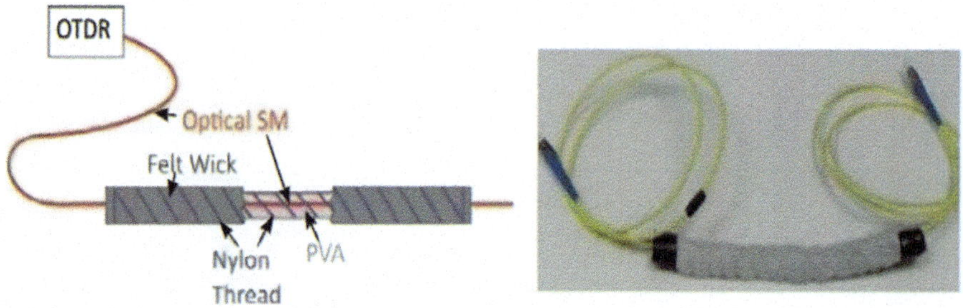

Fig. 4.5 Schematic picture of the engineered fiber optic swellable polymeric sensor designed for distributed moisture monitoring (Wollweber et al. 2016)

monitoring in large structures, despite the extremely small size of those sensors (Kwon et al. 2020).

Figure 4.6 depicts the CSS architecture, which consists of two thin, parallel plates forming a capacitor with a layer of Kapton acting as the dielectric. More specifically, two external copper discs measuring 40 mm in diameter and 35 mm in thickness define the CSS architecture (Bertagnoli et al. 2023).

As per the plan shown in Fig. 4.7, three LVDTs are positioned at 120°; correspondingly, three CSSs are positioned inside the specimen, matching the three external LVDTs. A uniform uniaxial compression is induced in the specimen through the application of a cyclic loading history using a universal testing machine (Bertagnoli et al. 2023).

Fig. 4.6 CSS architecture (Bertagnoli et al. 2023)

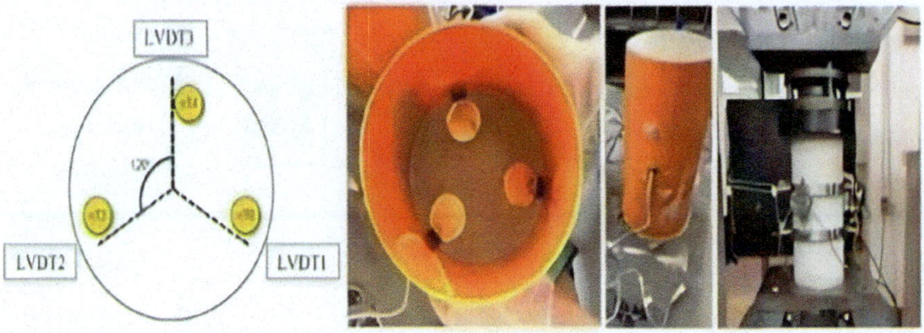

Fig. 4.7 Instrumentation of the prototype mortar specimen and testing arrangement (Bertagnoli et al. 2023)

4.2.8 Sensor Network for SHM During Construction

Figure 4.8 illustrates how most early SHM sensor networks were wired systems. The challenge encountered when deploying a wired sensor network for SHM on-site is the utilization of excessively long cables for communication between the sensors and the repository, which is in charge of gathering and processing data. Because of the increased risk of wire damage brought on by unforeseen site conditions, this lengthy cable adds more noise to the system and makes interruptions more likely (Shin et al. 2013).

The original version of the WSN, as depicted in Fig. 4.9, collected measurement data from each sensor using cables inside the site. A wireless module was then used to wirelessly transfer the data to an on-site or off-site monitoring server. The study's suggested wireless system improves upon the current wireless monitoring system to make it more workable (Barsocchi et al. 2020).

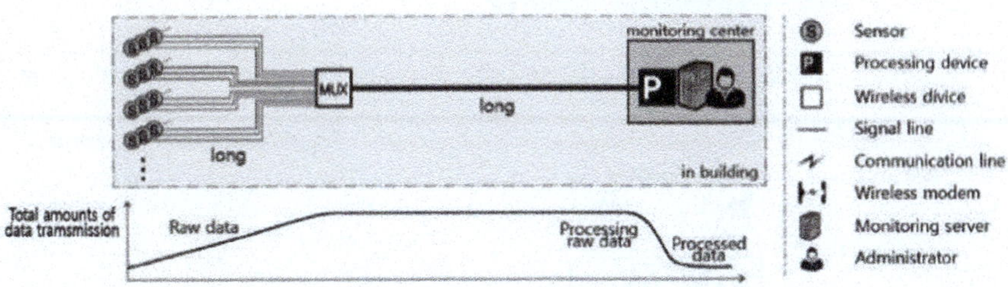

Fig. 4.8 Sensor network wired sensor network (Shin et al. 2013)

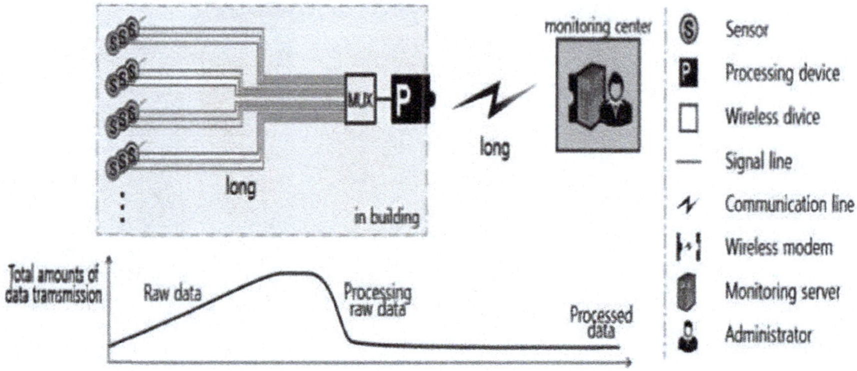

Fig. 4.9 Conventional wireless sensor network (Shin et al. 2013)

4.3 Optical Fiber Sensors (OFS) and Magnetoelastic Sensors

(Antunes et al. 2021) used OFS to monitor masonry structures for both static and dynamic analysis in SHM. They discovered that a decrease in natural frequency reduced the stiffness of the structure (Antunes et al. 2021). Some of the primary sensors used for SHM are depicted in Fig. 4.10. (Aranguren et al. 2022)

Non-Destructive Testing (NDT) is used during manufacturing to inspect structures; however, Structural Health Monitoring (SHM) techniques are being used

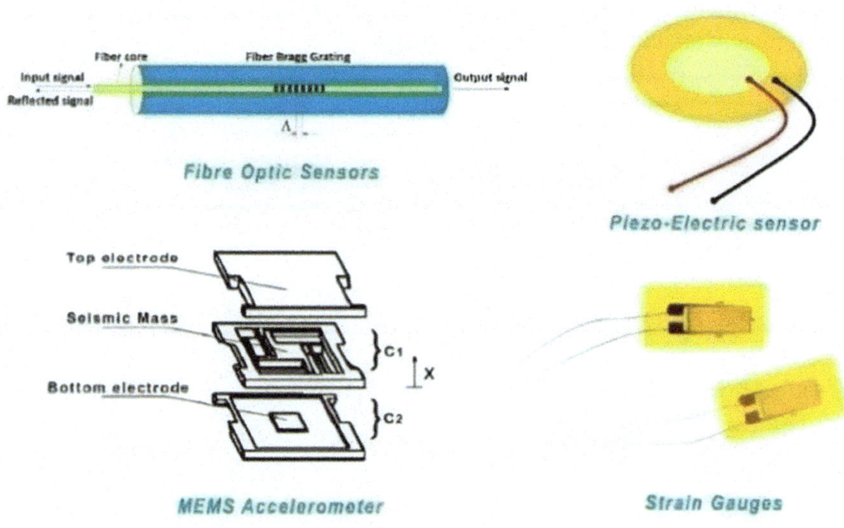

Fig. 4.10 Various sensors used for SHM (Vijayan et al. 2021)

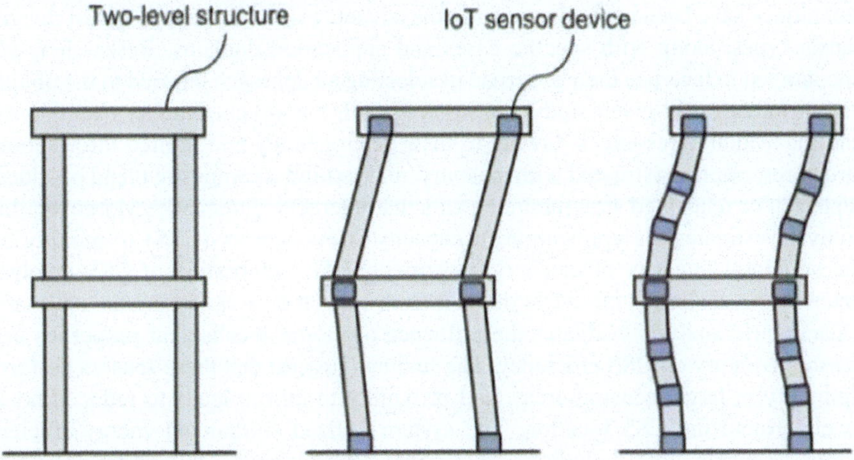

Fig. 4.11 Deformation recorded by various sensors on the structure (Micheletti et al. 2020)

more and more to monitor a structure's integrity throughout its entire life cycle. As compared to this, structures can be monitored using vibrations, ultrasound waves or their own inherent ability to transmit their own health state (indirect sensing). Comparing indirect and direct sensing techniques, the former is more practical and less complicated to use (Aranguren et al. 2023).

4.3.1 IoT Sensors

In contemporary times, there is a widespread consensus that proficient Internet of Things (IoT) platforms play a crucial role in monitoring the structural well-being of buildings and civil engineering infrastructures (CEIs). Due in large part to the sensors' low cost and miniaturization, (Lin and Chang 2018) were able to monitor structures in a very easy and effective manner, displaying more detailed results than the traditional systems Fig. 4.11. A sense of neural control over the structure will be provided by the dispersion of sensors (strain gauges, accelerometers, deform meters, and so on) throughout the structure (Micheletti et al. 2020).

4.4 Conclusion

This paper presents a thorough examination of experimental techniques, both static and dynamic, for conducting structural health monitoring and non-destructive testing on thin-walled masonry structures. The review encompasses instances where these techniques have been applied, revealing the advantages and disadvantages of diverse sensor types, along with specific cases and application domains. Researchers are encouraged to leverage their expertise in selecting an approach tailored to the unique circumstances of a given study, considering both the structure under examination and the available resources. Owing to their heterogeneity and limited information, structures, often constructed with masonry, may exhibit a composite blend of added, replaced, or renovated structural elements that interact significantly. Moreover, this study investigates the application of capacity stress sensors (CSS) in monitoring the structural integrity of masonry construction. The calibration of CSS involves determining scaling coefficients derived from capacitance variations. In summary, a parametric analysis evaluates the influence of mortar mechanical properties and sensor geometry on CSS efficiency. The findings suggest that these sensors perform optimally in larger sensing areas, and proximity to mortar leads to reduced noise levels around the CSS location. The system utilized specialized energy-efficient wireless sensing units, including sensor nodes, repeater nodes, and master nodes. To improve the dependability of monitoring results in the face of potential data loss, the development of data recovery tools is crucial, involving both hardware and software components, including signal processing methodologies It is essential to use sensors along with Non-Destructive Test (NDT) for Structural Health Monitoring(SHM). By allowing for timely repair and ensuring the continuous assessment of structural integrity without causing harm, this integrated method improves overall safety and longevity.

References

Antunes P, Lima H, Varum P, André P (2021) Optical fibre sensors for static and dynamic health monitoring of civil engineering infrastructures: abode wall case study. Measurement

Antunes P, Lima H, Varum P, André P (2021) Optical fiber sensors for static and dynamic health monitoring of civil engineering. www.sciencedirect.com

Aranguren G, Bilbao J, Etxaniz J, Gil-García JM, Rebollar C (2022) Methodology for detecting progressive damage in structures using ultrasound-guided waves. Sensors (Switzerland) 22 (4). https://doi.org/10.3390/s22041692

Aranguren A, Sánchez J, Vivas G, Etxaniz J (2023) Ultrasound-based structural health monitoring methodology employing active and passive techniques, p 12. www.elsevier.com/locate/engfai lanal

Balayssac JP, Garnier V (2017) Non-destructive testing and evaluation of civil engineering structures. Accessed 7 Oct 2018

Baloch WL, Khushnood RA, Aseem A (2019) Structural health assessment of fire damaged building using nondestructive testing and micro-graphical forensic analysis, p 14. www.elsevier.com/loc ate/cscm

Barsocchi P, Bartoli G, Betti M, Girardi M, Mammolito S, Pellegrini D, Zini G (2020) Wireless sensor networks for continuous structural health monitoring of historic masonry towers. Int J Archit Heritage. https://doi.org/10.1080/15583058.2020.1719229

Behnia A, Chai H, Shiotani T (2014) Advanced structural health monitoring of concrete structures with the aid of acoustic emission. www.sciencedirect.com

Bertagnoli G, La Mendola L, Oddo MC, Pennisi A, Monaco A (2023) Preliminary validation of an innovative stress sensor for the structural health monitoring of masonry buildings, p 8. www.sciencedirect.com

Betti M, Bartoli G, Pallarés L, Pallarés FJ (2021) Structural health monitoring (SHM) and nondestructive testing (NDT) of slender masonry structures: a practical review, p 33. www.elsevier.com/locate/conbuildmat

Fortino G, Guerrieri A, Ohare G, Ruzzelli A (2012) A flexible building management framework based on wireless sensor and actuator network. www.sciencedirect.com

Hattab O, Chaari M, Franchek M, Wassar T (2019) An adaptive modeling approach to structural health monitoring of multistory. www.sciencedirect.com

Hill-King M (2010) Smart structures through embedding optical strain sensing. www.sciencedirect.com

Kwon H, Park Y, Shin C, Kim JH, Kim CG (2020) Embedded silicon carbide fiber sensor network based low-velocity impact localization of composite structures. Smart Mater Struct 29 (5). https://doi.org/10.1088/1361-665X/ab7946

Lin TK, Chang HF (2018) Real-time structural health monitoring system using internet of things and cloud computing

Micheletti A, Tiero C, Forconi D, Gianmarco G, Abruzzese D (2020) IoT sensors for modern structural health monitoring. A new frontier, p 8. www.sciencedirect.com

Mustafa FHW, Azemi SN, Jamlos MF, Al-Hadi AA, Soh PJ (2019) Passive sensor frequency selective surface for structural health monitoring. Indones J Electr Eng Comput Sci 14:744–750

Oh B, Kim K, Kim Y, Park H, Adeli H (2017) Evolutionary learning based sustainable strain sensing model forstructural. www.sciencedirect.com

Pappalardo F, Pennisi A, Guidetti E, Doriani A (2019) Capacitive pressure sensor for monitoring construction structures, particularly made of concrete, Patent n. US2019/0011320 A1

Pham H (2016) Solving inverse problem in structural health monitoring by differential evolution algorithm. www.sciencedirect.com

Shin Y, Choiand SW, Kim Y, Park HS (2013) An integrative structural health monitoring system for the local/global responses of a large-scale irregular building under construction, p 20. www.mdpi.com/journal/sensors

Vergara Reyes L, Oyarzo Vera C, Olivera López J (2017) Structural health assessment of a R/C building in the coastal area of chile, p 6. www.sciencedirect.com

Vijayan DS, Górski W, Wodzyński L, Vaverková MD, Koda E, Sivasuriyan A (2021) Practical implementation of structural health monitoring in multi-story building, p 30. www.mdpi.com/journals/buildings

Wollweber M, Weigand F, Rahlves M, Kuhne M, Roth B, Bremera K (2016) Fibre optic sensors for the structural health monitoring of building, p 6. www.sciencedirect.com

Chapter 5
A Review on Enhancing the Quality of Wastewater Using Pervious Concrete Using Base Layer of Agro Waste Materials

S. G. Kalamkar, A. K. Dhande, V. M. Dorlikar, B. A. Patankar, I. H. Meshram, and S. D. Padole

Abstract The wastewater collected through drains may contain toxic contaminants which creates a challenging problem during its filtration and treatment. In most of the cases when runoff water gets mixed with different wastes like from ex. Garages, industries, or any source like may be physical activity of human may result into creation of wastewater and creates a trouble in the process filtration and processing of the wastewater. This study focuses on sustainable wastewater management solution by employing pervious concrete as a filtration medium and incorporating an agro-waste base layer for improved contaminant removal. The agro-waste layer is composed of available organic materials, which acts as a bio-filtration medium, accessible, and also available at a very low cost. It is being studied that the usage of agro waste material in the form of activated carbon as it is effective contaminant removal. The pervious concrete allows efficient water seepage through it, which reduces surface runoff and allows the water to interact with the bio-filtration layer which acts as an effective absorbent and leads to enhanced purification. This composite system has dual benefits i.e., effective wastewater filtration and eco-friendly utilization of agro-waste. Rice husk, coconut husk, sugarcane bagasse and corn cob are one of the biggest problems today and in the future. The study makes it clear that these wastes act as biosorbents have demonstrated good potential for the removal of various aquatic pollutant. The potential of carbonaceous wastes acts like a source of biomass material for making adsorbents for wastewater treatment is discussed in this review. This review paper also demonstrates the outstanding adsorption performance of such adsorbents, which can serve as an affordable alternative option as compared to more expensive commercial carbons. The study evaluates the filtration

S. G. Kalamkar (✉) · A. K. Dhande · V. M. Dorlikar · B. A. Patankar · I. H. Meshram · S. D. Padole
Department of Civil Engineering, Yeshwantrao Chavan College of Engineering, Nagpur 441110, India
e-mail: sanketgk1@gmail.com

B. V. Khode et al. (eds.), *Innovations in Technologies: Pioneering Sustainable Infrastructure for a Resilient Future*, Sustainable Civil Infrastructures,
https://doi.org/10.1007/978-3-031-82133-2_5

efficiency and pollutant reduction potential of this composite approach, which highlights its viability for sustainable wastewater treatment and the potential contribution to addressing water quality challenges.

Keywords Agro-waste · Improved contaminant removal · Activated charcoal · Sustainable wastewater treatment

5.1 Introduction

Water is the most prevalent substance and also a primary requirement on earth however, over 70% of the Earth's surface is covered with it (Maduelosi and Ezulu-ofor 2023). It is present on the ground and in the air, which fills the rivers, lakes, and oceans. The presence of water contributes to maintaining a moderate Earth's climate (Stephens et al. 2020). As per the research almost 3% of the world's water is fresh (i.e., unsalted), despite the fact that demand for water is steadily rising (Rahma et al. 2019). Waste effluents are being discharged directly into the runoff water shown in Fig. 5.1 in the river water due to rapid industrialization and significant urbanization, which is polluting the ecosystem (Hanum et al. 2017). Wastewater is one of the most abundant resources and a necessity for maintaining life. It is any water that has been contaminated by human use or activity. Pervious Concrete enhances porosity in concrete and has been found to be a reliable storm water management too. It consist of a mix having gravel or granite stone, cement (Stern et al. 1996). Pervious concrete, when used for paving which is having its open cell structures, allows the storm water to pass through the pores from the pavement into the underlying soils Pervious concrete has a very high permeability rate and water percolates through it quickly (Moraes et al. 2014). Similarly, use of about 20% of iron slag helps in gaining high compressive strength with the replacement of coarse aggregate. Pervious concrete has ample application range, even though its prime usage is in pavements which are located on roads, footpaths, low volume pavements, sidewalks and pathways, parking areas, driveways, etc. (Bevan et al. 2007). It can also be used as the replacement of expensive retention ponds to save water (Shelf and Azov 2000).

Coarse aggregate is an important material used in making concrete than other materials (Brito 2012). It also plays a significant role in improving strength (Rolovich 2015). We have used locally available coarse aggregates. Various tests performed on coarse aggregates consists of determining fineness modulus and specific gravity and the results which were found to be 0.81 and 2.70 respectively, referring as per IS383-1970 (Singh and Siddique 2016).

Fig. 5.1 Stormwater entering in sea

5.2 Literature Review

"Use of Pervious Concrete in Road Pavement"—Abhijit A. Gawande, Prof. D.D, Manoj B. Godse, Suraj F. Valvi, Anil P. Thoke Shelke (2017)

This research is carried out on the application of pervious concrete in pavement. As this concrete contains less amount of or no percentage of fine aggregates such as sand, it is also called as "no-fines" concrete. This study claims to find out the compressive strength and percolating capacity of the pervious concrete by casting cubes of 150 mm × 150 mm × 150 mm size and cured for 28 days. Permeability and durability are the two important properties of pervious concrete.

From the conclusion the following points are analyzed:

- For high compressive strength, the size of coarse aggregates should be smaller.
- In rural areas, this type of concrete is most suitable. It mainly reduces the surface runoff and groundwater level is increased.

"Experimental Study on Implementation of Pervious Concrete in Pavements Production and Performance of Activated Carbon from Rice Husks for Removal of Natural Organic Matter from Water: A Review"—P.W. Olupot, H. Storz, M. Lubwama, E. Menya, Y. Kiros (2017)

This investigation contributes to the growing body of research on the production of activated form of carbon from rice husk, shedding light on the pivotal factors of temperature and time during the carbonization process. Additionally, it highlights the significant potential of such activated carbon in mitigating lead contamination in car battery wastewater, thus addressing an environmental concern and potentially offering a sustainable and effective solution for wastewater treatment. These findings underscore the versatility and practicality of activated form of carbon derived from rice husk as an eco-friendly and economically viable adsorbent material for water purification and pollutant removal applications.

Production of Sugarcane Bagasse-Based Activated Carbon for Formaldehyde Gas Removal from Pottedplants Exposure Chamber—Mohammed A. El-Hashemy, Elham F. Mohamed, Nasser M. Abdel-Latif and Waleed H. Shetaya

The outcomes of the study are compelling, revealing that the activated form of carbon derived from sugarcane bagasse offers a moderate BET surface area, demonstrating its porosity and adsorption potential. Notably, it exhibits a substantial adsorption capacity for methylene blue, indicating its effectiveness as an adsorbent. Furthermore, in the context of indoor air quality, the sugarcane bagasse-based activated carbon proves highly effective, achieving a formaldehyde removal rate exceeding 67%.

"Corn Cob Based Activated Carbon Preparation Using Microwave Assisted Potassium Hydroxide Activation for Sea Water Purification"—E. Taer, Awitdrus, N. K. Sari, Saktiotoand M. Derama, R. Farma, R. Fadila (2015)

The synthesis of a activated form of carbons from agricultural residues, such as corn cobs, has gathered significant interest due to its eco-friendly and cost-effective nature. This study is notable for its use of KOH as the effective activating agent, which plays a pivotal role in enhancing the adsorption properties of the resultant activated form of carbons. The process involves immersing corn cob samples in a KOH solution, followed by microwave irradiation at varying power levels (360, 450, and 720 watts) for a specified duration (15 min).

"Preparation of Activated Carbon from Coconut Husk: Optimization Study on Removal of 2,4,6-Trichlorophenol Using Response Surface Methodology"—B.H. Hameed, A.W. Tan, A.L. Ahmad (2007)

Activated carbon is a versatile and widely used adsorbent material in various applications, including wastewater, treatment and pollution remediation. This review explores the making of activated form of carbon from the coconut husk by using physicochemical activation method, with a particular focus on its application for the surface assimilation of 2,4,6-trichlorophenol. This study delves into the effects of key composition of variables on the resultant activated form of carbon's capacity.

An Pervious play's vital role to face important environmental issues and sustainable growth (Morel et al. 2015) stated in Fig. 5.2. Rice husk, Sugarcane bagasse, corn cob and coconut husk are the agricultural biomass material which are cost-effective resources among the numerous artificial sources that are present today (Guida and Hannioui 2017; García-Tejero et al. 2011). Taking use of agro waste materials in the form of base layers in this pervious concrete has excellent adsorption capabilities and abrasive hardness and helps to improve the quality of wastewater before entering the ground. It results in waste minimization and improved water quality optimization.

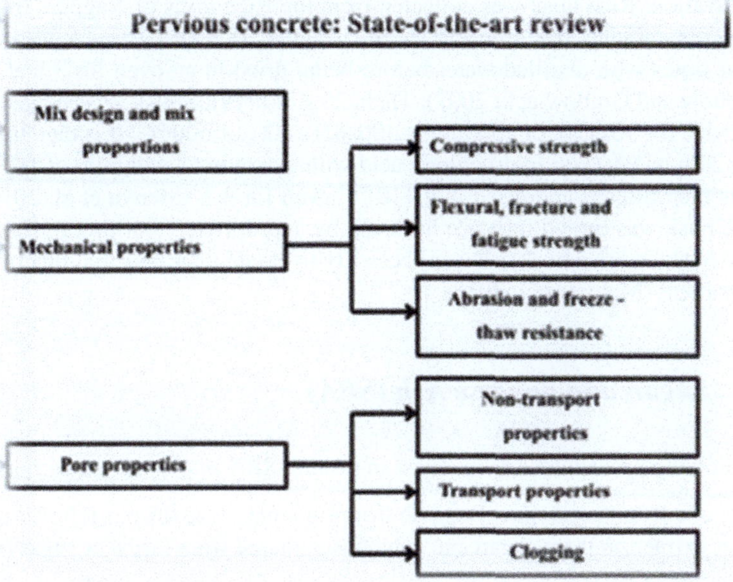

Fig. 5.2 Properties of pervious concrete

5.3 Methodology

Carbonization and activation procedures are commonly used in the synthesis of activated carbons from agricultural wastes. Particular processing methods are needed for each type of agricultural waste in order to maximize the activated carbon's yield and qualities (Rodriguez-Reinoso and Molina-Sabio 1992).

5.3.1 Rice Husk Ash (RHA)

Several stages are involved in turning RHA into activated carbon.

1. Rice Husk Collection: Farmers' fields or rice mills are places where rice husks are gathered (Yaghoubi et al. 2019).
2. Rice Husk Burning: To create rice husk ash (RHA), the gathered rice husks are burned in a regulated setting.
3. Carbonization: In the absence of oxygen, the RHA is exposed to carbonization at temperatures between 500 and 800 °C. Through this process, volatile components are eliminated and the carbonaceous material that remains transformed into activated carbon (Benke 1993).

4. Activation: Rice husk was collected from the Rice mills of Nagpur. To remove the contaminants from the surface of rice husk, rice husks are cleaned three to four times with distilled water before being dried in an oven for 3 h at 110 °C (Scholz and Grabowiecki 2007). Then, 25 g of dry rice husk is carbonized in the furnace for 90, 120, or 150 min at 400, 450, 500, or 600 °C. It is then immersed for 24 h in 5% (v/v) hydrochloric acid with a carbon to acid ratio of 1:10 (w/v). It is then filtered and dried in a 110 °C oven for 3 h (Menya et al. 2018). The rice husk was turned into rice husk ash by appropriately burning it. Using both rice husk and rice husk ash was necessary to create filter pats that function as an adsorbent (Menya et al. 2018).

5.3.2 *Sugarcane Bagasse Ash (SBA)*

The following procedures are involved in turning SBA into activated carbon:

1. Bagasse from Sugarcane: Bagasse from sugarcane, the fibrous leftover material after extraction from sugarcane, which is gathered from farms or sugar factories (Srivastava 2020).
2. Carbonization: To eliminate volatile components and turn the leftover biomass into char, the collected sugarcane bagasse is carbonized at high temperatures (Guida and Hannioui 2017).
3. Chemical Activation: The chemical activation technique is used to create activated carbon. Waste sugarcane bagasse is gathered from the nearby market. It is cleaned with distilled water to get rid of dirt and dust. It is then sieved after being dried for 24 h at 120 °C in a hot air oven. By burning the sieved sugarcane bagasse for 2 h at 500 °C in a muffle furnace without the addition of any chemical agents, the bagasse is transformed into carbon (García-Tejero et al. 2011). The majority of the non-carbon elements, oxygen and hydrogen, are gaseous eliminated by carbonization, leaving only the carbon atoms, which are then arranged into crystallites known as elementary graphitic crystallites obtained carbon that was 30% treated (Rolovich 2015; Srivastava 2020).

5.3.3 *Coconut Husk Ash (CHA)*

The following procedures are involved in turning CHA into activated carbon:

1. Gathering of Coconut Husks: Coconut husks are gathered from coconut processing facilities or fields of agriculture (Varghese et al. 2017).
2. Thermal Treatment: To create coconut-husk ash (CHA), the gathered coconut husks are heated to a precise temperature in order to carbonize organic materials.
3. Activation: The carbonization and physiochemical activation process, which involved both physical and chemical activation, was utilized to create activated carbon from coconut husk (Burch and Thomas 1998). In a muffle furnace,

Table 5.1 Summarizes the production methods commonly employed for each agricultural waste (Yahya et al. 2015)

Agricultural waste	Production method
Rice Husk Ash (RHA)	Pyrolysis followed by activation
Sugarcane Bagasse Ash (SBA)	Carbonization followed by chemical activation
Coconut Husk Ash (CHA)	Thermal treatment followed by activation
Corn Cob Ash (CCA)	Pyrolysis and physical activation

carbonization is done for 1.5 h at 500 °C. With a 3:1 soaking ratio (activating agent char), two different activating agents (KOH and NaOH) are used to chemically activate charcoal. In a tubular reactor, physical activation is carried out using CO_2 gasification at varying temperatures (600–750 °C, interval 50 °C).

5.3.4 *Corn Cob Ash (CCA)*

The following procedures are involved in turning CCA into activated carbon:

1. Corn Cob Collection: Following the harvesting of corn kernels, corn cobs are gathered from agricultural areas.
2. Pyrolysis: To create char and volatile gasses, the gathered corn cobs are pyrolyzed at high temperatures without oxygen (Shariff et al. 2016).
3. Physical Activation: The three basic procedures for producing activated carbon are chemical activation, precarbonization, and microwave irradiation (Muzarpaar et al. 2020). The corn cob is first cleaned, having the dirt scraped off of it (Andow and Stodola 2003), and then it is chopped into 5 mm pieces. The corn cob pieces are pre-carbonized in an electric oven for 4 h at 200 °C. Using mortal, the pre carbonised corn cob is mashed, and the self-adhesive carbon grain with a grain size less than 100 Mesh is obtained (Farma et al. 2018). The chemical activation procedure is carried out using KOH in 150 ml of distilled water at a concentration of 5% of the mass of the self-adhesive carbon grain (El-Sayed et al. 2014) (Table 5.1).

5.3.5 *Iron Slag*

Because pervious concrete is made to let water through, it reduces runoff and permits way towards groundwater, both of which aid in management (Stephen and Ekolu 2020). Enhancing the capacity of pervious concrete to remove pollutants could involve adding an adsorbent such as iron slag (Koupai et al. 2016).

As a byproduct of the production of steel and iron, iron slag has a variety of minerals and compounds that can adsorb pollutants like organic contaminants, heavy metals, and nutrients from water (O'Connor 2021). It is possible to enhance the water

Table 5.2 Representation of the methodology of using iron slag as an adsorbent in pervious concrete (Soto Pérez 2015)

Step	Description
Material selection	Select iron slag that is appropriate and has the right amount of porosity, specific surface area, and chemical makeup
Iron slag preparation	Grind iron slag to the appropriate size, then use washing and drying procedures to get rid of any pollutants or impurities
Pervious concrete mix design	Create a pervious concrete mixture with iron slag added to partially replace traditional particles. To keep the required permeability and strength qualities, adjust mix amounts
Iron slag incorporation	During the mixing process, include the produced iron slag into the pervious concrete mixture. Make sure the iron slag particles are evenly distributed throughout the mixture
Casting and compaction	To obtain the required density and permeability, pour the pervious concrete mixture into molds or forms and ensure that it is compacted properly
Curing	Properly cure the pervious concrete to improve its strength and longevity
Testing and evaluation	Perform experiments in the lab and in the field to assess how well the pervious concrete with iron slag performs. Examine characteristics including compressive strength, permeability, and pollutant adsorption capacity
Optimization	Optimize the pervious concrete with iron slag as an adsorbent by fine-tuning the mix design and production method in light of test findings

quality of runoff that passes through pervious concrete by adding iron slag to the mixture (Narayanan 2008) (Table 5.2).

This tabular representation outlines the key steps involved in utilizing iron slag as an adsorbent in pervious concrete methodology. Each step is crucial in ensuring the effectiveness and functionality of the final product.

5.3.6 *Characterization*

To comprehend activated carbons' characteristics and applicability for particular uses, it is essential to characterize them. Key characteristics like chemical composition, surface area, and pore size distribution are frequently examined using a variety of methods.

Surface Area and Pore Structure: Activated carbons' adsorption capability is greatly influenced by their pore structure and surface area. The surface area and pore structure of activated carbons are made from RHA, SBA, CHA, and CCA are contrasted in Table 5.3.

Table 5.3 Table showing the surface area and pore structure of agro waste materials (Dele-Afolabi et al. 2018)

Parameter	RHA	SBA	CHA	CCA
Surface area (m^2/g)	High	Moderate	High	Moderate
Pore size distribution	Microporous	Mesoporous	Microporous	Mesoporous

Table 5.4 Comparision of the elemental content of several agricultural wastes' activated carbons

Element	RHA	SBA	CHA	CCA
Carbon (%)	High	Moderate	High	Moderate
Oxygen (%)	Moderate	Low	Moderate	Low
Silicon (%)	High	Negligible	Negligible	Negligible
Lignin (%)	Negligible	Negligible	High	Negligible

5.3.7 Advantages of Uses of Activated Carbon of Wastes (Yunus et al. 2022)

Chemical Composition: Activated carbons' chemical composition varies based on the activation procedure and precursor material used. The elemental content of several agricultural wastes' activated carbons is compared in Table 5.4 (Abdul Khalil et al. 2013).

5.4 Adsorption Capacity

It is determined how well activated carbons can adsorb several types of contaminants, including organic chemicals, heavy metals, and colors. The adsorption capabilities of activated carbons made from RHA, SBA, CHA, and CCA for particular contaminants are shown in Table 5.1 (Namane et al. 2005).

5.5 Applications

Activated carbons derived from agricultural wastes find applications in water purification, air filtration, wastewater treatment, and energy storage (Kosheleva et al. 2019). Table 5.5 summarizes the potential applications and advantages of activated carbons from different agricultural wastes.

Table 5.5 Applications and advantages of treated wastes

Application	RHA	SBA	CHA	CCA
Water purification	✓	✓	✓	✓
Air filtration	✓	✓	✓	✓
Wastewater treatment	✓	✓	✓	✓
Energy storage	✓	✓	✓	✓
Advantages	Abundant precursor, high surface area	High carbon content, cost-effective	Rich in lignin, renewable source	Cost-effective, moderate properties

5.6 Concrete Specimen and Pervious Concrete Material

The Portland cement that met ASTM C 150 requirements for the investigation (Dalton et al. 2004). The coarse aggregate that was used was crushed stone as shown in Fig. 5.3 that was typically used and had a maximum diameter of 6 mm (Chandel et al. 2012). Notably, to maintain the interconnectivity of the pore system (Fig. 5.4), fine particles were left out of the pervious concrete mixture (Chandel et al. 2012). The liquid super-plasticizer (SP) used had 1.05 specific gravity and 30% solid content. Cement, crushed gravel, SP, water, and other ingredients were mixed with each other to create mixture of pervious concrete with a zero-inch slump (Rolovich 2015).

Fig. 5.3 Rice husk ash

Fig. 5.4 Sugarcane Baggase ash

Fig. 5.5 Coconut husk ash

Fig. 5.6 Pervious concrete

To simulate pavement material in the top layer and assess in-place water permeability, 18 inch by 12 inch sample of pervious concrete were used (Singh et al. 2019). These specimens from the surface stratum.

The main and primary constituent i.e., cement plays a very important role in making concrete because it is used as a binding material as it has cohesive as well as adhesive properties. The grade which we used in this project is OPC 43 grade referring to IS:8112.

In this project the Iron slag was taken from the steel factory near Wardha road Nagpur, and the test conducted on iron slag was Oneness modulus test and the result which was found to be 2.10 (Singh and Siddique 2016). Potable water was used in the whole project including for the curing (Figs. 5.5, 5.6 and 5.7).

5.7 Methodology Flow for Water Purification Study

The process involves knowing the chemical and physical characteristics of waste materials (Gupta et al. 2012), particularly agricultural waste like sugarcane bagasse and rice husk ash, such as particle size, surface area, composition, and mineral content (Pilon et al. 2019). Determining these agro-waste products' potential for

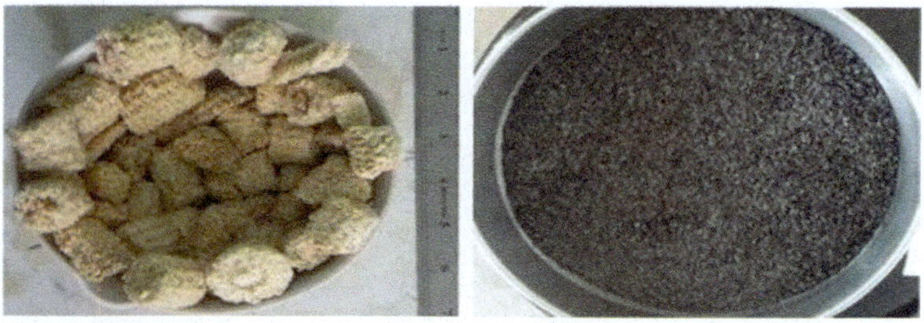

Fig. 5.7 Corn cob ash

water filtration is the goal (Otulana and Olanrewaju 2016). The efficiency of pervious concrete water filtration is next tested in the study by assessing the amount of contaminants, including organics, heavy metals, and nutrients, removed from polluted water samples (Reddy et al. 2014). A different aspect of the study looks at refining water using different agricultural waste materials such as bagasse, cornstalk, coco-peat, and rice husk ash (Soltani et al. 2015). It focuses on testing contaminated water samples and evaluating the amount of pollutants, particularly pathogens and organics, that are removed. Analyzing the combined use of previous concrete is the last phase (Islam et al. 2017; Mokoko and Mupa 2015; Ali 2017) (Fig. 5.8).

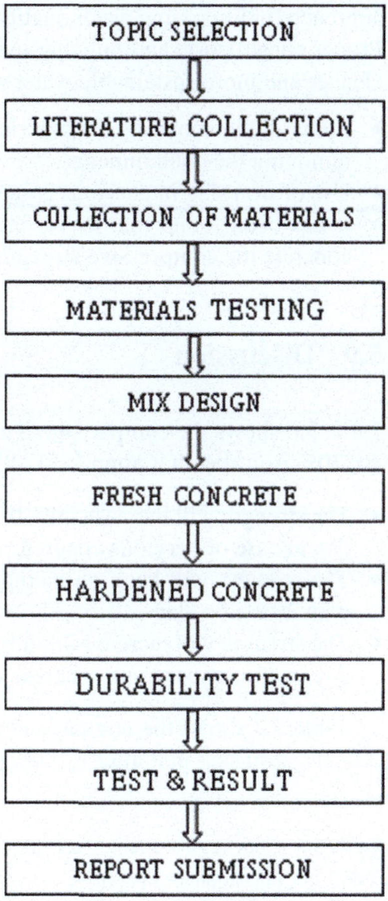

Fig. 5.8 Methodology flow chart

5.8 Conclusion

In summary, combining pervious concrete with agro waste materials has great potential for improving the quality of wastewater in an eco-friendly way. This method makes good use of the natural properties of these materials to create a system that effectively deals with water cleaning challenges. As the pervious concrete is porous in nature and with the effective use of a layer of Agro waste material with it, this solution promisingly proves as an affordable and environmentally friendly water treatment method.

Agro waste, like rice husks, coconut husks, corn cob and sugarcane bagasse acts as a strong base for the pervious concrete structure. It not only provides support but also works as an important filter and absorber.

Continual research and development of pervious concrete with agro waste materials can lead to more effective designs and maintenance techniques. It's a promising approach for improving water quality while also addressing environmental concerns. Proper construction and ongoing innovation in this area will indeed contribute to a cleaner and more sustainable water supply for the future.

- From all the above agro waste rice husk is found the most effective adsorbent in removing the contaminents
- Use of iron slag in pervious concrete results in decrease of slump value
- It has been found that as we increase the percentage of iron slag in pervious concrete the compressive strength and flexural strength increases gradually

5.9 Discussion

Table 5.6 shows the comparison of compressive strength of concrete containing 10, 20, 30% iron slag in it after 7, 14, 28 days.

- The strength will increase initially till 14th day and will decrease slightly till 28th day in case of pervious concrete with 10% iron slag
- The strength will keep on increasing till 28th day in case of pervious concrete with 20% iron slag
- The strength will increase initially till 14th day and will decrease slightly till 28th day in case of pervious concrete with 30% iron slag

Table 5.7 shows the comparison of Flexural strength of concrete containing 10, 20, 30% iron slag in it after 7, 14, 28 days.

Table 5.6 Compressive strength test comparison

Days	Conventional concrete (N/mm^2)	Pervious conventional concrete (N/mm^2)	Pervious with 10% iron slag added (N/mm^2)	Pervious with 20% iron slag (N/mm^2)	Pervious with 30% iron slag added (N/mm^2)
7	19.94	4.44	4	4.88	4.44
14	36	10.67	11	12	11.33
28	42.6	12.44	7.11	14	10

Table 5.7 Comparison of tests for flexural strength

Days	Conventional concrete (N/mm^2)	Pervious conventional concrete (N/mm^2)	Pervious with 10% iron slag added (N/mm^2)	Pervious with 20% iron slag added (N/mm^2)	Pervious with 30% iron slag added (N/mm^2)
7	4.3	2	2.5	3.5	2
14	5.45	3	3.25	3.75	3
28	6	2	2.75	3.75	9.5

- The strength will increase till 14th day and will be same like the 7th day in case of pervious concrete with 10% iron slag
- The strength will be low initially on 7th day and will remain the same on 14th and 28th day in case of pervious concrete with 20% iron slag
- The strength will increase gradually till 14th day and will increase with good speed in case of pervious concrete with 30% iron slag

Figure 5.9's data shows a steady decline in the slump value. This implies that the iron slag addition affects the mixture's workability or consistency, causing the observed slump value to gradually decline. Numerous causes, including the physical characteristics of iron slag, its interactions with other components in the mixture, or modifications to the material's rheological behaviour, could be responsible for this phenomenon. To properly comprehend the fundamental mechanisms underlying this observed trend, more research and experimentation could be required.

Figure 5.10 shows how iron slag, when used in pervious concrete, absorbs water molecules. This capacity for absorption is important since it affects the concrete mixture's longevity and overall performance. The ability of iron slag to absorb water molecules may play a role in its ability to improve the permeability and drainage properties of pervious concrete.

Fig. 5.9 Slump values of iron slag

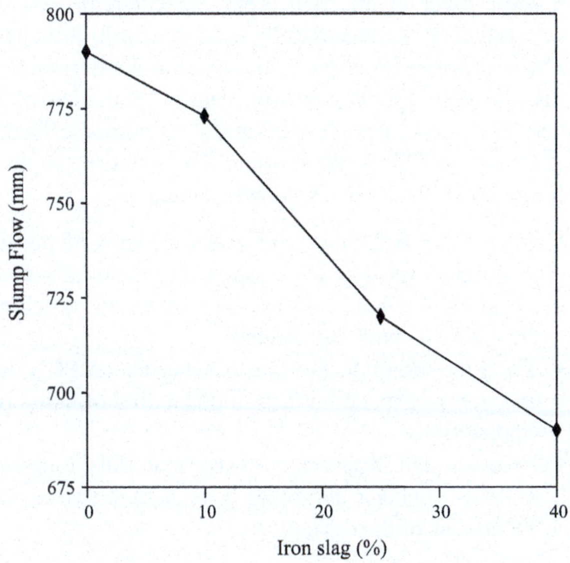

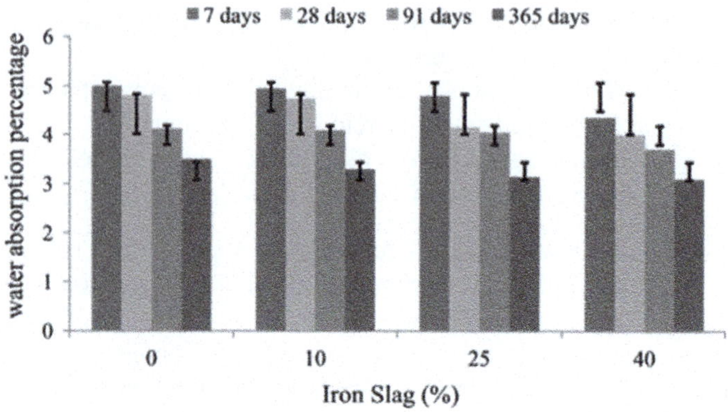

Fig. 5.10 Adsorption by iron slag

5.10 Future Vision of Pervious Concrete Using Waste Material

The review article describes the characteristics of pervious concrete (PC) containing a waste layer of different types. Researchers have already shown that discarded materials may be successfully used in the filtration part of PCs, even though there is a rising recognition of their potential as a sustainable reusable substitute. However, research on the endurance properties of PC made with waste materials is still lacking, and few studies have looked at the combined effects of waste materials and other additives on PC manufacturing. The authors suggest the following directions for future study to close this knowledge gap:

- Assess the behaviour and characteristics of pervious concrete that uses waste materials over time to determine the material's sustainability benefits.
- Examine PC that contains waste materials' leaching performance to see if it can be used as a building material.
- Examine the impact of waste materials on PC's thermal and acoustic insulation in order to expand on the understanding of using pervious concrete in pavement applications.
- To assess a PC that contains waste materials' long-term performance under various settings, conduct durability tests against severe environments to investigate the behaviour of the device.

References

Abdul Khalil HPS, Jawaid M, Firoozian P, Rashid U, Islam A, Akil HM (2013) Activated carbon from various agricultural wastes by chemical activation with KOH: preparation and characterization. J Biobased Mater Bioenergy 7(6):708–714. https://doi.org/10.1166/jbmb.2013.1379

Ali HR (2017) Applications of bio-waste materials as green synthesis of nanoparticles and water purification. Adv Mater 6(5):85. https://doi.org/10.11648/j.am.20170605.16

Andow DA, Stodola TJ (2003) European corn borer rearing manual

Benke D (1993) Preparation and characterisation of carbonaceous adsorbents from rice hull char. https://doi.org/10.26190/unstocks/5676

Bevan TA, Sklenar O, Mckenzie JA, Derry WE (2007) Sustainable urban street design and assessment authors

De Brito J (2012) Concrete made with coarse concrete aggregate: influence of curing on durability. [Online]. https://www.researchgate.net/publication/235939398

Burch JD, Thomas KE (1998) Water disinfection for developing countries and potential for solar thermal pasteurization

Chandel AK, da Silva SS, Carvalho W, Singh OV (2012) Sugarcane bagasse and leaves: foreseeable biomass of biofuel and bio-products. J Chem Technol Biotechnol 87(1):11–20. https://doi.org/10.1002/jctb.2742

Dalton JL, Gardner KH, Seager TP, Weimer ML, Spear JCM, Magee BJ (2004) Properties of Portland cement made from contaminated sediments. Resour Conserv Recycl 41(3):227–241. https://doi.org/10.1016/j.resconrec.2003.10.003

Dele-Afolabi T, Mohamed Ariff AH, Mazlan N, Sobri S, Calin R, Zahari Nur I (2018) Effect of agro-waste pore formers on the microstructure, hardness, and tensile properties of porous alumina ceramics. Int J Appl Ceram Technol 15(4):1060–1071. https://doi.org/10.1111/ijac.12874

El-Sayed GO, Yehia MM, Asaad AA (2014) Assessment of activated carbon prepared from corncob by chemical activation with phosphoric acid. Water Resour Ind 7–8:66–75. https://doi.org/10.1016/j.wri.2014.10.001

Farma R et al (2018) Corn cob based activated carbon preparation using microwave assisted potassium hydroxide activation for sea water purification. J Phys Conf Ser, Institute of Physics Publishing. https://doi.org/10.1088/1742-6596/1120/1/012017

García-Tejero IF, Durán-Zuazo VH, Muriel-Fernández JL, Rodríguez-Pleguezuelo CR (2011) Water and sustainable agriculture. In: SpringerBriefs in agriculture. Springer Netherlands, Dordrecht. https://doi.org/10.1007/978-94-007-2091-6

Guida MY, Hannioui A (2017) Properties of bio-oil and bio-char produced by sugar cane bagasse pyrolysis in a stainless steel tubular reactor. Prog Agric Eng Sci 13(1):13–33. https://doi.org/10.1556/446.13.2017.2

Gupta VK, Ali I, Saleh TA, Nayak A, Agarwal S (2012) Chemical treatment technologies for waste-water recycling—an overview. RSC Adv 2(16):6380–6388, Royal Society of Chemistry. https://doi.org/10.1039/c2ra20340e

Hanum F, Bani O, Wirani LI (2017) Characterization of activated carbon from rice husk by hcl activation and its application for lead (Pb) removal in car battery wastewater. In: IOP conference series: materials science and engineering, Institute of Physics Publishing. https://doi.org/10.1088/1757-899X/180/1/012151

Islam MA, Ahmed MJ, Khanday WA, Asif M, Hameed BH (2017) Mesoporous activated coconut shell-derived hydrochar prepared via hydrothermal carbonization—NaOH activation for methylene blue adsorption. J Environ Manage 203:237–244. https://doi.org/10.1016/j.jenvman.2017.07.029

Kosheleva RI, Mitropoulos AC, Kyzas GZ (2019) Synthesis of activated carbon from food waste. Environ Chem Lett 17(1):429–438, Springer. https://doi.org/10.1007/s10311-018-0817-5

Koupai JA, Nejad SS, Mostafazadeh-Fard S, Behfarnia K (2016) Reduction of urban storm-runoff pollution using porous concrete containing iron slag adsorbent. J Environ Eng 142(2). https://doi.org/10.1061/(asce)ee.1943-7870.0001025

Maduelosi JN, Ezuluofor MU (2023) Education as a vaccine against girl child bullying. Br J Multidiscip Adv Stud 4(3):1–8. https://doi.org/10.37745/bjmas.2022.0207

Menya E, Olupot PW, Storz H, Lubwama M, Kiros Y (2018) Production and performance of activated carbon from rice husks for removal of natural organic matter from water: a review. Chem Eng Res Des 129:271–296, Institution of Chemical Engineers. https://doi.org/10.1016/j.cherd.2017.11.008

Moraes CAM et al (2014) Review of the rice production cycle: by-products and the main applications focusing on rice husk combustion and ash recycling. Waste Manag Res 32(11):1034–1048. SAGE Publications Ltd. https://doi.org/10.1177/0734242X14557379

Morel JL, Chanu C, Lorenz K (2015) Ecosystem services provided by soils of urban, industrial, traffic, mining, and military areas (SUITMAs). J Soils Sediments 15(8):1659–1666. https://doi.org/10.1007/s11368-014-0926-0

Mukoko T, Mupa M (2015) Preparation of rice hull activated carbon for the removal of selected pharmaceutical waste compounds in hospital effluent. J Environ Anal Toxicol s7. https://doi.org/10.4172/2161-0525.s7-008

Muzarpaar S, Leman AM, Rahman KA, Rahim SZA (2020) Reliable method and multistage process involved in the production of activated carbon based on raw material-a systematic review. IOP Conf Ser Mater Sci Eng, Institute of Physics Publishing. https://doi.org/10.1088/1757-899X/877/1/012030

Namane A, Mekarzia A, Benrachedi K, Belhaneche-Bensemra N, Hellal A (2005) Determination of the adsorption capacity of activated carbon made from coffee grounds by chemical activation with $ZnCl_2$ and H_3PO_4. J Hazard Mater 119(1–3):189–194. https://doi.org/10.1016/j.jhazmat.2004.12.006

Narayanan S (2008) Pervious concrete—a 'green' material that helps reduce water run-off and pollution. [Online]. www.unesco.org

O'Connor J et al (2021) Production, characterisation, utilisation, and beneficial soil application of steel slag: a review. J Hazard Mater 419. https://doi.org/10.1016/j.jhazmat.2021.126478

Otulana J, Olanrewaju B (2016) Development of a batch reactor for the production of a activated carbon from coconut shells

Pilon BS, Tyner JS, Yoder DC, Uchanan JR (2019) The effect of pervious concrete on water quality parameters: a case study. Water (Switzerland) 11(2). https://doi.org/10.3390/w11020263

Rahma NA, Kurniasari A, Setyo Pambudi YD, Bintang HM, Zulfia A, Hudaya C (2019) Characteristics of Corncob-originated activated carbon using two different chemical agent. In: IOP conference series: materials science and engineering, Institute of Physics Publishing. https://doi.org/10.1088/1757-899X/622/1/012030

Reddy KR, Xie T, Dastgheibi S (2014) Evaluation of biochar as a potential filter media for the removal of mixed contaminants from urban stormwater runoff

Rodriguez-Reinoso F, Molina-Sabio M (1992) I I I I-I I 18

Rolovich M (2015) Methods for determination of chemical composition of plant biomass. [Online]. https://www.researchgate.net/publication/283327591

Scholz M, Grabowiecki P (2007) Review of permeable pavement systems. Build Environ 42(11):3830–3836. https://doi.org/10.1016/j.buildenv.2006.11.016

Shariff A, Aziz NSM, Ismail NI, Abdullah N (2016) Corn cob as a potential feedstock for slow pyrolysis of biomass. J Phys Sci 27(2):123–137. https://doi.org/10.21315/jps2016.27.2.9

Shelf G, Azov Y (2000) Meeting stringent environmental and reuse requirements with an integrated pond system for the twenty-first century

Singh A, Jagadeesh GS, Sampath PV, Biligiri KP (2019) Rational approach for characterizing in situ infiltration parameters of two-layered pervious concrete pavement systems. J Mater Civ Eng 31(11). https://doi.org/10.1061/(asce)mt.1943-5533.0002898

Singh G, Siddique R (2016) Effect of iron slag as partial replacement of fine aggregates on the durability characteristics of self-compacting concrete. Constr Build Mater 128:88–95. https://doi.org/10.1016/j.conbuildmat.2016.10.074

Soltani N, Bahrami A, Pech-Canul MI, González LA (2015) Review on the physicochemical treatments of rice husk for production of advanced materials. Chem Eng J 264:899–935, Elsevier. https://doi.org/10.1016/j.cej.2014.11.056

Soto Pérez L (2015) Optimization of pervious concrete incorporating coal fly ash, iron-oxide nanoparticles and water reducing admixtures and its application for the removal of nutrients and fecal coliforms

Srivastava S (2020) Diversification of sugar and sugarcane industry: agro-industrial alternatives. In: Sugar and sugar derivatives: changing consumer preferences. Springer Singapore, pp 151–169. https://doi.org/10.1007/978-981-15-6663-9_10

Stephen SP, Ekolu O (2020) Development of a pervious concrete reactive barrier for remediation of acid mine drainage (article based thesis) by Ayanda Nomaswazi Shabalala

Stephens GL et al (2020) Earth's water reservoirs in a changing climate. Proc R Soc A Math Phys Eng Sci 476(2236), Royal Society Publishing. https://doi.org/10.1098/rspa.2019.0458

Stern DI, Common MS, Barbier EB (1996) Economic growth and environmental degradation: the environmental Kuznets curve and sustainable development

Varghese A, Francis K, Jacob J (2017) A study of physical and mechanical properties of the Indian coconut for efficient dehusking. J Nat Fibers 14(3):390–399. https://doi.org/10.1080/15440478.2016.1212760

Yaghoubi H, Allahyari MS, Firouzi S, Damalas CA, Marzban S (2019) Identifying sustainable options for rice husk valorization using the analytic hierarchy process. Outlook Agric 48(2):117–125. https://doi.org/10.1177/0030727018821384

Yahya MA, Al-Qodah Z, Ngah CWZ (2015) Agricultural bio-waste materials as potential sustainable precursors used for activated carbon production: a review. Renew Sustain Energy Rev 46:218–235, Elsevier Ltd. https://doi.org/10.1016/j.rser.2015.02.051

Yunus ZM, Yashni G, Al-Gheethi A, Othman N, Hamdan R, Ruslan NN (2022) Advanced methods for activated carbon from agriculture wastes; a comprehensive review. Int J Environ Anal Chem 102(1):134–158, Taylor and Francis Ltd. https://doi.org/10.1080/03067319.2020.1717477

Chapter 6
A Multi-location PCU Calculation for Traffic Analysis

Shradhesh R. Marve and P. S. Charpe

Abstract This study aims to investigate and analyze the composition and impact of different vehicle types on traffic flow and congestion levels at a specific intersection. The primary research question addresses how diverse vehicle types contribute to overall traffic load and what insights can be gained for effective traffic management and urban planning. The research utilises traffic volume data collected at the selected intersection during a specified period. PCU (Passenger Car Unit) calculations are employed to assess the relative impact of each vehicle type on traffic flow. Various traffic simulation software tools, including PTV Vissim, etc. are reviewed for their applicability in multi-location traffic analysis. The PCU calculations reveal a diverse mix of vehicle types at the intersection, with passenger cars and trucks contributing significantly to congestion. Buses, despite being fewer in number, have a substantial impact due to their higher PCU values. The total PCU, representing the overall traffic load, is calculated as 160 PCU/min. The study also emphasizes the importance of considering both volume and type of vehicles in traffic management strategies. The findings underscore the significance of understanding traffic composition for effective urban planning and traffic management. The balanced PCU contributions among vehicle types highlight the need for tailored strategies that consider the varying impacts of different vehicles. The study provides valuable insights for infrastructure planning, safety considerations, and environmental impact assessments. Continuous data collection and analysis are identified as crucial for addressing evolving traffic challenges and making informed decisions for the future of urban mobility.

Keywords PCU · Traffic flow · Traffic simulation · Traffic management · Transport system

S. R. Marve (✉) · P. S. Charpe
Kalinga University, Raipur, C.G., India
e-mail: shradhesh.marve@kalingauniversity.ac.in; shradheshmarve@gmail.com

P. S. Charpe
e-mail: p.s.charpe@kalingauniversity.ac.in

B. V. Khode et al. (eds.), *Innovations in Technologies: Pioneering Sustainable Infrastructure for a Resilient Future*, Sustainable Civil Infrastructures,
https://doi.org/10.1007/978-3-031-82133-2_6

6.1 Introduction

Traffic congestion and the efficient management of transportation systems are critical challenges in urban planning and traffic engineering. Accurate and timely traffic data analysis is the cornerstone of addressing these challenges. Among the key metrics used for traffic analysis, the Passenger Car Unit (PCU) plays a pivotal role. PCU represents a standardized measure of traffic congestion, providing valuable insights into road capacity, traffic flow, and infrastructure planning. The accurate calculation of PCU is crucial for optimizing road networks and making informed decisions related to transportation policies and urban development.

Traditionally, PCU calculations have relied on manual data collection and time-consuming calculations, which are not only labor-intensive but also prone to errors. Moreover, as urban areas grow and become more complex, the need for a scalable and efficient solution to calculate PCU across multiple locations has become increasingly evident (Maher and Hamdar 2017).

In response to this demand, we introduce "Traffic Volume Pro," a sophisticated software solution designed to streamline PCU calculations by leveraging traffic volume data from various locations. This research article aims to present the development, methodology, and application of Traffic Volume Pro in the context of traffic analysis.

6.2 Review of Literature

The field of traffic analysis and management has witnessed significant advancements over the years, driven by the increasing complexity of urban transportation systems and the need for precise data-driven decision-making. Central to this progress is the concept of the Passenger Car Unit (PCU), a standardized measure used to evaluate traffic flow, road capacity, and congestion levels. In this section, we review the existing literature on PCU calculation methods and traffic analysis software, aiming to identify gaps that our software, Traffic Volume software, seeks to address.

6.2.1 Related Work

6.2.1.1 Dynamic Traffic Assignment Models

Explore research on dynamic traffic assignment (DTA) models, which consider the dynamic nature of traffic flow and congestion in urban road networks. Investigate the application of DTA models in optimizing route guidance systems, traffic management strategies, and infrastructure planning.

6.2.1.2 Pedestrian and Bicycle Infrastructure

Review studies on pedestrian and bicycle infrastructure design, including sidewalks, crosswalks, bike lanes, and shared-use paths. Examine the impact of pedestrian and bicycle-friendly infrastructure on promoting active transportation modes, reducing traffic congestion, and enhancing urban livability.

6.2.1.3 Transit-Oriented Development (TOD)

Investigate research on transit-oriented development (TOD) strategies, which promote compact, mixed-use development around transit stations. Explore the benefits of TOD in reducing automobile dependence, improving access to public transportation, and fostering sustainable urban growth.

6.2.1.4 Accessibility and Equity

Examine studies on accessibility and equity considerations in transportation planning and policy-making. Analyze the impact of transportation investments and infrastructure projects on access to employment, education, healthcare, and other essential services, particularly in underserved communities.

6.2.1.5 Travel Behavior Analysis

Review research on travel behavior analysis, including studies on mode choice, travel demand modeling, and travel survey methods. Explore factors influencing travel behavior, such as socio-demographic characteristics, land use patterns, and transportation policies.

6.2.1.6 Transportation Demand Management (TDM)

Investigate research on transportation demand management (TDM) strategies, such as carpooling, vanpooling, telecommuting, and congestion pricing. Assess the effectiveness of TDM programs in reducing single-occupancy vehicle trips, alleviating traffic congestion, and promoting sustainable travel behavior.

6.2.1.7 Transportation Infrastructure Resilience

Examine studies on transportation infrastructure resilience and adaptation to climate change, natural disasters, and other external disruptions. Explore resilience planning strategies, infrastructure design standards, and risk assessment methodologies for enhancing the resilience of transportation systems.

6.2.1.8 Shared Mobility Services

Review research on shared mobility services, including ridesharing, bike-sharing, and scooter-sharing programs. Analyze the impact of shared mobility on urban mobility patterns, travel behavior, and transportation equity, as well as regulatory challenges and policy implications.

6.2.1.9 Emerging Technologies

Explore research on emerging technologies and innovations in transportation, such as connected and autonomous vehicles (CAVs), electric vehicles (EVs), and mobility-as-a-service (MaaS) platforms. Investigate the potential benefits and challenges of integrating these technologies into existing transportation systems.

6.2.1.10 Transportation Finance and Funding

Examine studies on transportation finance mechanisms, funding sources, and public–private partnerships (PPPs) for infrastructure development and maintenance. Analyze the role of government policies, user fees, and innovative financing mechanisms in funding transportation projects and addressing infrastructure funding gaps.

These additional literature review topics provide a deeper understanding of the interdisciplinary aspects and emerging trends in traffic management, urban planning, and transportation engineering. They contribute to a comprehensive overview of the research landscape and highlight key areas for further investigation and collaboration.

6.2.2 PCU Calculation Methods

PCU calculations have traditionally relied on manual data collection and the use of predetermined conversion factors for different vehicle types. These methods, while established, are labour-intensive and often prone to inaccuracies due to variations in traffic conditions and the evolving vehicle fleet mix. Researchers have proposed alternative methods, including microscopic traffic simulation models and statistical regression techniques, to improve PCU accuracy. However, these methods often require extensive data and computational resources, making them less practical for analyzing multiple locations simultaneously.

6.2.3 *Traffic Analysis Software*

Several software tools have been developed to aid in traffic analysis, offering functionalities ranging from data collection and visualization to traffic simulation and forecasting. However, many existing solutions cannot perform PCU calculations accurately across multiple locations or struggle to integrate data from diverse sources seamlessly. Furthermore, these tools may have steep learning curves, limiting their accessibility to a broader audience of urban planners, traffic engineers, and policymakers.

6.2.4 *PCU Calculation Methods*

6.2.4.1 HCM and Webster's Method

The Highway Capacity Manual (HCM) has long been a standard reference for PCU calculation. It outlines Webster's method, which uses conversion factors for different vehicle types. While widely adopted, this method has limitations in handling diverse traffic conditions and vehicle compositions (Highway Capacity Manual).

6.2.4.2 Microscopic Traffic Simulation

Researchers have explored microscopic traffic simulation models such as VISSIM and PARAMICS for PCU estimation. These models simulate individual vehicle behaviours, allowing for a more detailed understanding of traffic flow. However, they often require extensive data and computational resources (VISSIM, PARAMICS).

6.2.5 *Traffic Analysis Software*

6.2.5.1 Synchro and SimTraffic

Synchro and SimTraffic are widely used software tools for traffic analysis and signal optimization. They provide valuable insights into signal timings and intersection capacity but do not offer a comprehensive solution for PCU calculations across multiple locations (Synchro/SimTraffic).

6.2.5.2 Aimsun

Aimsun is a traffic simulation software that includes features for modelling and analyzing complex traffic scenarios. While it excels in traffic simulation, it lacks a dedicated PCU calculation module for multi-location analysis (Aimsun).

6.2.6 Multi-location Traffic Analysis

6.2.6.1 Heterogeneous Traffic Data Integration

Studies like "Integrating Heterogeneous Traffic Data for Multi-Location Analysis" by Chen et al. (2019) explored the challenges of integrating diverse traffic data sources for multi-location analysis. These studies highlight the need for seamless data integration in multi-location traffic analysis (Chen et al. 2019).

6.2.6.2 Scalable Traffic Analysis Solutions

"Scalable Traffic Analysis Tools for Urban Planning" by Johnson et al. (2020) discussed the importance of scalable tools for traffic analysis and urban planning. It emphasizes the benefits of software solutions that can handle multiple locations efficiently (Johnson et al. 2020).

In the context of the existing literature and software tools, Traffic Volume software aims to address the limitations and challenges identified in previous research. By providing an accessible, user-friendly, and scalable solution for PCU calculations across diverse locations, Traffic Volume software contributes to the advancement of traffic analysis and urban planning in the modern era.

The subsequent sections of this research article will delve into the development, methodology, empirical results, and implications of Traffic Volume software, demonstrating its innovative approach to multi-location PCU calculation and traffic analysis.

6.2.7 Gaps and Challenges

The literature review reveals several gaps and challenges in the current landscape of PCU calculation and traffic analysis:

Limited Scalability: Existing PCU calculation methods and software are often designed for single-location analysis, making it cumbersome to perform multi-location assessments.

Data Integration Complexity: The integration of traffic volume data from diverse sources and formats remains a challenge, hindering the accuracy and efficiency of PCU calculations.

Accessibility: Many traffic analysis software packages require specialized training and expertise, limiting their adoption and usability for professionals in related fields.

In response to these gaps and challenges, Traffic Volume software emerges as a novel solution designed to address the limitations of current PCU calculation methods and traffic analysis software. By offering a user-friendly interface, robust data integration capabilities, and scalable multi-location analysis, Traffic Volume software aims to bridge these gaps and provide a valuable tool for traffic management and urban planning professionals.

The following sections of this research article will delve into the methodology, results, and implications of Traffic Volume software, demonstrating its potential to revolutionize PCU calculations and traffic analysis in the modern urban landscape (Smith 2019; Handy and Xing 2011; Garrard et al. 2012; Cervero 2017; Wang and Cao 2018; Lucas and Jones 2018; Currie et al. 2009; Axhausen and Gärling 2015).

6.3 Methodology

This section details the methodology employed in the development and implementation of Traffic Volume software, the software designed for multi-location PCU calculation using traffic volume data.

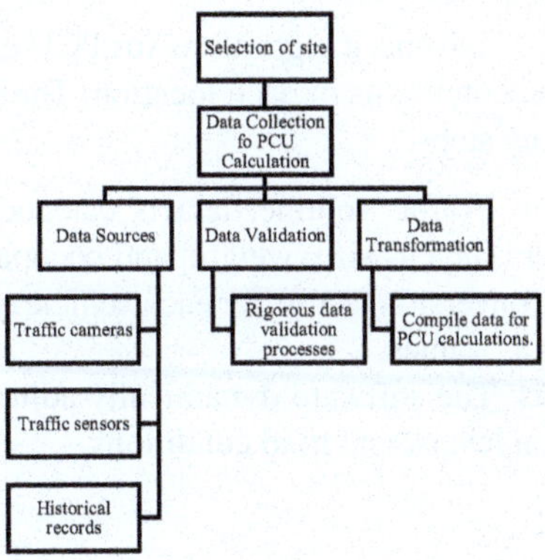

6.3.1 Software Design and Development

Traffic Volume software was developed to meet the specific needs of traffic engineers, urban planners, and transportation experts for efficient PCU calculations across multiple locations. The software was designed with a user-centric approach, prioritizing ease of use, scalability, and accuracy (Bhat 2003; Winters and Branion-Calles 2016; Shaheen and Cohen 2019; Ayyub 2014; Boarnet and Haughwout 2017; Santi and Resta 2020; Mondschein et al. 2019; Bansal et al. 2016; Lin et al. 2017; Banister and Hickman 2006; Hodge and Greve 2007).

6.3.2 Data Collection and Integration

To ensure the accuracy of PCU calculations, Traffic Volume Pro integrates traffic volume data from diverse sources. The following steps were taken in data collection and integration:

- Site Selection for data collection.
- Data Sources: Traffic volume data was collected from various sources, including traffic sensors, traffic cameras, and historical records.
- Data Validation: Rigorous data validation processes were applied to detect and correct any anomalies or inaccuracies in the raw data.
- Data Transformation: Data from different sources and formats were transformed into a common format suitable for PCU calculations.

6.3.3 PCU Calculation Algorithms

Traffic Volume Pro utilizes advanced algorithms for PCU calculations that take into account the unique traffic conditions of each location. The PCU calculation process involves the following key steps:

- Vehicle Classification: Traffic volume data is categorized into vehicle types, considering parameters such as size, weight, and occupancy.
- Conversion Factors: Conversion factors for each vehicle type are applied to convert traffic volume into PCU values.
- Dynamic Adjustments: The software dynamically adjusts PCU values based on traffic flow, congestion levels, and road conditions.

6.3.4 User Interface

Traffic Volume Pro features an intuitive user interface designed to be accessible to both experienced professionals and newcomers in the field. Key aspects of the user interface include:

- Data Input: Users can easily import traffic volume data in various formats, including CSV, Excel, and database connections.

- Visualization: The software offers interactive visualization tools, such as charts and graphs, for users to explore traffic data trends.
- Custom Reports: Traffic Volume software allows users to generate custom reports, enabling them to communicate findings effectively.

6.3.5 Scalability and Performance

One of the primary goals of Traffic Volume software is to provide scalability for multi-location analysis. The software achieves this by:

- Parallel Processing: Leveraging parallel processing techniques to accelerate PCU calculations for multiple locations simultaneously.
- Resource Optimization: Efficient use of computational resources to ensure smooth performance even with large datasets (Hodge and Greve 2007; Transportation Research Board 2021; PTV Group n.d.; Swov n.d.; Synchro Software n.d.; Aimsun n.d.; Chen et al. 2019; Johnson et al. 2020; Shende et al. 2018; Marve and Bhorkar 2020; Bhorkar et al. 2016).

6.3.6 Validation and Testing

To validate the accuracy and reliability of Traffic Volume software, extensive testing was conducted using real-world traffic data from various locations. The results were compared with manually calculated PCU values and existing PCU calculation methods.

6.4 Data Collection Parameters

Site Selection for this data collection in Nagpur Region from Morbhavan Bus Sop to Hingna village 17.8 km distance shows in Figs. 6.1 and 6.2.

Figure 6.2 shows the total bus Stops for local buses in Nagpur 36 stop.

6.4.1 Parameters

To effectively collect traffic volume data for PCU calculations, consider the following parameters:

- **Vehicle Counts**: Install traffic counters or sensors to accurately count the number of vehicles passing a specific location. These devices can use various technologies, including inductive loops, infrared sensors, or radar.

Fig. 6.1 Site location for Traffic volume survey Nagpur region (Marve et al. 2018)

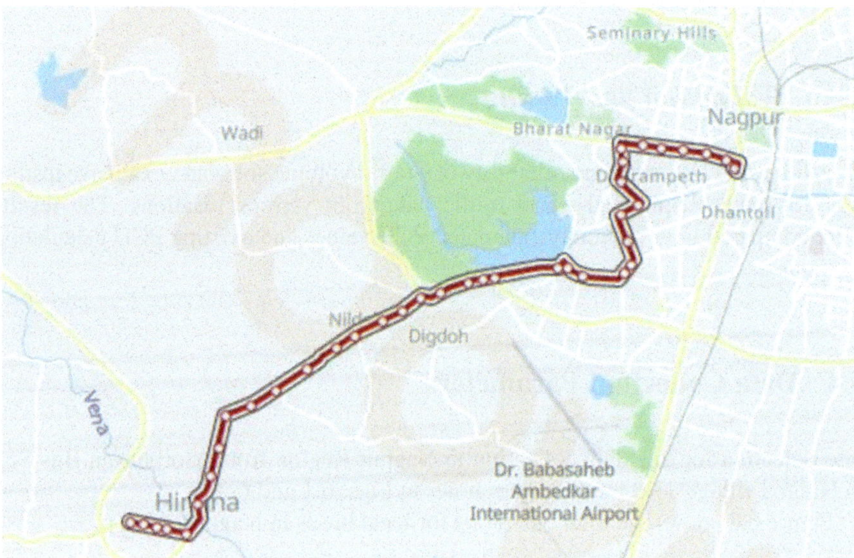

Fig. 6.2 Private bus stop location (Marve and Bhorkar 2020; Bhorkar et al. 2016; Barde et al. 2022)

- **Vehicle Classification**: Implement systems that classify vehicles into different categories, such as cars, trucks, buses, and motorcycles. This classification helps determine the appropriate PCU values for each vehicle type.
- **Speed**: Collect data on the speed at which vehicles are travelling. This information can be valuable for analysing traffic flow and congestion levels.

Table 6.1 The parameters to consider when collecting traffic volume data for PCU calculations (Barde et al. 2022)

Parameter	Description
Vehicle counts	– Use traffic counters or sensors to count vehicles passing a specific location
	– Technologies: Inductive loops, infrared sensors, radar, video cameras, etc.
Vehicle classification	– Implement systems to classify vehicles (e.g., cars, trucks, buses, motorcycles)
Speed	– Collect data on vehicle speeds to analyze traffic flow and congestion levels
	– Speed measurement devices: Radar guns, laser detectors, GPS, etc.
Direction	– Record the direction of vehicle movement (inbound or outbound) to understand traffic flow patterns
	– Important at intersections and road segments with bidirectional traffic

- **Direction**: Record the direction of vehicle movement, whether inbound or outbound. Understanding traffic flow patterns is essential for PCU calculations, especially at intersections.

Table 6.1 represents various parameters and their descriptions.

6.4.2 Calculation

Calculation Part of PCU (Passenger Car Unit)

Calculation part of PCU (Passenger Car Unit) based on the traffic volume data you've collected. PCU calculations involve assigning a standardized value to each vehicle type to represent its effect on traffic flow. Here's an example calculation using the data provided.

Example PCU Calculation

In this example, we'll calculate PCUs for each vehicle type (cars, trucks, buses, motorcycles) using the collected traffic volume data.

Step 1: Vehicle Classification

First, classify the vehicles based on the data collected:

Cars: 60 vehicles/min.
Trucks: 30 vehicles/min.
Buses: 10 vehicles/min.
Motorcycles: 20 vehicles/min.

Step 2: Assign PCU Values

Assign PCU values to each vehicle type. PCUs represent the relative capacity of each vehicle type compared to a standard passenger car, which is typically assigned a PCU of 1.0:

Cars: PCU = 1.0.

Trucks: PCU = 2.0 (as an example, trucks are often considered to have twice the impact of cars).

Buses: PCU = 3.0 (buses occupy more road space and often have a greater impact on traffic).

Motorcycles: PCU = 0.5 (motorcycles are smaller and have less impact on traffic).

Step 3: PCU Calculation

Now, calculate the PCU for each vehicle type based on their counts and assigned PCU values:

PCU for Cars = (60 vehicles/min) × (1.0 PCU/vehicle) = 60 PCU/min.

PCU for Trucks = (30 vehicles/min) × (2.0 PCU/vehicle) = 60 PCU/min.

PCU for Buses = (10 vehicles/min) × (3.0 PCU/vehicle) = 30 PCU/min.

PCU for Motorcycles = (20 vehicles/min) × (0.5 PCU/vehicle) = 10 PCU/min.

Step 4: Total PCU Calculation

To calculate the total PCU for the intersection, sum the PCUs for all vehicle types:

Total PCU = PCU for Cars + PCU for Trucks + PCU for Buses + PCU for Motorcycles.

Total PCU = 60 PCU/min + 60 PCU/min + 30 PCU/min + 10 PCU/min = 160 PCU/min.

In this example, the total PCU for the intersection during the specified period is 160 PCU/min. This value represents the overall impact of different vehicle types on traffic flow at the intersection, considering both their counts and PCU assignments.

These PCU calculations are crucial for assessing traffic capacity, congestion levels, and the effectiveness of traffic management strategies at the intersection. They provide valuable insights for urban planning and transportation decision-making. Following Table 6.2 shows the values for PCU Calculations.

"Vehicle Type" lists the categories of vehicles (cars, trucks, buses, motorcycles).

Table 6.2 Summarizes the PCU calculation for different vehicle types based on the example data provided earlier (*Source* Data collection explained) (Transportation Research Board 2021)

Vehicle type	Vehicle count (per minute)	Assigned PCU Value	PCU calculation
Cars	60	1.0	60 × 1.0 = 60
Trucks	30	2.0	30 × 2.0 = 60
Buses	10	3.0	10 × 3.0 = 30
Motorcycles	20	0.5	20 × 0.5 = 10
Total			160

"Vehicle Count (per minute)" represents the collected traffic volume data for each vehicle type.

"Assigned PCU Value" shows the PCU value assigned to each vehicle type for the calculation.

"PCU Calculation" demonstrates the calculation for each vehicle type by multiplying the vehicle count by the assigned PCU value.

"Total" sums up the PCUs for all vehicle types, representing the total PCU for the specified period.

This table provides a clear overview of the PCU calculations, allowing you to assess the impact of different vehicle types on traffic flow and congestion levels at the intersection.

From the PCU (Passenger Car Unit) calculation example provided, several insights can be derived to assess the traffic conditions and implications at the intersection during the specified period:

1. **Overall Traffic Load**: The total PCU of 160 PCU/min indicates the total traffic load at the intersection during the observed time frame. This value represents the combined effect of different vehicle types on traffic flow.
2. **Traffic Composition**: By analyzing the PCU contributions of individual vehicle types, you can understand the composition of traffic at the intersection. In this example, cars and trucks contribute equally to the total PCU, while buses have a lower but significant impact, and motorcycles have the least impact.
3. **Congestion and Capacity**: The total PCU value provides insights into congestion levels at the intersection. A higher total PCU suggests greater congestion and reduced traffic flow capacity, while a lower total PCU indicates smoother traffic flow.
4. **Transportation Planning**: PCU calculations are essential for transportation planning and infrastructure development. They help assess the impact of different vehicle types on road capacity and aid in making informed decisions regarding road expansions, signal timing adjustments, or the need for additional lanes or alternate routes.
5. **Traffic Management**: Traffic managers and engineers can use PCU calculations to optimize traffic signal timings, implement traffic control measures, or develop strategies to alleviate congestion based on the specific composition of traffic at the intersection.
6. **Safety Considerations**: Understanding the types and volumes of vehicles at an intersection is critical for safety assessments. For instance, intersections with a high percentage of trucks may require specialized road design or signage to enhance safety.
7. **Environmental Impact**: PCU calculations can also be used to estimate the environmental impact of traffic, including fuel consumption and emissions. Heavier vehicles, such as trucks and buses, contribute more to environmental pollution compared to lighter vehicles like cars and motorcycles.

In summary, PCU calculations provide a comprehensive assessment of traffic conditions, composition, and impact at a specific intersection. These insights are valuable for traffic management, urban planning, safety evaluations, and environmental considerations, ultimately contributing to more efficient and sustainable transportation systems.

Table 6.3 represents data collection of the site for peak hours for 7 days and Fig. 6.3 shows its graphical representations.

6.5 Results and Conclusion

- The balance of PCUs among different vehicle types indicates that both passenger cars and trucks contribute significantly to traffic congestion.
- Buses, while less numerous, have a substantial impact due to their high PCU values, highlighting the importance of considering vehicle type in congestion assessments.
- Motorcycles, despite being numerous, have a relatively low impact on congestion due to their low PCU values.
- The total PCU of 160 PCU/min provides a quantitative measure of congestion and traffic load at the intersection, which can inform traffic management decisions and urban planning efforts.

These results are valuable for traffic engineers, urban planners, and policymakers as they provide insights into the composition and impact of different vehicle types on traffic flow and congestion levels at the intersection. These insights can guide decision-making for traffic management and infrastructure improvements (Marve and Baitule 2016).

Conclusion:

In this study, we conducted PCU calculations based on traffic volume data collected at a specific intersection during a specified period. The results of our analysis provide critical insights into the composition and impact of different vehicle types on traffic flow and congestion levels. These findings have significant implications for traffic management, urban planning, and decision-making.

Key Findings:

- **Traffic Composition**: The PCU calculations revealed a diverse mix of vehicle types at the intersection, with passenger cars and trucks being prominent. Buses, although fewer in number, have a substantial impact on congestion due to their high PCU values.
- **Traffic Management**: Understanding the varying impacts of different vehicle types is crucial for effective traffic management. It underscores the need for tailored strategies that consider both the volume and type of vehicles on the road. These strategies can include optimizing signal timings, implementing dedicated lanes, and enhancing safety measures.

Table 6.3 PCU calculation for different locations in Peak hours morning as well as evening (Currie et al. 2009)

Days	Vehicle characteristics	Volume	Total	Vehicle %	Passenger car equivalent (PCE) value /factor	PCU (N X PCE)	Total volume (in PCU)
Day 1	Motor cycles	2328		44.98	0.5	1164	4828.00
	Cars	960		18.55	1	960	
	Bus	416	5176	8.04	3	1248	
	Trucks	288		5.56	3	864	
	Autos	736		14.22	0.5	368	
	NMV	448		8.66	0.5	224	
Day 2	Motor cycles	2496		48.22	0.5	1248	5832.00
	Cars	1320		25.50	1	1320	
	Bus	448	5664	8.66	3	1344	
	Trucks	488		9.43	3	1464	
	Autos	424		8.19	0.5	212	
	NMV	488		9.43	0.5	244	
Day 3	Motor cycles	2344		45.29	0.5	1172	6200.00
	Cars	1688		32.61	1	1688	
	Bus	520	5912	10.05	3	1560	
	Trucks	440		8.50	3	1320	
	Autos	448		8.66	0.5	224	
	NMV	472		9.12	0.5	236	
Day 4	Motor cycles	2360		45.60	0.5	1180	6276.00
	Cars	1416		27.36	1	1416	
	Bus	520	5896	10.05	3	1560	
	Trucks	528		10.20	3	1584	
	Autos	528		10.20	0.5	264	
	NMV	544		10.51	0.5	272	
Day 5	Motor cycles	2120		40.96	0.5	1060	7932.00
	Cars	1768		34.16	1	1768	
	Bus	784	6456	15.15	3	2352	
	Trucks	744		14.37	3	2232	
	Autos	520		10.05	0.5	260	
	NMV	520		10.05	0.5	260	
Day 6	Motor cycles	2144		41.42	0.5	1072	8136.00
	Cars	1688		32.61	1	1688	
	Bus	960	6424	18.55	3	2880	

(continued)

Table 6.3 (continued)

Days	Vehicle characteristics	Volume	Total	Vehicle %	Passenger car equivalent (PCE) value /factor	PCU (N X PCE)	Total volume (in PCU)
	Trucks	672		12.98	3	2016	
	Autos	552		10.66	0.5	276	
	NMV	408		7.88	0.5	204	
Day 7	Motor cycles	1688		32.61	0.5	844	5088.00
	Cars	1136		21.95	1	1136	
	Bus	424	4680	8.19	3	1272	
	Trucks	448		8.66	3	1344	
	Autos	472		9.12	0.5	236	
	NMV	512		9.89	0.5	256	

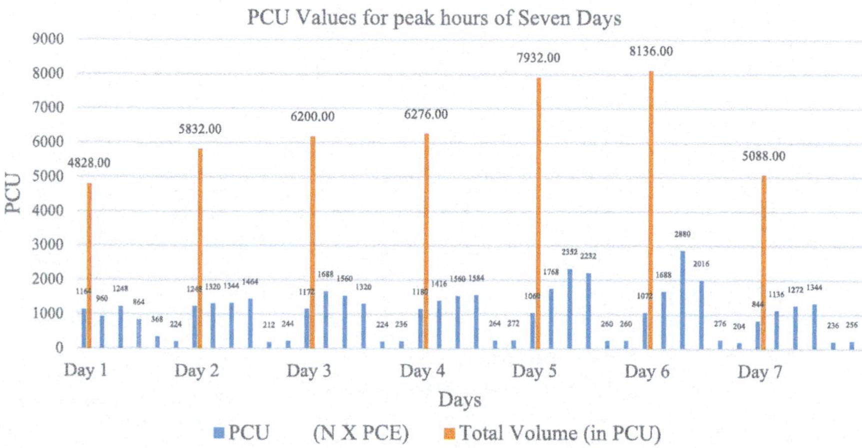

Fig. 6.3 PCU peak hour traffic volume data collection (Marve and Baitule 2016)

- **Infrastructure Planning**: The study highlights the importance of infrastructure planning and development to accommodate the unique characteristics of each vehicle type. Road expansions, lane additions, and alternative routes may be necessary to ensure efficient traffic flow.
- **Safety Considerations**: Safety remains a top priority in traffic management. The data on vehicle types and volumes aid in identifying safety concerns, especially at intersections with specific vehicle types that require specialized road designs and signage.
- **Environmental Impact**: The environmental footprint of traffic is another critical consideration. Heavier vehicles, such as trucks and buses, have a more significant

impact in terms of fuel consumption and emissions. This information can inform sustainability initiatives and the promotion of eco-friendly transportation modes.

- **Data-Driven Decision-Making**:

This study underscores the importance of data-driven decision-making in traffic management and urban planning. Access to accurate and up-to-date traffic volume data is fundamental to addressing traffic challenges and optimizing transportation systems (Bhorkar et al. 2016).

- **Conclusion**:

In conclusion, the PCU calculations presented in this study serve as a valuable tool for understanding and managing traffic dynamics at a specific intersection. The insights gained contribute to safer, more efficient, and environmentally sustainable transportation systems. As cities continue to grow and traffic patterns evolve, data-driven approaches will play a pivotal role in shaping the future of urban mobility and ensuring the well-being of communities.

- **Future Directions**:

As traffic patterns evolve, continuous data collection and analysis will be essential for making informed decisions. Future research can focus on more extensive and updated data collection, allowing for refined PCU calculations and the development of more effective traffic management strategies.

References

Aimsun. (n.d.) Aimsun. https://www.aimsun.com/

Axhausen KW, Gärling T (2015) Travel behaviour: a conceptual review. In: De Palma A, Lindsey R (eds) Handbook in transport economics, vol 1. Edward Elgar Publishing, pp 113–146

Ayyub BM (2014) Infrastructure resilience metrics: a review. Struct Saf 48:1–7

Banister D, Hickman R (2006) Transport futures: thinking the unthinkable. Transp Policy 13(4):325–339

Bansal P, Kockelman KM, Singh MP (2016) Assessing public opinions of and interest in new vehicle technologies: an Austin perspective. Transp Res Part A Policy Pract 94:397–409

Barde SM et al (2022) A review of parking management system at SSCET campus. Int J Res Publ Rev 3(7):2996–2999. [Online]. https://doi.org/10.55248/gengpi.2022.3.7.15

Bhat CR (2003) Simulation estimation of mixed discrete choice models using randomized and scrambled Halton sequences. Transp Res Part B Methodol 37(9):837–855

Bhorkar M, Marve SR, Baitule P (2016) A survey on environmental impacts due to traffic congestion in peak hours. IJSTE Int J Sci Technol Eng 2(08):2009–2012. [Online]. http://www.ijste.org/articles/IJSTEV2I8054.pdf

Boarnet MG, Haughwout A (2017) Evolving paradigm of resilience in transportation planning: a synthesis. Transp Res Part A Policy Pract 96:298–307

Cervero R (2017) Transit-oriented development's ridership bonus: a product of niche market appeal or station area quality? J Transp Land Use 10(1):213–234

Chen X et al (2019) Integrating heterogeneous traffic data for multi-location analysis. J Transp Eng Part A Syst 145(7):04019030. https://doi.org/10.1061/JTEPBS.0000213

Currie G et al (2009) A tale of two cities: factors affecting car use in Melbourne and Sydney. Urban Stud 46(7):1543–1568
Garrard J et al (2012) Active transport, urban design, and public health. Epidemiol Rev 29(1):27–43
Handy S, Xing Y (2011) Factors correlated with bicycle commuting: a study in six small US cities. Int J Sustain Transp 5(2):91–110
Hodge G, Greve C (2007) Public–private partnerships: an international performance review. Public Adm Rev 67(3):545–558
Johnson L et al (2020) Scalable traffic analysis tools for urban planning. Transp Res Rec 2674(6):186–197. https://doi.org/10.1177/0361198120904706
Lin J et al (2017) Autonomous vehicles: a bibliometric analysis of interdisciplinary research. Transp Res Part C Emerg Technol 85:205–222
Lucas K, Jones P (2018) The social distribution of transport resources: a review. Transp Rev 38(6):739–767
Maher M, Hamdar S (2017) Dynamic traffic assignment models: a review. Transp Res Part C Emerg Technol 79:276–291
Marve SR, Shende SR, Chalkhure AN (2018) Public transportation system in Chandrapur City. Int J Sci Res Sci Eng Technol 4(10):306–312. https://doi.org/10.32628/18410IJSRSET
Marve SR, Baitule MBP (2016) Traffic congestion minimization study for Hingna area of Nagpur City, MS, India. Int J Eng Res Technol ISSN 4(30):1–4. [Online]. https://www.ijert.org/research/traffic-congestion-minimization-study-for-hingna-area-of-nagpur-city-ms.-india-IJERTCONV4IS30024.pdf
Marve SR, Bhorkar MP (2016) Analysis of traffic congestion of Hingna region in Nagpur City. Int Res J Eng Technol 03(04):2593–2598. [Online]. https://www.irjet.net/archives/V3/i4/IRJET-V3I4514.pdf
Marve SR, Bhorkar MP (2020) Traffic congestion study for a developing area of Nagpur City, pp 1–4
Mondschein A et al (2019) The travel and environmental implications of shared mobility: evidence from two North American cities. J Transp Geogr 78:96–107
PTV Group (n.d.) PTV Vissim. https://www.ptvgroup.com/en/solutions/products/ptv-vissim/
Santi P, Resta G (2020) Shared mobility systems: a survey. IEEE Trans Intell Transp Syst 21(8):3335–3348
Shaheen S, Cohen A (2019) Carsharing and personal vehicle services: worldwide market developments and emerging trends. Int J Sustain Transp 13(10):685–695
Shende SR, Pathan SR, Marve SR, Jumnake AG et al (2018) A review on design of public transportation system in Chandrapur City. J Res 04(01):41–47. [Online]. http://www.journal4research.org/articles/J4RV4I1007.pdf
Smith MJ (2019) Dynamic traffic assignment: a critical review and a look forward. Transp Res Rec 2673(8):472–484
Swov (n.d.) Parameterized microscopic simulation (Paramics). https://www.swov.nl/en/research/facts-and-figures/factsheets/parameterized-microscopic-simulation-paramics
Synchro Software (n.d.) Synchro/SimTraffic. https://www.synchrosoftware.com/
Transportation Research Board (2021) Highway capacity manual. https://trb.org/main/public/blurbs/169998.aspx
Wang J, Cao X (2018) Transit-oriented development and its travel behavior effects: a review of empirical evidence from the United States and China. J Plan Lit 33(3):273–288
Winters PL, Branion-Calles M (2016) The effectiveness of transportation demand management strategies: a review of the literature. J Plan Lit 31(4):399–415

Chapter 7
Traffic Data Analysis by Using Automatic Traffic Counter and Classifier

Kunal Chandra and Bhalchandra V. Khode

Abstract The purpose of traffic planning is to promote the safe and timely movement of people and goods by creating plans and methods of road use. Without proper traffic controls or effective plans, current road infrastructure will not be able to support the needs of cities. An increase in population results in an increase in the number of vehicles on the road. This study investigates the characteristics and smoothness of traffic in Mathura City. To conduct the traffic survey, an Automatic Traffic Counter and Classifier (ATCC) are used at several locations in the selected study area. Traffic patterns can be studied better by using the data that we will obtain through traffic surveys. The research focuses mainly on the characteristics affecting the management of traffic flows to maintain rules and regulations in traffic systems. The results of this research will be used to develop an efficient road system that can handle unpredictable traffic congestion.

Keywords Automatic Traffic Counter and Classifier (ATCC) · Traffic count · Traffic flow · Pneumatic tube

7.1 Introduction

Traffic may be defined as people's movement in motorized or non-motorized vehicles from one place to another. While designing highways, bridges, or any other road infrastructure, traffic analysis plays a vital role (Karuppanagounder and Muneera 2017). Traffic analysis is the first step to be taken while designing a road, as according to the result, the number of lanes that are required in the particular stretch is decided. It can be conducted in two ways: manual or automatic (Mansoori and Achar 2018). In today's world, due to enhancements in technology, we have several automatic

K. Chandra (✉)
Transportation Engineering, G. H. Raisoni College of Engineering, Nagpur, India
e-mail: kunalchandra263@gmail.com

B. V. Khode
G. H. Raisoni College of Engineering, Nagpur, India

B. V. Khode et al. (eds.), *Innovations in Technologies: Pioneering Sustainable Infrastructure for a Resilient Future*, Sustainable Civil Infrastructures,
https://doi.org/10.1007/978-3-031-82133-2_7

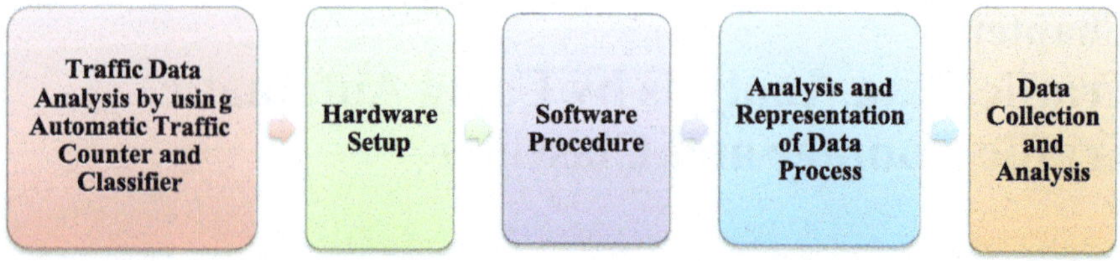

Fig. 7.1 Process for analyzing traffic data using ATCC

approaches that can be used for our traffic analysis (Rao 2013). These approaches collect and study the incidental details and give an accurate result. Inductive loop detectors and pneumatic sensors are some of the devices that were used for collecting traffic data (Chandra and Sikdar 2000). Data that is collected through these tools is converted into usable data by using software such as Metro Count (Bertini 2005). By using an automatic traffic counter classifier, we get information on various aspects, such as types of vehicles, traffic frequency, vehicular speed, etc., according to which we can predict the traffic forecast for later years (Prasad 2009).

7.2 Methodology: (Rao 2013)

Process Flow:

Figure 7.1, shows the process for analyzing traffic data using ATCC. In this process for analyzing the traffic data, the first thing that is required is the hardware setup (Justo and Khanna 2011). Then after that, software process came into play. With the help of these two processes data was collected, and from that collected data analyses were made.

7.3 Setup for Instrument

In two stages initiation process is completed.

7.3.1 Hardware Setup

- **Equipment Used**:

Road tube, Anchors, Duct Tape, Nails, Tape Measure, Sledge Hammer, Gloves, Counter.

Fig. 7.2 Installation of pneumatic tube at NH 530B (Photograph taken by Author)

- **Installation Procedure**:

As shown in Fig. 7.2, for installation of pneumatic tube, that we used in NH 530B, fasteners are used to attach the road tube to a road so that it sticks to its place while vehicles move across it (Bibinahake et al. 2024). If the stretch is very wide, center clamps or duct tape are used (Uddin 2009). Nails are used to attach the fasteners to the asphalt roads, and according to the guidelines, the nail head should be 1/4" above the pavement.

1. On the opposite side of the road, install an anchor with the help of a nail and a small hammer. If you are using more than two pipes, then measure the distance between them and place the anchor accordingly (Thakur et al. 2024).
2. After attaching the anchors on both sides of a road, stretch the pipe over the road and secure it with an anchor.
3. Make sure that pipes are perpendicular to the road, and they should be placed in such a way that both wheels of vehicles should collide on the same axle at the same time. If these conditions are fulfilled, then you can attach the pipe with the anchor and hammer it into the pavement with an iron nail (Gusain et al. 2024).
4. According to the manual, tighten the tube to 10–20% and put some strong tape on it so that it doesn't loosen enough at the time of examination when heavy vehicles such as trucks, etc. move across it (Mendhe et al. 2022).
5. Now, as one end of the pipe is attached to the pavement, put another end tube on the automatic classifier counter. Make sure that the counter is supported by something fixed, such as a tree, fence, or something similar (Ghodmare et al. 2021).

7.3.2 Software Procedure

The procedures for starting the Road Side Unit are as follows.

1. Connect the counter to your laptop by using a USB cable.
2. Select the Metro Count software. Click on the RSU button to connect as the software opens.

Fig. 7.3 Automatic traffic counter and classifier (photograph taken by Author)

3. A dialog box appears that contains information about the pipes that are connected to the counter.
4. After that, click on the configure button, which will take you to the setting files.
5. In the setting files, provide the location name, traffic count starting time, number of intervals, etc. It is up to you when you have to start traffic reading.
6. If the counter is flashing a red light per second, then the counter is properly connected to the pneumatic tubes, and it is working properly. Now it will save the traffic data until the specific time that is mentioned in the setting file.

As shown in Fig. 7.3, this photograph was clicked by the author during installation process. Under this box is an automatic traffic counter classifier machine, in which pneumatic tubes are connected. With the use of this machine, we can count traffic volume and classify the type of vehicles automatically.

7.3.3 Analysis and Representation of Data Process

1. The recorded data obtained by the software is analyzed by the MC Report program.
2. This program gives the reading of vehicle counts at different intervals, and apart from that, it provides the same data in graphs and diagrams.
3. It also provided us with information regarding types of vehicles, traveling speed, number of axles, etc.

7.4 Data Collection and Analysis

An automatic traffic counter was used to collect traffic data from Mathura at NH-530B (Mathura–Bareilly). A 12-h data set was collected at this site. Figure 7.4, describes the map of Mathura–Bareilly highway where the experiment was carried out.

Fig. 7.4 Map of Mathura–Bareilly NH-530B

Following data is collected from the site which includes vehicle type such as car, jeep, cycles etc. that passes from the particular stretch in the given time. All the data is collected through automatic traffic counter and classifier. As shown in Table 7.1, that describes the volume count for particular vehicles at NH 530B. Tables 7.1 and 7.2 describes commercial vehicles per day at NH 530B.

From the Tables 7.1 and 7.2, data was collected, and the following graphs were made, which show the peak hour and non-peak hour flow of traffic at NH 530B.

Table 7.1 Classified volume count

Vehicle type	Mathura–Hathras road (NH-530B)
Car, jeep, van	4033
2-Axle	371
3-Axle	556
MAV	1424
SCV	385
Agri. tractor without trailer	151
Cycle	93
2 Wheeler	5331
3 Wheeler	984
Bullock cart	0
Bus	194
Total vehicles	13,199
Total PCU's	15,831

Table 7.2 Commercial vehicles per day

	HCV			Tractor			Bus		
	Day-1	Day-2	Day-3	Day-1	Day-2	Day-3	Day-1	Day-2	Day-3
Average CVPD	1735			163.33			914.33		
Total CVPD	2812.66								
Consider total CVPD	2813								

7.5 Result

- The total vehicle counted was 15,831.
- The 62.3% vehicles were the cars and jeeps, 27.23% vehicles were the two wheelers and rest are the other vehicles.
- The 85th percentile speed was 48 km/h with minimum of 12 km/h and maximum of 70 km/h
- Most of the vehicles travelled within speed of 30 to 60 km/h.

7.6 Conclusion

Earlier, traffic data was collected manually, which could result in high human error, but as time passes and due to technological enhancements with the use of automatic traffic counter classifiers, the collection of traffic data is quick and easy. This device provides us with incidental details that we couldn't achieve manually. Apart from that, we can collect a vast amount of data from it, such as on a weekly, monthly, and annual basis.

The traffic volume study for different vehicle classes in district Mathura is shown in the Tables 7.1 and 7.2. From this study, it can be concluded that whether it is a two-wheeled or a four-wheeled vehicle, it could create a traffic disturbance during peak hours in Mathura, as shown in Figs. 7.5, 7.6, 7.7, and 7.8. In the following figures, traffic flow is shown during peak and non-peak hours for two and four wheeled vehicles in Mathura, which could lead to traffic congestion during office hours. Cities like Mathura would have similar peak-hour traffic figures, which will help in future street projects and road management.

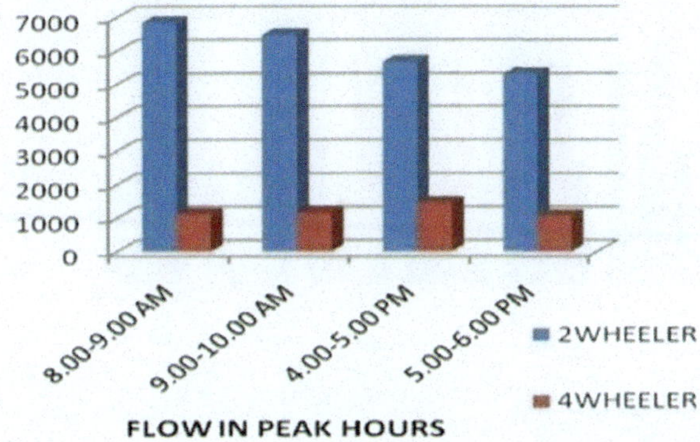

Fig. 7.5 Peak hour flows of two-wheeler and four-wheelers

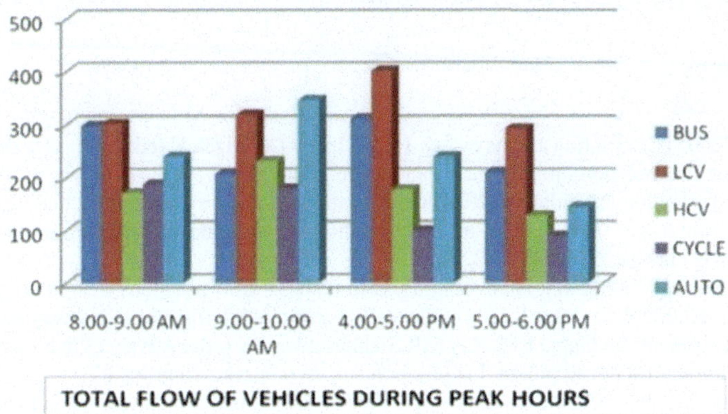

Fig. 7.6 Total peak hour flows at NH 530B

Fig. 7.7 Non peak hour flows of two-wheeler and four-wheelers

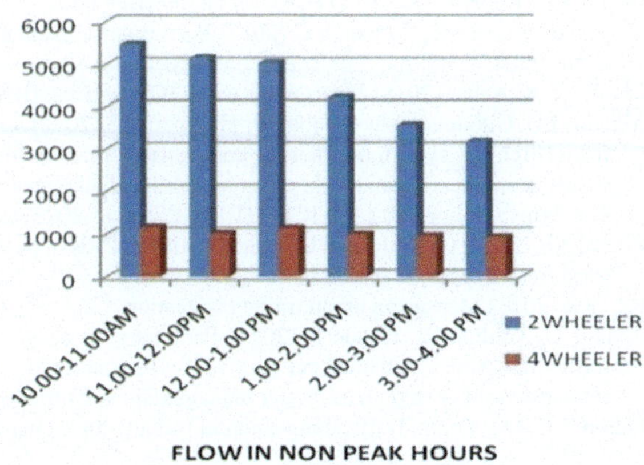

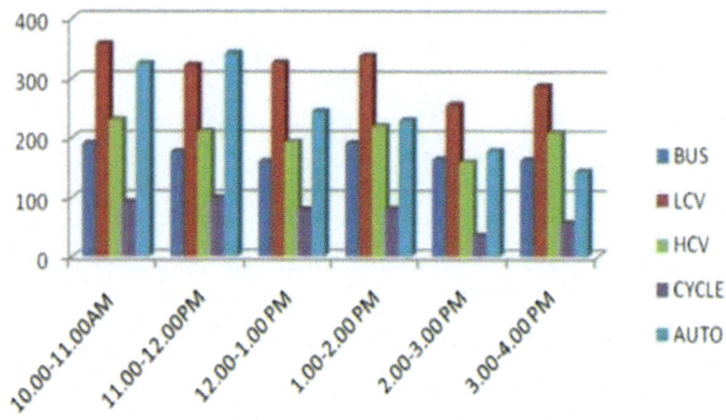

Fig. 7.8 Non peak hour flows at NH 530B

References

Bertini RL (2005) Congestion and its extent. In: Levinson D, Krizek K (eds) Access to destinations: rethinking the transportation future of our region. Elsevier

Bibinahake Y, Wadhai P, Vaidya N (2024) Comparative study on smart road technology. In: International conference on recent advancement in infrastructural development, water management and climate change (IC-RAIWC-22)

Chandra S, Sikdar (2000) Factors affecting PCU in mixed traffic situations in urban roads

Ghodmare S, Khode B, Ladekar S (2021) The roll of artificial intelligence in industry 4.0 and smart city development. In: Gupta LM, Ray MR, Labhasetwar PK (eds) Advances in civil engineering and infrastructural development. Lecture notes in civil engineering, vol 87

Gusain I, Sharma AK, Mishra N, Khode BV (2024) Experimental analysis of concrete replaced by different percentage of demolished concrete in a concrete mix. In: International conference on recent advancement in infrastructural development, water management and climate change (IC-RAIWC-22)

Justo CEG, Khanna SK (2011) Highway engineering book

Karuppanagounder K, Muneera C (2017) Performance evaluation of urban links under heterogeneous traffic condition. Eur Transp 65:1–10

Mansoori A, Achar (2018) Smart roads using IOT devices. IRJET 5(06)

Mendhe PD, Ghodmare SD, Wadhai PJ, Titarmare AS (2022) Sustainable transportation using automated highway system for intercity connectivity. In: 10th International conference on emerging trends in engineering and technology—signal and information processing (ICETET-SIP-22), pp 1–6. https://doi.org/10.1109/ICETET-SIP-2254415.2022.9791761

Prasad NV (2009) Central Road Research Institute (CRRI), New Delhi to determine the PCU value for different types of vehicles

Rao KR (2013) Measuring urban traffic congestion

Thakur D, Wadhai PJ, Khode B (2024) Parametric evaluation of highway redesign models for traffic flow from a statistical perspective. In: International conference on recent advancement in infrastructural development, water management and climate change (IC-RAIWC-22)

Uddin A (2009) Draft. Traffic congestion in Indian cities: challenges of a rising power

Chapter 8
Failure Analysis and Maintenance Recommendations for Delhi-Kolkata Highway

Kunal Chandra and Bhalchandra V. Khode

Abstract Transport infrastructure is very important for the development of the country. The purpose of this article is to assess the types of pavement damage and choose optimal maintenance to overcome them on the Delhi-Kolkata Highway also known as (Old NH-2). Road surfaces can be damaged in a number of ways, including potholes, subsidence, ruts, shear damage, etc. heavy loads, insufficient drainage and heavy rains are the main reasons that cause damage to flexible paving. People suffer many problems in their daily life due to broken pavement. Road surface damage not only causes problems for people, but also increases maintenance costs. This study examines the failure categories of flexible pavements, their causes and their repair along with related maintenance measures.

Keywords Highway failure · Highway maintenance · Flexible pavement · Rigid pavement · Potholes · Cracks · Raveling · Rutting · Repair

8.1 Introduction

Transport is essential for the social, economic and industrial growth of a country. Therefore, transport is classified into three types: air, rivers and roads. The main purpose of this research is to find out the many defects of elastic pavement and their causes, and how to solve them with the recommendations of maintenance methods. It is clear that people's favorite mode of transport is road transport. Such transport is necessary to connect villages to cities or to connect cities to regions of other states, but great difficulties can arise if the transport cover is destroyed. Pavement failure can include cracks, potholes, holes, bowing and other problems. These pavement

K. Chandra (✉)
Transportation Engineering, G. H. Raisoni College of Engineering, Nagpur, India
e-mail: kunalchandra263@gmail.com

B. V. Khode
G. H. Raisoni College of Engineering, Nagpur, India

B. V. Khode et al. (eds.), *Innovations in Technologies: Pioneering Sustainable Infrastructure for a Resilient Future*, Sustainable Civil Infrastructures,
https://doi.org/10.1007/978-3-031-82133-2_8

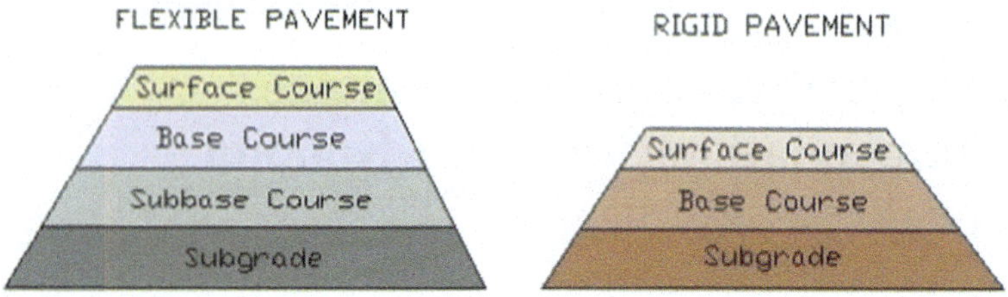

Fig. 8.1 Layers of flexible and rigid pavement

problems can be caused by poor drainage infrastructure, poor construction materials or heavy rainfall (Khaing and Htwe 2014).

As shown in Fig. 8.1, pavements are divided into two categories: flexible pavement and rigid pavement (Hasan and Sobham 2020).

- **Flexible Pavement**—This type of pavement is mostly constructed from different materials. The top layer transfers the weight to each layers, which then distributes it equally to subsequent layers. The flexible road surface is less stressed this way. Such a road surface curves when loaded with tires (Khanna and Justo 1991).
- **Rigid pavement**—This type of paving is made of reinforced concrete or cement concrete slabs. These pavements become strong enough to withstand the vehicle loads caused by traffic. Because it has a high modulus of elasticity and stiffness, it distributes the vehicle load over a large base area (Khanna and Justo 1991).

In India, flexible pavement is mostly taken into account for building. It is crucial to maintain roads after they are built so that people may utilize them for a longer amount of time. Another factor is that India has a tough environment, necessitating upkeep to minimize damage to the roads during a single season. The creation of potholes and fissures on the bituminous surface is a sign of flexible pavement collapse.

8.2 Literature Review

Kishore and Ramu (2020), suggested proper drainage system design, maintenance and construction. They also focus on maintenance frequency. Road repair is necessary if the road is damaged. Various methods are used in this study, including highway planning and design, which includes steps such as highway construction, geometric design, and the elements that influence these models. In addition, they talked about the numerous defects of the road surface and their maintenance. In addition, various maintenance measures that should be considered when implementing these maintenance works were discussed, including banks, bridges, traffic systems, etc. Khare and Verma (2018), suggested, damage to the road surface, which is due to a certain impairment of operability, manifests itself in cracks, wrinkles, etc. If these factors are not considered, serious consequences can occur. The main purpose of their paper is to properly maintain the pavement so that it remains in good condition for future use. Their job is to identify the problem, its causes and proper maintenance to ensure smooth traffic. The research has a limitation or research gap in that they do not emphasize the practical side because they only used theory as the basis of their research. Khaing and Htwe (2014), explained and suggested some ways to extend pavement life. According to him, there are several techniques to extend the life of the coating, including routine maintenance, periodic maintenance and emergency repairs. The work focuses on failure patterns and visual maintenance using various techniques including IRC formulas and three-layer soil mechanics. Routine, occasional and emergency repairs are part of visual maintenance, which also includes maintenance of bitumen and asphalt pavements. Overall, their main goal is to improve the road surface on the Magway-Yangon route in Myanmar. They also used the IRC method to estimate the coating thickness. This results in a maximum achievable coverage of 6 inches, while the pellets have a maximum coverage of 12 inches. **Parveen et al.** insist that things like hidden marks, holes, broken edges, faulty design and other things should not be ignored because they can lead to catastrophic accidents. In this study, they looked at possible problems with a flexible roof, their causes and maintenance. In the Fatehpur district of Uttar Pradesh, they used 10 km of alignment in the test, using 8 km of flexible and 2 km of rigid pavement. They also pay attention to factors that contribute to pavement defects such as excessive traffic, improper mixing and others. According to this study, the maintenance technique is 60% identical to the traditional procedure. Data are collected through field surveys. As a consequence of their field study, they have identified multiple failures in various locations around the Fatehpur district. According to their failure situations, they also divided their maintenance kinds. They came to the conclusion that both urgent and regular maintenance are carried out in the Uttar Pradesh district of Fatehpur. The study manuscript's shortcoming or research gap is that it focuses on maintenance types rather than the approaches that might be taken to address these failure-related issues. **Praveen Kumar, Ankit Gupta et al. (2010)**, outlines the solutions to reduce pavement failure. Pavement collapse can be caused by a variety of factors. Applying the maintenance process to the pavement might lengthen its lifespan. This essay discusses the issues related to Indian pavement failure. This document describes numerous preventative methods for extending the life of a pavement based on experience. For the maintenance In addition, they highlight the value of pavement modals in predicting pavement strength and service life. Pavement modals include surface property models, models based on environmental factors, pavement classification models, and other models. The lack of a defined standard, because it varies according to need, is a limitation of scientific publications or a lack of research. These models are not suitable for seasonal traffic job, the effort is centered on pavement performance modals.

8.3 Causes and Factors Contributing to Pavement Failure: (Khanna and Justo 1991)

- There are errors in the materials used
- Errors in quality assurance and construction methodology
- Inadequate drainage system
- Movement of heavy trucks or increase in load repetitions due to increased traffic
- Poor sealing of the pavement in the initial phase of construction
- Elements such as rain, rising ground water, etc.

The following measures are aimed at solving problems related to pavement damage, such as compaction, cutting, filling, etc. Road surface maintenance is divided into three categories: (Rashid 2017)

- **Regular repairs**—This includes patching potholes, repairing curbs, fixing road signs, etc.
- **Periodic maintenance**—which includes the replacement of the wear layer of the coating and the maintenance of various components.
- **Special repairs**—such as resurfacing, road widening and environmental damage. Flexible pavement is characterized by low bending strength and flexible response under load. There are deformations in the flexible cover of the last layer. The figure shows many layers of flexible and rigid pavements (Fig. 8.1).

8.4 Methodology

The following steps are taken to resolve pavement related issues. The Delhi-Kolkata highway was damaged at many places due to heavy rains or movement of heavily loaded trucks on the pavement causing inconvenience to the pavement. There were several defects including cracks and holes. Maintenance is often done to solve this problem as shown in Table 8.1. Some pavement defects and their care are discussed below (Hasan and Sobham 2020).

Pavements must be stable enough to withstand the proposed traffic under current subsurface and climatic conditions to be successfully maintained. Covering an old pavement with a new one can strengthen it by increasing the thickness of the pavement in one or more layers to an appropriate level.

Types of overlay (Khanna and Justo 1991):

- Flexible overlay over flexible pavement
- Rigid overlay over rigid pavement
- Rigid overlay over flexible pavement
- Flexible overlay over rigid pavement
- **Flexible overlay over flexible pavement**—Depending on the intended traffic and existing subsurface conditions, flexible pavement over flexible pavement is

Table 8.1 Pavement failures and their maintenance

S. no	Type of failure	Description	Remedial measure
1.	Potholes	The most frequent kind of pavement collapse is this one. The pavement is prone to this kind of breakdown in all areas. The result was structural breakdown and a bumpy road These sorts of failure were created by the movement of big load vehicles and rainfall	At shallow depths, pit filling is one measure that can be considered. Otherwise, if there are more cavities than usual, we consider filling with pre-mix
2.	Alligator cracking	This kind of pavement breakdown, which results from the relative motions of the pavement layer materials, is frequently observed. When large vehicles are used often, the pavement might expand or shrink due to moisture variations or fatigue failure. Another reason for this disaster is that the foundation course was built improperly, which caused the surface course to break	Different sorts of care are needed for different types of alligator cracking, such as carpeting for common alligator cracking but patching for effective alligator cracking
3.	Block cracking	Block cracking is another type of damage that can be seen in pavement. Such failures are caused by unstable foundations and poor construction	A coating of sealant can be applied to prevent such defects
4.	Slippage cracking	The road surface became rough due to such cracking. Poor drainage and unstable treads cause these cracks	The methods of applying the sealing layer are considered when repairing various broken carpets
5.	Transverse cracking	This sort of cracking typically develops as a result of inadequate mix design or heavy load vehicle movement on the pavement	Such cracks are repaired by surface treatment such as carpeting
6.	Longitudinal cracking	This type of cracking is usually caused by poor mix design or heavy vehicle traffic on the pavement	Such cracks are repaired by surface treatment such as carpeting

(continued)

Table 8.1 (continued)

S. no	Type of failure	Description	Remedial measure
7.	Raveling	Pavement irregularities are due to such damage. This occurs when the compaction process is done incorrectly during construction and the pavement binder cannot hold the aggregate	Such cracks are repaired by surface treatment
8.	Corrugation and shoving	Such damage causes unevenness and bumps in the pavement. This is due to a combination of poor design, heavy traffic and inappropriate links, among others	Premix filling, rolling and compaction are possible remedies for such defects
9.	Depression	Another form of failure is pavement depression caused by heavy rains and inadequate drainage systems	To eliminate this type of failure, the damaged area of the surface layer is removed and replaced with a premix filler
10.	Rutting	These types of failures are caused by the movement of heavy vehicles and poor construction practices that lead to clogged water troughs that cause hydroplaning in moving cars	The primer should be used to restore the damaged area. The premix is then compacted to secure the coating

required. The IRC takes into account the CBR pavement design approach when determining the total thickness. The current thickness of the coating is determined using test pits dug in the coating.

$$h_o = h_d - h_e$$

where, h_o = overlay thickness required
h_d = total design thickness required
h_e = total thickness of existing pavement

- **Rigid overlay over rigid pavement**—Since the slab of the new and old pavement do not match perfectly, the rigid pavement behaves like a monolithic pavement when built on top of the existing rigid pavement. Making the front surface uneven or adding a thin coating of bitumen between the slabs can also solve the problem.

$$h_o = (h_d^a - x h_e^b)^n$$

where, h_o = overlay thickness required
h_d = total design thickness required

Table 8.2 Rigid overlay design factors

Agency	Existing pavement condition	X		
Crops of engineers and, PCA	Good condition	1.00		
	Initial cracking	0.75		
	Badly cracked	0.35		
Agency	Construction method	a	b	c
Crops of engineer	Poured direct on old pavement	1.40	1.40	1/1.4
	Levelling course	2.20	2.00	1/3
PCA	As(1) above	1.87	2.00	1/2
	As(2) above	2.00	2.00	1/2

h_e = total thickness of existing pavement

Values of a, b, x and n depends on the type of pavement and method of overlay, as shown in Table 8.2.

- **Rigid overlay over flexible pavement**—By laying down firm pavement, the thickness for the stiff overlay is calculated. In order to determine K value, a plate bearing test on flexible pavement is performed.
- **Flexible overlay over rigid pavement**—The wheel load is distributed over a wider area when a flexible pavement is applied to a rigid pavement. In addition, as a result of the bituminous coating, the temperature difference of the rigid surface is reduced. A properly designed and constructed and timely installed bituminous pavement can significantly extend the life of a rigid pavement.

$$h_f = 2.5(Fh_d - h_e)$$

where, h_f = overlay thickness required
h_d = design thickness of rigid pavement
h_e = total thickness of existing pavement
F = factor depending upon modulus of rigid pavement

8.5 Study Area

One of the busiest highways in India is the Delhi-Kolkata highway. It starts from Delhi and ends at Kolkata. The length of the road is approximately 1465 kilometers. A 20 km survey will be organized to identify the various existing pavement damages and the maintenance techniques needed to repair them. Figure 8.2, is the study area where this work was carried out.

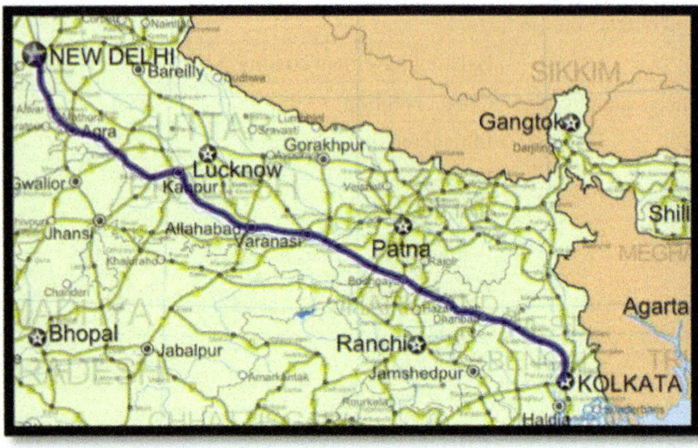

Fig. 8.2 Map of study area

8.6 Results and Analysis

Based on the information collected on several types of failures, all maintenance processes are classified under maintenance. All the maintenance data is obtained from the authorities.

8.7 Conclusion

Based on various pavement failure studies. The conclusions are as follows:

- Weather factors such as heavy rain, the type of material used, the amount of traffic on the pavement and humid air on the ground floor are the main causes of damage to flexible covers.
- From Table 8.3, it can be concluded that routine maintenance must be done, but in some areas it must be urgent. The Delhi-Kolkata Expressway needs urgent maintenance to keep traffic flowing; otherwise, being one of the busiest highways in India, traffic jams may occur.
- Though the work was carried out for 20 km, the maximum percentage was coved by cracks and pothole as shown in Fig. 8.3. Due to which the percentage of urgent maintenance is higher as shown in Fig. 8.4, comparing rest of maintenance frequency.
- For the maintenance of potholes and cracks, we basically use bitumen, but insist using bitumen if we use modified bitumen (a mixture of bitumen and crumb rubber) it not only enhance the pavement life but also reduce the maintenance cost.

Table 8.3 Indicates fault category and maintenance frequency

Failure	Failure occurrence (location)	Maintenance
Potholes in pavement	14	Urgent maintenance
Cracks in pavement	12	Routine maintenance
Raveling in pavement	8	Periodic maintenance
Corrugation and shoving in pavement	6	Urgent maintenance
Rutting in pavement	6	Periodic maintenance
Depression in pavement	5	Urgent maintenance

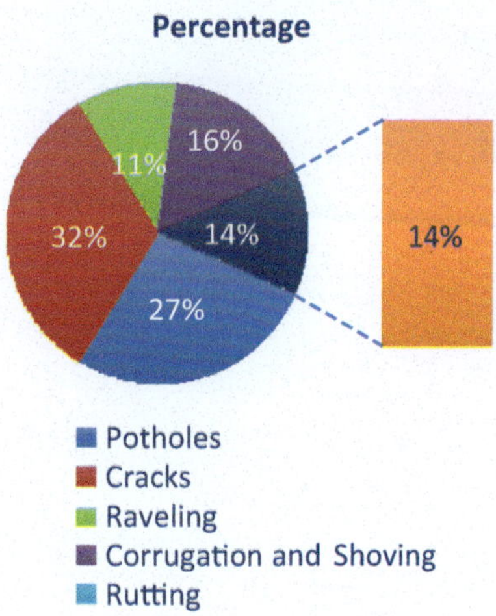

Fig. 8.3 Failure occurrence

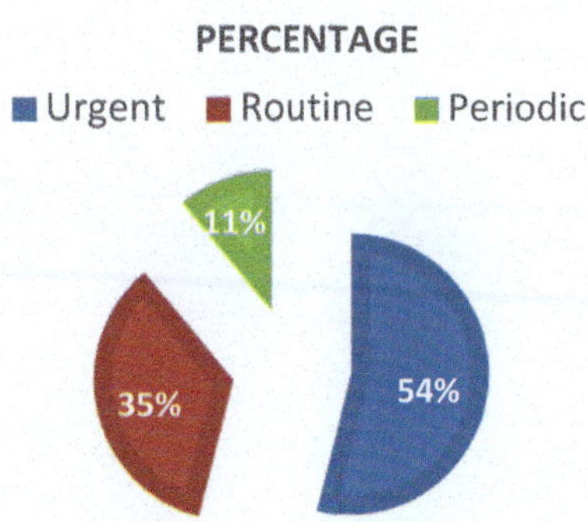

Fig. 8.4 Maintenance frequency

- But if standardization and reinforcement are not properly monitored, it can lead to repeated road maintenance of the same road.

References

Hasan M, Sobham M (2020) Highway failure and their maintenance. Int J Sustain Transp Technol 03

Khaing H, Htwe T (2014) Study on failures and maintenance of flexible pavement. Int J Sci Eng Technol Res 3(14):2984–2990

Khanna SK, Justo CEG (1991) Highway engineering book

Khare S, Verma T (2018) Defect and maintenance of road. Int J Adv Res Sci Eng 07

Kishore G, Ramu K (2020) Evaluation of highway failure and their maintenance. Int Res J Eng Technol

Chapter 9
Identification and Analysis of Black Spots on NH-353D (Nagpur-Armouri Highway)

Radhika D. Chaudhari and Bhalchandra V. Khode

Abstract Nowadays, accidents become a major cause of death. The main root cause of accidents is the increase in congestion on the roads. According to an examination of road accident statistics from 2020, there are 42 accidents and the deaths of 15 people each hour on Indian roads, for a daily average of 1003 accidents and 360 fatalities. Attention must be paid to the fact that more than 56 percent of accidents and more than 61 percent of deaths occurred on highways, which make up around 5% of the nation's overall road network. This paper is based on a thorough investigation of black spots and traffic safety. A sizable portion of these incidents occurs on national highways, which has severely stunted economic growth and caused enormous losses. The primary purpose of this study is to locate dark spots and conduct in-depth research for evaluation and correction. The strategies and procedures utilized for selecting the project study spots (where a large number of accidents happened), collecting accident data, dividing the data for better observation, analyzing black spots, and suggesting mitigation measures are all included in this research. This work describes how to identify black spots by applying the Weighted Severity Index (WSI). This study concludes the rectification methods according to short-term, long-term, reduction, or improvement of the geometric layout as determined by existing road situations.

Keywords Black spots · National highways · Fatalities

9.1 Introduction

Traffic accidents are a prominent source of disability, hospitalization, and fatality in the globe, notably in India. According to the World Health Organization (WHO), around one in ten people who died on roads worldwide are from India. Due to escalating transportation, a growing population, and more vehicular access across the entire length of the nation, casualties and fatalities from road crashes are becoming

R. D. Chaudhari (✉) · B. V. Khode
Transportation Engineering, G. H. Raisoni College of Engineering, Nagpur, India
e-mail: cradhika1996@gmail.com

B. V. Khode et al. (eds.), *Innovations in Technologies: Pioneering Sustainable Infrastructure for a Resilient Future*, Sustainable Civil Infrastructures,
https://doi.org/10.1007/978-3-031-82133-2_9

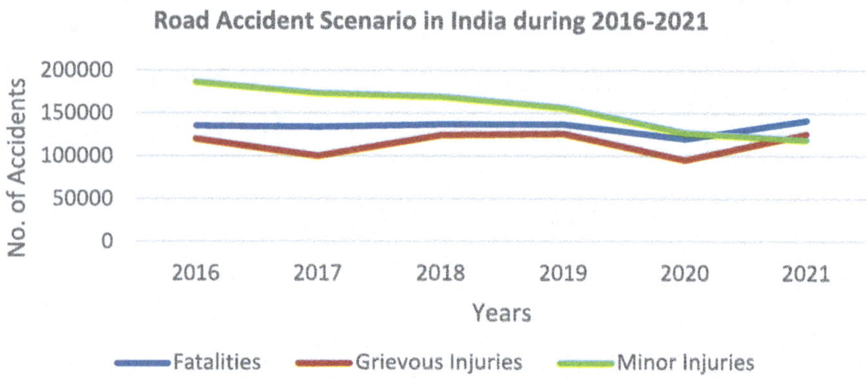

Fig. 9.1 Road accident scenario (*Source* Road Accidents in India 2021, MoRTH.)

matters of extreme concern. The top contributory factor of mortality among persons aged 15 to 49 across the world is a result of traffic-related damages. Road accidents in India claimed over 1.3 lakh lives and injured over 3.4 lakh people in the year 2020 (Road Accident Data 2020).Vehicle accidents have serious economic as well as social repercussions; if nothing is done, it is predicted that by 2030 there will be approximately 2.4 million vehicle fatalities annually worldwide. The overall number of fatal crashes in the nation did, however, decline in 2020 compared to the year before (Fig. 9.1), in part because of a variety of causes, including improved traffic management, the introduction of the newly enacted Motor Vehicle Act, and the COVID-19-induced shutdown. Since nine out of ten of the global populace now resides in lower and middle-income nations, the majority of these occurrences will take place there. In the context of overall fatalities on the world's roadways, motorcycle riders, cyclists, and pedestrians will make up close to half of them. Finding the positions of the black spots is covered in the first part and the road safety audit in the second part of this research. Roadway "Black Spots" are locations with a maximum number of incidents that frequently happened in the past. An accidental spot, according to MoRTH, is a section of a national highway with a length of about 500 m where both 5 crashes (in all the previous three years set collectively including deaths and injuries/grievous injuries) or ten casualties (in 3 years placed in tandem) occurred during the last three calendar years. "Black spots" are sites with accident numbers and/or rates higher than a set minimum or locations that have more fatalities than other similar regions on the transport network.

9.2 Objectives

The research's primary objective is:

1. To investigate and pinpoint the accidental situation.

2. To propose corrective actions for the smooth and secure movement of vehicular traffic.
3. To lower the frequency and severity of crashes on highway networks.
4. To identify each hazardous location based on the scope, nature, and severity of the mishaps.

9.3 Literature Review

Mendhe et al. (2022) proposes a five-layer concept to provide an automated interstate structure for the metro political regions (Nagpur and Mumbai). Automobiles and roadways work together to coordinate vehicle growth, avoid roadblocks, and enhance traffic flow, therefore increasing safety and reducing congestion. Nuli et al. (2022) picked a rural route spanning from Rallaguda Flyover to Vardhamana Engineering Institute of Hyderabad City for highway security reviewing in this study. The additional purpose aims to discover the high-risk zones under the jurisdiction of Rajiv Gandhi International Airport (RGIA) Police Station, Hyderabad. Incident-prone locations are recognized using historical accident data to calculate weighted severity and accident severity indexes. Rodke et al. (2022) seek to investigate the approaches implemented at various sites for calculating the rate at which saturation occurs while taking into account variable traffic characteristics. Nimburkar et al. (2022) address several elements of signal timing based on previous study findings. It is also attempted to determine the proper characteristics and methods for the Indian traffic environment. Yadav and Khode (2022) focus on reducing energy consumption by utilizing an optimized speed profile. The MMRDA's metro line 1 (Versova-Ghatkopar) is being evaluated for this purpose. Sanghai and Pawade (2022) explain strategies for developing an abrasion reducer that include the soil's structural damage. The sliding stress of damping friction was computed by assuming a set and elastic base. Yadav et al. (2021) employed the concept of categorizing incidents based on the degree of injuries. Accident data including the overall amount of occurrences for every single year along NH 66 section of Alappuzha District in Kerala was received from the PWD, NHAI, and Road Department during 2017 and 2019. These incident statistics were classified depending on the nature, kind, and causes of the accident, type of injury, road condition, and weather condition. The average severity index (ASI) was determined after rating the blackspots, and correction procedures for chosen spots were offered. Gupta et al. (2021) investigate current traffic situations. The study will be based on the maximum wheel load that the existing road can carry with the current traffic volume. The IRC traffic flow principles will also be considered for this investigation. Metre et al. (2021) investigate Nagpur, India, as a compact city. The study emphasises the compact city concepts and its implications for equitable growth in India (Metre et al. 2021). Bari and Khode (2021) investigate the diffusion of particulate matter (PM10) at an unsignalized crossroads on metropolitan roadways. It was discovered that the early-morning peak has an elevated level of PM10, which gradually decreases near off-peak and then climbs again in the late afternoon peak. Ghodmare et al. (2021) attempted to concentrate on the trajectory of industrial and economic growth as urbanization progressed. It attempts to consolidate factual knowledge linked to advances in technology all over the globe. Ghodmare and Gulghane (2021) illustrate how a multi-attribute utility theory-based technique is used to evaluate sustainability for Nagpur, an expanding metropolitan metropolis, using a variety of individual performance metrics. The results are provided as scaled values for several of the performance measures considered in this article. By collecting incidental data from Bidar Police Station, Goyal et al. (2020) chose Hulasur-Basavakalyana SH for the identification and study of blackspots. The data was analyzed using the Crash Analysis Method, which was based on MoRTH criteria, and a total of six blackspots were detected (Goyal et al. 2020). Short-term remedies were proposed for each detected cause, and if the indicated remedial actions were implemented, the accident rates would be reduced to the greatest degree practicable. Pitale and Katre (2019) discovered that several research on the detection and analysis of blackspots have been undertaken. The approaches and procedures employed for project analysis area identification (wherein frequent injuries happened), accident data gathering system, data splitting for easier investigation, and evaluation of black zones are presented. This research concludes the detection of black spots using various approaches such as the weighted severity index and average severity index. Bisht and Tiwari (2020) illustrate the present condition of the country's policy concerning black spots. This study evaluates the effectiveness of strategies aimed at identifying and correcting high-risk zones on National Highways Furthermore, research outlines the policies' assets along with drawbacks depending on the findings of an assessment done to assess the state of roadway security in India (Bisht and Tiwari 2020). Long-lasting initiatives, involving building of a vehicular underpass, by-pass, flyover, and carriageway enlargement, can be suggested. Pitale et al. (2019) chose a 97-km stretch of NH 3 from Mumbai to Nashik, and accident data from the previous three years (2014 to 2017) were classified based on the type of the mishap, Reasons of the mishap, Kind of Damage (mortal, disquieting & inconsequential), Road Feature, Road Condition & Climate Condition. Using an Average Severity Index Method, accidental severity was assessed, and priority was assigned based on the weightage of the spot, as well as corrective measures. Arun et al. (2016) report the discovery of black spots on NH-21A between Baddi and Nalagarh. The severity of the accidental places was determined using data gathered from Nalagarh and Baddi Police Stations and PWDs. Collision diagrams were created for each detected blind spot to determine the type and sort of incident (Arun et al. 2016). Amle and Ghodmare (2016) investigated National Highway 6, which connects Nagpur and Kondhali. Accident data from the previous 3 to 5 years have been gathered and analyzed. Based on the information gathered, a key segment is selected. This might help the country's highway safety audits pinpoint specific places on national roads based on the geometric aspect that is targeted to enhance as part of safeguarding enhancement initiatives (Amle and Ghodmare 2016). Bobade et al. (2015) focused on identifying unexpected risk patches on highways as well as on motorways. This paper recommended that the hazardous zones be discovered by rating qualities based on severity as well as generating the severity score (Bobade et al. 2015). The overall harshness indices are obtained through summing the weights. The Accidental Severity Index was generated by adding an amount of importance for each current parameter and dividing it by the overall severity. By Sorate et al. (2015), a 34-km section of the Mumbai-Bangalore NH within the Pune municipality boundary experienced 110 deadly mishaps, taking 111 lives in the previous 3 years. Therefore, the major goal of the research is to find hazardous spots on NH-4, which spans about 14.5 km through the New Katraj Terminal to Chandani Chowk, as well as provide corrective solutions. This research focuses on structural mistakes along with how they are linked to additional defects (Sorate et al. 2015). Shaik and Reshma (2015) chose the Kanakpura Highway (NH 209) traverses through Bangalore to locate accidental zones in the current research. On a 15-km length, 8 black patches were noticed. Accidents are commonly caused by dense traffic, poor road geometry, excessive speed, short view lengths, and an absence of traffic etiquette (Shaik and Reshma 2015).

9.4 Methodology

1. The project stretch is the National Highway (NH353D), passing through Nagpur-Umred-Bhiwapur-Nagbhid-Brahmapuri-Armori having a length of 137 km.
2. The location of the study area was obtained from Google Earth.
3. Accidental statistics according to the spots will be gathered from the Public Works Department (PWD), or road agency, including the kind of accident, the reasons for the accident, and the categories of the accidents, the road characteristics, and the climatic conditions.
4. Sort the accident records by the number of accidents that occurred at a certain location.
5. To identify numerous traffic and highway-related components that contribute to accidents.
6. Separate crash records for 2020–2022 by kind of crash, reasons for the crash, kind of damage (deadly, minor and major), and Traffic characteristics along with traffic quality &climate status.
7. The data collected at (chosen location) is then analyzed by applying WSI formula based on the relevance of the parameter (WSI).
8. Assign the location the highest ranking and the most weightage based on the number of accidents that have taken place.
9. Examination of the top-ranking accidental locations.
10. The percentages were determined by assigning ranking and weightage, and on an accidental basis.
11. The accidental hotspots were recognized using the Weighted Severity Index (WSI).
12. Conduct a detailed examination of the indicated accidental spots and investigate the reason for the assessment of each black spot.

13. Alternatives for accidental locations are also proposed.
14. Set forth the Geometric Design Improvement activities.
15. Evaluation of appropriate steps for rectification and execution.
16. Black Zones will be corrected with the use of crash information on these areas and will be significantly decreased after the introduction of immediate actions.

9.5 Data Collection and Analysis

The following information should be gathered and analyzed to identify the accident-prone location in our chosen hazardous spots.

a. Location-wise incident information will be gathered from PWD, or the transportation authority together with the accident's nature, its causes, how it was classified, its characteristics, and its state.
b. Vulnerable locations where accidents are likely to happen, as advised by the police department.
c. Separating of Accident Records

The last three years (from 2020 to 2022) as well as the current year's accident data are analyzed at the PWD office and police station and divided into categories based on the accident's nature (Table 9.3), causes, kind of damage (deadly, minor and major) (See Table 9.1), pavement characteristics, condition as well as climate conditions (See Table 9.2).

According to the graph (Fig. 9.2), most individuals suffered fatalities in 2022, there were more grievous injuries in 2022, and there were more minor injuries in 2022. In 2022, the overall number of accidents has increased.

According to the graph (Fig. 9.3), the majority of Continuous Stretch accidents happened in 2022, the most Slight Bend injuries occurred in 2022, the most Sharp

Table 9.1 Accident classification based on the kind of injury

Year	Injury type			Total accident
	death	Grievous	Minor	
2020	26	72	44	142
2021	28	96	56	180
2022	38	112	62	212

Table 9.2 Accident classification based on condition of road

Year	Continuous stretch	Slight bend	Sharp bend	Tapered bend	Total accident
2020	46	32	22	42	142
2021	72	45	29	34	180
2022	85	53	30	44	212

Table 9.3 Accident classification based on accident nature

Year	Hit from side	Hit from back	Hit pedestrian	Head on collision	Skidding	Total
2020	41	32	18	26	25	142
2021	48	43	21	35	33	180
2022	66	51	26	43	26	212

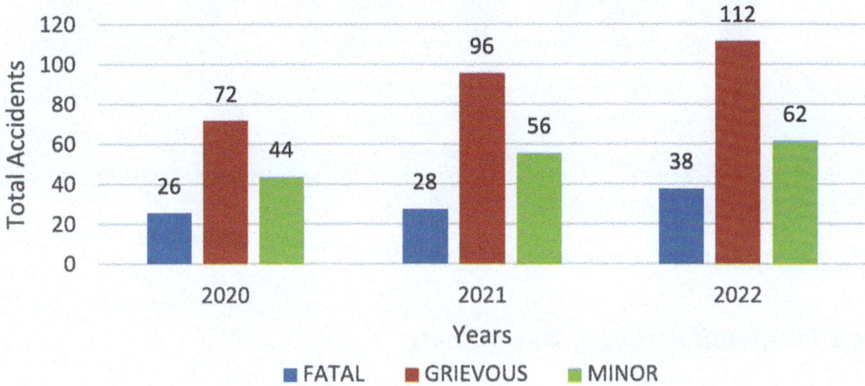

Fig. 9.2 Accident classification based on kind of injury

Bend incidents occurred in 2022, and the most Tapered Bend accidents occurred in 2022. A substantial amount of incidents happen in 2022.

The graph (Fig. 9.4) illustrates that striking from side incidents occurred frequently in 2022, Passenger strikes in 2022, striking back in 2022, slipping injuries in 2021, and Head-on wrecks in 2022.

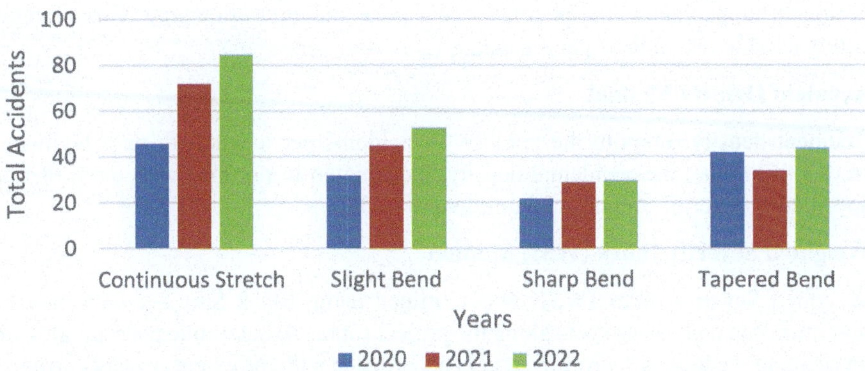

Fig. 9.3 Accident classification based on condition of road

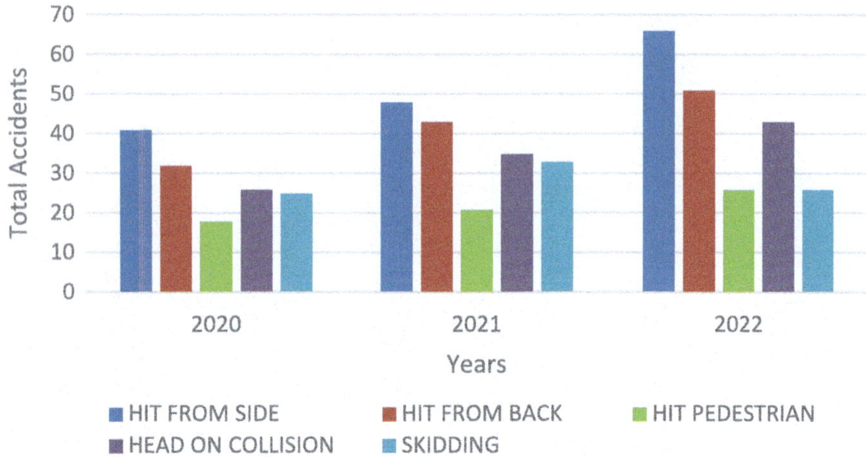

Fig. 9.4 Accident classification based on accident nature

9.5.1 Identification of Black Spots

The black spot assessment procedure is a technique for locating the most hazardous points on road sections, commonly referred to as black spots. Black spots are determined by the percentage of fatalities or serious injuries that occurred in a certain location. Methods used for determination of accidental hotspots are as follows:

Ranking Method

Using this method, the factors that result in mishaps are initially observed. The factor that continuously impacts accidents happening based on logical analysis is then given the superior ranking, while the factor that doesn't add significantly to accidents happening in that specified area is given the inferior ranking. This strategy is restricted to use in that specific area.

Accident Density Method

Accident density refers to the number of incidents per unit length on a highway. In this technique, the accidental density is estimated in portions with more than a specified number of events and a unit length of 500 m.

Weighted Severity Index (WSI) Method

Weighted Severity Index (WSI) is determined using Black Spot accident data to determine the dangerous spots along the project route. To determine the criticality of an accident location, a scoring system in accordance with the extent of an occurrence was implemented. The rating is known as WSI (Weighted Severity Index). The WSI is a non-dimensional metric that indicates the amount of damage on a road section. The below formula was employed:

$$\text{Weighted Severity Index, } (WSI) = (41 * K) + (4 * GI) + (1 * MI) \quad (9.1)$$

where, K = number of deaths
GI = several grievous injuries
MI = several minor injuries

9.5.2 Analysis of Blackspots

Accident Data are analyzed using the Weighted Severity Index (WSI) Method. Using the above formula, accidental zones or patches are identified as sites having WSI greater than 40. (Srinivasan et al. 2011).

9.6 Results

9.7 Conclusion

The present research aims to locate blackspots on National Highway (NH-353D), also known as the branch of NH53. The incident locations were rated through the Weighted Severity Index (WSI) (See Table 9.4). This technique was found to be efficient in recognizing blackspots. According to the study and results, the average WSI is 44.05 which is above the Benchmark value i.e. above 41 (Fig. 9.5). It has been found that most of the identified locations are junctions and built-up areas linking to NH (Fig. 9.6). The reason for the occurrence of mishaps is overspeeding and poor road geometrics. We offered short-term measures for blackspots while the long-term remedies are implemented.

Table 9.4 WSI calculations

Sr. No	Black spot location	Total no. of fatal accidents	Total no. of grievous accidents	Total no. of minor accidents	Weighted severity index (WSI)
1	A	1	3	1	54
2	B	1	4	2	59
3	C	0	2	7	15
4	D	1	1	2	47
5	E	1	3	2	55
6	F	0	2	3	11
7	G	2	3	0	94
8	H	1	4	2	59
9	I	0	2	1	9
10	J	1	1	3	48
11	K	2	2	2	92
12	L	2	0	5	87
13	M	1	1	3	48
14	N	1	3	2	55
15	O	0	2	2	10
16	P	0	1	6	10
17	Q	1	4	4	61
18	R	0	1	8	12
19	S	1	0	4	45
20	T	0	2	2	10

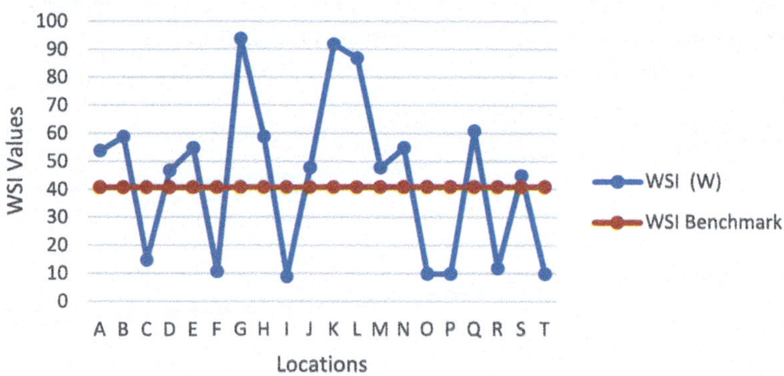

Fig. 9.5 Weighted severity index and benchmark

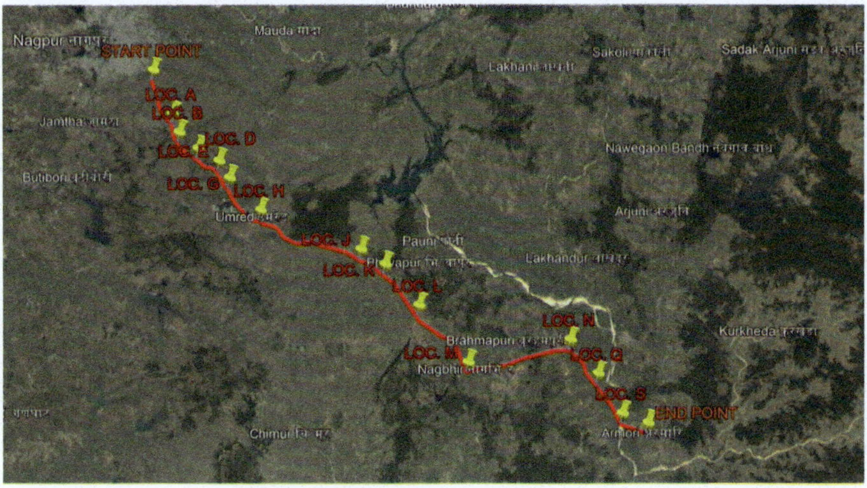

Fig. 9.6 Map showing black spot locations

References

Amle S, Ghodmare S (2016) Accident analysis and crash prediction modeling on national highways. Indian J Sci Technol 12(3):969–978

Arun K, Chauhan A, Thakur A, Singh K, Tiwari A (2016) Black spot analysis on NH-21A. Indian J Sci Technol 9(2):1–5

Bari C, Khode B (2021) PM10 dispersion modelling at unsignalised intersection using caline4—a case study. Adv Civ Eng Infrastructural Dev 87:531–539

Bisht L, Tiwari G (2020) Assessing the black spots focused policies for indian national highways. Transp Res Procedia 48:2537–2549

Bobade S, Patil J, Sorate R (2015) Identification of accidental black spots on national highways and expressways. IOSR J Mech Civ Eng 12(3):61–67

Ghodmare S, Gulghane A (2021) An attempt for sustainability evaluation considering implementation of rapid transit system. In: 2nd International conference on sustainable materials and practices for built environment, Material Today Proceedings, Jaipur, India, pp 448–451

Ghodmare S, Khode B, Ladekar S (2021) The role of artificial intelligence in industry 4.0 and smart city development. Adv Civ Eng Infrastructural Dev 87:591–604

Goyal A, Choudhary M, Nagar B (2020) Accident analysis and blackspots identification of Hulasur-Basavakalyana Road-Karnataka-a case study. Int Res J Eng Technol 7(4):7183–7187

Gupta M, Ghodmare S, Khode B (2021) Feasibility study on widening and designing of pavement. Int J Sci Res Sci Technol 8(3):1044–1048

IRC 67:2022, Code of practice for road signs (4th revision)

IRC SP: XXX 2020, Guidelines for identifying and treating black spots

Mendhe P, Ghodmare S, Titarmare A, Wadhai P (2022) Sustainable transportation using automated highway system for intercity connectivity. In: 10th International Conference on Emerging Trends in Engineering and Technology. IEEE, Nagpur, India, pp 1–6

Metre K, Baghel H, Suman G, Batra M, Ghodmare S (2021) Compact city and related impact on sustainable development in urban areas. Adv Civ Eng Infrastructural Dev 87:523–530

Nimburkar A, Ghodmare S, Tenpe A (2022) Traffic signal synchronization system for highway corridor in urban area. In: 10th International Conference on Emerging Trends in Engineering and Technology. IEEE, Nagpur, India, pp 1–6

Nuli S, Rohan K, Joshnavi C (2022) A study on road safety audit and black spot identification. Adv Sustain Mater Infrastruct 1086:1–8

Patel D, Umrigar F, Mishra C, Vankar A (2013) Road safety audit of selected stretch from Umreth junction to Vasad junction. Int J Sci Mod Eng 1(6):2319–2386

Pitale N, Katre N (2019) Analysis of blackspots and its rectification-a review. Int J Res Appl Sci & Eng Technol 7(4):1794–1796

Pitale N, Katre N, Bobade S (2019) Analysis of black spots on NH-3 and its rectification. J Transp Syst 4(2):16–32

Road accident data 2020

Rodke S, Ghodmare S, Deshmukh A (2022) Critical analysis of saturation flow rate and parameters affecting it. In: 10th International Conference on Emerging Trends in Engineering and Technology. IEEE, Nagpur, India, pp 1–6

Sanghai S, Pawade P (2022) Performance evaluation of friction dampers for building with soil-structure interaction. Mater Today: Proc 60(1):194–210

Shaik S, Reshma E (2015) Black spots studies on national highway 209 Kanakpura road Bangalore for a stretch of 15km. Int J Eng Res & Technol 2(7):1353–1359

Sorate R, Kulkarni R, Bobade S, Patil M, Talathi A, Sayyad I, Apte S (2015) Identification of black spots on national highway 4 (New Katraj Tunnel to Chandani Chowk). IOSR J Mech Civ Eng 12(3):61–67

IRC SP 88 Manual on road safety audit Indian roads congress New Delhi

Srinivasan S, Iyer S, Chand M, Srinath K (2011) Scientific identification and improvement of accident prone locations on national highways in Kerala. J Indian Road Congr 48(3):1–10

Yadav D, Ghodmare S, Kumar N (2021) Mitigation of Black-Spot on highways by the application of a safe system approach. Mater Today: Proc 52(3):1228–1235

Yadav D, Khode B (2022) Metro follows an optimized speed trajectory to save energy. In: 10th International Conference on Emerging Trends in Engineering and Technology. IEEE, Nagpur, India, pp 1–6

Chapter 10
Characterization and Micro Structural Analysis of Concrete Using Industrial By-Products and Incinerated Ash

Pranita S. Dambhare, S. A. Dhale, A. Das, K. M. Tajne, and A. Yadav

Abstract For the last vicennial, scientists have been focusing attention on mitigating the risks of environmental contamination brought on by the manufacturing of industrial waste and cementitious materials. Finding efficient waste to effectively incorporate into concrete production is the most feasible approach. Incinerated ash, an industrial by-product, offers unique qualities and long-term advantages that make it suitable as a building material. As a result, the characteristics of waste material are thoroughly covered in this paper to assess how effective it is as filler or cement substitute in concrete. This study examines the impact of replacing silica fume in concrete with incinerated ash on the material's strength and penetration capabilities. To achieve this, four concrete mixes containing four different percentages of silica fume (6, 8, 10, and 20%) and incinerated ash (4, 6, 8, and 10%) were prepared, along with a control concrete mix. It was noted that the control-mix concrete was easier to work than the waste-mix concrete. These investigations verified that 4% incinerated ash replacement combined with 8% silica fume results in improved mechanical properties and elevated pozzolanic reactivity. In the meanwhile, further study is required, particularly in the area of durability details, to assess the comprehensive performance of concrete made from industrial and incinerated ash.

P. S. Dambhare
Ph.D Research Scholar, Department of Electronics and Computer Science, R T M Nagpur University, Nagpur, India
e-mail: pranita.bhandari@pcenagpur.edu.in

S. A. Dhale · A. Das (✉)
Principal & Professor, Priyadarshini College of Engineering, Nagpur, India
e-mail: alakadaspce@gmail.com

K. M. Tajne
Assistant Professor, Priyadarshini College of Engineering, Nagpur, India

A. Yadav
Associate Professor, Government College of Engineering, Nagpur, India

P. S. Dambhare
Assistant Professor, Yashwantrao Chavan College of Engineering, Nagpur, India 440033

B. V. Khode et al. (eds.), *Innovations in Technologies: Pioneering Sustainable Infrastructure for a Resilient Future*, Sustainable Civil Infrastructures,
https://doi.org/10.1007/978-3-031-82133-2_10

Keywords Concrete · Incinerated ash · Silica fume · SEM · XRD

10.1 Introduction

The construction industry emphasizes the use of alternative materials to promote sustainability and minimize CO_2 emissions because of the rising urbanization and excessive usage of concrete. The use of industrial byproduct in concrete is sustainable and has advantages for the environment, using waste or recycled materials in concrete is thought to be the most practical option. Solid waste ash is mostly utilized as industrial processing by-products (Juenger et al. 2019; Gnanaraj et al. 2021). The generation of waste was noted at 2.01 billion in 2016 in the world and due to industrialization and economic growth it is believed this will increase to 3.40 billion tonnes by 2050 (Kaza et al. 2018). Incinerated ash (BMWA), Municipal Solid Waste Ash, Fly ash has been frequently used as a functional mineral additive. This waste material is adopted in concrete to decrease the environmental effects connected to concrete manufacturing. The requirement to reduce the heat intensity of the cement industry and promote low-carbon technology adoption has made compliance with sustainable development (Thomas et al. 2021). Since the creation of concrete, OPC (Ordinary Portland Cement) has served as the main binding component. The use of OPC in concrete has drawn a lot of attention in recent decades due to its detrimental effects on the environment brought on by excessive limestone extraction and CO2 emissions (Miller et al. 2019; Vassilev 2010; Vassilev et al. 2013). In the past few years, the concrete industry has widely incorporated various by-products like metakaolin, incinerated ash, silica fume, ground granulated blast furnace slag and agro-industrial by-products like rice husk and fuel ash of palm oil, to prevent their uncontrolled accumulation and burning in landfills that would pollute air, water and land (Agrela et al. 2019). Adding industrial by-products a replacement of cement has been used as a mineral supplement to lessen the effects of conventional methods of producing concrete. It was discover that the pore structure of the concrete matrix becomes more refined with the addition of wastes like silica fume, incinerated ash, fly ash, tiles powder, a byproduct of iron and steel industry (GGBS) and electric arc furnace slag. In comparison to control mix concrete, the ITZ (Interfacial Transition Zone) is denser. So the harmful compounds' entry into concrete becomes smaller (Praseeda and Rao 2022). The use of trash material in concrete reduces the impact of the production of conventional concrete. The industrial trash used (by-products) in concrete becomes the primary concern of numerous research studies in both developed and developing nations, but there is scant research on its utilization of incinerated ash in concrete as a mineral admixture, especially in developing nations like India. This could be the result of ignorance regarding the uses of incinerated ash in the building sector. The unique aspect of this study is the simultaneous reporting of XRF and XRD data on cement, industrial by-products, and incinerated ash. This allows the identification of toxic elements and chemical form as well as its textural distribution and relationship to other elements. The same study sample has also been subjected to SEM–EDS analysis. To better understand and interpret SEM examination of hardened concrete has been done with and without waste material. The current study's goal can be summed up as:-

1. Examining and evaluating earlier research on materials from waste that are able to be made sustainable concrete, in addition to their attributes, and choosing the best ones based on the requirements and environment of the case studies.
2. Analysis of the industrial waste by-product and compare it with previous research studies. The outcomes of experimental work and earlier studies are used to compile data on waste materials, produced sustainable concrete mixes, and their properties.
3. To ascertain the chemical form, textural distribution, and compatibility of other elements in concrete

10.2 Materials and Its Characterization

In the present study cement (OPC), aggregate, industrial by- product, incinerated medical ash, and potable water are used for making concrete. The chemical and physical characteristics of materials play a crucial role in determining their potential applications. The qualities of the by-product determine how well it works as a substitute for conventional building materials. This section reviews the physical, chemical, and morphological properties of cement, silica fume, and burned ash that are relevant to the study of their use as building materials.

10.2.1 Ordinary Portland Cement (OPC)

OPC is used for the study after being confirming from IS: 12,269–1987 and IS: 269–2015. The following table enumerates cement's physical and chemical properties (Table 10.1):

SEM images displayed the texture and geometry of the surface of cement tiny bit shown in Fig. 1a, b, respectively. Studies and SEM examination of cement particles revealed that the material had rough surface textures. Some cement particles are oblong, while most are angular. In the manufacturing process cement is cooled very slowly. Under ambient atmospheric conditions, a heat exchange takes place between the solid material and vertically incoming fresh air during clinker cooling. The numerous crystalline phases that can form under these slow cooling conditions are reflected in the intricate XRD patterns displayed in Fig. 10.2. It was feasible to determine distinct high-intensity peak mineral phases as well as some overlapping low-intensity peak mineral phases.

The characteristics pick at 33.1, 45.249 having phase value 2.7, 2.0024 indicates calcium carbonate presence in the sample. The cement pattern's most intense peaks'

Table 10.1 Cement's physical and chemical characteristics

Oxide	%
CaO	60.57
SiO_2	17.91
Al_2O_3	4.55
Fe_2O_3	4.84
MgO	0.88
Alkalies (1 < 20, Na_2O)	0.11
SO_3	2.4
Loss of ignition	2.8
Fineness (m^2/kg)	373
Setting Time [Initial] (min.)	74
Soundness (mm)	2.6
Setting Time [Final] (min.)	378

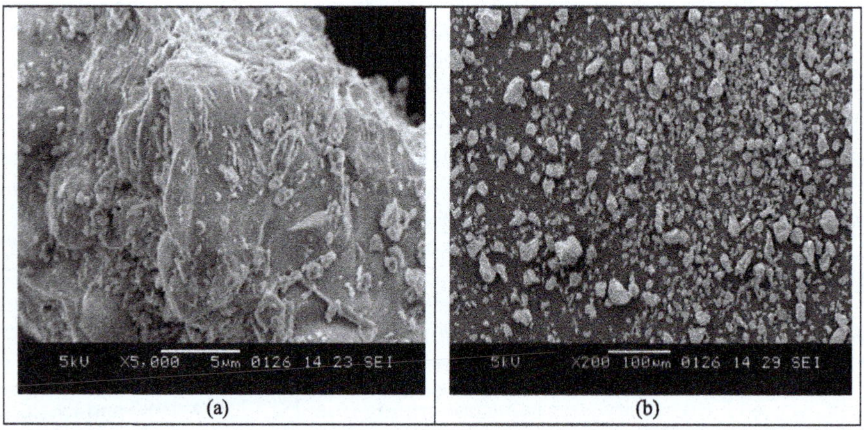

Fig. 10.1 SEM micrographs of cement sample **a** Surface texture **b** Particle shape

diffraction angle and d- spacing values are indicated in Fig. 10.2 above. The primary phases are determined by these values. Diffraction peak at 34.397 and 41.284 indicates the presence of sodium carbonate crystals. The diffraction peak at 47.30 indicates the presence of silicon. The peak at 34.397 shows that C3S and C2S hydration reactions also result in the production of calcium hydroxide or portlandite, which is crystalline but a lower peak indicates a small amount. The highest intensity peak at 31.04 shows that sodium silicate is the major component.

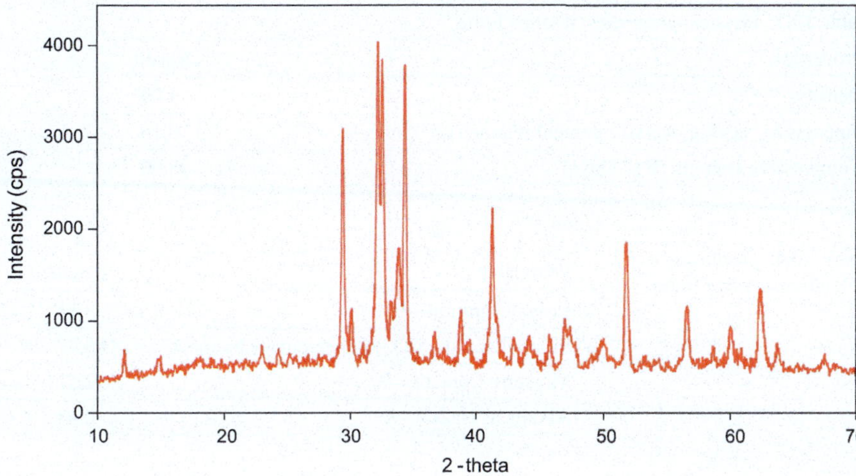

Fig. 10.2 X-ray diffraction patterns for cement

Table 10.2 Properties of aggregate sample [Fine]

Test	Result
Specific gravity	2.60
Bulking of Sand	25%
Silt Content	3.17%
Fineness modulus	2.76
Water absorption	1.0%
Bulk density (kg/m3)	1.65

10.2.2 Aggregate

In the present study, graded coarse aggregate (10 and 20 mm) nominal sizes was employed, following IS: 383—1970. The specific gravity of coarse aggregates is 2.95 with angular shape. Concrete is made with natural river sand or fine aggregate. It is obtained locally from the nearby river of specific gravity 2.6 and after being filtered through a 4.75 mm screen, it complies with IS 383—1978 Zone II. The table below displays fine aggregate's numerous attributes (Table 10.2).

10.2.3 Silica Fume (SF)

In electric arc furnaces, coke and high-grade quartz are reduced to create silicon and ferrosilicon alloys, from which Silica fume (SF) comes out as a byproduct. The silica fume sample was obtained from Black Cat Industries located in Nagpur. While SF has

Table 10.3 Physical properties of silica fume

Properties	Result
Density	2.09
Fineness by sieving oversize percent retained on 45mic. %	0.16
Compressive strength % (7 Days)	95.00

Table 10.4 Chemical properties of silica fume

Properties	Result
Loss of Ignition—%	3.81
SiliconDioxide(SiO2) in percentage by mass	86.37
Moisture Content—%	2.57
Alkalis as Na2O—%	0.43

many benefits when added to concrete mixtures, it is still one of the building materials. It also has drawbacks that several researchers have noted. The need for water is increased when SF is used in concrete because of its fineness; therefore, a superplasticizer must be used to obtain the required flow ability. Silica fume enhances lime-consuming activity, spherical, small variety of pore sizes, compressive strength and reduced heat release are all positively impacted by the elevated pozzolanic reactivity in (Hamada et al. 2023). A table illustrating the various characteristics of silica fume is provided below (Tables 10.3, 10.4 and 10.5).

According to the SEM study, silica fume shows a high concentration of tiny spherical (SiO2) silicon dioxide particles. The application of silica fume reduces concrete cracking. Microstructure study of silica fume particles show that the material had smooth surface textures and spherical in shape (Fig. 10.3 and 10.4).

Above figure highlights the values of d-spacing and diffraction angle of the cement pattern's most intense peaks. The primary phases are determined by these values. Diffraction peak at 30.671 and 31.681 with spacing 2.9126 and 2.8220 shows the presence of cubic and orthorhombic tri-calcium aluminate and 45.71 indicates the presence of silica crystals.

10.2.4 Incinerated Ash

Hospitals, medical research facilities, and other medical sources produce biomedical waste after incinerating biomedical waste, incinerated ash (BMWA) is produced. The most effective method of handling biomedical waste is incineration. In this procedure the handling of biomedical waste, carried out by burning it in the designed incinerators. There is a lot of oxygen present when incinerated ash is incinerated.

Table 10.5 Element composition of incinerated ash table

Compound	Na	Mg	Al	Si	S	Cl	K	Ca	Ti	V	Cr	Mn	Fe	
Conc	2.007	2.246	2.446	4.19	3.030	7.450	0.813	67.202	3.526	149.3	0.200	0.164	3.360	
Unit	%	%	%	%	%	%	%	%	%	ppm	%	%	%	
Compound	Ni	Cu	Zn	Ga	As	Br	Sr	Zr	Sn	Sb	Te	Ba	Ir	Pb
Conc	405.5	0.168	2.369	42.8	0.0	713.1	0.131	497.5	620.3	390.9	476.7	0.212	31.7	0.153
Unit	ppm	%	%	ppm	ppm	ppm	%	ppm	ppm	ppm	ppm	%	ppm	ppm

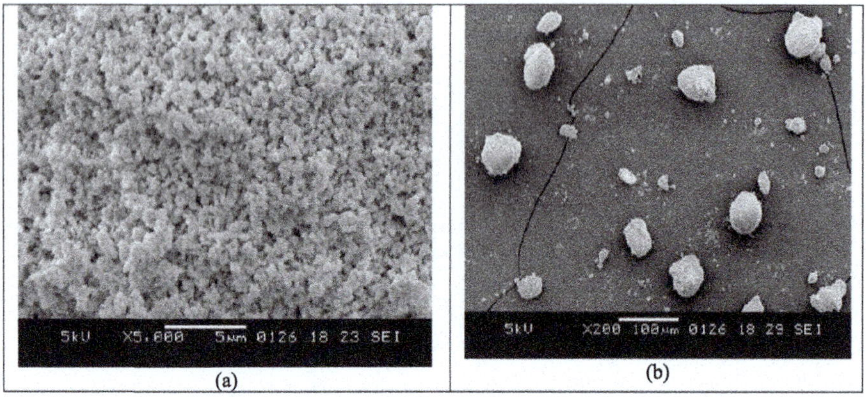

Fig. 10.3 SEM micrographs of silica fume sample **a** Surface texture **b** Particle shape

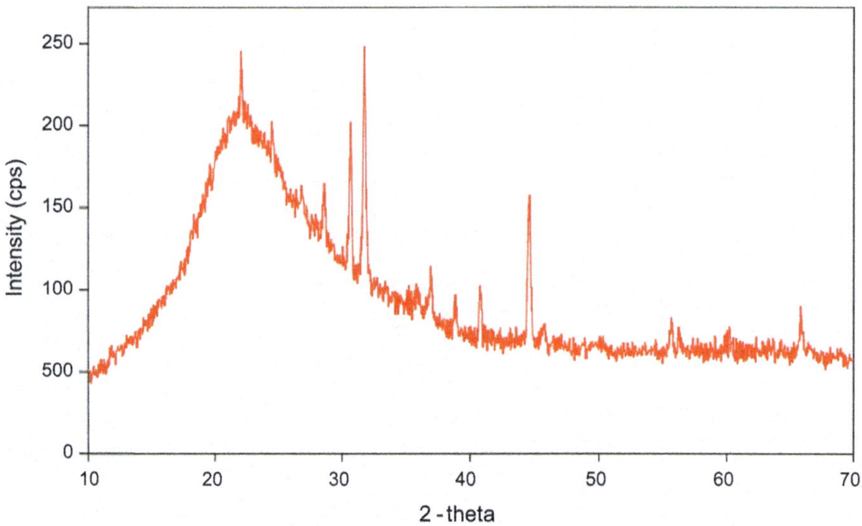

Fig. 10.4 X-ray diffraction of Silica fume

Incineration removes 90% of contaminants and reduces weight by 75% (Kaur et al. 2023).

10.2.5 The Oxide Content of Incinerated Ash

Compound	Na2O	MgO	Al2O3	SiO2	SO3	Cl	K2O	CaO	TiO2	V2O2	Cr2O3	Na2O	Ir
Conc	2.252	3.040	3.699	7.062	5.530	5.530	0.692	62.729	3.583	161.0	0.177	2.252	31.7
Unit	%	%	%	%	%	%	%	%	%	ppm	%	%	%
Compound	Fe2O3	NiO	CuO	ZnO	Ga2O3	As2O3	Br	SrO	ZrO2	SnO2	Sb2O3	TeO2	PbO
Conc	2.895	309.5	0.126	1.760	34.5	0.0	421.3	912.1	396.8	473.5	282.1	360.6	976.1
Unit	%	ppm	%	%	ppm	ppm	ppm	ppm	ppm	ppm	ppm	ppm	%

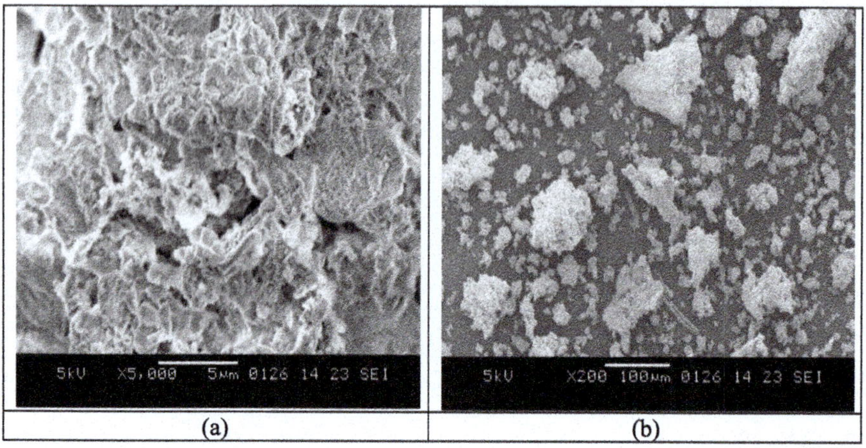

Fig. 10.5 SEM micrographs of incinerated ash sample **a** Surface texture **b** Particle shape

Microstructure analysis of incinerating ash is crucial for enhancing performance and forecasting their behavior in a blended cognitive system. Couple of researchers have examined BMWA utilizing SEM or scanning electron microscopy. As seen in Fig. 10.5a, b, the majority of published work has said that three different types of particles existed in untreated BMWA: irregular particles, rough and angular (Kaur et al. 2019; Akyıldız et al. 2017). Based on the BMWA SEM images the angular particle size and texture are bumpy. Better interlocking is provided by the angular shape and a superior mechanical bond develops with the cement paste by rough surface texture. Due to rough surface texture blended concrete may produce less workability than control concrete.

The cement pattern's most intense peaks' diffraction angle and d-spacing values are shown in the above figure. These values determine the initial phases. The primary phases are determined by these variables. The diffraction peak at 29.377 with spacing 3.03 shows the presence of elite (M) and 39.372 indicates the presence of calcite crystals. As a function of temperature stabilization Elite has 7 polymorphs: 1 Rhombohedra, 3 Monoclinic (M1, M2, M3) and 3 Triclinic (T1, T2 and T3) (Fig. 10.6).

10.3 Microstructural Analysis of Concrete

Concrete's microstructure can be more thoroughly examined using scanning electron microscopy, which can also be used to pinpoint the factors influencing the material's durability and mechanical qualities. The microstructure of waste-filled concrete and control concrete can be analyzed using scanning electron microscopy. This method

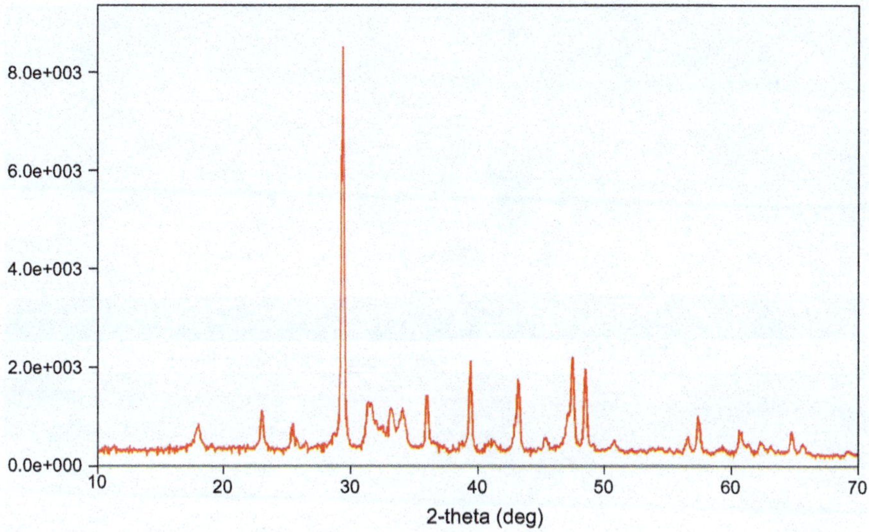

Fig. 10.6 X-ray diffraction of bio medical waste ash

validates the mechanism put forward by compressive strength. Strength tests were conducted after 28 days, and tiny cursed pieces were pull out from the.

The integrated system that constitutes the concrete microstructure is made up of an ITZ between the aggregate and cement hydration products as well as calcium hydroxide, calcium silicate hydrate (CSH) gel, coarse and fine aggregate and calcium sulfoalumininate hydrate (ettringite and monosulfate). In 8% silica fume concrete more CSH gel formation is seen. SEM image of control mix concrete is shown in Fig. 10.7a. After a 28-day curing period, a modification in the specimens' microscopic structure was noted in each replacement. In figures b, c, d the primary hydration product calcium silicate hydrate gel accountable for the higher compressive strength is visible in the SEM image. The pozzolanic reaction and the development of CSH gel were noticed after 28-day healing period. Figure 10.7a depicts the porosity and voids in standard concrete. Figure 10.7b shows that concrete made by 8% silica fume replacement makes the structure denser having fewer voids and micro-cracks. The calcium silicate hydrate gel forms a good bond with incinerated ash & silica fume concrete and the addition of these materials to the concrete decreased the voids and cracks. In comparison to control mix concrete, ITZ is denser when silica fume and incinerated ash are added. The concrete specimens' compressive strength was enhanced by the calcium silicate hydrate (CSH) and dense structure that was produced as a result of the hydration reaction. Ettringite formations in control mix concrete were found which describes that as compared to incinerated ash and silica fume concrete, the generation of calcium silicate hydrate gel was less. Hence the compressive strength is less as compared to other concrete with waste material. CSH gel formation is good in 8% replacement of silica fume concrete as depicted

Fig. 10.7 SEM analysis of **a** Control mix concrete **b** Concrete with cement & 8% silica fume **c** Concrete with cement & 4% incinerated ash **d** Concrete with cement & 8% silica fume & 4% incinerated ash

in Fig. 10.7b Consequently, the strength is higher than reference concrete. The 4% replacement of incinerated ash also gives good compressive strength, SEM image 2.7c shows the denser microscopic structure. SEM image 2.7d shows a very denser structure where ettringite formation is negligible and also the calcium silicate hydrate gel formation is more. Therefore, this combination gives more strength as compared to control mix concrete.

10.4 Conclusion

Taking into consideration, the aforementioned studies have provided an overview of how different alternative materials might be utilized as binder materials for concrete manufacture, to maximize their effectiveness. Concrete's durability and strength could be improved by using fly ash, GGBS, Silica Fume, and incinerated ash as precursor materials. Workability, strength, and different properties of concrete depend on several variables, including precursor materials, the dosage of solutions, curing conditions, and more. Concrete has strong mechanical strength and durability if the right curing technique, binder materials, and proportion, solution, and ratio are chosen with care. Therefore, the development of environmentally friendly concrete for construction can be facilitated by the use of industrial products and incinerated ash which reduce soil pollution and greenhouse gas emissions thereby increasing the greener impact on the environment due to materials. The precursor materials are thoroughly analyzed in this work, along with their physical and chemical properties, microstructure analysis (SEM), and detailed chemical compositions (XRD).

The intent of this study was to examine the effects of silica fume and burned ash as precursor materials on the characteristics of both fresh and cured concrete. The main conclusions of the study are as follows:

- Characterization of incinerated ash reveals that it has a range of shapes and sizes of particles, making it suitable for use as filler in concrete. The rough surface texture ensures better interlocking and a superior mechanical bond with the cement paste.
- When 8% silica fume replacement concrete is analyzed using an SEM, it reveals a dense microstructure and strong constituent material bonding because of the formation of CSH gel, similarly, the disjointedness of constituent particles and micro cracks are observed.
- These investigations verified that 4% incinerated ash replacement combined with 8% silica fume results in improved mechanical properties and elevated pozzolanic reactivity.
- The XRD analysis of cement and burned ash indicates a crystalline nature, while the analysis of silica fume reveals an amorphous nature.
- The control mix concrete was more workable than the concrete made of waste material.
- By using silica fumes and incineration to partially replace cement in concrete, the issue of contamination of underground water table, limited space, and expensive land disposal is resolved, preventing pollution in the environment.

Investigators and those working in the building materials sectors are going to find this study useful in resolving the problems brought on by the rising demand for cement. It will also help find meaningful solutions to problems related to industrial by-products and incinerated ash.

References

Abdul-Rashid R, Frantz GC (1992) MSW incinerator ash as aggregate in concrete and masonary. J Mater Civil Eng ASCE 4(4):353–368

Agmuthu P (2009) Solidification/stabilization disposal of medical waste incinerator fly ash using cement. Malay J Sci 3:241–255

Agrela F, Cabrera M, Morales MM, Zamorano M, Alshaaer M (2019) Biomass fly ash and biomass bottom ash. Woodhead Publishing Series in Civil and Structural Engineering 23–58. https://doi.org/10.1016/B978-0-08-102480-5.00002-6

Akyildiz A, Kose ET, Yildiz A (2017) Compressive strength and heavy metal leaching of concrete containing medical waste incineration ash. Constr Build Mater 138:326–332

Akyıldız A, Köse ET, Yıldız A (2017) Compressive strength and heavy metal leaching of concrete containing medical waste incineration ash. Constr Build Mater 138:326–332. https://doi.org/10.1016/j.conbuildmat.2017.02.017

Al-Mutairi N, Terro M, Al-Khaleefi AL (2004) Effect of recycling hospital ash on the compressive properties of concrete: statistical assessment and predicting model. Build Environ 39:557–566

Al-Rawas AA, Hago AW, Taha R, Al-Kharousi K (2005) Use of incinerator ash as a replacement for cement and sand in cement mortars. Build Environ 40:1261–1266

Altin S, Altin A, Elevli B, Cerit O (2003) Determination of hospital waste composition and disposal methods: a case study, Polish. J Environ Stud 12:251–255

Anastasiadou K, Christopoulos K, Mousios E, Gidarakos E (2012) Solidification/stabilization of fly ash and bottom ash from medical waste incineration facility. J Hazard Mater 93:201–208

Aubert JE, Husson B, Vaquier A (2004) Use of municipal solid waste incineration fly ash in concrete. Cem Concr Res 34:957–963

Augustine UE (2016) Hospital ash waste-ordinary Portland cement concrete. Sci Res 4:72–78

Azni I, Katayon S, Ratnasamy M, Johari MMNM (2005) Stabilization and utilization of hospital waste as road and asphalt aggregate. J Mater Cycles Waste Manage 7:33–37

BIS.: 2386 (Part III) - 1963, 'Methods of test for aggregates for concrete - part 1: particle size and shape. Bureau of Indian Standards, New Delhi, India

BIS.: 10262–1982, 'Recommended guidelines for concrete mix design. Bureau of Indian Standards, New Delhi, India

BIS.: 1199–1959, Indian standard methods of sampling and analysis of concrete. Bureau of Indian Standards, New Delhi, India

BIS.:383–1970,'Specification for coarse and fine aggregate from natural sources for concrete. Bureau of Indian Standards, New Delhi

BIS.:516–1959,'Methods of test for Strength of Concrete. Bureau of Indian Standard, New Delhi-1999, India

BIS.:5816–1999,'Methods of test for Splitting Tensile Strength of Concrete. Bureau of Indian Standard, New Delhi-1999, India

BIS.:8112–1989, 'Indian standard 43 Grade ordinary Portland cement – specification. Bureau of Indian Standards, New Delhi, India

Christopoulos K, Anastasiadou K, Mousios E, Gidarakos E (2012) Solidification/stabilization of fly and bottom ash from medical waste incineration facility. J Hazard Mater 207:165–170

CPCB, (2008), "Hazardous Waste Management, Chapter 9. Highlights 2008. Central Pollution Control Board, India. http://www.cpcb.nic.in/Highlights/ 2008/ chapter-9.pdf."

Gagg CR (2014) Cement and concrete as an engineering material: an historic appraisal and case study analysis. Eng Fail Anal 40:114–140. https://doi.org/10.1016/j.engfailanal.2014.02.004

Global Ready mix Concrete (RMC) Market worth over USD US$ 624.82 Bn by 2025: QY Research, Inc—Press Release—Digital Journal, (n.d.). http://www.digitaljournal.com/pr/4087041. Accessed 11 August 2020

Gnanaraj SC, Chokkalingam RB, Thankam GL, Pothinathan SKM (2021) Durability properties of self-compacting concrete developed with fly ash and ultra fine natural steatite powder. J Mater Res Technol 13:431–439. https://doi.org/10.1016/j.jmrt.2021.04.074

Hamada HM, Abed F, Katman HYB, Humada AM, Al Jawahery MS, Majdi A, Yousif ST, Thomas BS (2023) Effect of silica fume on the properties of sustainable cement concrete. J Mater Res Technol 24:8887–8908. https://doi.org/10.1016/j.jmrt.2023.05.147

Juenger MCG, Snellings R, Bernal SA (2019) Supplementary cementitious materials: new sources, characterization, and performance insights. Cem Concr Res 122:257–273. https://doi.org/10.1016/j.cemconres.2019.05.008

Kaur H, Siddique R, Rajor A (2019) Influence of incinerated ash on the properties of concrete. Constr Build Mater 226: 428–441. https://doi.org/10.1016/j.conbuildmat.2019.07.239

Kaur H, Siddique R, Rajor A (2023) Effect of bacteria on strength properties and toxicity of incinerated incinerated ash concrete. Environ Technol 44(12). https://doi.org/10.1080/09593330.2021.2012265

Kaza S, Yao L, Bhads-Tata P, Van Woerden F (2018) What a waste. Manager. https://doi.org/10.1680/imuen.2000.32747

Manjunatha M, Preethi S, Mounika HG, Niveditha KN (2021) Life cycle assessment (LCA) of concrete prepared with sustainable cement-based materials. Mater Today Proc. https://doi.org/10.1016/j.matpr.2021.01.248

De Matteis C, Pollastri S, Mantovani L, Tribaudino M (2024) Potentially toxic elements speciation in bottom ashes from a municipal solid waste incinerator: a combined SEM-EDS, μ-XRF and μ-XANES study. Environ Adv 15:100453. https://doi.org/10.1016/j.envadv.2023.100453

Miller SA et al (2019) Rice-based ash in concrete: a review of past work and potential environmental sustainability. Resour Conserv Recycl

Ohemeng EA, Ekolu SO (2020) Comparative analysis on costs and benefits of producing natural and recycled concrete aggregates: a South African case study. Case Stud Constr Mater 13. https://doi.org/10.1016/j.cscm.2020.e00450

Praseeda D, Rao KS (2022) Durability performance and microstructure analysis of nano engineered blended concrete. Clean Mater 6:100155. https://doi.org/10.1016/j.clema.2022.100155

Samarakoon SSM, Ruben P, Pedersen JW, Evangelista L (2019) Mechanical performance of concrete made of steel fibers from tire waste. Case Stud Constr Mater 11. https://doi.org/10.1016/j.cscm.2019.e00259

Seth D, Shrivastava RL, Shrivastava S (2016) An empirical investigation of critical success factors and performance measures for green manufacturing in cement industry. J Manuf Technol Mang 27:1076–1101. https://doi.org/10.1108/JMTM-04-2016-0049

Tangadagi RB, Manjunatha M, Bharath A, Preethi S (2020) Utilization of steel slag as an eco-friendly material in concrete for construction. J Green Eng 10:2408–2419

Thomas BS, Yang J, Mo KH, Abdalla JA, Hawileh R, Ariyachandra E (2021) Biomass ashes from agricultural wastes as supplementary cementitious materials or aggregate replacement in cement/geopolymer concrete: a comprehensive review 40:102332. https://doi.org/10.1016/j.jobe.2021.102332

Usahanunth N, Tuprakay S, Kongsong W, Tuprakay SR (2018) Study of mechanical properties and recommendations for the application of waste Bakelite aggregate concrete, case Stud. Constr Mater 8:299–314. https://doi.org/10.1016/j.cscm.2018.02.006

Vassilev SV et al (2010) An overview of the chemical composition of biomass. Fuel

Vassilev S et al (2013) An overview of the composition and application of biomass ash. Part 1, Phase– mineral and chemical composition and classification. Fuel

Chapter 11
Effect of HDRB Base Isolation System on RCC Framed G + 5 Irregular Plan with Mass Unsymmetric Building

Anil Ninawe and Sanket Sanghai

Abstract Considering the perpetual nature of progress, it is clear that modern development has reached a new level. Buildings that can withstand seismic loads must be designed and built by architects and engineers. The passive control method known as "base isolation" must be used to prevent the building from collapsing due to heavy horizontal and vertical loads. Its ability to sustain energy rakishness caused by parallel seismic loads and to encourage litheness at the foot of the structure is one of the most significant features of the foot Isolation Structure System. In this review paper, we investigate the performance of base isolation, which is a popular and efficient passive control method that resists these forces by keeping the superstructure and substructure apart. Base isolated buildings and fixed base buildings are evaluated for high-rise structures with vertical irregularities. The impact of LRB base isolators for both regular and irregular high-rise and low-rise buildings is demonstrated in the current study. At the moment, a structure's dynamic analysis can be carried out via time history analysis. The results are in the form of tale drift, period, base shear, and displacement using SAP2000 software.

Keywords High damping base-isolation system · Time history analysis · Seismic zone · R.C.C Frame structure · Mass irregularity

11.1 Introduction

The adoption of base isolation has emerged as an innovative approach to address challenges faced by structures in seismic zones. Present-day building engineers widely embrace this technique, appreciating its versatility across various structural components and its ability to yield desired outcomes. Seismic isolation, therefore, stands as a time-tested architectural concept that effectively shields buildings, specific

A. Ninawe · S. Sanghai (✉)
G. H. Raisoni College of Engineering, Nagpur, Maharashtra 440010, India
e-mail: sanket.sanghai@raisoni.net

B. V. Khode et al. (eds.), *Innovations in Technologies: Pioneering Sustainable Infrastructure for a Resilient Future*, Sustainable Civil Infrastructures,
https://doi.org/10.1007/978-3-031-82133-2_11

sections of structures, or even installed equipment from the destructive repercussions of ground accelerations. This approach showcases its adaptability and reliability in modern engineering practices, emphasizing its integral role in creating earthquake-resistant designs. Recent seismic events have underscored the importance of structuring buildings with straightforward, regular layouts, as irregularities can heighten vulnerability to damage. (Deringöl & Güneyisi 2021) Irregularities in a structure expose it to heightened loads and deformations at points of discontinuity, potentially leading to structural collapse and dissatisfaction among occupants (Jangid 2023; Balaji et.al. 2012; Reddy et.al. 2021).

The base-isolation techniques are a fascinating design tool used nowadays to decouple the building after the destructive impacts of ground accelerations in seismic events (Reddy et.al. 2021; Satyanarayana & Gopal 2018). The use base-isolation has emerged as a new way to deal with the issues that structures in seismic zones confront. Today's engineers who are employed in building design accept base isolation. The technique is well-liked in the field of study since it may be applied broadly in different portions of the structure and provide desired results. The methods are popular because they produce successful results the method used in the past has been to extend Initially, the building was in existence. Relative to the important period of the traditional fixed base building. Moving a structure's fundamental frequency away from the dominating frequency of earthquake motion and the fundamental relative frequency of the permanent foundation structure is one of the objectives of seismic isolation. The 2nd purpose of an isolation system is to provide an extra means of energy dissolution, hence decreasing the received modification of the structure. First base-isolation system considered is a hybrid base- isolation system (HDRB + FS) consisting of a low- friction Friction Slider (FS) operating parallel to an HDRB. Seismic isolation is a time-tested architectural idea that isolates a building, a section of a building, or even equipment installed inside from the damaging effects of ground accelerations (Majage & Phadatare 2018; Patel & Jamani 2017, Chaurasia & Pal 2019). Structures with standard geometry and regular distributed mass and consistency in plan and height withstand far less damage than those with Plan-irregular arrangements (Rai et.al. 2020; HTUN et.al. 2014; Sahu et.al. 2020) . Regardless, the needs of the contemporary world and the growing populace have compelled draft and architects to create unconventional plans. The data was produced using the reaction spectrum technique, because it is generally known that torsion negatively affects both conventional and base isolated(Thomas & Mathai 2016) A base- isolation system should have the following characteristics: (1) horizontal flexibility, except in sandy soils, to prolong the structural period and decrease the effects of earthquakes. (2) The uptake of energy like damping in the building to reduce displacements; and (3) effective primary rigidity in small displacements to deal with environmental loads when in service. This mass asymmetry may be caused by a water tank at the top of the structure, a heavy machine at any level, etc. Torsion in the building grows as this eccentricity does.

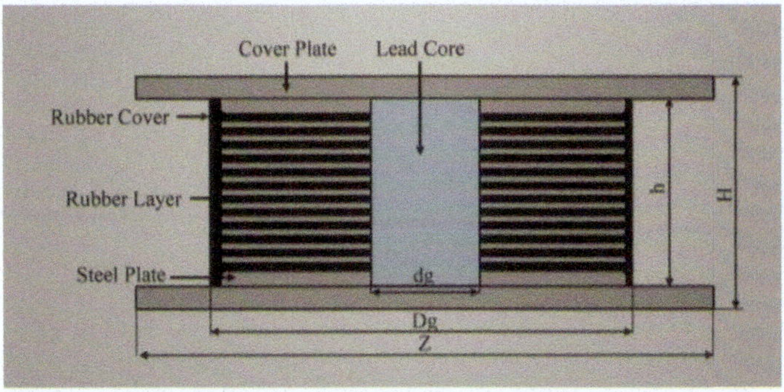

Fig. 11.1 Rubber bearing system

11.1.1 Rubber Bearing System

In 1975, New Zealand marked a significant milestone by being the first country to implement lead rubber bearings (HDRB) in structural engineering. Subsequently, this technology gained widespread adoption in Japan, United States and New Zealand, The LRB is built using a lead core whose diameter ranges varies 15 to 33% of the bonded diameter & LDE. Along with the ones composed of rubber and neoprene, These systems are also available in lead nucleus and steel laminated rubber forms (Bandhekar & Dabhekar 2019; Sanghai & Pawade 2022; Gorle et.al. 2018; Nimje et.al 2018). Elastomeric bearings represent a later evolutionary stage, evolving from the use of natural and synthetic rubber bearings in bridge applications. These bearings are frequently employed as seismic isolators. Thin steel plates are vulcanized to rubber plates to create the rubber laminated isolators. The more advanced of such are lead- nucleated laminated rubber kinds. Lead Laminated Rubber Bearing Systems are extremely advanced seismic isolators made up of layers of having a lead nucleus in the middle and steelrubber plasticcoating (Fig. 11.1).

11.1.2 Friction Pendulum Bearing

Seismic isolation systems, such as the sliding friction pendulum (FPS), aim to dissipate vibration energy during earthquakes by prolonging the operational period away from the dominant occurrence of ground acceleration motion (NDMA report, 2014; Mosalam et.al. 2018; Kalam et.al. 2017; Williams et.al. 2008). A type of base isolation system suitable for medium to large-scale buildings is the sliding friction pendulum base isolation structure. In recent years, the sliding friction pendulum system has gained widespread acceptance as an effective seismic isolation tool. The primary objective is to safeguard buildings from ground tremors during seismic events (Figs. 11.2, 11.3 and 11.4).

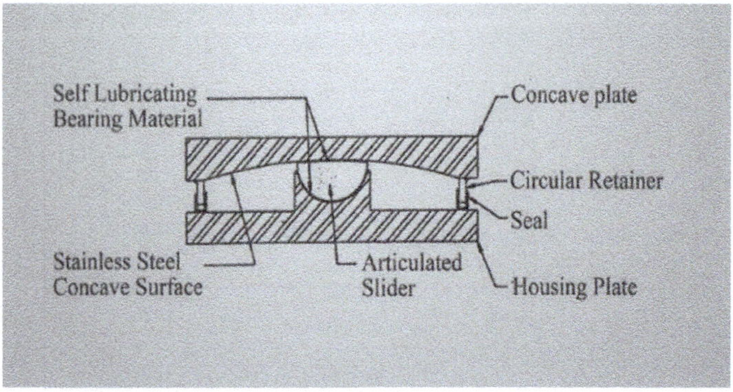

Fig. 11.2 Friction pendulum bearing

Fig. 11.3 Building plan

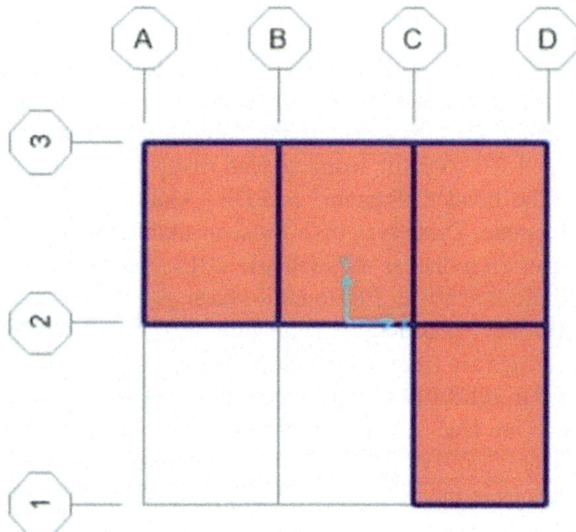

11.1.3 High Damping Rubber Bearing (HDR)

A damping influence is one that has the power to lessen, restrict, or even stop oscillations in an oscillatory system (Nahkies 2015; Tripathi, et.al. 2021; 22.IS 15988 2013; 23.IS 875 1987; Lou et.al. 2022; IS 456 2000). It belongs to the seismic bearing isolation and is comparable HDR. It is made of layer's of steel between special rubber with high damping properties. As a result, high damping rubber bridge bearings have outstanding vibration abatement capabilities. The exceptional characteristics of high damping rubber bearings enable them to absorb a portion of the seismic input energy before it reaches the substructure. Their remarkable flexibility and high restoration

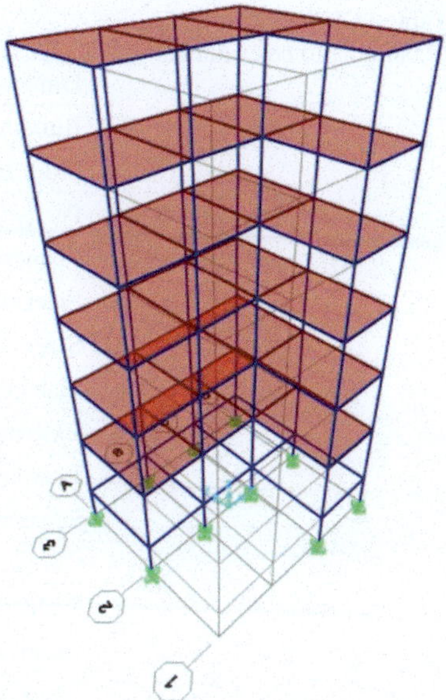

Fig. 11.4 Building 3D model generated in SAP2000

capacity allow for the subsequent restoration of this absorbed energy after the earthquake event. HDRB exhibits notable characteristics such as high vertical loading capacity, rapid recovery ability, and effective damping capacity. Its inherent hysteric nature allows HDRB to generate a seismic isolation effect across a spectrum of earthquake magnitudes, including small, medium, or large events. The self-sufficiency of HDRB in producing the seismic isolation function eliminates the necessity for installing supplementary devices. Consequently, this leads to reduced maintenance and repair expenses. There are various uses for these bearings since temperature effects on their flexibility and restraining power are so minimal (Tables 11.1 and 11.2).

The laminated High damping rubber bearing vertical stiffness is represented by

$$K = \frac{EA}{tr} \quad (11.1)$$

where,

A = the bearing's surface area.

tr = the total thickness of rubber in the bearing.

E = the composites' instantaneous compression modulus. According to Kelly (1997), the immediate compression modulus for a bearing with a square plan is given by

$$E = 6.7352G \quad (11.2)$$

Table 11.1 Data for analysis of building to be examined

Plane Details	Dimensions
Dimensions of Plane	9m × 8m
Building's total height	19.5 m
Story Height	3.0 m
Parapet-Wall Height	0.9 m
Foundation Depth	1.5 m
Sizes of beams	B1—300 mm × 365 mm
Sizes of columns	C1—520mm × 520 mm
Slab depth	125 mm
External walls thickness	230 mm
Internal walls thickness	150 mm
Seismic zone	III
Condition of Soil	Medium class
Reduction Factor of Response	5
Factor of Importance	1.20
Floor finish	1.00 k N/m2
Live-load for all floors	3.00 k N/m2
Live-load for mass asymmetry	3.00 kN/m2
Concrete's Grade	M25
Steel Grade	Fe500 & Fe415
Concrete's Density	25 kN/m3
Brick Masonry Density	19 kN/m3

** according to IS 1893–2016 (part 1). In comparison to their next floor, we must increase the live load by between 150 and 200 percent

Table 11.2 Earthquake parameter

Sr. No	Model description	Data
1	Zone	III
2	Zone factor	0.16
3	Type of building	Residential
4	Response reduction factor	5 (SMRF)
5	Importance Factor	1.2
6	Building height	19.5
7	Soil condition	Medium
8	Damping Ratio	5%

where,

G = the bearing's shear modulus, which is often influenced by the hardness of the rubber.

S = the shape factor, or the ratio of the loaded area to the force- free area of the rubber layer.

The laminated rubber bearings' horizontal rigidity is represented by

$$K_h = \frac{GA}{tr} \tag{11.3}$$

11.2 Time History Method

The analysis of the dynamic repose of the R.C.C structure at each step of time during which the structural foundation is subjected to a certain ground motion acceleration time history, or as one would put it, during an earthquake.

**We have used San Fernando time history data with magnitude of 6.61, Castaic—Old Ridge Route Station.

11.2.1 Aim & Objectives

The aim and objectives of the presented study is:

- To investigate how a building's asymmetrical mass affects its seismic performance.
- To study & design base-isolation system for R.C.C irregular Building with mass unsymmetrical and irregular structure.To evaluate the outcomes of base- isolation system and base-free model construction.
- To compare the results of building models with and without base-isolation system.
- To look into how the asymmetrical mass of a building impacts its seismic performance.

11.3 Methods of Analysis

Using SAP 2000 software, the displacement and stresses of numerous key structural constituents are determined. In order to analyse the structure for earthquakes, IS 1893-2016 (part (1) is used. The building will be examined using the SAP-2000 programme, and this study will be based on the method time history analysis (Nibasumba 2001; Alhan et.al. 2016; Shoaei & Mahsuli 2019). The structures basic time period is computed in accordance with the code's instructions.

11.4 Analysis Data

We have calculate the different structural component as per IS 1893:2016 part. The following calculation as shown below:

11.4.1 Load Calculation for Slab

A	Dead Load	25*(0.125*4*3) = 37.5 KN
B	Live Load	3*4*3 = 36 KN
C	Floor Finish	1*3*4 = 12
D	Water Proofing	0.7*3*4 = 8.4
	TOTAL FACTORED LOAD	1.5(37.5 + 36 + 12 + 8.4) = 144.9 kN

Slab:

A	Dead Load	25*(0.125*4*3) = 37.5 KN
B	Live Load	3*4*3 = 36 KN
C	Floor Finish	2*3*4 = 24
	TOTAL FACTORED LOAD	1.5(99) = 148.9 kN

Slab for Asymmetry Floor:

A	Dead Load	25*(0.13*4*3) = 37.5 KN
B	Live Load	(3 + 2.4)*4*3 = 64.8 KN
C	Floor Finish	2*3*4 = 24
	TOTAL FACTORED LOAD	1.5(99) = 191.7 kN

Total Load act on slab = 144.9 + 191.7 + 148.5 + 148.5 + 191.7 + 148.5 + 148.5 = 1079.1 KN.

11.4.2 Load Calculation for Beam

A	Dead Load	25*(0.365*0.23*12) = 11.885 KN *6 = 151.11
B	wall Load	19*0.23*(3-0.365)*12 = 138.1794*6 = 829.0764
	TOTAL FACTORED LOAD	1.5(151.11 + 829.0764) = 1471.311 kN

Total factored load on beam = 1471.311.

11.4.3 Load Calculation for Column

Assume size of column 0.5*0.5 m for trial.

A Column self-weight = 25*0.5*0.5*19.

= 118.75 KN.

Total factored Load = 178.125 KN.

$L_{eff} = 0.65*l$.

$= 0.65*(3–0.365)$.

= 1.71 m.

$\frac{L_{eff}}{D} = \frac{1.71}{0.5} = 3.42 < 12$.

Short column.

We have general equation;

$P_u = 0.4* f_{ck}* (A_g- A_{sc}) + 0.67* f_y* A_{sc}$.

(Assuming % of steel, p = 1%).

$P_u = 15.23*A_g$.

$(2728.536*10^3) = 15.23* A_g$.

$A_g = 268{,}733.027\ mm^2$

Considering a square column.

Depth(D) = Width(B) = $\sqrt{Ag}$ = 518.39 mm.

Assume = 520 mm. ∴ Adopt size of column = B* D = 520 *520 mm.

11.5 Analysis of Building for an Earthquake

11.6 Load Combination

When multiple different types of loads act on the structure, a load combination is created. In the course of analysis, several load cases and load combination cases in accordance with IS 1893:2016 are taken into consideration to determine the most critical element stresses in the structure (Figs. 11.5, 11.6 and 11.7)(Tables 11.3 and 11.4).

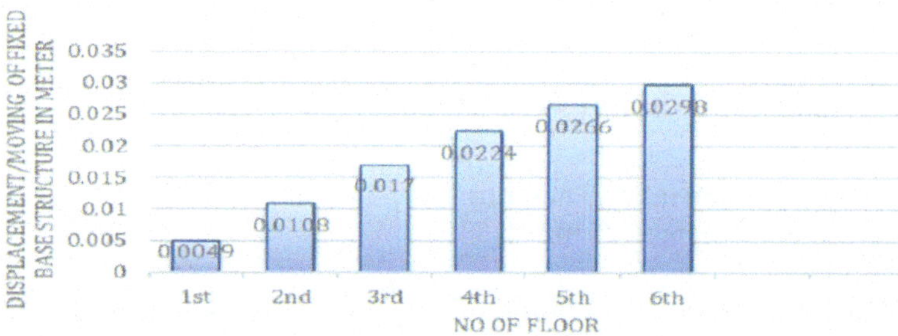

Fig. 11.5 Displacement of fixed base structure in meter

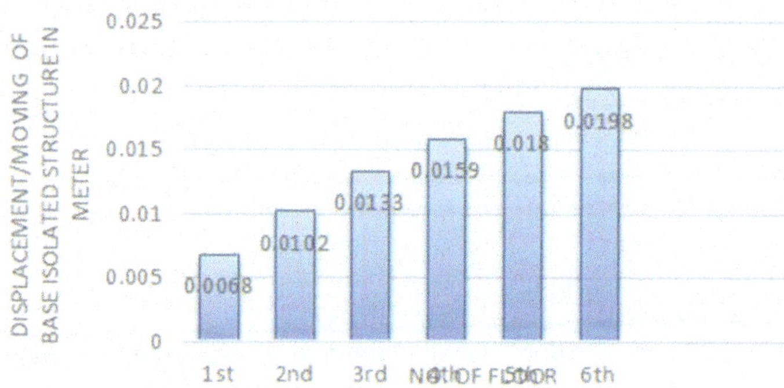

Fig. 11.6 Displacement of base isolated structure in meter

Fig. 11.7 Difference between fixed base and base isolated structure

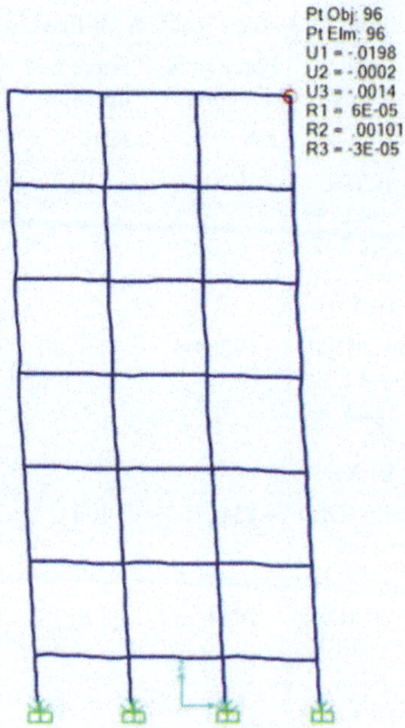

a. Fixed Isolated Structure

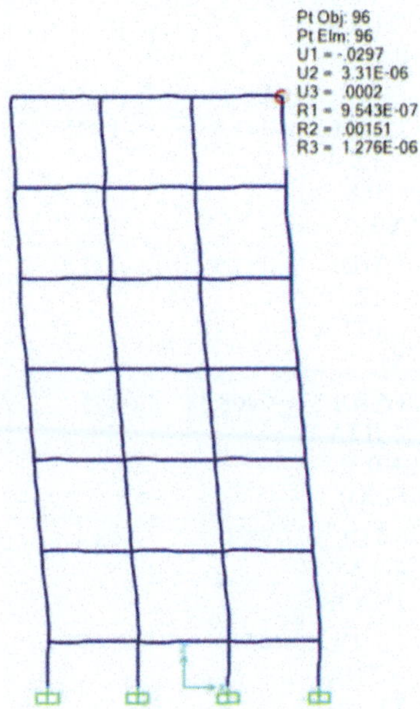

b. Base Isolated Structure

Table 11.3 Base reaction for fixed base building frame

Load Cases	Force in X direction	Force in Y direction	Force in Z direction	Moment in X direction	Moment in Y direction	Moment in Z direction
Unit	KN	KN	KN	KN-m	KN-m	KN-m
1.2(DL + LL + (ELX + 0.3 ELY))	−1920.067	−110.20	16,160.802	889.7900	−29,366.326	−129.598
1.2(DL + LL + (ELY + 0.3 ELX))	−1920.067	−110.20	16,160.802	889.7900	−29,366.326	−129.598
1.5 (DL + 0.3 THA X)	−134	−134	14,845.277	1120.561	−1474.135	−144.0458
1.5 (DL + (ELX + 0.3 ELY))	−2444.334	−140	14,845.277	1120.561	−36,608.942	−157.9880
0.9 DL + 1.5 (ELX + 0.3 ELY))	−2346.734	-87.4	8977.160	672.3403	−36,087.2897	−99.369
1.2 (DL + LL + (THA X))	−117.5	−117.5	16,160.072	885.766	−1119.397	−117.2366
1.2 (DL + LL + 0.3 THA X)	−117.5	−117.5	16,160.072	885.766	−1119.397	−117.2366
1.5 (DL + (ELY + 0.3 ELX))	−2408.34	−145	14,848.277	1125.5071	−36,650.9471	−157.973
0.9 DL + 1.5 THA X	−88.4	−88.4	8904.166	674.303	−844.4061	−86.4275
1.5 (DL + THA X)	-145	−145	14,745.774	1120.5671	−1415.145	−144.0458

(continued)

Table 11.3 (continued)

Load Cases	Force in X direction	Force in Y direction	Force in Z direction	Moment in X direction	Moment in Y direction	Moment in Z direction
Unit	KN	KN	KN	KN-m	KN-m	KN-m
0.9 DL + 0.45 THA X	-87.4	−87.4	8907.166	675.350	−851.488	−86.4275
0.9 DL + 1.5 (ELY + 0.3 ELX)	−2346.734	−87.4	8907.166	675.350	−36,099.2897	−99.357

Table 11.4 Base reaction for base isolated building frame

Load casse	Force in X direction	Force in Y direction	Force in Z direction	Moment in X direction	Moment in Y direction	Moment in Z direction
Unit	KN	KN	KN	KN-m	KN-m	KN-m
1.2 (DL + LL + (ELX + 0.3 ELY))	−1925.752	−118.2	16,060.601	880.2360	−29,295.125	−129.580
1.2 (DL + LL + (ELY + 0.3 ELX))	−1925.752	−118.2	16,060.601	880.2360	−29,295.125	−129.580
1.5 (DL + 0.3 THA X)	−143	−143	14,766.40	1123.935	−1418.4690	−147.088
1.5 (DL + (ELX + 0.3 ELY))	−2408.430	−142	14,766.40	1123.935	−36,640.540	−159.980
0.9 DL + 1.5 (ELX + 0.3 ELY))	−2340.838	−88.4	8850.588	670.551	−36,070.740	−97.366
0.9 DL + 1.5 (ELY + 0.3 ELX))	−2340.838	−88.4	8850.588	670.551	−36,070.740	−97.373

(continued)

Table 11.4 (continued)

Load casse	Force in X direction	Force in Y direction	Force in Z direction	Moment in X direction	Moment in Y direction	Moment in Z direction
1.2 (DL + LL + (THA X))	−112.2	−112.2	16,060.61	877.262	−1110.734	−113.260
1.2 (DL + LL + 0.3 THA X)	−112.2	−112.2	16,060.61	877.262	−1110.734	−113.260
1.5 (DL + (ELY + 0.3 ELX))	−2400.438	−142	14,760.301	1121.925	−36,639.55	−157.9836
0.9 DL + 1.5 THA X	−85.8	−85.8	8850.521	673.551	−846.684	−85.475
1.5 (DL + THA X)	−142	−142	14,754.301	1120.9251	−1414.4674	−144.0458
0.9 DL + 0.45 THA X	−85.8	−85.8	8850.521	673.551	−846.684	−85.475

11.7 Results

11.8 Conclusion

The comparison made on the irregular structure in mass and in plane behavior having fixed and isolated base using High Damper rubber isolation.

- As compare with Base Isolated Structure Frame, Fixed Base Structure Frame exhibits greater shear.
- The result indicates that the normal base structure frame has a larger displacement than the High Damper rubber isolated structure frame. The 5th floor motion of structural models with isolation systems is found to be lower than the corresponding motion without isolation systems, according to the resultant graph. Consequently, the base isolation system is effectively work in reducing the dynamic response on the structural system.
- Base isolated structure give the RC structure horizontal flexibility.
- According to the Base Reactions table, Fixed Moment on isolated base structure is less than as compare to the with fixed base systems. Able in supporting the structure against earthquake load

References

Alhan C, Öncü-Davas S (2016) Performance limits of seismically isolated buildings under near-field earthquakes. Eng Struct 116:83–94

Balaji CDR, Vasanth K, Premkumar R (2012) Analytical study of seismic and wind response of symmetric and asymmetric base insolated building. International Conference on Advances in Architecture and Civil Engineering Proceedings (AARCV 2012) 21, pp 299

Bandhekar R, Dabhekar K (2019) Seismic analysis of various shapes of building

Chaurasia R, Pal A (2019) Comparative analysis of multi-storey rc frame building with and without floating column using base-isolation in seismic zone. Int J Adv Eng Res Sci 602–604

Deringöl AH, Güneyisi EM (2021) Effect of using high damping rubber bearings for seismic isolation of the buildings. Int J Steel Struct 21(5):1698–1722

Gorle T, Suryawanshi S, Kedar S, Mange U, Charde S, Khode V (2018) Seismic analysis of a multi-storey (G+ 8) RCC frame building for different seismic zones in Nagpur. (MH). Int J Eng Res Technol (IJERT), ISSN:2395–0056, (MH)

National Disaster Management Guidelines, Seismic Retrofitting Of Deficient Buildings and Structures 2014

Htun T, Aung DKM (2014) Investigation on the structural response of reinforced concrete building without and with base- isolation system

IS 456:(2000) Plain and reinforced concrete, Bureau of Indian Standards, New Delhi

IS 15988:(2013) Seismic evaluation and strengthening of existing reinforced concrete buildings—Guidelines, Bureau of Indian Standards, New Delhi

IS 875:1987(Part 2), Design loads for buildings and structures BIS, New Delhi

Jangid RS (2023) Seismic performance of supplemental inerter and spring with on-off effects for base-isolated structures. J Infrastruct Intell Resil 2(2):100038

Kalam et al (2017) Methods of seismic retrofitting by using Etabs software

Lou X, Huang Y, Lv Y, Huang Q (2022) Seismic performance of the series seismic isolation systems designed by the procedures of GB50011-2010 and ASCE/SEI 7–16. Case Stud Constr Mater 17:e01184

Majage SC, Phadatare NP (2018) Design of high damping rubber isolator for RC multistoried structures and its comparative seismic analysis. Int Res J Eng Technol (IRJET) 5

Mosalam KM et al (2018) Performance-based engineering and multi- criteria decision analysis for sustainable and resilient building design

Nahkies PB (2015) Seismic retrofitting—Testing feasibility

Nibasumba PT (2001) Singhua University, China, Strengthening of a reinforced concrete column by SFRC

Nimje RP et al (2018) Enhancing the strength of existing building using retrofitting techniques

Patel S, Jamani A (2017) Effect of base-isolation on seismic performance of RC irregular buildings. Int Res J Eng Technol e-ISSN:2395–0056

Rai A, Sahu SK, Shrivastava V (2020) Static analysis of asymmetric reinforced concrete structure with lead rubber bearing

Reddy KA, Prasad JSR, Malagavalli V (2021) Comparative study of multi-storey RC building with and without base-isolation. Ilkogretim Online 20(6):8–16

Sahu SK, Rai A, Shrivastava V (2020) Review on seismic response of asymmetric reinforced concrete structure with lead rubber bearing

Sanghai S, Pawade P (2022) Performance evaluation of friction dampers for building with soil-structure interaction. Mater Today: Proc 60:194–210

Satyanarayana GVV, Gopal J (2018) Comparative study on fixed base and base-isolation having regular and irregular RC frames (G+5 & G+20) By using Etabs

Seranaj A, Garevski M (2015) Analysis of reinforced concrete buildings with different location of seismic isolation system

Shoaei P, Mahsuli M (2019) Reliability-based design of steel moment frame structures isolated by lead-rubber bearing systems. In Structures 20:765–778 Elsevier

Thomas T, Mathai A (2016) Study of base-isolation using friction pendulum bearing system. J Mech Civil Eng 19–23
Tripathi S et al (2021) Literature review on seismic retrofitting of Rc frame structure using energy dissipation device
Williams RJ et al (2008) Decision analysis for seismic retrofit of structures

Chapter 12
Arduino Based Traffic Actuated Signal System for Traffic Management System

Trupti G. Dongarwar and N. H. Pitale

Abstract In a lot of the nation's major metropolitan area, gridlocked traffic is becoming a significant problem. This serious issue of congestion on the roadways has been caused by the unpredictable failure of traffic lights, subpar law enforcement, and inadequate transportation planning. A novel approach utilizing RFID and Arduino has been developed to deal with the crucial problem of vehicle and time congestion at traffic lights. Currently, emergency vehicles are stuck in gridlock because of the blockage. This rush hour causes human casualties because the EMV cannot get to the destination in time. This study has improved an innovative concept to traffic management by employing the order of priority of the vehicles to circumvent it. In this approach, emergency vehicles like ambulances are given priority so they can get to their destination on time and reduce travel time. The suggested system is made to receive data regarding any emergencies, including ambulances, by using RF identification technology. We are utilizing RFID technology and one Arduino for implementation. For such emergencies, the technology has the capability of opening a whole lane. This project's primary goal is to provide an innovative way of tracking and controlling traffic congestion on roadways and emergency service vehicles. The most suitable spots for radio frequency readers have been selected so that the reader can read the radio frequency tags on ambulances easily. When the EV arrives, the specific approach signal will change from green to blue. As a result of the current traffic system's shortcomings, it reduces traveling duration and conserves thousands of lives. The present traffic system will be strengthened by this project. Corrective actions are taken and applied in our system to regulate traffic.

Keywords RFID tags · Traffic signals · RFID reader

T. G. Dongarwar (✉) · N. H. Pitale
G. H. Raisoni College of Engineering, Nagpur, India
e-mail: dongarwartrupti@gmail.com

B. V. Khode et al. (eds.), *Innovations in Technologies: Pioneering Sustainable Infrastructure for a Resilient Future*, Sustainable Civil Infrastructures,
https://doi.org/10.1007/978-3-031-82133-2_12

12.1 Introduction

Population growth is accelerating in emerging nations like India. This leads to a variety of issues, including complicated traffic laws and occasionally even accidents. In addition, rush hour traffic causes lengthy delays, resource waste, as well as financial loss. Traffic jams in particular are a factor in the high emission rate that affects the local population's health. Wildlife and shuttle services. Traffic jams are frequently linked to obstructions in emergency vehicles' paths, making human life an extremely precious commodity for any nation (Gulati and Srinivasan 2019). Due to the frequently incidents and disasters happening, like crashes, explosions, and medical crises, EVs must arrive quickly to prevent serious human casualties. As a result, medical emergencies and fire departments are dispersed throughout the city to speed up response times in situations involving emergencies. Cities have seen a very quick increase in population, which has led to a huge increase in localized traffic. Additionally, the frequency of fatalities caused by emergency vehicle arrival delays has increased significantly in recent years. Emergency services must thus arrive promptly to prevent the loss of human life, which is extremely important for any nation (Liu and Shen 2018). The primary objective for developing an effective traffic management system is to reduce traffic jams. The topic of traffic safety control systems is the subject of several research projects. Additionally, all emergency vehicles must perform crucial duties such as giving the vehicle precedence and adjusting traffic signals to enhance road safety methods. Assisting an emergency vehicle as it attempts to escape gridlock is crucial in the present traffic condition. In today's world, particularly in metropolitan areas, there is an increase in vehicle traffic. Future traffic management will become a major concern due to the on-going rise in road users. Traffic congestion is becoming a major problem. Long periods are spent by numerous vehicles at the signal. Because of increased consumption and several other issues, people are having a difficult time getting to work, and some are working in businesses. Before being fixed for some time, traffic lights are readied, and only then will they be changed to another signal (Kanungo et al. 2014). They don't have a system in place for ambulances, fire trucks, police cars, or other emergency vehicles to get precedence. Even before they reach the intersection, such automobiles require smart traffic signals to open the route for them. Due to the traditional traffic signals' inability to assist in the provision of dynamic and adaptable services for business activities, there are several additional disadvantages to traditional traffic control systems. The traffic lights are created in advance and are only fixed for a short period before switching to another signal (Samir et al. 2021). In India, where the population is growing daily and traffic is correspondingly increasing, traffic lights require good coordination to keep traffic flowing during peak hours. Currently, all systems operate automatically, but when an ambulance arrives in India, it can be operated manually at the intersection by a traffic officer. This is no longer the case. As a result, we have suggested a technique known as "emergency vehicle clearance using blue mode (Harshith et al. 2021). Here, we have considered a two-lane road. A radio frequency reader will also be positioned on the side of the road to scan the radio frequency tag of an ambulance. For implementation, we are using RFID technology and the Arduino-Uno board. The device can open a whole lane in an emergency like this. On the EV, there is an RFID tag. A few meters before the main signal, an RFID Reader is used to detect the emergency vehicle. A traffic signal is informed of an approaching ambulance thanks to an RFID reader. RFID TAGS are pre-programmed with the code that allows the main module (ARDUINO). Microcontroller to adjust traffic lights (Gunda et al. 2018). Blue light is only utilized when an emergency vehicle arrives at the signal system to signify open, whereas red LED signals CLOSED and a green LED indicates OPEN signal. The aim of this project is to produce an innovative traffic control system by resolving the drawbacks of the existing one. This can recognize emergency vehicles and give them the highest priority so they can arrive at their location in a short period (Dubey and Rane 2017).

12.2 Related Work

Various research work done by authors on traffic actuated signal system is discussed below a five-layer concept to provide an automated interstate structure for the metro political regions (Nagpur and Mumbai). Automobiles and roadways work together to coordinate vehicle growth, avoid roadblocks, and enhance traffic flow, therefore increasing safety and reducing congestion (Mendhe et al. 2022). The approaches implemented at various sites for calculating the rate at which saturation occurs while taking into account variable traffic characteristics (Rodke et al. 2022). The address several elements of signal timing based on previous study findings. It is also attempted to determine the proper characteristics and methods for the Indian traffic environment (Nimburkar et al. 2022). Focus on reducing energy consumption by utilizing an optimized speed profile. The MMRDA's metro line 1 (Versova-Ghatkopar) is being evaluated for this purpose (Yadav and Khode 2022). The explanation strategies for developing an abrasion reducer that include the soil's structural damage. The sliding stress of damping friction was computed by assuming a set and elastic base (Sanghai and Pawade 2022) to investigate how drivers perceive the level of service at Unsignalised junctions. Infrastructure authorities will find it easier to prioritize the important aspects of infrastructure development with the aid of the suggested model (Suprava and Dutta 2022). The Diffusion of particulate matter (PM10) at an unsignalised crossroads on metropolitan roadways. It was discovered that the early-morning peak has an elevated level of PM10, which gradually decreases near off-peak and then climbs again in the late afternoon peak (Bari and Khode 2021) the investigate Nagpur, India, as a compact city. The study emphasizes the compact city concepts and its implications for equitable growth in India (Metre et al. 2021). Attempted to concentrate on the trajectory of industrial and economic growth as urbanization progressed. It attempts to consolidate factual knowledge linked to advances in technology all over the globe (Ghodmare et al. 2021). Illustrates how a multi-attribute utility theory-based technique is used to evaluate sustainability for Nagpur, an expanding metropolitan metropolis, using a variety of individual performance metrics. The results are provided as scaled values for several of the performance measures considered in this article (Ghodmare and Gulghane 2021).The RFID technology is used for the comparison analysis, which shows that the system is six times more effective than the traditional way when 100 documentation are checked for accuracy in 500 s as opposed to 3000 s when done manually (Prasad 2021). Vehicle density -based System. This system can open and close the gate mechanically. The buzzer sound suggests that the gate is closed (Bura and Korra 2020) in “Image Processing in Intelligent Traffic Management” explored real-world issues regarding applications and the execution of traffic surveillance and control activities. They have included the several types of detectors used for traffic detection, as well as their capabilities, strengths, and shortcomings. They also covered the processing of images and development for vehicle recognition and categorization, vehicle counting, and density estimate using different strategies such as background removal, edge detection, algorithm methodology, and so on (Gulati and Srinivasan 2019). Created a stream of traffic light supervisor that, as soon as an EV approaches the signal, grants priority to that vehicle. The shortest route for an EV can be found by the controller. This study correspondingly intentions to improve traffic signal systems to ensure that controller may automatically turn on the green light for the roadways when it detects the movement of EVs or pedestrians. However, it is not stated whether a traffic signal will turn green if an EV approaches it without experiencing an emergency. Keep in mind that EVs do not receive any preference when they aren’t required to assist with incident recovery (Naseer et al. 2018). developed a mass-based traffic control system based on the ATmega8 microcontroller, in which IR sensors are used to identify the level of traffic density presenting each road at an intersection, and in the case of emergency vehicles, their approach was identified by presenting a unique RFID card for the same (Pandey et al. 2017).sophisticated intelligent traffic light management system utilizing IR sensors and a PICmicroprocessor that alienates information collected by the sensors by using an XBee Tran receiver running ZigBee firmware to transfer information from the devices to the microprocessor (Ghazal et al. 2016). This traffic light control system is not working based on real time. Processors and controllers are the basis of this system. These ITLCS have constraints because they perform based on established hardware that is set by the information system and lacks real-time modification flexibility. In order to calculate density, it also includes an infrared sensor. The processor uses this information to make decisions. However, one of its main drawbacks is that each vehicle must mount a single metal plate, which drives up the expense of the system (Padmini et al. 2015). He is proposed a green wave theory to give any EV passage through converting entirely of the red lights in its path to green, so giving the targeted vehicle a full green wave. The main drawback of green waves is that if they are disturbed, they can lead to traffic issues that can be made worse by coordination. When this happens, the line of cars in the green wave gets so long that some of the cars can’t make it to the intersections in time and have to stop (Sundar et al. 2014).

Role of components required in this project:

RFID READER: In our implementation, the EV was detected at the junction of the signal section using an RFID card. The RFID tag is the only thing the EV needs to be functional. Every RFID tag has a unique ID. These complete IDs were kept in a traffic control data base. One RFID reader will be installed in a part of a traffic signal, and it was this RFID reader that was utilized to collect the identification code. Then use an analogue connection to communicate the ID data to the Arduino. An RFID reader is a tool used to extract data from an RFID tag. Data is transmitted from a tag to a reader using radio waves. Typically, an RFID reader is used to find the automobiles that have RFID tags installed on them. We are utilizing the reader there to find the emergency vehicles that are in risk. After that, the RFID reader will read the code assigned to those vehicles (Gunda et al. 2018).

RFID TAG: The TAG often remains optimal in the absence of a source of electricity. An electronic tag called an RFID tag uses radio frequency waves to transmit and receive data using an RFID reader. At least two primary components make up an RFID tag in its entirety. Receiving radio frequency (RF) waves via the antenna. The next step is the integrated circuits (IC), which are used to process, store, modulate, and demodulate the radio wave that the antenna has picked up. A RFID chip is another name for an RFID Tag (Prakash et al. 2020).

ARDUINO: A compact Arduino board called the Nano is built around an ATmega328P or ATmega628 microcontroller. The connection is the same for the Arduino UNO board. The Arduino Nano, which depends on a microcontroller, has 16 digital pins that can be used for many different kinds of purposes Developed by Arduino.cc, the Arduino Nano is an open-source, breadboard-friendly microcontroller module based on the Microchip ATmega328P that was first made accessible in 2008.There are 30 male I/O headers on the board known as the Arduino Nano. The Arduino Nano is organized using the Arduino IDE. Mini USB and the IDE for Arduino are needed to launch projects utilizing the Arduino Nano board. Installing the Arduino IDE Programmed on the earlier mentioned laptop or desktop is required. Through the tiny USB port on the PC, the Arduino Nano board gets the code (Gunda et al. 2018).

LED: A light-emitting diode is a semiconductor that emits light. By electrical current passes through it.Photons are produced as the voids in the semiconductor interact with the ions, indicating that the vehicle is prepared to proceed, slow down, or proceed (Porter et al. 2023).

Power Supply (5 V Battery): The USB connector can provide sufficient power for Arduino boards to function properly. An external power source can be used to power it exterior socket connector, or PC port and provides 5 V DC voltage for use. Using 5 V because, with current technology, it offers the best balance of noise immunity, utilization of energy, and rate. In order to minimize the need for additional power equipment, combining networks like detection devices and other gadgets naturally attempted to use the same voltage (Ritchie et al. 2023).

12.3 Research Methodology

A combination of user-friendly hardware and software, Arduino boards is an open-source technology framework that can read components, such as a sensor's light, a button's push, or an indication and transform it into a result. Lots of projects have been inspired by Arduino across the years, and the community's contributions have amassed an amazing amount of readily available information around the technology that can be very beneficial to both beginners and experts. The primary areas of expertise include sensors and controlling devices with remotes. Within the range of electromagnetic radiation, the near-infrared, mid-infrared, and far-infrared areas comprise the infrared section. For short-term programs, optical wireless connectivity uses infrared transmission of information. When an infrared sensor senses its environment, it either emits or detects infrared light. RFID Reader and RFID Tag are the two parts of a Radio frequency identification system. An antenna on the RFID reader emits high-frequency electromagnetic waves to a tag. RFID tags come in two different forms: active and passive.RFID tags that are passive are fuelled through the influence of the reader's probing electromagnetic waves, whereas tags that are active are powered by batteries. If a tag is found, the RFID reader will scan the surroundings for them, it reads the tag's unique ID. Additionally, the Arduino will toggle the LED state if the Unique ID matches the Unique ID in the code. Electric electricity is required for all Arduino boards to operate. An electrical supply, which might be an AC connector, USB (Universal Serial Bus) connection and the power source, or controlled electrical gadgets, is utilized to supply electricity to the boards. The most popular method for powering the Arduino is by using a USB port.

The primary approach of the suggested model is to allow unrestricted traffic flow in order to prevent obstruction lanes for emergency vehicles during emergencies. Because priority vehicle clearance is not effectively controlled by the current model's congestion. Thus, radio frequency wireless communication technology is used in this project. The work involved interprets the methods that users use to cross intersections, such as automated assembly, ambulance clearance, etc. These roles are completed with the assistance of an Arduino, a Radio frequency identification reader as well as RFID tag. As the block diagram illustrates (Fig. 12.1), one Arduino is shown in the block diagram linked to an RFID reader. The emergency vehicles' RFID tags are detected and scanned by this roadside RFID reader, which is positioned 15 m ahead of the signalling framework at the desired square. An emergency vehicle's signal will turn blue when it enters the emergency lane of traffic and the RFID reader detects and scans the RFID tag on the vehicle. An emergency lane is suggested to be located 15 m ahead of the intersection in order to prevent the delay. To allow an emergency vehicle to travel freely through this lane without being delayed, other vehicles are not permitted to pass through it.

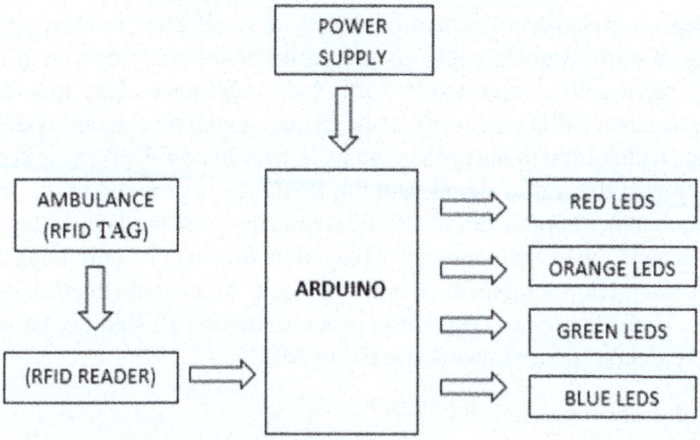

Fig. 12.1 Block diagram for proposed system

12.4 Proposed Methodology

Summary of the suggested algorithm for controlling traffic signals in real time.

The suggested model's algorithm is created in Embedded C and tested using the Arduino IDE. Generally, applications based on microcontrollers are developed using Embedded C. A established and standardized universal language for programming is embedded C. Compiler, a platform-specific program, is utilized for the C programs are converted to binary files that are particular to the target CPU. To run the programs needed to operate the suggested system, we utilize the Arduino IDE. The open-source Arduino IDE platform is used to program microcontrollers to carry out particular tasks. We are utilizing the Arduino UNO program for this task. Languages such as C and C + + are accommodated by the Arduino UNO The Arduino Software includes a message section, a text control panel, an interface with icons for basic tasks, and an editor for text for programming. In order to upload and interact with programs, it establishes a connection with the Arduino hardware. The setup () and loop () routines are the two primary Arduino functions used to execute code. It is simpler to modify code for running programs in the frameworks (Williams et al. 2023).

12.5 Proposed Model

With the help of an Arduino, emergency vehicles such as ambulances may free flow through traffic signals. The RFID reader is situated beside the traffic signal over the routes. When an EV moves into the emergency lane of traffic and the RFID reader recognizes and scans the car's RFID tag, the emergency vehicle's signal will turn blue. To avoid the delay, it is advised to place an emergency lane 15 m in advance of the intersection. As a result, the ambulance uses the traffic signal to pass. In this research, we suggested a novel method that is more effective in clearing traffic for ambulances at traffic signals while also requiring less time to operate n innovative solution is suggested for minimizing traffic jams and unnecessary time delays. We used RFID tags and RFID readers to create a smart ambulance as one example of this cutting-edge technology in the project we were working on. The reader is positioned at the junction of the traffic signal, and the RFID tag is fastened to the ambulance. When the ambulance approaches the traffic signal, the scanner scans the tag to retrieve the input data and the buzzer massage is played to inform other commuters about the entry of an emergency vehicle on emergency track. And handle the task by utilizing an Arduino Micro Processor type Kit in order to modify the Red Signal into a Blue Signal at the intersection, allowing the EV to pass.

12.6 Experimental Result

The prototype of the suggested system (Fig. 12.2) depicts an ambulance's crossing at a traffic signal. This model has RFID tags and readers that will collect data from the ambulance when the tags are fastened to the vehicle. The RFID reader at the few meters before receives a response from this input. At the intersection, we make use of RFID tags and readers. Upon arriving at the traffic signals, the ambulance's reader scans the tag for input and transfers it to the Arduino microprocessor for the intended action.

Arduino an Arduino microprocessor is utilized to gather and analyze the input data, resulting in an output that changes the LED lights from red to blue in accordance with the specific route on which the ambulance arrives at a traffic signal. RFID

Fig. 12.2 Prototype of proposed model

tags and readers are used in this system. We employ the Embedded C programming language to implement the suggested system's operations, which are contingent upon the Arduino Microcontroller's operating conditions. Typically, applications based on microcontrollers are developed using Embedded C. The efficiency of the connected device is improved by this programming format, which enhancing software commands for the supplied information. The suggested system is coded in Embedded C. because it offers the Arduino board a great deal of programming flexibility for developing systems with embedded applications. In addition, compared to other an elevated level programming language, it requires the least amount of hardware time required for the entire operation.

12.7 Advantages

Efficient Traffic Flow: The system facilitates a smoother flow of traffic by giving priority to specific cars based on RFID tags, such as emergency vehicles or public transportation, enabling these vehicles to pass through crossings or obstacles more quickly.

Reduced Congestion: It is possible to reduce overall traffic congestion on roadways by allocating priority to Emergencies vehicles. This is especially helpful in cities where transportation congestion is a major problem.

Enhanced Safety: When faced with dire circumstances, emergency vehicles like fire trucks and ambulances can arrive at their destinations faster and possibly save lives. Systems that are based on priorities make sure that typical traffic doesn't cause delays for these vehicles.

Improved Public transportation: By providing a more rapid and reliable travel than private vehicles, prioritizing public transportation vehicles may encourage more people to take buses or trains.

Customizable Prioritization: Different vehicle kinds can be prioritized by the system according to predetermined parameters. For instance, it can give buses precedence during rush hours or prioritize ambulances over other emergency vehicles.

Scalability: Systems based on Arduino are easily expanded to cover greater regions or handle more vehicles as needed because they are very scalable.

Cost Effectiveness: Comparing Arduino boards and RFID technology to other sophisticated traffic management systems, they are comparatively less expensive. This lowers the cost of implementing a priority-based system, particularly for emerging or smaller communities.

Flexibility: As traffic patterns differ over time, the system can simply reprogrammed or changed to accommodate a variety of traffic circumstances.

Integration with Existing Infrastructure: Without requiring significant changes, RFID technology can be easily linked with the current traffic infrastructure, such as traffic lights or toll booths.

12.8 Limitation

- Implementing IT'S into practice can be rather expensive at first.
- The potential for human entry mistakes, which can happen if the RFID tag is not scanned correctly or if the database of the system is having problems.
- Arduino-based traffic control systems need to be maintained on a regular basis to ensure optimal performance, just like any other electronic device.

12.9 Conclusion

Our goal with this project is to use an Arduino, an RFID module, create an autonomous signal clearing system for emergency vehicles that is prioritized based on traffic management. With the help of the Arduino Processing Kit's simulation scheme, the system could operate automatically, and the ambulance could smoothly free flow at intersection. The system's goal is to provide traffic managers more freedom. The RFID techniques for unrestricted emergency vehicle movement are presented in this paper. Every city in the nation is experiencing worsening traffic conditions every day. The suggested action is very beneficial as technology and digitization become more widespread. Therefore, we can conclude that there will be a shorter wait for emergency services vehicles to arrive at their destination when using this system. Emphasizing emergency vehicles as they enter the lane is the main objective in order to reduce Lag in time durations. This is because ambulances stuck in traffic for extended periods of time have resulted in numerous fatalities. The suggested system operates continuously and manages EVs at intersections with greater accuracy.

12.10 Future Scope

It can be enhanced from a hardware perspective by numerous new and current technologies. However, from an application perspective, alternative techniques for detecting traffic violations can take the place of current methods. These algorithms could monitor other traffic offenses not involved in this model or enhance the suggested algorithms.

The following are potential upgrades to the suggested system:

- Updating additional traffic to the offense algorithms.
- Developing quicker versions of suggested fundamental algorithms.
- Utilizing the recommended framework with wireless sensing networks.
- Introducing neural networks that can adjust settings in response to changes in traffic density in real time.

- Including extra features in the framework, like a computerized locking and unlocking mechanism.

References

Bari C, Khode B (2021) PM10 dispersion modelling at unsignalised intersection using Caline4—a case study. Adv Civ Eng Infrastructural Dev 87:531–539

Bura V, Korra S (2020) Traffic control system for vehicles on Indian roads using raspberry Pi. IOP Conf Ser Materials Sci Eng 981

Dubey A, Rane S (2017) Implementation of an intelligent traffic control system and real time traffic statistics broadcasting. In: International conference of electronics, communication and aerospace technology (ICECA) vol 2. IEEE, pp 33-37

Ghazal B, ElKhatib K, Chahine K, Kherfan M (2016) Smart traffic light control system. In: Third international conference on electrical, electronics, computer engineering and their applications (EECEA), pp 140–145

Ghodmare S, Gulghane A (2021) An attempt for sustainability evaluation considering implementation of rapid transit system. In: 2nd international conference on sustainable materials and practices for built environment. Mater Today Proc. Jaipur, India, pp 448–451

Ghodmare S, Khode B, Ladekar S (2021) The role of artificial intelligence in industry 4.0 and smart city development. Adv Civ Eng Infrastructural Dev 87:591–604

Gulati I, Srinivasan R (2019) Image processing in intelligent traffic management. Int J Recent Technol Eng 8(2S4):213–219

Gunda H, Waghunde R, Malwatkar S, Desai A, Baviskar P (2018) Smart traffic management system using arduino and RFID tags. IJCRT 6(2)

Harshith N, Pasha S, Amrutha MR, Bharti N, Ravi DJ (2021) Zero traffic for emergency vehicles. Int J ProgIve Res Sci Eng 2

Kanungo A, Sharma A, Singla C (2014) Smart traffic lights switching and traffic density calculation using video processing. In: Recent advances in Engineering and computational sciences (RAECS). IEEE, pp 1–6

Kaushik PG, Dahake VD (2015) Design of intelligent traffic light controller. Int J Eng Sci Manag 5:120–129

Liu J, Shen J (2018) Secure traffic light control using fog computing. Futur Gener Comput Syst 78(2):817–824. Elsevier B.V

Mendhe P, Ghodmare S, Titarmare A, Wadhai P (2022) Sustainable transportation using automated highway system for intercity connectivity. In: 10th international conference on emerging trends in engineering and technology. IEEE, Nagpur, India pp 1–6

Metre K, Baghel H, Suman G, Batra M, Ghodmare S (2021) Compact city and related impact on sustainable development in urban areas. Adv Civ Eng Infrastructural Dev 87:523–530

Naseer F, Khan KB, Khaliq MS, Raheel M (2018) Smart roadlights and auto traffic-signal controller with emergency override. In: International conference on ITA. Springer

Nimburkar A, Ghodmare S, Tenpe A (2022) Traffic signal synchronization system for highway corridor in urban area. In: 10th international conference on emerging trends in engineering and technology. IEEE, Nagpur, India, pp 1–6

Pandey S, Jethva M, Jangam S, Panwar Y (2017) Density based traffic control system. Int J Sci Res Dev (IJSRD) 576–578

Porter RJ et al. (2023) Emergency vehicle preemption systems. 47(1):87–94. IEEE Transaction

Prakash N, Udaykumar E, Kumar N (2020) Arduino based traffic congestion control with automatic signal clearance for emergency vehicles and stolen vehicle detection. In: 2020 International Conference on Computer Communication and Informatics (ICCCI-2020), Jan. 22−24, Coimbatore, INDIA.

Prasad K (2021) Design and implementation of RFID based -document verification system. In: Third international conference on inventive research in computing applications (ICIRCA). IEEE
Ritchie SG et al. (2023) A rule based expert system for emergency vehicle signal preemption. 1(1):33–40. IEEE Transaction
Rodke S, Ghodmare S, Deshmukh A (2022) Critical analysis of saturation flow rate and parameters affecting it. In: International Conference in Emerging Trends in Engineering and Technology-Signal and information Proceeding, pp 1–6
Samir A, Elsagheer M, AlShalfan KA (2021) Intelligent traffic management system based on the internet of vehicles (IoV). J Adv Transp 4037533
Sanghai S, Pawade P (2022) Performance evaluation of friction dampers for building with soil-structure interaction. Mater Today Proc 60(1):194–210
Sundar R, Hebbar S, Golla V (2014) Implementing intelligent traffic control system for congestion control, ambulance clearance, and stolen vehicle detection. IEEE SensS J 15:1109–1113
Suprava J, Dutta M (2022) Assessment of service quality at unsignalised intersections using traffic simulation and computational intelligence, 365–377
Williams BM et al. (2023) Traffic signal peemption for emergency vehicles. 45(3):453-464. IEEE Transaction
Yadav D, Khode B (2022) Metro follows an optimized speed trajectory to save energy. In: 10th international conference on emerging trends in engineering and technology. IEEE, Nagpur, India, pp 1–6

Chapter 13
Development of a Roundabout at Margao, Goa Using Civil 3D

Anuj A. Varhokar and P. J. Wadhai

Abstract Using civil 3D, popular civil engineering software, this research study analyses and designs a rotary (roundabout) with five legs. Roundabouts have gained popularity as an effective traffic control measure due to their ability to improve safety and traffic flow. The study focuses on the design process, traffic analysis, geometric considerations, and safety assessment for a rotary with five legs, aiming to provide insights into the advantages and challenges of such intersections. The present research focuses on creating an effective and efficient roundabout arrangement using civil 3D software. Through the examination of traffic patterns, road geometry, and the software assists in designing a roundabout that maximizes traffic flow while reducing congestion.

Keywords Civil 3D · Roundabout · Conflict Point · Intersection · Traffic Flow

13.1 Introduction

Many people in India are starting to own their cars these days due to the country's rapidly growing population and growing traffic issues. Congestion has resulted on the city's squares and highways as a result. In our lives, highways are crucial for transportation. It is crucial for all kinds of travel, including business travel, travel for pleasure, and the transportation of commodities. Many businesses use roads to transport their goods from one location to another. It supports human development in many spheres, including social, medical, and scientific. Separate the roadway into lanes; depending on the design location, we can supply two, three, or four lanes. Rotating intersection at a location with a minimum of 500 PCU/hr and a maximum of 3000 PCU/hr, as per IRC 65-1976 rules (IRC 1976).

The majority of Margao City's urban regions have severe traffic congestion issues. These issues, together with traffic flow on the roads, are the primary obstacles at the

A. A. Varhokar (✉) · P. J. Wadhai
G H Raisoni College of Engineering, Nagpur, India
e-mail: anuj.varhokar.mtechts@ghrce.raisoni.net

B. V. Khode et al. (eds.), *Innovations in Technologies: Pioneering Sustainable Infrastructure for a Resilient Future*, Sustainable Civil Infrastructures,
https://doi.org/10.1007/978-3-031-82133-2_13

crossroads. Traffic jams are caused by cars speeding on the roadway, which also contributes to traffic jams and accidents. To lessen the issue of traffic and accidents, we provide intersections. Because of the heavy traffic in the area where this paper is being written, AutoCAD Civil 3D is being used to design intersections in a way that minimizes traffic. So there is a lot less of a traffic issue than there was previously. A road is divided into two sections by an intersection so that a car can turn left or right with ease. On a highway, it is extremely difficult to offer an intersection since cars are moving at such high speeds and in numerous directions. Every traffic flow's efficiency is dependent upon intersections. Another name for a rotary intersection is a traffic rotary. It is quite big, and cars pass through it continuously at slow speeds. Every vehicle leaves Central Island by one of several paths and travels in one direction before exiting onto the road.

The problem of traffic congestion is getting worse every day. Vehicle servicing is expensive and there is a delay in getting somewhere because of the traffic issue. Owing to the intersection, the cars will enter it more slowly and exit in the appropriate direction. Autodesk developed AutoCAD Civil 3D, software for civil engineering design and documentation. Roads, trains, and building projects are designed using AutoCAD Civil 3D software (Singh et al. 2019). Professionals and civil engineers utilize AutoCAD Civil 3D. Any project can be simply detailed in one location with the help of this program. We can design better intersections.

Objective

- To address the challenges unique to five-legged rotaries and provide practical recommendations for their successful implementation.
- This paper intends to serve as a valuable resource for practitioners involved in the design of intricate rotary intersections, guiding them toward effective solutions that harness the benefits while mitigating the challenges of such configurations.

Advantages of Using Civil 3D for Rotary Design

Parametric Design. Civil 3D's parametric design capabilities allow you to quickly adjust and modify design elements while maintaining relationships between components.

Data Integration. Civil 3D integrates various design data, including survey data, land development data, and geospatial information, into a single platform, streamlining the design process.

3D Visualization. Visualize the roundabout design in 3D, allowing for better communication with stakeholders and a clearer understanding of the final product.

Dynamic Updating. As design changes are made, civil 3D automatically updates related elements, ensuring consistency throughout the project.

Collaboration. Civil 3D supports collaboration through data sharing, allowing multiple team members to work on the same project concurrently.

Efficiency. The specialized tools in civil 3D streamline the roundabout design Process, reducing design time and potential errors.

13.2 Literature Review

Hemavathy and Malarvizhi (2023), the study focuses on designing a roundabout at a four-armed intersection in Chennai to lessen traffic congestion. Discovered that elevated roundabout bridges, as compared to flyovers, are a better approach to reduce traffic congestion at intersections. Dash (2021), reduced traffic congestion can be achieved by putting alternate routes and a thorough parking strategy into practice. Roundabout conversions can improve traffic flow, volume, speed, and safety. Gaurav (2020), the planning and assessment of roundabouts at various crossroads have been the subject of numerous studies. Researchers have concentrated on enhancing traffic conditions and using manual traffic surveys to assess roundabout capacity. The process of designing a roundabout involves choosing a location based on predetermined standards and measuring traffic volume during peak hours. Research has also shown how crucial it is to comprehend how roundabout traffic flows, including how to weave, merge, and diverge. Zacharia et al. (2019), Explain an innovative geometric idea for creating rotaries at traffic interchanges. explains the various aspects of rotary design, such as weaving length, width, design speed, and radius provides a software suite designed for creating turn signals at intersections. Highlights the significance of slowing down near rotaries to guarantee secure vehicle movement. Vasantha Kumar et al. (2017), at a complicated multi-leg crossroads in Roya Pettah, Chennai, designing a circular intersection can assist reduce traffic jams during rush hour. The paper outlines strategies for resolving traffic congestion brought on by India's fast urbanization. When compared to grade separators, implementing a rotary intersection based on the study's design may be a more affordable option. Mahajan et al. (2013), the impact of geometric components on roundabout entry capacity under heterogeneous traffic conditions has not been extensively studied. Due to the inadequacy of current empirical entrance capacity models for varied traffic situations, models considering roundabout geometry are required. Pratelli (2006), experimented with offering solutions to improve and ease the flow of traffic at two of the intersections. The signal durations have been modified for peak flow in the afternoon. The optimal cycle length of a transmission is derived using traffic quantity and saturation quantities [Webster technique]. The first junction, Jyoti Talkies, should be improved by expanding the road. The capacity of the rotary is assessed at the other junction known as the "Vallabh Bhawan roundabout" to determine if it is within its allowed limits or not. The use of a synchronized rotary is recommended. Rodke et al. (2022), The findings of the research showed that motorcycles have a significant impact on traffic flow and should be considered while designing signalized intersections geometrically and operating systems. Mendhe et al. (2022), we look at the focus on introducing fresh methods for managing urban transit. The Automated Highway System will create tools to assist cities in overcoming obstacles that prevent them from using five-layer theoretical creative frameworks. Nimburkar et al. (2022), additionally, an effort is made to determine the proper approach and parameters for the traffic situation in India. It has been discovered that several crucial factors include traffic volume, phase sequence, offsets, saturation flow, cycle length, and traffic mix. It has been noted that when building a signal system in an urban region, consideration should be given to sign synchronization and related factors. Pathareet et al. (2023), compared to a thermoplastic composite body panel, a conventional bus with an aluminium covering and supporting steel bars demonstrated exceptional weight savings of more than 55%. According to the research, these kinds of transportation are perfect for land-side airport applications. Additionally, a list of possible advantages for airside operations is supplied (Yadav et al. 2022). Yadav and Khode (2022), Optimal speed, acceleration, tractive effort, and work done are obtained for the entire journey and it is found that 58.67% of energy is saved for the Versova to Ghatkopar journey and 28.44% energy is saved for the Ghatkopar to Versova journey. This output will be sent to input via a feedback mechanism. Additionally, surgeries are offered (Yadav et al. 2022). Yadav et al. (2022), both short-term and long-term actions have been implemented based on the accident severity. Eventually, 10 black spot locations and an estimate were determined by analysis of short-term measures. Jena et al. (2022), the most important factor affecting the ALOS of unsignalized junctions is the state of the pavement. The suggested methodology will assist transportation officials in setting priorities for the most important aspects of infrastructure development funding (Sanghai and Pawade 2022).

Benefits of Rotary Intersections

Traffic Flow Efficiency. Rotaries promote continuous traffic flow, reducing the need for complete stops and allowing vehicles to merge seamlessly, thereby decreasing congestion and delays.

Reduced Collision Severity. The design of rotaries eliminates high-speed, right-angle collisions, leading to a significant reduction in severe accidents. Collisions that do occur tend to be at lower speeds, reducing the risk of serious injury.

Energy Savings and Environmental Impact Rotaries reduce idling time, resulting in lower fuel consumption and emissions. This contributes to improved air quality and reduced carbon footprint.

Lower Long-Term Costs. Rotaries have lower maintenance costs compared to signalized intersections, as they require fewer electrical components and experience less wear and tear.

Challenges of Five-Legged Rotaries

Geometry and Layout. Designing an efficient layout that accommodates all five legs while maintaining appropriate curvature and lane configurations is demanding.

Traffic Control. Managing the intricate interactions between vehicles from different directions requires precise traffic control measures and clear signage.

Pedestrian and Cyclist Integration. Ensuring the safety and ease of movement for pedestrians and cyclists becomes more intricate in a five-legged configuration.

Conflict Points. The higher number of potential conflict points increases the need for thorough analysis to minimize the chances of collisions.

Conflict Points Analysis

Conflict Points. Identify potential conflict points where vehicle paths intersect. These points include vehicle-to-vehicle, vehicle-to-pedestrian, and vehicle-to-bicycle interactions.

Reduction of Conflicts. Roundabouts typically have fewer conflict points compared to traditional intersections with traffic signals. However, analyzing conflict points helps identify any potential safety concerns and allows adjustments in design to mitigate them.

13.3 Methodology

13.3.1 Study Area

The location selected for designing the roundabout is situated at Jio Tower Margao—Bypass road, Margao Goa. At the intersection, there are five roads named Davali Rd, Margao Bypass, St. Joseph Road, and railway Station road roads with 6 m carriageway. Due to rapid growth and development, the population of this area has increased in the last decade which also led to an increase in the number of vehicles a traffic. All five roads have commerce signals of 30 s allowance which is not efficiently handling the business so to overcome this problem we think that a roundabout will be a more efficient option and hence we are designing the roundabout for the said intersection (Table 13.1 and Fig. 13.1).

Table 13.1 Intersection arm details

Arms	Road name
ARM-1	Margao Bypass Road
ARM-2	Davali Road
ARM-3	Power House Road
ARM-4	Tempel Garden Road
ARM-5	Magao Railway Station Road

Fig. 13.1 Satellite image of existing roundabout

13.3.2 Data Collection

To design the geometry of highways and intersections, the ground surface data is necessary. The group that was managing the road's development provided the point CSV file surface file. The survey data is stored as a point CSV file with the point CSV file extension.

13.3.3 Traffic Analysis

Traffic analysis is a critical process to ensure the efficiency and safety of a roundabout design following (IRC 106:1990) guidelines. It involves evaluating various aspects of traffic flow and performance.

The technique of gathering vehicle data allows for the analysis of traffic conditions by measuring the amount of traffic at the intersection. To get the most data, it is necessary to determine the peak period for the specific intersection. The morning and evening peak hours for JIO Tower Square were respectively from 9:00 to 10:00 a.m. and 17:00 to 18:00 p.m. It was done to gather data (Table 13.2).

Table 13.2 Peak hour data

Period	Total peak hour traffic at junction, PCU/hr	Peak hour time
Morning peak hour period	566	10:00 to 11:00
Afternoon peak hour period	599	12:00 to 13:00
Evening peak hour period	518	17:00 to 18:00

13.4 Design Criteria

The horizontal geometry of the center line, as well as the profile and cross-section of the highway, were given the following design criteria depending on the Geometric Design Standards Manual (IRC: 65: 1976).

I. Design speed = 30 km/h
II. Number of Lanes = 5
III. Entry Radius = 20
IV. Exit Radius = 30 m
V. Radius of central Island = 10 m
VI. Entry Width = 10 m
VII. Exit Width = 12 m
VIII. Weaving Width = 11 m
IX. Weaving Length = 30 m

13.4.1 Design Procedure

- Create Surface.
- Insert CSV files (in Excel format containing northing, elevation, and easting) into the civil 3D by AutoCAD application.
- Using the alignment generation tool, generate alignment.
- In this assignment, adhere to the IRC specifications for design.
- Create a surface profile
- Using the profile generation tools, draw a grade level and vertical bends.
- Form the Assembly. Individual subassemblies are added to form the assembly (Fig. 13.2).

13.4.2 Horizontal Alignment Design

A road's horizontal alignment indicates its orientation and placement in plain view. Tangents (straight sections), circular curves, and transition spirals between tangents and curves are its three geometric components. Proper alignment design boosts roadway capacity while also improving design speed efficiency. Survey data in Excel

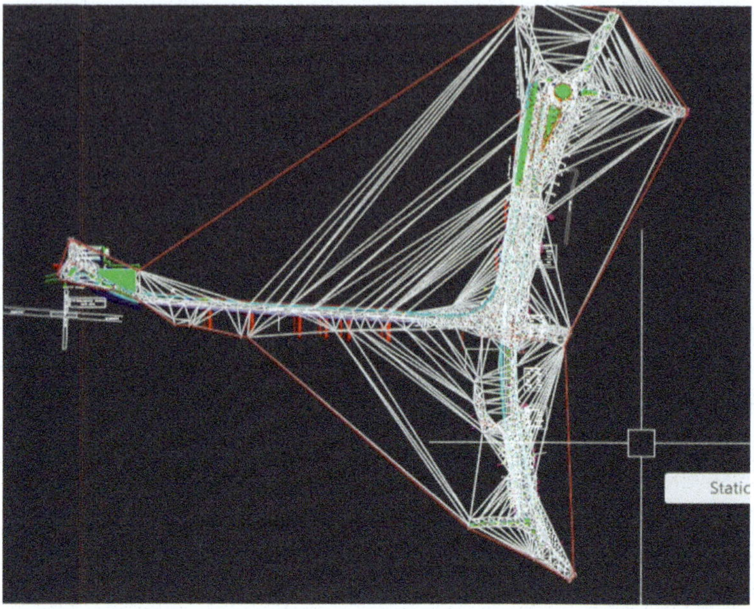

Fig. 13.2 Existing ground surface generated by AutoCAD Civil 3D

format, including easting, northing, and elevations, was imported into Civil 3D. This automatically generated the current ground surface. After determining the design requirements to be applied to the alignment, the alignment generation tool was used to create the road alignment. The spiral and circular curves were created using the Geometry Editor 'tool, and an alignment curve report was received (Fig. 13.3).

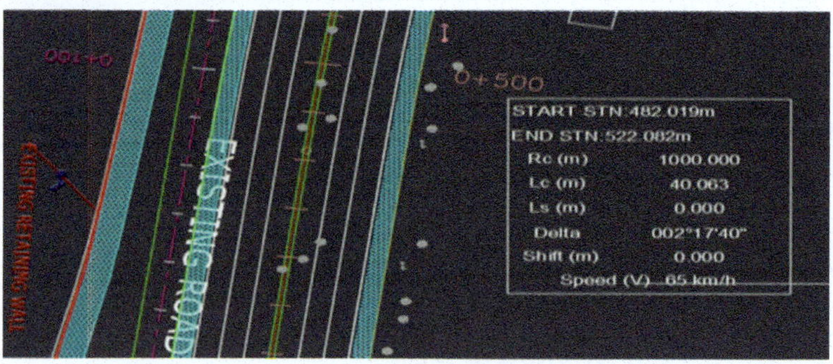

Fig. 13.3 Horizontal alignment generated by AutoCAD Civil 3D

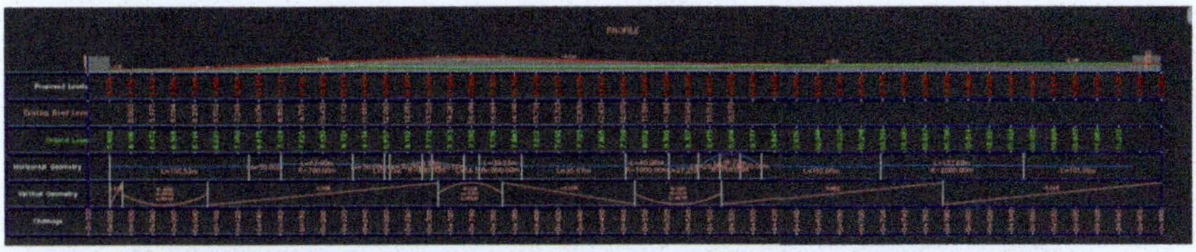

Fig. 13.4 Vertical alignment generated by AutoCAD Civil 3D

13.4.3 Vertical Alignment Design

The longitudinal section of a path is specified as a vertical alignment. It is made up of vertical curves and slopes, and it impacts vehicle convenience at high speeds, sight distance, and stopping distance. Vertical alignment may be produced once the horizontal alignment is fixed. To do this, the horizontal alignment ground profile is created first, followed by the vertical alignment on the profile's viewpoint. Profile geometry was selected by using minimal K design guidelines for vertical sag and crest curves that match the fundamental requirements for stopping sight distance, accessibility, and aesthetics. Vertical curvature lengths were manually determined by applying this approach (Fig. 13.4).

$$L = KA$$

where,

K = length required for a 1% change of grade.

L = vertical curve length.

A = change of grade in %.

13.5 Assembly

In the Civil 3D software, assembly is utilised to define the platform. The platform's "Subassembly" command designates a lane, shoulder, and other features needed to create a standard cross section of a road. The center of the platform has additional sections on the right and left with subassemblies. Using pre-existing subassemblies or building subassemblies from scratch are the two possibilities for assembly creation (Fig. 13.5).

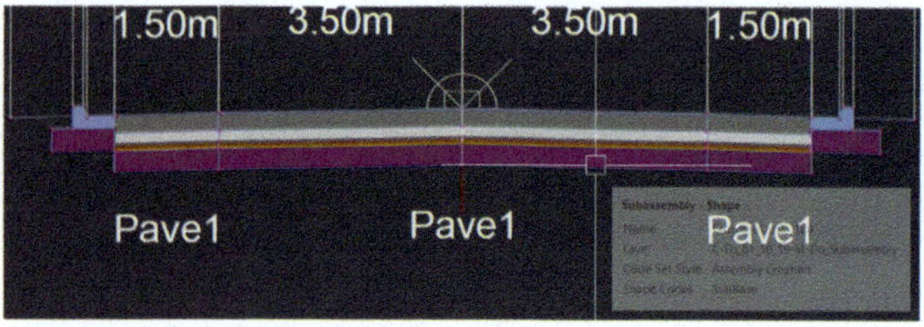

Fig. 13.5 Assembly generated by AutoCAD Civil 3D

13.6 Conclusion

AutoCAD Civil 3D saves much time and work and facilitates a comfortable and timely completion of the design process. In this study, AutoCAD Civil 3D is used to design crossings. Manual design is time- and money-consuming, and unreliable reports are accessible. AutoCAD Civil 3D makes it possible to create quality intersections more quickly. Every car approaching the rotary is gently pushed to slow down and keep moving at a slower pace. The IRC 65:1976 states that a junction can be created with a maximum traffic volume of 3000 pcu/hr and a minimum traffic volume of 500 pcu/hr.

The following has to be put into practice at Jio Tower, Margao Junction, to reduce the amount of conflict that results from traffic flowing from four different roads:

Margao Bypass Road, Davali Road, Power House Road, Tempel Garden Road and Margao Railway Station Road. Because we removed the central island of the road dividing Power House Road and Margao Railway Station Road to provide a suitable turning radius for vehicles flowing from Railway Station Road to Powerhouse Road, it also helps to prevent lines and minimise the crash rate. In addition, the queuing problem arises from the limited mobility on Margao Bypass Road, which is resolved if the central island Diameter is (10 m) and (ICD = 30 m) is as per 65:1976 (Fig. 13.6).

Key Findings:

- The IRC guidelines offer a structured framework for geometric design, traffic analysis, safety assessment, and other critical aspects of roundabout design.
- Proper data collection and analysis play a vital role in accurately determining traffic patterns, turning movements, and geometric requirements.
- Integrating pedestrian and cyclist accommodations, drainage design, and aesthetic considerations contribute to the holistic success of the roundabout design.

Recommendations:

- Continuously update the design to accommodate changing traffic patterns and future growth to ensure long-term efficiency.
- Consider using advanced traffic simulation software for more detailed analysis, especially for complex roundabouts with intricate traffic flows.

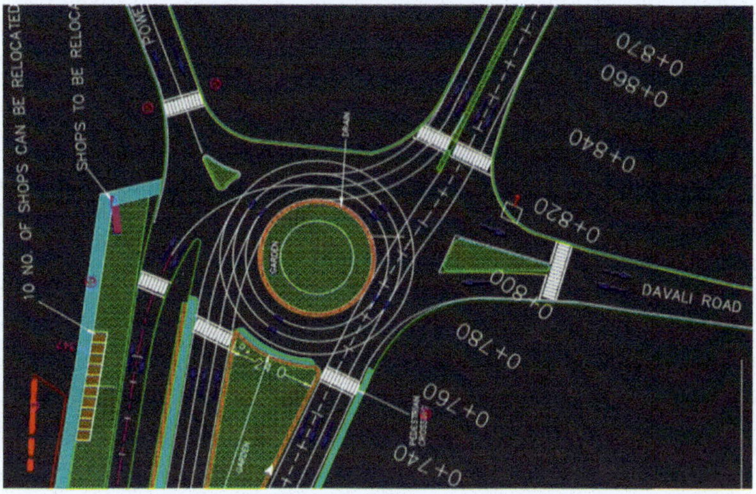

Fig. 13.6 Proposed round about generated by AutoCAD Civil 3D

- Collaborate closely with local transportation authorities to align the design with regional preferences and specific needs.
- Explore innovative solutions for pedestrian safety and accessibility within the roundabout design.

Limitations of the Study:

- The research focused on the utilization of civil 3D within the context of IRC guidelines. Different design standards and software platforms may yield varied results.
- The study's scope was limited to design considerations, excluding real-world implementation challenges, cost analysis, and post-construction assessments.
- The efficacy of the design was not tested in a live environment, and simulation outputs should be validated with actual traffic observations.
- The availability and accuracy of data significantly impact the precision of the design, and the study assumed complete and accurate data availability.
- External factors like local laws, regulations, and urban planning strategies could influence the practical implementation of the designed roundabout.

References

Dash DK (2021) Congestion and performance evaluation of roundabouts: case study at Bhubaneswar City; India. Des Eng 4:181–195. ISSN: 0011-9342. https://www.researchgate.net

Gaurav DH (2020). Int Res J Modernization Eng Technol Sci 2(06). e-ISSN: 2582–5208. https://www.irjmets.com

Hemavathy M, Malarvizhi G (2023) Designing a roundabout to overcome traffic congestion at a four-armed intersection. Eng Innov 7:95–103. ISSN: 2813–1002

IRC: 65–1976. Guidelines for planning and design of roundabouts. Indian Roads Congress, New Delhi

Jena S et al (2022) Assessment of service quality at unsignalized intersections using traffic simulation and computational intelligence. Transp Lett 14(4):365–377

Kewale P et al (2021) Design and implementation of RFID based -document verification system. In: 2021 third international conference on inventive research in computing applications (ICIRCA). IEEE

Mahajan SK, Umadekar A, Jethwa K (2013) New concept of traffic rotary design at road intersections. In: 13th COTA international conference of transportation professionals (CICTP 2013) Procedia—social and behavioral sciences, vol 96, pp 2791–2799

Mendhe PD, Ghodmare SD, Wadhai PJ, Titarmare AS (2022) Sustainable transportation using automated highway system for intercity connectivity. In: 2022 10th international conference on emerging trends in engineering and technology—signal and information processing (ICETET-SIP-22), pp 1–6. https://doi.org/10.1109/ICETET-SIP-2254415.2022.9791761

Nimburkar AS, Ghodmare S, Tenpe A (2022) Traffic signal synchronization system for highway corridor in urban area. In: 2022 10th international conference on emerging trends in engineering and technology—signal and information processing (ICETET-SIP-22), pp 1–6. https://doi.org/10.1109/ICETET-SIP-2254415.2022.9791577

Pathare A, Gulghane A, Ali Khan BKH (2023) Comparative analysis of personal rapid transit system for interconnecting metro station with airport. J Polym Compos 11(1). https://doi.org/10.37591/JoPC. ISSN: 2321–2810

Pratelli (2006) Design of modern roundabouts in urban traffic systems. Urban transport XII: urban transport and the environment in the 21st century, vol 89

Rodke SJ, Ghodmare SD, Deshmukh A (2022) Critical analysis of saturation flow rate and parameters affecting it. In: 10th international conference on emerging trends in engineering and technology—signal and information processing (ICETET-SIP-22), pp 1–6. https://doi.org/10.1109/ICETET-SIP-2254415.2022.9791785

Sanghai S, Pawade P (2022) Performance evaluation of friction dampers for building with soil-structure interaction. Mater Today: Proc 60:194–210

Singh AB, Pandey RK, Atul E, Pandey S (2019) Designing of rotary intersection using AutoCAD civil 3D software. Int J Tech Innov Modern Eng Sci (IJTIMES) Impact Factor 5.22 (SJIF-2017) 5(08). e-ISSN: 2455–2585

Talwekar MA, Tenpe A (2022) Design of rotary intersection at deep Banglaw chock. In: 2022 10th international conference on emerging trends in engineering and technology—signal and information processing (ICETET-SIP-22), pp 1–4. https://doi.org/10.1109/ICETET-SIP-2254415.2022.9791540

Vasantha Kumar S, Gulati H, Arora S (2017) Design of a rotary for an uncontrolled multi-leg intersection in Chennai, India. In: IOP conference series: materials science and engineering, vol 263, p 032030. https://doi.org/10.1088/1757-899X/263/3/032030

Yadav D, Khode B (2022) Metro follows an optimized speed trajectory to save energy. In: 2022 10th international conference on emerging trends in engineering and technology—signal and information processing (ICETET-SIP-22), pp 1–6. https://doi.org/10.1109/ICETET-SIP2254415.2022.9791629

Yadav DK, Ghodmare SD, Naveen Kumar N (2022) Mitigation of blackspots on highways by the application of safe system approach. Mater Today: Proc 52:1228–1235

Zacharia AB, Madhavan H, Anjaneyulu MVLR (2019) Geometric factors influencing entry capacity of roundabouts under heterogeneous traffic conditions. Arch Transp 49(1). e-ISSN (online): 2300–8830. https://doi.org/10.5604/01.3001.0013.2778

Chapter 14
Improvement of Access Point on NH Passing Through Industrial Area

Omkar A. Mogarkar and Sujesh D. Ghodmare

Abstract By meeting the journey and transport requirements of both individuals and goods, transportation has assisted in society's growth from the dawn of time. Deaths, on the other hand, have increased by a factor of two over the last ten years. As a result, study and planning are crucial for a road project, keeping in mind both the present and potential future needs of the area. The main issue of defining both qualitative and quantitative standards to move forward with a thorough road condition assessment is addressed in this study. The route is an 8-km section of the Nayara Energy Limited in the Gujarat, district of Jamnagar. The various entry and exit gates of Nayara Energy Ltd. face a problem of traffic congestion while entering and leaving the gates and also an issue of queuing for other commuters. To address this issue, various improvement plans were suggested based on traffic data, pavement inventory, topography of the region, etc. Road authorities may use this technique as a reliable tool in the framework of effective road upgrading management. The main goal of this article is to provide a thorough review of the plan for improving current roadways along with the pertinent technologies to address the issue of congestion and avoid accidents on the road. The four-lane road's paved shoulders being upgraded will speed up transportation, ensure a free flow of traffic, and encourage social contact in neighboring villages.

Keywords Road Safety · Traffic Accidents · Transport Environment

14.1 Introduction

The Ministry of Road Transport and Highways owns the system of major roads that make up India's national highways. One of the primary factors contributing to traffic accidents and the phenomenon of congestion is access. To avoid traffic in cities, towns, and villages, national roads include overpass access or some controlled

O. A. Mogarkar (✉) · S. D. Ghodmare
G. H. Raisoni College of Engineering, Nagpur, India
e-mail: om.mogarkar@gmail.com

B. V. Khode et al. (eds.), *Innovations in Technologies: Pioneering Sustainable Infrastructure for a Resilient Future*, Sustainable Civil Infrastructures,
https://doi.org/10.1007/978-3-031-82133-2_14

access, where entry and departure are through the side of the flyover. These highways are constructed as well for a speed of 100 km/h. There are interchanges on certain national roads, however, not all of them have completely regulated access. The Ministry of Highways and Road Transportation has decided to develop all-new national highways with an arrangement of divided 4-lane highways and above as access-controlled roadways as much as possible, as well as all green-field 4-lane national highway initiatives with the capability of closed tolling where access from and leave from a national highway is only allowed at a predetermined point to ensure effortless circulation of traffic, road safety, and that a road user (Figs. 14.2, 14.3 and 14.4). Therefore, service roads linking the two places must be provided by the state governments and other infrastructure development organizations at their own cost within the permitted right of way of any such access-controlled premises. Depending on the availability of Right of Way (RoW), it may or may not be practicable to provide service roads along highways that travel through built-up or populated regions, or along highways that run through open expanses. While the NH development authority must make sure that it properly plans features and structures to assimilate and slow down traffic at all connect points when the highway is being developed, the authority would struggle to continue meeting such incremental demands of any future growth unless such access to the highway is successfully regulated and controlled.

14.2 Literature Review

In this paper, Gautam et al. discusses how crucial road transport is to an economy's operations in emerging nations. This study provides data that may be used to identify regions of a country's road infrastructure stock that are poor or strong. According to this study's findings, there is a covariate relationship between infrastructure for roads and the economy's growth (Quiroz and Gautama 1992). The impact of improved road infrastructure on traffic deaths and injuries has been covered by Noland et al. The author of this study came to the following conclusions: Changes in demographic groups, greater usage of seat belts, lower alcohol intake, and increased use of health services to reduce deaths overall (Robert 2003). Flahaut et al. has talked about how the local environment and the road system might contribute to dangerous driving. The author has concluded that local conditions and the state of the roads constantly affect auto accidents (Flahaut 2004). Road infrastructure safety evaluation is a process that Appleton et al. have addressed. With regard to road safety, this method has controlled and tracked the performance of the road authority. It depended on global study linking infrastructure future to determining feature risk ratings. This shows that the safety evaluation of the road infrastructure was conducted correctly (Mulyono et al. 2008). According to Ahmed et al., there are various aspects involved in the design research of highways, and these factors are crucial in determining the probability of traffic accidents. The goal of the highway or roadside design is to reduce or eliminate the danger of traffic accidents. For the betterment of every motorist, the author might concentrate on enhancing the standard of the road system's preservation and care, as well as its infrastructure (Appleton 2009). In their discussion of technologies for improving highway operation, investment, and pricing, Winston et al. also noted that these innovations would increase travel time, dependability, and safety while lowering the cost of maintaining the roads. The author has concluded that the employment of technology or policymakers would lessen traffic jams and delay time for drivers, as well as increase safety on the roads (Ahmed 2013). The development and construction stages of the road infrastructure have been covered by Tunde et al. Descriptive and inferential statistics are used in this study to decide on data collection and evaluation. The author of this study has drawn the following conclusions: respondents' characteristics in terms of organization, titles, educational background, and years of professional expertise (Jamroz et al. 2014). This article by Chepchieng et al. discusses how road safety is the biggest issue with transit in emerging economies. The fundamental objectives of the transportation system on the roads are to develop or create a bit more effective, to enable easy movement on the road, and to reduce road traffic accidents. The study's findings, according to the author, contrasted different types of road accidents, the characteristics of those incidents, the characteristics of the road itself, and the impact of road safety (Lokesha and Mahesha 2016). Persia et al. have described road infrastructure safety management (RISM) in this paper. The highway agency will make its choice by a set of RISM processes, which include system execution, assessment of impact, efficiency assessments, and road safety inspections (Blair et al. 2017). The author has concluded that the primary factor in increasing traffic safety is the management of the safety of the road network. In this study, Lokesha et al. explored road transport, emphasizing how important it is to the development of agriculture. The improvement of road infrastructure is crucial for agriculture and for the general economic expansion that raises living standards (Lipovac et al. 2015). The author of this study concluded that better road infrastructure on rural roads decreased transportation costs and increased marketing. In this study, Blair et al. spoke about possible solutions to improve the security of the road infrastructure. The author has concluded that both urban and rural arterial highways have effectively used these solutions (Tunde et al. 2014). Points of access are a crucial part of the vehicular transit system since they connect the public highway to side properties, private projects, and services. Yet, access is seen as one of the primary causes of traffic accidents and congestion. Their position, design, and sight aspects are frequently connected with several concerns that are incompatible with the essential safety requirements for automobile and pedestrian movement. Furthermore, their poor security record is commonly attributed to left-turn motions, either inside or outside. The focus of this study is on uncontrolled approaches, which are entry points that lack traffic signals, stop signs, access rights, yield signs, or right-turn acceleration/deceleration lanes. The lack of suitable geometrical layout and access management techniques results in these sites being mishap spots.

14.3 Study Area

The chosen case study is the National Highway (NH151A), which is an essential component of the Indian road network. Access nodes of various locations are described below (Figs. 14.1, 14.2, 14.3 and 14.4).

Traffic volume counts at the study locations are given below (Tables 14.1, 14.2, 14.3):

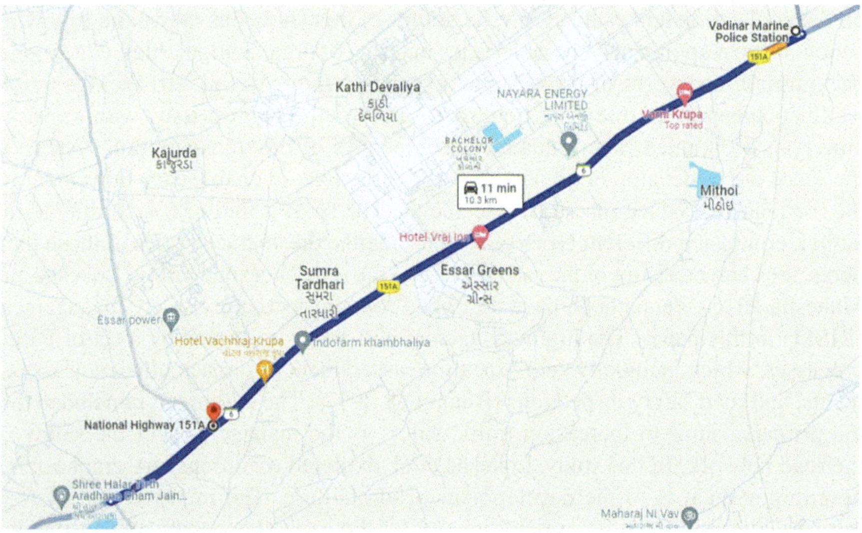

Fig. 14.1 Location of project site

Fig. 14.2 Nayara main gate at CH. 73 + 770

Fig. 14.3 Nayara material gate at CH. 72 + 760

Fig. 14.4 Nayara boundary gate (near the end of Nayara Boundary Wall) at CH. 69 + 590

14.3.1 Spot Speed Study

- The spot speed investigations (Table 14.4) are essential for designing and analyzing highway systems by monitoring various variables and components.
- Recorded speeds are used for capacity evaluation, geometric design, safety precautions, speed trends, and evaluation. Speed studies serve in decision-making and evaluating the success of roadway improvements. They also identify the need for further regulation, realignment, or rebuilding.
- Spot speed records (Fig. 14.5) may be used for several safety purposes, such as: Identifying current traffic operations and evaluating traffic control devices, developing highway design components, and assessing the performance of traffic control devices or initiatives, such as signage and symbols, traffic operational adjustments, and speed enforcement campaigns.

Table 14.1 ADT (turning movement) at Nayara main gate

Types		IN				OUT			
		ADT	ADT %	PCU	CVPD	ADT	ADT %	PCU	CVPD
Fast passenger vehicles	Two wheeler	87.00	16.76	0.75	65.25	75.00	14.94	0.75	56.25
	Auto rick	0.00	0.00	2.00	0.00	0.00	0.00	2.00	0.00
	Car/jeep /TAXI	355.00	68.40	1.00	355.00	350.00	69.72	1.00	350.00
	Trip van (Tata magic)	2.00	0.39	1.00	2.00	2.00	0.40	1.00	2.00
	Mini bus	2.00	0.39	3.70	7.40	2.00	0.40	3.70	7.40
	School bus	0.00	0.00	3.70	0.00	0.00	0.00	3.70	0.00
	Govt bus	0.00	0.00	3.70	0.00	0.00	0.00	3.70	0.00
	Private bus	42.00	8.09	3.70	155.40	42.00	8.37	3.70	155.40
Fast commercial vehicles	Mini LCV	31.00	5.97	2.00	62.00	31.00	6.18	2.00	62.00
	LCV	0.00	0.00	2.00	0.00	0.00	0.00	2.00	0.00
	2 Axle Truck	0.00	0.00	3.70	0.00	0.00	0.00	3.70	0.00
	3 Axle Truck	0.00	0.00	3.70	0.00	0.00	0.00	3.70	0.00
	MAV	0.00	0.00	5.00	0.00	0.00	0.00	5.00	0.00
	Tractor	0.00	0.00	2.00	0.00	0.00	0.00	2.00	0.00
	Tractor + trailer	0.00	0.00	5.00	0.00	0.00	0.00	5.00	0.00
Slow modes	Cycle	0.00	0.00	0.50	0.00	0.00	0.00	0.50	0.00
	Cycle rick	0.00	0.00	2.00	0.00	0.00	0.00	2.00	0.00
	Animal drawn	0.00	0.00	2.00	0.00	0.00	0.00	2.00	0.00
Toll exempted vehicles	Car	0.00	0.00	1.00	0.00	0.00	0.00	1.00	0.00
	Bus	0.00	0.00	3.70	0.00	0.00	0.00	3.70	0.00
	Truck	0.00	0.00	3.70	0.00	0.00	0.00	3.70	0.00
Total		519.00	100.00		647.05	502.00	100.00		633.05

Table 14.2 ADT (turning movement) at Nayara material gate

Types		IN				OUT			
		ADT	ADT %	PCU	CVPD	ADT	ADT %	PCU	CVPD
Fast passenger vehicles	Two wheeler	5.00	0.87	0.75	3.75	3.00	0.46	0.75	2.25
	Auto rick	0.00	0.00	2.00	0.00	21.00	3.19	2.00	42.00
	Car/ jeep/taxi	245.00	42.46	1.00	245.00	243.00	36.93	1.00	243.00
	Trip van (Tata magic)	4.00	0.69	1.00	4.00	10.00	1.52	1.00	10.00
	Mini bus	10.00	1.73	3.70	37.00	12.00	1.82	3.70	44.40
	School bus	3.00	0.52	3.70	11.10	1.00	0.15	3.70	3.70
	Govt bus	0.00	0.00	3.70	0.00	9.00	1.37	3.70	33.30
	Private bus	31.00	5.37	3.70	114.70	38.00	5.78	3.70	140.60
Fast commercial vehicles	Mini LCV	139.00	24.09	2.00	278.00	165.00	25.08	2.00	330.00
	LCV	25.00	4.33	2.00	50.00	25.00	3.80	2.00	50.00
	2 Axle Truck	49.00	8.49	3.70	181.30	47.00	7.14	3.70	173.90
	3 Axle Truck	44.00	7.63	3.70	162.80	42.00	6.38	3.70	155.40
	MAV	9.00	1.56	5.00	45.00	26.00	3.95	5.00	130.00
	Tractor	4.00	0.69	2.00	8.00	4.00	0.61	2.00	8.00
	Tractor + trailer	9.00	1.56	5.00	45.00	12.00	1.82	5.00	60.00
Slow modes	Cycle	0.00	0.00	0.50	0.00	0.00	0.00	0.50	0.00
	Cycle rick	0.00	0.00	2.00	0.00	0.00	0.00	2.00	0.00
	Animal drawn	0.00	0.00	2.00	0.00	0.00	0.00	2.00	0.00
Toll exempted vehicles	Car	0.00	0.00	1.00	0.00	0.00	0.00	1.00	0.00
	Bus	0.00	0.00	3.70	0.00	0.00	0.00	3.70	0.00
	Truck	0.00	0.00	3.70	0.00	0.00	0.00	3.70	0.00
Total		577.00	100.00		1185.65	658.00	100.00		1426.55

Table 14.3 ADT (turning movement) at Nayara boundary gate

Types		IN				OUT			
		ADT	ADT %	PCU	CVPD	ADT	ADT %	PCU	CVPD
Fast passenger vehicles	Two wheeler	338.00	73.80	0.75	253.50	354.00	75.80	0.75	265.50
	Auto rick	21.00	4.59	2.00	42.00	20.00	4.28	2.00	40.00
	Car/jeep/taxi	66.00	14.41	1.00	66.00	55.00	11.78	1.00	55.00
	Trip van (Tata magic)	6.00	1.31	1.00	6.00	12.00	2.57	1.00	12.00
	Mini bus	1.00	0.22	3.70	3.70	1.00	0.21	3.70	3.70
	School bus	2.00	0.44	3.70	7.40	2.00	0.43	3.70	7.40
	Govt bus	0.00	0.00	3.70	0.00	0.00	0.00	3.70	0.00
	Private bus	0.00	0.00	3.70	0.00	0.00	0.00	3.70	0.00
Fast commercial vehicles	Mini LCV	11.00	2.40	2.00	22.00	12.00	2.57	2.00	24.00
	LCV	8.00	1.75	2.00	16.00	6.00	1.28	2.00	12.00
	2 Axle Truck	1.00	0.22	3.70	3.70	0.00	0.00	3.70	0.00
	3 Axle Truck	0.00	0.00	3.70	0.00	1.00	0.21	3.70	3.70
	MAV	0.00	0.00	5.00	0.00	0.00	0.00	5.00	0.00
	Tractor	1.00	0.22	2.00	2.00	3.00	0.64	2.00	6.00
	Tractor + trailer	2.00	0.44	5.00	10.00	1.00	0.21	5.00	5.00
Slow modes	Cycle	1.00	0.22	0.50	0.50	0.00	0.00	0.50	0.00
	Cycle rick	0.00	0.00	2.00	0.00	0.00	0.00	2.00	0.00
	Animal drawn	0.00	0.00	2.00	0.00	0.00	0.00	2.00	0.00
Toll exempted vehicles	Car	0.00	0.00	1.00	0.00	0.00	0.00	1.00	0.00
	Bus	0.00	0.00	3.70	0.00	0.00	0.00	3.70	0.00
	Truck	0.00	0.00	3.70	0.00	0.00	0.00	3.70	0.00
Total		458.00	100.00		432.80	467.00	100.00		434.30

Table 14.4 Average speed and 85th percentile speed at particular locations

Sr. no	Direction	Time	Average speed (Kmph)	85th percentile speed (Kmph)
1	Vadinar Jn. towards Dwarka	09:20 to 09:50	44.0	44.2
2	Vadinar Jn. towards Jamnagar	13:40 to 14:10	38.7	39.1
3	Vadinar Jn. towards Vadinar	14:20 to 14:50	28.5	29.8
4	Mithoi Jn. towards Dwarka	10:15 to 10:45	52.7	53.0
5	Mithoi Jn. towards Jamnagar	13:40 to 14:10	59.3	59.7
6	Mithoi Jn. towards Mithoi	13:40 to 14:10	28.4	28.9
7	Nayara main gate towards Dwarka	13:40 to 14:10	51.3	51.2
8	Nayara main gate towards Jamnagar	13:40 to 14:10	46.9	47.7
9	Nayara material gate towards Dwarka	13:40 to 14:10	53.4	53.8
10	Nayara material gate towards Jamnagar	13:40 to 14:10	52.9	53.3
11	Devariya Jn. towards Dwarka	13:40 to 14:10	46.4	46.8
12	Devariya Jn. towards Jamnagar	13:40 to 14:10	54.8	54.4
13	Devariya Jn. towards Devariya	13:40 to 14:10	29.7	30.2
14	Essar main gate towards Dwarka	13:40 to 14:10	63.7	64.6
15	Essar main gate towards Jamnagar	13:40 to 14:10	64.0	64.4
16	Essar material gate towards Dwarka	13:40 to 14:10	46.7	47.8
17	Essar material gate towards Jamnagar	13:40 to 14:10	50.4	51.2
18	Vadaliya Sinhan towards Dwarka	13:40 to 14:10	52.0	53.2
19	Vadaliya Sinhan towards Jamnagar	13:40 to 14:10	51.5	52.3

Fig. 14.5 Images showing spot speed studies at the study location

14.3.2 Problem Statement

The At Nayara intersection, NH-151A connects Jamnagar and Dwarka, and the main entrance gate, material gate, and boundary gate of Nayara is the critical junction on the stretch. Traffic delays, the risk to pedestrian life, and the lack of proper facilities like symbols, signals, and marking are major problems at this junction.

Nayara's main gate as shown in the figure lies at CH. 73 + 770 and at latitude 22.32 and longitude 69.55. The collapsible gate placed at the median (in front of Nayara Main Gate) is opened only at peak hours of morning and evening for the entry and exit of Nayara Staff Buses. The entry and exit of staff buses at the peak hours causes queue formation on the road as well as traffic congestion.

Nayara Material Gate as shown in the figure lies at CH. 72 + 760 and at latitude 22.31 and longitude 69.74. Temporary barricading is done on a median in front of Nayara Material Gate. A retaining wall on both sides of the gate is seen. The usage of road bollards is done for safe entry and exit of trucks and trailers handling materials. During the movement of trucks and trailers at the material gate, the traffic situation is worst causing the other commuters to wait for half to one hour daily.

Nayara Boundary Gate as shown in the figure lies at CH. 69 + 590 and at latitude 22.30 and longitude 69.72. This node is the end of the Nayara Energy Boundary wall and the starting point of Essar Power Plant. The end gate of the Nayara boundary creates traffic disturbances to the adjoining Essar power plant. Present research aims to locate blackspots on National Highway (NH-353D), also known as the branch of NH53. The incident locations were rated through the Weighted Severity Index (WSI) (see Table 14.4). This technique was found to be efficient in recognizing blackspots. According to the study and results, the average WSI is 44.05 which is above the Benchmark value i.e. above 41 (Fig. 14.5). It has been found that most of the identified locations are junctions and built-up areas linking to NH (Fig. 14.6). The reason for the occurrence of mishaps is overspeeding and poor road geometrics. We offered short-term measures for blackspots while the long-term remedies are implemented.

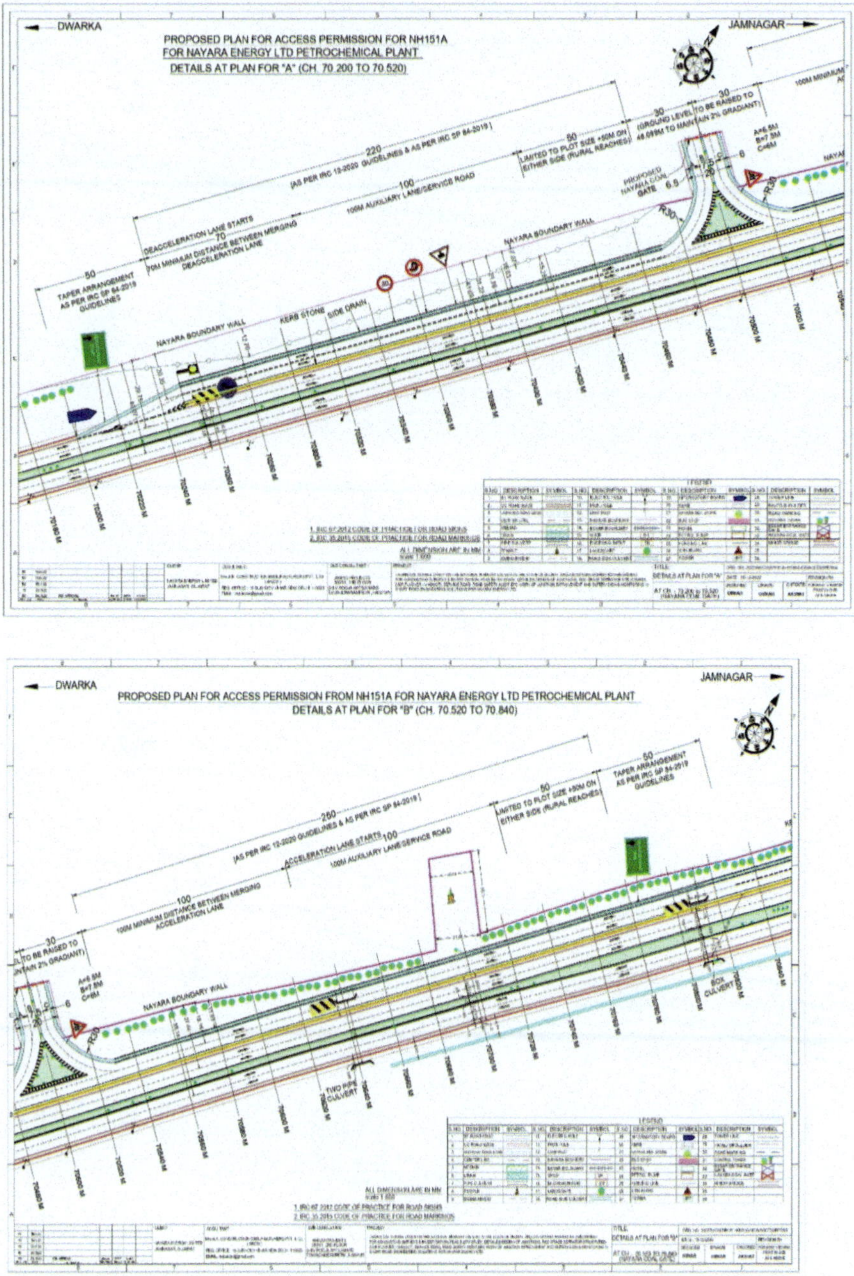

Fig. 14.6 Strip plan of proposed service road

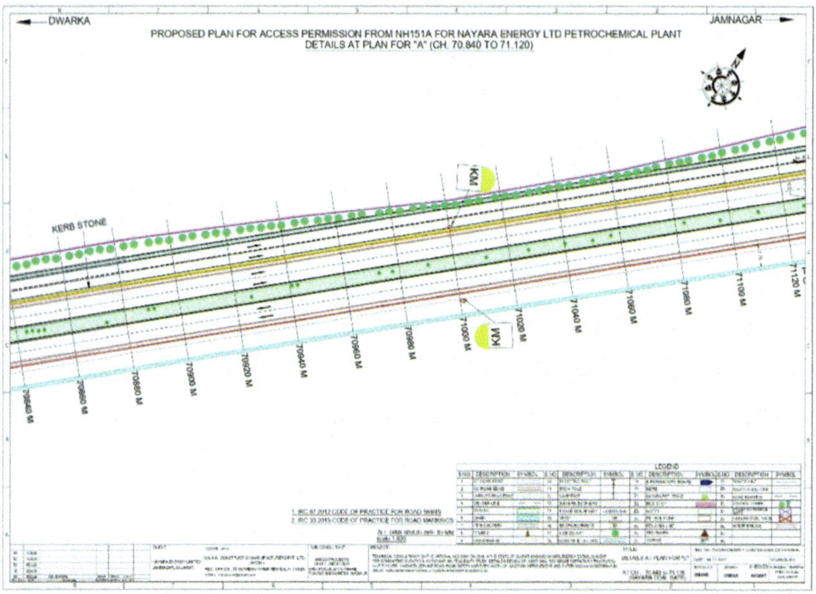

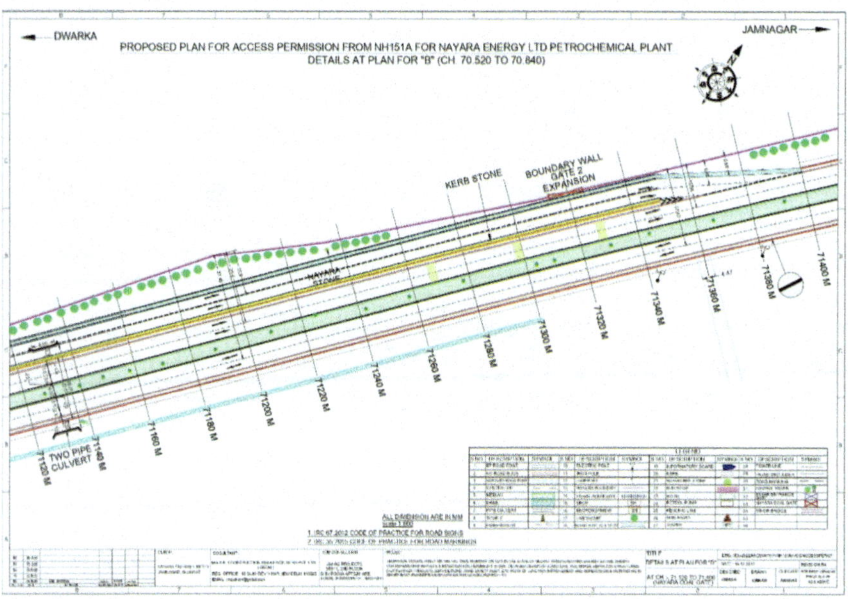

Fig. 14.6 (continued)

14.3.3 Solution

For this problem, we have four alternatives flyover, provision of service road, signalize intersection, and divert traffic of Nayara on another road. Out of these four alternatives construction of a flyover is advantageous at this intersection. Increasing population growth of about 15.5% per decade and an average of 48,157 no. of vehicles increasing per year so rapid construction of flyover at this intersection. Another option is the provision of a service road for the separate movement of vehicles heading towards various Gates of Nayara Refinery Plant.

14.3.4 Recommendations

1. **At Grade**
    - Signalization is one of the most widely used solution to counteract the issue of congestion on intersections. Proposal of signalization is given at Mithoi Junction heading towards Nayara's gate discussed above.
    - Proposal of Service road adjacent to NH151A from Nayara Boundary Gate to Nayara Material gate.
    - Proposal of separate lane for Nayara Energy Ltd.'s commuters.
    - Diversion of Local traffic from NH.
2. **Grade Separated**
    - Construction of flyover is suggested from Mithoi Junction to Devariya Junction to eliminate the conflict of vehicles passing from Mithoi to Nayara Gates and vice versa with local traffic.

14.4 Conclusion

This research offers a complete approach to the management of uncontrolled access points, i.e., access points lacking traffic signals, stop signs, right-of-way, yield signs, or right-turn acceleration/deceleration lanes (Fig. 14.6). These road network components guarantee that the public highway is linked to side properties, private projects, and facilities. However, because to insufficient circumstances and constraints in their position, construction, and visibility, they are frequently connected with a number of safety hazards.

Various options were suggested to officials of Nayara Energy Ltd. such as construction of flyover, Diversion of existing traffic, designing a separate lane for Nayara's traffic, service road, etc. Construction of service road is taken forward by the officials as a feasible option taking into account current traffic scenario of both local traffic and Nayara's traffic. The proposed service road is shown in figure below.

Taken together, the findings of this study highlight the need of a holistic strategy that considers a variety of criteria related to the physical and functional design of uncontrolled access sites. Further data collection would assist confirm and validate the trend and findings of this research, resulting in more accurate and comprehensive criteria targeted at enhancing the operational and safety efficiency of these road network parts.

References

Ahmed I (2013) Road infrastructure and road safety. Transp Commun Bull Asia Pac
Appleton I (2009) Road infrastructure safety asses. Proceeding of Fourth IRTAD, Seoul, Korea
Blair T, Chris J, Tariro M (2017) Treatments that need to be used for providing safe road infrastructure. Aust NZ Res Pract 26(03)
Flahaut B (2004) Impact of road infrastructure and local environment on road unsafety. Accid Anal Prev 36(06):1055–1066
Jamroz K, Budzynski M, Kustra W, Michalski L, Gaca S (2014) Tools for road infrastructure safety management. Transp Res Procedia 3:730–739
Lipovac K, Tesic M, Maric B, Deric M (2015) Self-reported and observed seat belt use. Accident analysis and prevention
Lokesha MN, Mahesha M (2016) Impact of road infrastructure on agriculture development and rural road infrastructure development programmes. Int J Humanit Soc Sci Invention 5(11):1–7
Mulyono AT, Kushari B, Agustin J (2008) Monitoring and evaluating infrastructure safety deficiencies towards integrated road safety improvement. In: Australasian road safety research, policy and education conference.
Quiroz C, Gautama S (1992) Road infrastructure and economic development. Western Africa department and infrastructure and urban development
Robert BN (2003) How road infrastructure improvement can affect traffic facilities and injuries. Accid Anal Prev 35(04):599–611
Tunde SOB, Perera S, Udeaja C, Zhou L (2014) Challenges of implementing Infrastructure mega projects through PPP. Int J Eng Constr 3:142–154

Chapter 15
Study and Analysis of Traffic for Planning of National Highway Development: A Case Study

Yashaswi K. Thakre and P. Y. Pawade

Abstract The study looked at the issues with traffic congestion and what led to them at particular road crossings in Jamnagar, Gujarat. Additionally, the traits of crossings that make them more prone to congestion issues and the spatial distribution of traffic congestion at road intersections were determined. Additionally, estimates of traffic volume, delays, and delay causes were made. Data on intersection characteristics, traffic volume and composition, traffic delay and causes, and land use activities were gathered by direct field surveys. Additionally, information was gathered from previous study reports, journals, books, atlases, and the Internet. Traffic flows and delays at the analyzed intersections exhibit spatiotemporal fluctuations, according to the analysis of the data gathered. Additionally, it was discovered that parking issues and traffic wardens were the main reasons for delays at city intersections. To lessen traffic congestion and delays at the intersection in the city, the study suggests that the crossroads be signalized and that parking be tightly restricted.

Keywords Traffic Congestion · Traffic Volume Count · Spot Speed

15.1 Introduction

Traffic congestion is one of the most important issues with urban transportation. It occurs when the demand for urban transportation networks exceeds the supply of those networks (Hao et al. 2017). Today, traffic congestion affects almost all cities in both developed and developing nations. It primarily shows itself as frequent lines, hold-ups, and time waste that commuters encounter along key networks, particularly during peak hours (Huang et al. 2016). Traffic congestion has become an unsolvable issue in metropolitan centers due to constant population growth, rising household

Y. K. Thakre (✉) · P. Y. Pawade
G. H. Raisoni College of Engineering, Nagpur, India
e-mail: yashaswi.thakre.mtechts@ghrce.raisoni.net

B. V. Khode et al. (eds.), *Innovations in Technologies: Pioneering Sustainable Infrastructure for a Resilient Future*, Sustainable Civil Infrastructures,
https://doi.org/10.1007/978-3-031-82133-2_15

incomes, and resulting rise in car usage, along with bad land-use planning, poor transport design, and planning (Bonela and Raghuram Kadali 2022).

Urban transportation is severely hampered by traffic congestion. It is a nuisance that has built itself into daily life in practically every urban location on the planet (Kumar and Ranjitha 2013). More importantly, traffic congestion makes travel times unpredictable, forcing urban commuters to prepare for these issues by leaving home early simply to don't be late (Mankar and Khode 2016a).

At road intersections, the issue of traffic congestion in metropolitan areas is severe. No other location on city highways may be as congested as a road intersection, in particular. It was discovered that there are more potential and actual confrontations at road crossroads than everywhere else in the street and highway systems (Marve and Bhorkar 2020). This is due to the fact that at intersections, many traffic flows from various approaches that are making left-turn, though, or right-turn movements compete for the same physical area at once. Additionally, to these traffic movements, pedestrians also attempt to cross the street in this area, increasing the already problematic traffic situation (Mankar and Khode 2016b).

Urban traffic issues get considerably worse by the centralized location of the majority of workplaces in the same areas (often in the city centers), making morning and evening rush hour traffic largely unidirectional. This study looks at the latter issue, which causes local variance in congestion in metropolitan locations (Marve and Bhorkar 2016). This study's main goal is to investigate the characteristics of traffic congestion at a road intersection in Nagpur, as well as to pinpoint potential reasons and provide remedies.

Like many other Indian state capitals, Nagpur experiences major daily attention from the issue of traffic congestion (Sayre et al. 2021). Movements within cities for work, recreation, markets, stores, and schools are becoming more and more challenging, uncomfortable, delayed, and a waste of time, energy, and resources (Ladhi et al. 2018). The issue is especially severe in the morning and evening rush hours, when vehicles are stuck in lengthy lines, stressing commuters and cutting into their workdays. Nagpur's situation is still under control, but there are already some routes that show indicators of possible jams.

15.2 Objectives

1. To research the current traffic conditions at a few chosen intersections.
2. To study the traffic volumes, delay and its causes for a chosen intersection.
3. To determine the operation's traffic performance at intersections.
4. To carry out traffic-improvement suggestions for the chosen intersections.

The research efforts were restricted to identifying the city's congested intersections conducting studies to determine the root reasons of the congestion, and proposing solutions to enhance traffic flow.

15.3 Literature Review

Nannan et al. established relative importance of the index of assessment through the suggested approaches of fuzzy analytic hierarchy procedure. Following this, grey methods of correlation for ranking preferences by resemblance to an optimal approach were merged to create the combined decision-making approach. The efficiency of the combination technique was confirmed by a real-life study of traffic congestion at junctions using five assessment indices. The assessment findings of the various ways demonstrate that the suggested method, by analogy to an optimal option and analytical hierarchy process-gray correlation, solves the polarization of analytical dominance process-techniques for ranking preference. As a result, the suggested combination decision-making approach offers a more precise and trustworthy way of determining the level of traffic congestion. Huang et al. study offers a fresh approach to reducing traffic congestion. The authors use VISSIM to represent and investigate an issue with urban traffic congestion for this purpose. Two different traffic light sequences are also included. On the other hand, their unique congestion control approach is evaluated using an urban traffic structure with three different types of intersections. Finally, they show how three different stream of traffic scenarios—low, intermediate, and heavy—work to compare them to the multiline control techniques, the standard light regulation rules, and their unique control strategy. The results of the simulation demonstrate that the newly suggested traffic control approach is superior than the others. It is important to highlight that their new suggested approach not only successfully reduces the problem of congestion but also stops additional traffic congestion from occurring in the community. Someshwara et al. critically reviewed Surrogate Safety Measures (SSM) and how they are used at unsignalized intersections with a concentration on the T-intersection, in this work. The T-intersection safety assessment with SSM in a poor country environment is also highlighted in this work. The results of this study are more beneficial to identify which are appropriate safety criteria for assessing roadway security at T-intersections. It is essential that they point out that their new recommended approach not only successfully reduces the problem of congestion but also stops additional traffic obstructions occurring in the community. Sunil et al. examining the appropriate approaches to increase the traffic capacity and seeks to improve on the operational performance of the signalized crossings. This investigation showed that the chosen traffic structure is now experiencing severe deterioration that is leading to unsafe situations. As a result, there must be right away consideration given to improving the LOS by offering multiple options. Pratik et al. research variety of traffic elements required for planning, designing, and operating transportation facilities. They suggested the need of more capable highway infrastructure to handle an increase in vehicle traffic. Estimating the volume of traffic and capacity of highway infrastructure under diverse traffic conditions is fairly challenging for the capacity study of roadways. The variables determining the allowable capacity of roadways and the characteristics that impact it are examined and analyses in this paper, which also discusses and reviews the capacity calculation of roadways under different traffic circumstances by a number of authors. To enhance the state of the road and the flow of traffic, this research may be used to analyze various techniques for assessing the capacity of roads. Mankar et al. investigate the significance of lane width as it relates to the traditional concept (green shield) used to analyze the capacity of urban roadways. The Green Shield Model is implemented to determine the capacity of urban roadways, and its outcomes are assessed with a Microscopic Simulation Model. The quick rise in lane width on the roadway is investigated, and the outcomes indicate that as lane width grows, so does the capacity of the roadway. Marve et al. Goals to evaluate the Nagpur's congestion at selected intersections and noise pollution by conducting surveys at particular distances in order to gather such huge quantities of information. They have suggested provision of bus bay at a junction might be lessened the congestion. By using a variety of techniques, including roadway expansion, strengthening roadways, limiting the routes taken by rickshaws and hawkers, and the use of fly overs in future metros, traffic congestion can be improved. Kumar et al. examining the appropriate approaches to increase the traffic ability, the current study seeks for improved efficiency of the signalized intersections. The study's findings showed that the chosen traffic system is now experiencing substantial deterioration that is leading to unstable situation. As a result, there must be quick attention given to enhancing the LOS by offering several options. Shradhesh et al. study to determine how much noise and particulate matter (PM) from different sources, including two-wheelers, vehicles, trucks, and industries, is affecting the air's quality. The findings will reveal the levels of suspended particulate, in the atmosphere and will be compared to the permitted concentrations for an industrial area. Precautions must be taken to decrease the influence degree of these pollutants in accordance with the standards set by the CPCB. Sourabh et al. study the negative social, environmental, and economic repercussions of traffic congestion. They suggested a rail-based transit system is needed to replace passengers use of private and public transportation on roadways. This study compares the current traffic scenario with the anticipated congestion scenario against the installation of metro rail, taking into account factors such as queue length, LOS, delay, pollutants and fuel usage. Saurabh et al. research focus in a restructuring of the current signaling system and a change to the segments, which not only results in a shorter line but also shorter travel times and less fuel usage. Transportation productivity will increase as a result of this modification.

15.4 Theories of Traffic Flow and Congestion Issues

Several congestion measurements have been established taking into account various performance parameters in order to evaluate the amount of congestion. According to these standards, there are five different types of congestion measures: speed, trip time, delay, level of services (LoS), and congestion indices. The issues with traffic flow are what cause traffic congestion. Individual drivers' and vehicles' movements between two points and their interactions with one another are referred to as traffic flow. Unfortunately, it is challenging to research traffic flow since it is impossible to forecast driver behavior with absolute precision. Fortunately, traffic streams tend to have some consistency and may be somewhat mathematically represented because drivers typically operate within a range that is fairly stable. The three primary variables in traffic flow for which mathematical correlations can be constructed are flow, density, and velocity. These connections aid in the development, creation, and maintenance of roadway facilities.

The density or concentration is the number of vehicles per unit length of the highway, the flow is the number of vehicles crossing a specific point on the highway in a given amount of time, and the speed is the space mean speed.

15.5 Methodology

15.5.1 Selection of Site

A single homogenous portion of the project road is chosen to run along the Nayara Refinery boundary in Jamnagar, Gujarat, following a thorough site visit and reconnaissance survey. At intersections like Vadinar, Mithoi, Mondh, Vadaliya Sinha, and Devariya junction, the volume of traffic is measured. The turning action of traffic towards SH has been considered at Vadinar Junction. Additionally, NH151A's LHS and RHS have been computed. Vehicles coming into and going out of Mithoi Junction are recorded. The LHS and RHS of NH151A have also been taken into account. The Mandha Junction's vehicle entry and exit counts are computed. The LHS and RHS of NH151A have also been taken into account. The turning action of traffic along Rural Lane has been considered for Devariya Junction. The LHS and RHS of NH151A have also been taken into account. The Vadaliya Sinhan Junction's vehicle entry and exit counts are computed. The LHS and RHS of NH151A have also been taken into account.

15.5.2 Data Collection

Data on intersection attributes, the traffic volumes, and intersection characteristics are all needed for this research. Both primary and secondary sources were used to gather this information. Direct field observation and field data were used to gather the primary information. The secondary information was gathered from books, journals, websites and other studies.

At the chosen intersection, a reconnaissance investigation of the study area was initially conducted to determine the factors that make them subject to high traffic. The next step was a traffic census, which was manually carried out using a straightforward hand-tally approach to determine the volume and make-up of vehicles at the intersections. We kept a running tally of every vehicle that went through the intersections, broken down by class or category. We were focus on one lane at a time and count each vehicle as they passed through the intersections. This is so that turning actions may be addressed completely before cars leave junctions.

15.5.3 Survey

To fulfill this objective, the following traffic surveys were carried out traffic volume count survey at one location for 24 Hours.

Video graphic survey for 24 h at one location (Table 15.1).

Turning movement survey for 24 hours at 1 location (Table 15.2).

Straight movement survey for 3 days at 1 location (Table 15.3).

Table 15.1 Schedule of traffic volume count survey

Location	Survey dates
Vadinar Junction	25, 26, 27 August 2022
Mithoi Junction	02, 03, 04 August 2022
Mandha Junction	01, 02, 03 August 2022
Vadaliya Sinhan	02, 03, 04 August 2022
Devariya Junction	01, 02, 03 August 2022

Table 15.2 Schedule of turning movement survey

Sr. no	Location	Survey dates
1	Vadinar Junction	26 August 2022
2	Mithoi Junction	04 August 2022
3	Mandha Junction	01 August 2022
4	Vadaliya Sinhan	02 August 2022
5	Devariya Junction	02 August 2022

Table 15.3 Schedule of straight movement survey

Sr. no	Location	Survey dates
1	Vadinar Junction	25, 26, 27 August 2022
2	Mithoi Junction	02, 03, 04 August 2022
3	Mandha Junction	01, 02, 03 August 2022
4	Vadaliya Sinhan	02, 03, 04 August 2022
5	Devariya Junction	01, 02, 03 August 2022

15.6 Approach and Methodology for Traffic Survey

The site of the traffic survey has been decided upon, and its layouts have been developed in accordance with the purpose of this research and the data that will be collected there. The survey was carried out according to the guidelines in the TOR. To perform the survey, a special team consisting of a traffic engineer and a data collector was put together. Among the pursuits were.

15.6.1 Training of Enumerators

Prior beginning of work on surveys, counters who were to be deployed for the traffic survey received training on how to identify different vehicle groups and fill out forms.

15.6.2 Survey Teams

Each team was consisting of one supervisor and adequate enumerator, depending on location and scope of the survey.

15.6.3 Selection of Traffic Survey Location Along the Section

The sites for the traffic surveys were chosen based on homogeneous highway segments. Using the results of the reconnaissance study, the highway's portions with uniform traffic volumes are located. Officials helped determine the final site for the project highway.

15.6.4 Classified Volume Count Survey

The counts were taken down every 60 min and were done by hand. Three days were dedicated to conducting the survey (straight motions). The TOR/IRC code-suggested vehicle classification was adhered to. The templates for the Traffic Study reflect this as well. Each finding has been shown both graphically and tabularly. MS-EXCEL tool was used to computerize the acquired data.

Picomixer STA Software.

Picomixer STA is a vehicle computation and effective traffic testing software application which uses the effective artificial intelligence and image processing algorithms.

Identification, Detection, Counting, The categorization of various vehicles types, its average velocity, estimating highway flow volumes, etc. are a few of the most crucial characteristics of Picomixer STA Software.

STA functional capabilities:

- Vehicle count and record data every hour of the day and night.
- Separating and classifying vehicles in 5 groups:
  - Car
  - Mini bus
  - Truck
  - Bus
  - Trailer
- Also separating and classifying vehicles in 2 groups:
  - Lightweight Vehicles
  - Heavyweight Vehicles
- Calculates average traffic volume and record data every hour of the day and night.
- Calculates average vehicle speed and record data every hour of the day and night.
- Support moving in both directions simultaneously (Figs. 15.1, 15.2 and 15.3).

Classified Traffic Volume Counts

Traffic Characteristics on Project Road.

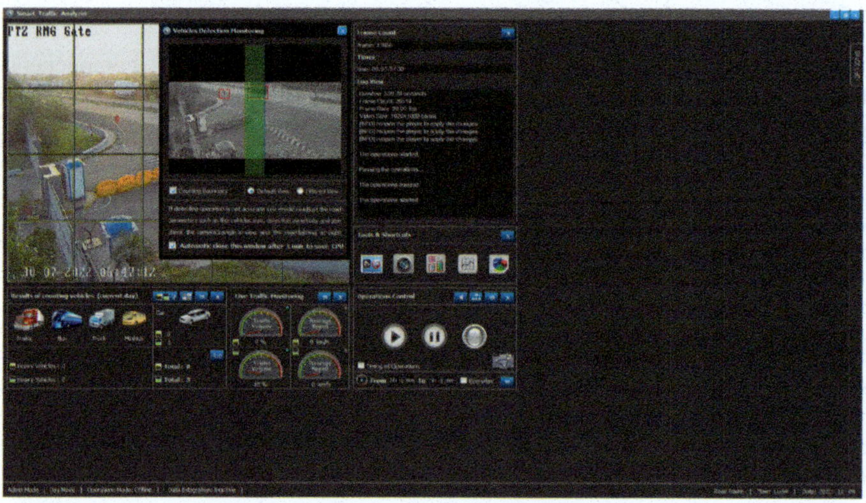

Fig. 15.1 Working window of Picomixer STA Software (Mandha Junction)

Fig. 15.2 Vadinar and Devariya Junction

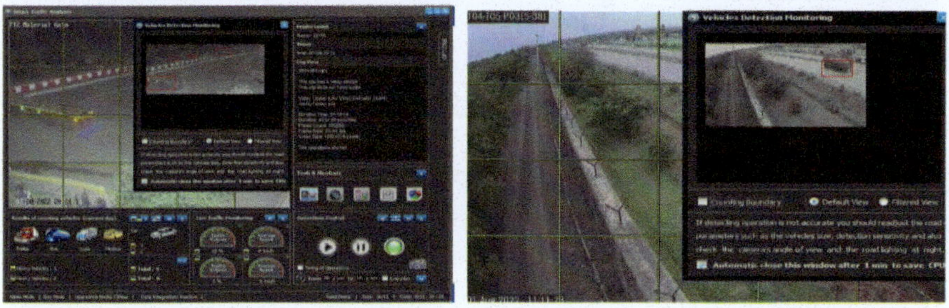

Fig. 15.3 Vadaliya Sinhan and Mithoi Junction

A tabular summary of all the data gathered from the traffic survey is shown for every site along the Project Highway. The table provides an overview of the Average Daily Traffic (ADT) for the month of August in 2022.

Passenger Car Unit (PCU):

The diverse traffic conditions on the proposed Road lead to intricate interactions amongst different types of vehicles. IRC has suggested unifying various forms of transportation into a single "Passenger Car Unit (PCU)" in order to cope with congestion. In accordance with the interface parameter value of each vehicle type, equivalent PCUs are generated (Table 15.4).

Maximum and Minimum Traffic in a day:

The maximum and minimum traffic observed at the survey location is as follows (Tables 15.5, 15.6, 15.7, 15.8, 15.9, 15.10, 15.11 and 15.12):

Maximum and Minimum Traffic at a straight stretch (Tables 15.13, 15.14, 15.15, 15.16, 15.17):

Average Daily Traffic (ADT) and PCU on Turnings (Tables 15.18, 15.19, 15.20, 15.21 and 15.22):

Table 15.4 Recommended PCU factor for various types of vehicle

Sr. no	Vehicle type	Urban roads		Rural road
		% Composition of vehicle type in traffic stream		
		5%	10% and above	
Fast vehicles				
1	Two wheelers	0.5	0.75	0.5
2	Passenger cars, pick up vans	1	1	1
3	Auto rickshaw	1.2	2	1
4	Light commercial vehicles	1.4	2	1.5
5	Truck or bus	2.2	3.7	3
6	Agricultural tractor without trailer	1.4	2	1.5
7	Agricultural tractor with trailer, MAV			
Slow vehicles				
8	Cycle	0.4	0.5	0.5
9	Cycle rickshaw	1.5	2	1.5
10	Tonga (horse drawn vehicle)	1.5	2	6
11	Bullock cart*			8
12	Hand cart	2	3	3

Source IRC 64-1990 & IRC-SP-41 1994

Table 15.5 Vadinar Junction

Survey location	Vadinar Junction
Maximum traffic time	08:00–09:00
No. of vehicles	524
Minimum traffic time	03:00–04:00
No. of vehicles	13

Table 15.6 Mithoi Junction (IN)

Survey location	Mithoi Junction (IN)
Maximum traffic time	18:00–19:00
No. of vehicles	49
Minimum traffic time	00:00–01:00
No. of vehicles	01

Average Daily Traffic (ADT) and PCU on Straight Stretch (Tables 15.23, 15.24, 15.25, 15.26 and 15.27)

Spot Speed:

Table 15.7 Mithoi Junction (OUT)

Survey location	Mithoi Junction (OUT)
Maximum traffic time	08:00–09:00
No. of vehicles	89
Minimum traffic time	00:00–01:00
No. of vehicles	01

Table 15.8 Mandha Junction (IN)

Survey location	Mandha Junction (IN)
Maximum traffic time	08:00–09:00
No. of vehicles	226
Minimum traffic time	03:00–04:00
No. of vehicles	01

Table 15.9 Mandha Junction (OUT)

Survey location	Mandha Junction (OUT)
Maximum traffic time	18:00–19:00
No. of vehicles	154
Minimum traffic time	03:00–04:00
No. of vehicles	01

Table 15.10 Vadaliya Sinhan Junction (IN)

Survey location	Vadaliya Sinhan Junction
Maximum traffic time	08:00–09:00
No. of vehicles	115
Minimum traffic time	01:00–02:00
No. of vehicles	02

Table 15.11 Vadaliya Sinhan Junction (OUT)

Survey location	Vadaliya Sinhan Junction
Maximum traffic time	17:00–18:00
No. of vehicles	113
Minimum traffic time	00:00–01:00
No. of vehicles	02

Table 15.12 Devariya Junction

Survey location	Devariya Junction
Maximum traffic time	08:00–09:00
No. of vehicles	756
Minimum traffic time	02:00–03:00
No. of vehicles	06

Table 15.13 Vadinar Junction

Survey location	Vadinar	
	LHS	RHS
Maximum traffic time	08:00–09:00	18:00–19:00
No. of vehicles	602	611
Minimum traffic time	23:00–24:00	23:00–24:00
No. of vehicles	52	73

Table 15.14 Mithoi Junction

Survey location	Mithoi	
	LHS	RHS
Maximum traffic time	08:00–09:00	18:00–19:00
No. of vehicles	602	509
Minimum traffic time	23:00–24:00	04:00–05:00
No. of vehicles	52	30

Table 15.15 Mandha Junction

Survey location	Mandha Junction	
	LHS	RHS
Maximum traffic time	09:00–10:00	17:00–18:00
No. of vehicles	791	613
Minimum traffic time	04:00–05:00	04:00–05:00
No. of vehicles	65	32

Table 15.16 Vadaliya Sinhan Junction

Survey location	Vadaliya Sinhan Junction	
	LHS	RHS
Maximum traffic time	10:00–11:00	08:00–09:00
No. of vehicles	415	404
Minimum traffic time	23:00–24:00	03:00–04:00
No. of vehicles	52	33

Table 15.17 Devariya

Survey location	Devariya	
	LHS	RHS
Maximum traffic time	18:00–19:00	08:00–09:00
No. of vehicles	457	404
Minimum traffic time	02:00–03:00	03:00–04:00
No. of vehicles	62	33

Table 15.18 Vadinar Junction

Types		ADT	ADT %	PCU	CVPD
Fast passenger vehicles	Two wheeler	2498	45.42	0.75	1873.5
	Auto rick	226	4.11	2	452
	Car/jeep/taxi	852	15.49	1	852
	Trip van (Tata magic)	285	5.18	1	285
	Mini bus	39	0.71	3.7	144.3
	School bus	3	0.05	3.7	11.1
	Gov. bus	1	0.02	3.7	3.7
	Private bus	30	0.55	3.7	111
Fast commercial vehicles	Mini LCV	60	1.09	2	120
	LCV	147	2.67	2	294
	2 Axle truck	115	2.09	3.7	425.5
	3 Axle truck	536	9.75	3.7	1983
	MAV	687	12.49	5	3435
	Tractor	11	0.2	2	22
	Tractor + trailer	5	0.09	5	25
Slow modes	Cycle	1	0.02	0.5	0.5
	Cycle rick	0	0	2	0
	Animal drawn	0	0	2	0
Toll exempted vehicles	Car	0	0	1	0
	Bus	1	0.02	3.7	3.7
	Truck	0	0	3.7	0
Total		5500	100		10,056.5

- Spot speed studies are essential for evaluating many parameters and components required in the planning and evaluation of the transportation network.
- For capacity analysis, geometric design, safety precautions, speed patterns, and evaluations, reported speeds are considered. Processes for making choices and before-and-after analyses to evaluate the successful implementation of road changes are aided by speed analyses.
- Additionally, traffic studies aid in determining whether new law enforcement, realignment, or reconstruction are required for the highway.
- There are several safety functions via spot speed data, which include the following: identifying current traffic activities, evaluating traffic management devices, creating highway design components, and gauging the success of traffic management initiatives, such as speed monitoring initiatives, signs and markings, and operational modifications to the traffic flow (Table 15.28).

Table 15.19 Mithoi Junction

Types		(IN)				(OUT)			
		ADT	ADT %	PCU	CVPD	ADT	ADT %	PCU	CVPD
Fast passenger vehicles	Two wheeler	338	73.8	0.75	253.5	354	75.8	0.75	265.5
	Auto rick	21	4.59	2	42	20	4.28	2	40
	Car/jeep/taxi	66	14.41	1	66	55	11.78	1	55
	Trip van (Tata magic)	6	1.31	1	6	12	2.57	1	12
	Mini bus	1	0.22	3.7	3.7	1	0.21	3.7	3.7
	School bus	2	0.44	3.7	7.4	2	0.43	3.7	7.4
	Gov. bus	0	0	3.7	0	0	0	3.7	0
	Private bus	0	0	3.7	0	0	0	3.7	0
Fast commercial vehicles	Mini LCV	11	2.4	2	22	12	2.57	2	24
	LCV	8	1.75	2	16	6	1.28	2	12
	2 Axle truck	1	0.22	3.7	3.7	0	0	3.7	0
	3 Axle truck	0	0	3.7	0	1	0.21	3.7	3.7
	MAV	0	0	5	0	0	0	5	0
	Tractor	1	0.22	2	2	3	0.64	2	6
	Tractor + trailer	2	0.44	5	10	1	0.21	5	5
Slow modes	Cycle	1	0.22	0.5	0.5	0	0	0.5	0
	Cycle rick	0	0	2	0	0	0	2	0
	Animal drawn	0	0	2	0	0	0	2	0
Toll exempted vehicles	Car	0	0	1	0	0	0	1	0
	Bus	0	0	3.7	0	0	0	3.7	0
	Truck	0	0	3.7	0	0	0	3.7	0
Total		458	100		432.8	467	100		434.3

15.7 Discussion

At intersections like Vadinar, Mithoi, Mandha Junction, Vadaliya Sinhan Junction, and Devariya Junction, the volume of traffic is measured. The peak hour timing for Vadinar Junction is 08:00 to 09:00 with traffic volume of 524 vehicles for turnings. For straight stretch, peak hour duration is from 08:00 to 09:00 with traffic volume of 602 for Vadinar LHS & 18:00 to 19:00 with traffic volume of 611 for Vadinar RHS. Through Mithoi, vehicles leaving the junction are more for span of 08:00 to 09:00 i.e., 89 and number of vehicles entering the junction are 49 from 18:00 to 19:00. Number of vehicles are 617 from 08:00 to 09:00 & 447 from 18:00 to 19:00 for LHS & RHS respectively. The number of vehicles entering the Mandha Junction are 226 from 08:00 to 09:00 and leaving the gate are 154 from 18:00 to 19:00. For Mandha

Table 15.20 Mandha Junction

Types		(IN)				(OUT)			
		ADT	ADT %	PCU	CVPD	ADT	ADT %	PCU	CVPD
Fast passenger vehicles	Two Wheeler	87	16.76	0.75	65.25	75	14.94	0.75	56.25
	Auto rick	0	0	2	0	0	0	2	0
	Car/jeep/taxi	355	68.4	1	355	350	69.72	1	350
	Trip van (Tata magic)	2	0.39	1	2	2	0.4	1	2
	Mini bus	2	0.39	3.7	7.4	2	0.4	3.7	7.4
	School bus	0	0	3.7	0	0	0	3.7	0
	Gov. bus	0	0	3.7	0	0	0	3.7	0
	Private bus	42	8.09	3.7	155.4	42	8.37	3.7	155.4
Fast commercial vehicles	Mini LCV	31	5.97	2	62	31	6.18	2	62
	LCV	0	0	2	0	0	0	2	0
	2 Axle truck	0	0	3.7	0	0	0	3.7	0
	3 Axle truck	0	0	3.7	0	0	0	3.7	0
	MAV	0	0	5	0	0	0	5	0
	Tractor	0	0	2	0	0	0	2	0
	Tractor + trailer	0	0	5	0	0	0	5	0
Slow modes	Cycle	0	0	0.5	0	0	0	0.5	0
	Cycle rick	0	0	2	0	0	0	2	0
	Animal drawn	0	0	2	0	0	0	2	0
Toll exempted vehicles	Car	0	0	1	0	0	0	1	0
	Bus	0	0	3.7	0	0	0	3.7	0
	Truck	0	0	3.7	0	0	0	3.7	0
Total		519	100		647.1	502	100		633.1

Junction, Number of vehicles are 792 from 09:00 to 10:00 & 613 from 17:00 to 18:00 for LHS & RHS respectively. Number of vehicles entering the Devariya Junction are 115 from 08:00 to 09:00 & leaving the Junction are 113 from 17:00 to 18:00. LHS & RHS traffic count of Devariya Junction are 457 from 18:00 to 19:00 and 404 from 08:00 to 09:00 respectively.

For Vadaliya Sinhan Junction, number of vehicles entering the gate are 115 from 08:00 to 09:00 and leaving the gate are 113 from 17:00 to 18:00. LHS & RHS traffic count of Vadaliya Sinhan Junction are 421 from 10:00 to 11:00 and 407 from 08:00 to 09:00 respectively.

Table 15.21 Vadaliya Sinhan Junction

Types		(IN)				(OUT)			
		ADT	ADT %	PCU	CVPD	ADT	ADT %	PCU	CVPD
Fast passenger vehicles	Two wheeler	5	0.87	0.75	3.75	3	0.46	0.75	2.25
	Auto rick	0	0	2	0	21	3.19	2	42
	Car/jeep/taxi	245	42.46	1	245	243	36.93	1	243
	Trip van (Tata magic)	4	0.69	1	4	10	1.52	1	10
	Mini bus	10	1.73	3.7	37	12	1.82	3.7	44.4
	School bus	3	0.52	3.7	11.1	1	0.15	3.7	3.7
	Gov. bus	0	0	3.7	0	9	1.37	3.7	33.3
	Private bus	31	5.37	3.7	114.7	38	5.78	S3.7	140.6
Fast commercial vehicles	Mini LCV	139	24.09	2	278	165	25.08	2	330
	LCV	25	4.33	2	50	25	3.8	2	50
	2 Axle truck	49	8.49	3.7	181.3	47	7.14	3.7	173.9
	3 Axle truck	44	7.63	3.7	162.8	42	6.38	3.7	155.4
	MAV	9	1.56	5	45	26	3.95	5	130
	Tractor	4	0.69	2	8	4	0.61	2	8
	Tractor + trailer	9	1.56	5	45	12	1.82	5	60
Slow modes	Cycle	0	0	0.5	0	0	0	0.5	0
	Cycle rick	0	0	2	0	0	0	2	0
	Animal drawn	0	0	2	0	0	0	2	0
Toll exempted vehicles	Car	0	0	1	0	0	0	1	0
	Bus	0	0	3.7	0	0	0	3.7	0
	Truck	0	0	3.7	0	0	0	3.7	0
Total		577	100		1185.65	658	100		1426.6

Table 15.22 Devariya Junction

Types		ADT	ADT %	PCU	CVPD
Fast passenger vehicles	Two wheeler	1995	49.13	0.75	1496.25
	Auto rick	54	1.33	2	108
	Car/jeep/taxi	1333	32.82	1	1333
	Trip van (Tata magic)	20	0.49	1	20
	Mini bus	32	0.79	3.7	118.4
	School bus	7	0.17	3.7	25.9
	Gov. bus	0	0	3.7	0
	Private bus	60	1.48	3.7	222
Fast commercial vehicles	Mini LCV	0	0	2	0
	LCV	198	4.88	2	396
	2 Axle truck	57	1.4	3.7	210.9
	3 Axle truck	83	2.04	3.7	307.1
	MAV	105	2.59	5	525
	Tractor	38	0.94	2	76
	Tractor + trailer	0	0	5	0
Slow modes	Cycle	7	0.17	0.5	3.5
	Cycle rick	71	1.75	2	142
	Animal drawn	1	0.02	2	2
Toll exempted vehicles	Car	0	0	1	0
	Bus	0	0	3.7	0
	Truck	0	0	3.7	0
Total		4061	100		4986.05

Table 15.23 Vadinar Junction

Types		LHS				RHS			
		ADT	ADT %	PCU	CVPD	ADT	ADT %	PCU	CVPD
Fast passenger vehicles	Two wheeler	1365.67	22.30	0.75	1024.25	1608.00	24.72	0.75	1206.00
	Auto rick	193.00	3.15	2.00	386.00	230.67	3.55	2.00	461.33
	Car/ jeep/ taxi	2243.00	36.63	1.00	2243.00	2196.00	33.76	1.00	2196.00
	Trip van (Tata magic)	677.67	11.07	1.00	677.67	844.33	12.98	1.00	844.33
	Mini bus	51.00	0.83	3.70	188.70	28.33	0.44	3.70	104.83
	School bus	9.00	0.15	3.70	33.30	8.00	0.12	3.70	29.60
	Gov. bus	19.67	0.32	3.70	72.77	12.33	0.19	3.70	45.63
	Private bus	285.33	4.66	3.70	1055.73	282.67	4.35	3.70	1045.87
Fast commercial vehicles	Mini LCV	178.33	2.91	2.00	356.67	157.33	2.42	2.00	314.67
	LCV	280.67	4.58	2.00	561.33	270.33	4.16	2.00	540.67
	2 Axle truck	138.67	2.26	3.70	513.07	135.33	2.08	3.70	500.73
	3 Axle truck	276.67	4.52	3.70	1023.67	249.67	3.84	3.70	923.77
	MAV	394.00	6.43	5.00	1970.00	468.00	7.20	5.00	2340.00
	Tractor	0.67	0.01	2.00	1.33	2.00	0.03	2.00	4.00
	Tractor + trailer	5.00	0.08	5.00	25.00	10.00	0.15	5.00	50.00
Slow modes	Cycle	5.33	0.09	0.50	2.67	1.00	0.02	0.50	0.50
	Cycle rick	0.00	0.00	2.00	0.00	0.00	0.00	2.00	0.00
	Animal drawn	0.00	0.00	2.00	0.00	0.00	0.00	2.00	0.00
Toll exempted vehicles	Car	0.00	0.00	1.00	0.00	0.00	0.00	1.00	0.00
	Bus	0.00	0.00	3.70	0.00	0.00	0.00	3.70	0.00
	Truck	0.00	0.00	3.70	0.00	0.00	0.00	3.70	0.00
Total		6123.67	100.00		10,135.15	6504.00	100.00		10,607.93

Table 15.24 Mithoi Junction

Types		LHS				RHS			
		ADT	ADT %	PCU	CVPD	ADT	ADT %	PCU	CVPD
Fast passenger vehicles	Two wheeler	1367.67	22.32	0.75	1025.75	1246.00	21.22	0.75	934.50
	Auto rick	192.33	3.14	2.00	384.67	189.33	3.22	2.00	378.67
	Car / jeep/ taxi	2242.33	36.65	1.00	2242.33	2232.00	38.01	1.00	2232.00
	Trip van (Tata magic)	677.67	11.06	1.00	677.67	494.67	8.42	1.00	494.67
	Mini bus	53.67	0.85	3.70	198.57	26.33	0.45	3.70	97.43
	School bus	8.67	0.15	3.70	32.07	5.67	0.10	3.70	20.97
	Gov. Bus	17.00	0.29	3.70	62.90	44.00	0.75	3.70	162.80
	Private bus	284.67	4.66	3.70	1053.27	258.67	4.40	3.70	957.07
Fast commercial vehicles	Mini LCV	175.67	2.88	2.00	351.33	371.00	6.32	2.00	742.00
	LCV	274.00	4.56	2.00	548.00	224.00	3.81	2.00	448.00
	2 Axle truck	142.67	2.27	3.70	527.87	215.33	3.67	3.70	796.73
	3 Axle truck	273.33	4.53	3.70	1011.33	338.67	5.77	3.70	1253.07
	Mav	390.33	6.44	5.00	1951.67	209.33	3.56	5.00	1046.67
	Tractor	1.00	0.02	2.00	2.00	6.33	0.11	2.00	12.67
	Tractor + trailer	6.00	0.10	5.00	30.00	11.00	0.19	5.00	55.00
Slow modes	Cycle	5.00	0.10	0.50	2.50	0.00	0.00	0.50	0.00
	Cycle rick	0.00	0.00	2.00	0.00	0.00	0.00	2.00	0.00
	Animal drawn	0.00	0.00	2.00	0.00	0.00	0.00	2.00	0.00
Toll exempted vehicles	Car	0.00	0.00	1.00	0.00	0.00	0.00	1.00	0.00
	Bus	0.00	0.00	3.70	0.00	0.00	0.00	3.70	0.00
	Truck	0.00	0.00	3.70	0.00	0.00	0.00	3.70	0.00
Total		6112.00	100.00		10,101.92	5872.33	100.00		9632.23

Table 15.25 Mandha Junction

Types		LHS				RHS			
		ADT	ADT %	PCU	CVPD	ADT	ADT %	PCU	CVPD
Fast passenger vehicles	Two wheeler	2056.33	29.18	0.75	1542.25	2085.00	29.49	0.75	1563.75
	Auto rick	256.00	3.63	2.00	512.00	222.00	3.14	2.00	444.00
	Car/ jeep/ taxi	2146.67	30.46	1.00	2146.67	2146.67	30.36	1.00	2146.67
	Trip Van (Tata magic)	859.33	12.19	1.00	859.33	829.00	11.73	1.00	829.00
	Mini bus	27.67	0.39	3.70	102.37	48.00	0.68	3.70	177.60
	School bus	6.00	0.09	3.70	22.20	16.67	0.24	3.70	61.67
	Gov. bus	23.33	0.33	3.70	86.33	71.33	1.01	3.70	263.93
	Private bus	238.33	3.38	3.70	881.83	211.33	2.99	3.70	781.93
Fast commercial vehicles	Mini LCV	218.00	3.09	2.00	436.00	194.33	2.75	2.00	388.67
	LCV	341.67	4.85	2.00	683.33	351.00	4.96	2.00	702.00
	2 Axle truck	142.33	2.02	3.70	526.63	188.33	2.66	3.70	696.83
	3 Axle truck	293.33	4.16	3.70	1085.33	328.67	4.65	3.70	1216.07
	MAV	402.33	5.71	5.00	2011.67	340.00	4.81	5.00	1700.00
	Tractor	15.33	0.22	2.00	30.67	8.67	0.12	2.00	17.33
	Tractor + trailer	21.33	0.30	5.00	106.67	28.67	0.41	5.00	143.33
Slow modes	Cycle	0.00	0.00	0.50	0.00	0.00	0.00	0.50	0.00
	Cycle rick	0.00	0.00	2.00	0.00	0.00	0.00	2.00	0.00
	Animal drawn	0.00	0.00	2.00	0.00	0.00	0.00	2.00	0.00
Toll exempted vehicles	Car	0.00	0.00	1.00	0.00	0.00	0.00	1.00	0.00
	Bus	0.00	0.00	3.70	0.00	0.00	0.00	3.70	0.00
	Truck	0.00	0.00	3.70	0.00	0.00	0.00	3.70	0.00
Total		7048.00	100.00		11,033.28	7069.67	100.00		11,132.78

Table 15.26 Vadaliya Sinhan Junction

Types		LHS				RHS			
		ADT	ADT %	PCU	CVPD	ADT	ADT %	PCU	CVPD
Fast passenger vehicles	Two wheeler	1279.00	22.77	0.75	959.25	1201.33	25.12	0.75	901.00
	Auto rick	193.33	3.44	2.00	386.67	167.33	3.50	2.00	334.67
	Car/jeep/taxi	1887.00	33.60	1.00	1887.00	1667.00	34.86	1.00	1667.00
	Trip van (Tata magic)	679.00	12.09	1.00	679.00	483.33	10.11	1.00	483.33
	Mini bus	46.67	0.83	3.70	172.67	14.33	0.30	3.70	53.03
	School bus	9.00	0.16	3.70	33.30	5.00	0.10	3.70	18.50
	Gov. bus	17.33	0.31	3.70	64.13	35.00	0.73	3.70	129.50
	Private bus	263.00	4.68	3.70	973.10	180.67	3.78	3.70	668.47
Fast commercial vehicles	Mini LCV	149.00	2.65	2.00	298.00	178.33	3.73	2.00	356.67
	LCV	281.00	5.00	2.00	562.00	198.33	4.15	2.00	396.67
	2 Axle truck	136.33	2.43	3.70	504.43	167.67	3.51	3.70	620.37
	3 Axle truck	274.33	4.88	3.70	1015.03	296.33	6.20	3.70	1096.43
	MAV	389.67	6.94	5.00	1948.33	185.00	3.87	5.00	925.00
	Tractor	1.00	0.02	2.00	2.00	2.00	0.04	2.00	4.00
	Tractor + trailer	5.67	0.10	5.00	28.33	0.67	0.01	5.00	3.33
Slow modes	Cycle	5.33	0.09	0.50	2.67	0.00	0.00	0.50	0.00
	Cycle rick	0.00	0.00	2.00	0.00	0.00	0.00	2.00	0.00
	Animal drawn	0.00	0.00	2.00	0.00	0.00	0.00	2.00	0.00
Toll exempted vehicles	Car	0.00	0.00	1.00	0.00	0.00	0.00	1.00	0.00
	Bus	0.00	0.00	3.70	0.00	0.00	0.00	3.70	0.00
	Truck	0.00	0.00	3.70	0.00	0.00	0.00	3.70	0.00
Total		5616.67	100.00		9515.92	4782.33	100.00		7657.97

Table 15.27 Devariya

Types		LHS				RHS			
		ADT	ADT %	PCU	CVPD	ADT	ADT %	PCU	CVPD
Fast passenger vehicles	Two wheeler	1311.00	24.42	0.75	983.25	1203.00	25.17	0.75	902.25
	Auto rick	190.00	3.54	2.00	380.00	167.00	3.49	2.00	334.00
	Car/ jeep/taxi	1935.67	36.05	1.00	1935.67	1663.67	34.81	1.00	1663.67
	Trip van (Tata magic)	466.33	8.69	1.00	466.33	484.00	10.13	1.00	484.00
	Mini bus	15.67	0.29	3.70	57.97	15.67	0.33	3.70	57.97
	School bus	8.00	0.15	3.70	29.60	5.00	0.10	3.70	18.50
	Gov. bus	35.00	0.65	3.70	129.50	35.33	0.74	3.70	130.73
	Private bus	186.33	3.47	3.70	689.43	179.67	3.76	3.70	664.77
Fast commercial vehicles	Mini LCV	273.00	5.09	2.00	546.00	177.00	3.70	2.00	354.00
	LCV	194.67	3.63	2.00	389.33	197.33	4.13	2.00	394.67
	2 Axle truck	223.67	4.17	3.70	827.57	168.00	3.51	3.70	621.60
	3 Axle truck	292.67	5.45	3.70	1082.87	299.67	6.27	3.70	1108.77
	MAV	224.67	4.18	5.00	1123.33	181.33	3.79	5.00	906.67
	Tractor	8.33	0.16	2.00	16.67	2.00	0.04	2.00	4.00
	Tractor + trailer	3.67	0.07	5.00	18.33	1.00	0.02	5.00	5.00
Slow modes	Cycle	0.00	0.00	0.50	0.00	0.00	0.00	0.50	0.00
	Cycle rick	0.00	0.00	2.00	0.00	0.00	0.00	2.00	0.00
	Animal drawn	0.00	0.00	2.00	0.00	0.00	0.00	2.00	0.00
Toll exempted vehicles	Car	0.00	0.00	1.00	0.00	0.00	0.00	1.00	0.00
	Bus	0.00	0.00	3.70	0.00	0.00	0.00	3.70	0.00
	Truck	0.00	0.00	3.70	0.00	0.00	0.00	3.70	0.00
Total		5368.67	100.00		8675.85	4779.67	100.00		7650.58

Table 15.28 Average speed and 85th percentile speed at particular locations

Sr. no	Direction	Time	Average speed (Kmph)	85th percentile speed (Kmph)
1	Vadinar Jn. towards Dwarka	09:20 to 09:50	44.0	44.2
2	Vadinar Jn. towards Jamnagar	13:40 to 14:10	38.7	39.1
3	Vadinar Jn. towards Vadinar	14:20 to 14:50	28.5	29.8
4	Mithoi Jn. towards Dwarka	10:15 to 10:45	52.7	53.0
5	Mithoi Jn. towards Jamnagar	13:40 to 14:10	59.3	59.7
6	Mithoi Jn. towards Mithoi	13:40 to 14:10	28.4	28.9
7	Mandha Junction towards Dwarka	13:40 to 14:10	51.3	51.2
8	Mandha Junction towards Jamnagar	13:40 to 14:10	46.9	47.7
9	Vadaliya Sinhan Junction towards Dwarka	13:40 to 14:10	53.4	53.8
10	Vadaliya Sinhan Junction towards Jamnagar	13:40 to 14:10	52.9	53.3
11	Devariya Jn. towards Dwarka	13:40 to 14:10	46.4	46.8
12	Devariya Jn. towards Jamnagar	13:40 to 14:10	54.8	54.4
13	Devariya Jn. towards Devariya	13:40 to 14:10	29.7	30.2
14	Essar main gate towards Dwarka	13:40 to 14:10	63.7	64.6
15	Essar main gate towards Jamnagar	13:40 to 14:10	64.0	64.4
16	Essar material gate towards Dwarka	13:40 to 14:10	46.7	47.8
17	Essar material gate towards Jamnagar	13:40 to 14:10	50.4	51.2
18	Vadaliya Sinhan towards Dwarka	13:40 to 14:10	52.0	53.2
19	Vadaliya Sinhan towards Jamnagar	13:40 to 14:10	51.5	52.3

15.8 Conclusion

A major problem that most Indian metropolises are facing is the country's growing urbanization. There is a rise in the number of vehicles on highways as a consequence of the migration of numerous individuals from towns to the countryside. As an outcome, there is traffic congestion on metropolitan roadways due to the increasing development of vehicles and the lack of corresponding upgrades to infrastructure. Reduced use of public transportation is another aspect of the traffic problem.

References

Bonela SR, Raghuram Kadali B (2022) Review of traffic safety evaluation at T-intersections using surrogate safety measures in developing countries context. Int Assoc Traffic Saf Sci 46:307–321

Hao N, Feng Y, Zhang K, Tian G, Zhang L, Jia H (2017) Evaluation of traffic congestion degree: an integrated approach. Int J Distrib Sens Netw 13(7)

Huang Y-S, Weng Y-S, Wu W, Chen B-Y (2016) Control strategies for solving the problem of traffic congestion. IET Intell Transp Syst 10(10):642–648. ISSN 1751-956X

Sunil Kumar V, Ranjitha J (2013) Improvement of traffic operations in congested signalized intersections: a case study in Bangalore City. Int J Eng Res Technol (IJERT) 2(7). ISSN: 2278-0181

Ladhi SC, Ghodmare SD, Sayankar BB (2018) Sustainability assessment considering traffic congestion parameters in Nagpur City. 5(4). e-ISSN: 2395-0056, p-ISSN: 2395-0072

Mankar PU, Khode BV (2016) Capacity estimation of urban roads under mixed traffic condition. Int Res J Eng Technol (IRJET) 03(04). e-ISSN: 2395-0056, p-ISSN: 2395-0072

Mankar PU, Khode BV (2016) Comparative study of methods used for a capacity estimation of road. IJSTE Int J Sci Technol Eng 2(9). ISSN (online): 2349-784X

Marve SR, Bhorkar MP (2016) Analysis of traffic congestion of Hingna Region in Nagpur City. Int Res J Eng Technol (IRJET) 3(4). e-ISSN: 2395-0056, p-ISSN: 2395-0072

Marve SR, Bhorkar MP (2020) Traffic congestion study for a developing area of Nagpur City. Research Gate Publication

Sayre S, Wadhai P, Khode BV (2021) Study of traffic management on congested intersections, by redesign of signal. Int J Sci Res Sci Technol 8(2):301–311. p-ISSN: 2395-6011, e-ISSN: 2395-602X

Chapter 16
Study on PEB Structure and Conventional Steel Building: A Review

Abhishek R. Kuralkar and Isha P. Khedikar

Abstract The pre-engineering Building framework is a cutting-edge in steel structure design that produces affordable, sustainable, and environmentally friendly buildings. The pre-engineered Building conception is a revolutionary approach for single-story manufacturing structures. This approach's versatility stems from its lightweight and cost-effective construction as well as from its excellent prefabrication and predesigning. Compared to the conventional steel building (CSB) design, which calls for structures with roof trusses, this idea provides a number of advantages. The PEB and CSB concepts are the primary subjects of this essay. With the aid of Staad Pro, a warehouse structure is designed as both a CSB and PEB to complete the study. AISC 360:10 designs and analyses the PEB, while IS 800:2007 (LSM) designs and analyses the CSB.

Keywords Pre-engineered building · Prefabricated structure · Steel building systems · Economical frame · Cantilever frame · Staad-pro

16.1 Introduction

Pre-engineered building is ones that are designed, entirely manufactured in the industrial unit, and then conveyed to the construction location in a fully demolished state. All of the building's components are put together and constructed on-site using nut-bolts, which shortens the construction time. The building components were either produced to order or ready-made, but the drawings were already created. Taking use of industrial methods that allow for the economically viable quantity manufacture of components, these constructions were preplanned, or pre-engineered to have specified widths, span, bays, and heights. They also used details for flashing, roofing, gutters, windows, doors, and cladding, among other things.

A. R. Kuralkar (✉) · I. P. Khedikar
G.H. Raisoni College of Engineering, Nagpur, India
e-mail: abhishek.kuralkar.mtechstr@ghrce.raisoni.net

B. V. Khode et al. (eds.), *Innovations in Technologies: Pioneering Sustainable Infrastructure for a Resilient Future*, Sustainable Civil Infrastructures,
https://doi.org/10.1007/978-3-031-82133-2_16

16.2 Pre-engineering Buildings

The steel building systems used in pre-engineered buildings are prefabricated and predesigned. As the title suggests, this idea entails pre-engineering structural elements use a pre-established register of materials for construction and production processes it may effectively satiate a range of structural and artistic plan criteria. The PEB idea's fundamental idea is to only provide the section at site in accordance with the needs there. The sections may differ along their entire length in accord with the bending moment diagram.

16.3 Framing Mechanism

16.3.1 Conventional Steel Framework

Hot rolled sections (beams and columns) that are produced in mills are utilized in typical steel constructions. The highest possible internal tension in each member determines the size of that member. Due to the consistent depth of a hot rolled section, numerous portions of member—shown by the shaded area—are in locations with less stress within and exceed the specifications of the design.

16.3.2 Pre-engineered Steel Framework

Pre-engineered frameworks are constructed from a large stock of standard steel plates that are kept on hand by the PEB producer. PEB frames frequently include flanges and webs of varying thicknesses running along each individual component, and they are typically tapered. By matching the internal stress (bending moment) diagram's shape, the frame optimizes the use of materials and lowers the structure's overall weight. The secondary frame consists of wall roof purlins and girts in the shape of a "Z." Compared to the typical hot rolled "I" or "C" shaped segments seen in traditional steel constructions, they are lightweight (Fig. 16.1).

16.4 Elements of Pre-engineering Building

16.4.1 Main Elements

The principal elements of a pre-engineered building are as follows:

- Minor framework or Purlins, girts etc.

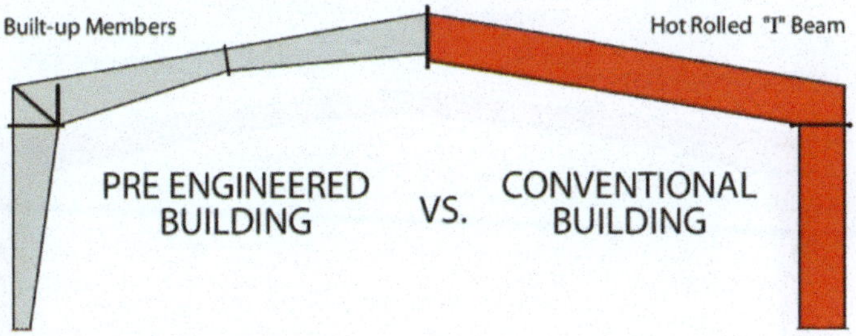

Fig. 16.1 Conventional and pre-engineering building. (*Source* https://www.epack.in/wp-content/uploads/2018/08/Peb-comparison.webp)

- Bracing structure
- Accessories like Turbo vents, Ridge Vents, Skylights etc. Mezzanine system
- Insulations
- Roof & Wall Sheeting
- Attachments like canopies, fascia etc.
- Windows, Doors, Ventilators Crane system (Figs. 16.2, 16.3 and 16.4).

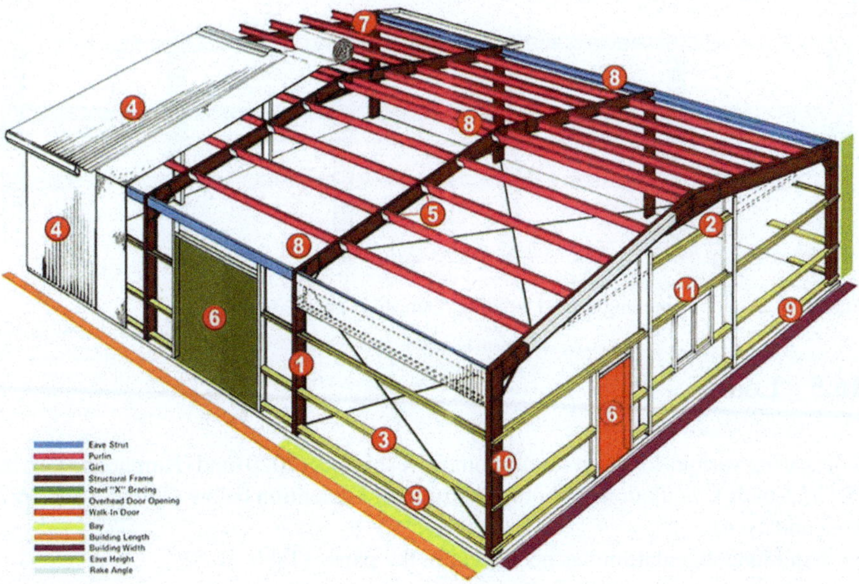

Fig. 16.2 Element of PEB building. (*Source* http://fj.chinatwsteelstructure.com/Content/upload/2017287246/20171213105120392797O.jpg)

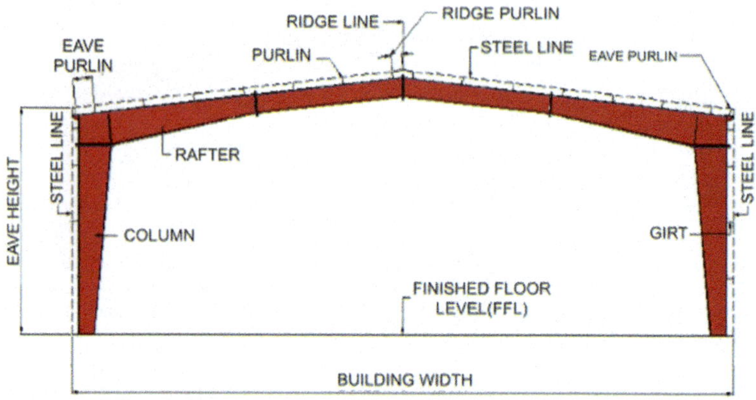

Fig. 16.3 New era PEB parameters. (*Source* https://i0.wp.com/bsiinternational.net/wp-content/uploads/2021/08/bsi-prod-struc-steel.jpg?w=1138&ssl=1)

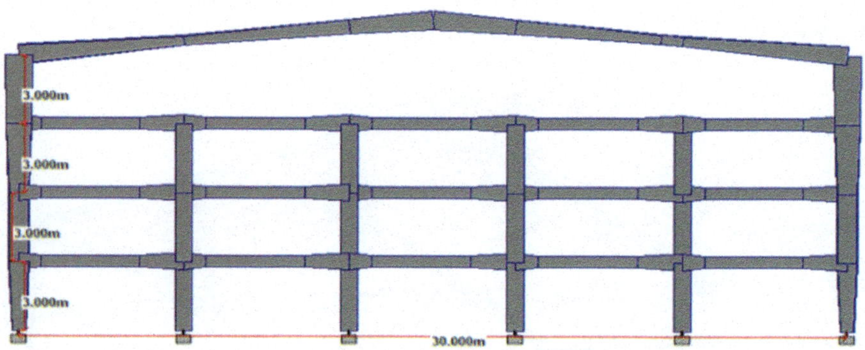

Fig. 16.4 Section of pre-engineered building (Qureshi et al. 2020)

16.5 Loads

Comparing various loads on frame from Qureshi et al. (2020) and (Kumar and Grover 2015). Loads Calculations of both the structure as mention below (Tables 16.1, 16.2, 16.3 and 16.4).

Wind load calculation has been used in the paper (Table 16.5).

Table 16.1 Load estimate on both the structure (Qureshi et al. 2020)

Loads	PEB frame	CSB frame	
	Rafter load	Load according to IS 875	
	According 875–1987	As intermediate panel point	At end panel point
DL	1.69 KN/m	2.58 KN	1.287 KN
LL	5.63 KN/m	7.57 KN	3.784 KN
	Load at mezzanine Beam	Load at mezzanine beam	
DL	23.44 KN/m	23.44 KN/m	23.44 KN/m
LL	22.51 KN/m	22.51 KN/m	22.49 KN/m

Table 16.2 Load estimate on both the structure (Kumar and Grover 2015)

Load-type	Value
Seismic load	Industrial building design is typically not governed by seismic load, hence it is not taken into consideration
Dead load	
Self-weight	Software created using model data
Top sheeting	60.0 N/m^2
Fixture	25.0 N/m^2
Facilities	100.0 N/m^2
Purlins	60.0 N/m^2
Live load	750.0 N/m^2
Wind-load	Wind load coefficients according to IS 875 (Part-3)
Load combinations	Combinations of loads according to IS 800:2007

Table 16.3 Wind load estimate on CSB structure (Qureshi et al. 2020)

Case	Column		CSB panel			
	Left (KN)	Right (KN)	Windward (KN)		Lee ward (KN)	
WL_{-1}	5.711	−0.322	−7.287	−3.637	−1.967	−0.978
WL_{-2}	−0.317	5.709	−1.967	−0.978	−7.287	−3.637
WL_{-3}	3.168	−2.848	−11.227	−5.607	−5.909	−2.957
WL_{-4}	−2.846	3.168	−5.907	−2.959	−11.229	−5.607
WL_{-5}	−1.901	−1.901	−5.907	−2.957	−2.257	−1.129
WL_{-6}	−1.901	−1.901	−2.257	−1.126	−5.907	−2.959
WL_{-7}	4.437	4.437	−9.849	−4.917	−6.207	−3.101
WL_{-8}	4.437	4.437	−6.204	−3.103	−9.846	−4.918

Table 16.4 Wind-load estimate on PEB structure (Qureshi et al. 2020)

Case	Column		PEB rafter	
	Left (KN)	Right (KN)	Windward (KN)	Lee ward (KN)
WL_{-1}	5.711	−0.308	−4.687	−1.267
WL_{-2}	−0.318	5.708	−1.267	−4.689
WL_{-3}	3.167	−2.848	−1.137	−3.807
WL_{-4}	−2.846	3.167	−3.809	−1.137
WL_{-5}	−1.901	−1.901	−3.809	−1.456
WL_{-6}	−1.901	−1.901	−1.456	−3.809
WL_{-7}	4.437	4.437	−6.348	−3.989
WL_{-8}	4.437	4.437	−3.989	−6.348

Table 16.5 Estimate of Rafter (Qureshi et al. 2020)

Sr. no	Descriptions	According (IS 800:2007) in CSB	According to (IS 800:2007) in
1	Distance (m)	30.00	30.00
2	Shift maximum (mm)	30.064	104.077
3	Axial force (K-N)	967.404	459.153
4	Shear force (sy) (K-N)	134.876	119.441
5	Bending moment (Mz)	281.984	521.231
6	Steel amount (KN)	84.794	19.836

16.6 Results

Here discussing structural analysis and design, particularly regarding the effects of different loads on steel consumption in a frame structure. Variations in loads such as wind load, dead load, and live load can indeed influence the design and material requirements of a structure. When analyzing a structure, engineers consider various loads that act on it.

Comparing results of research paper it's significantly found that, consumptions of steel in pre-engineering are 76.60% less in (see Fig. 16.5) and 40.23% in (see Fig. 16.6) as compared to Conventional steel building.

16.7 Literature Review

The pre-engineered building (PEB) concept represents a modern approach to constructing single-storey industrial buildings, characterized by its emphasis on pre-designing and prefabrication. Its lightweight nature and cost-effectiveness make it

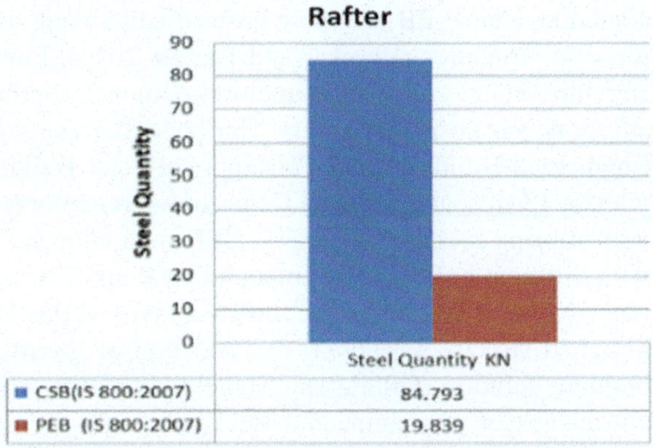

Fig. 16.5 Comparing steel consumption of CSB and PEB (Qureshi et al. 2020)

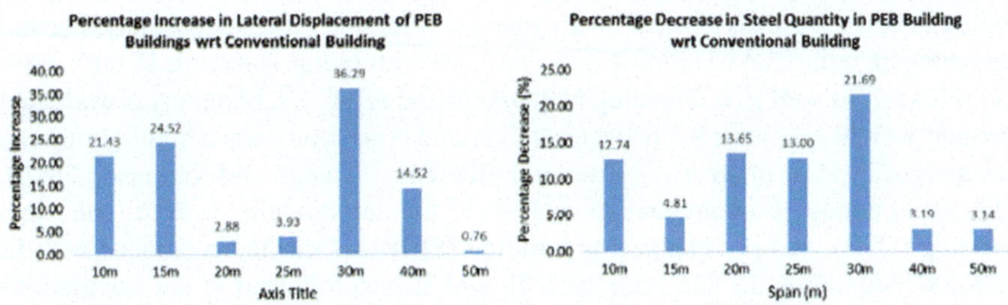

Fig. 16.6 Comparing steel consumption of CSB and PEB (Kumar and Grover 2015)

a versatile choice compared in contrast to the traditional steel building (CSB) idea, which relies on roof trusses. These paper aims to conduct a relative analysis between the PEB and C-SB concepts, highlighting their respective advantages and drawbacks (Wankhade and Pajgade 2014). Less material was used in the PEB instance than in the CSB case. This case study compares pre-engineered and conventional industrial buildings, focusing on bending moments using Staad Pro v.8 and Indian standard codes. It analyzes components like purlins, girts, and bracings, considering loads for Vadodara, Gujarat. Various truss systems were explored for roofing efficiency. The comparison assesses economy and time-saving (Sharma 2017). The paper highlights the benefits of pre-engineered buildings over conventional ones, emphasizing cost-effectiveness, time-saving, future potential, versatility, and economic advantages, particularly relevant in developing nations like India. Focused on industrial sheds, it draws from reviews and case studies, showcasing experimental and analytical research. Findings demonstrate the economical, energy-efficient, and adaptable nature of pre-engineered structures (Titiksh 2015). Pre-engineered buildings are prefabricated steel structures with predetermined designs. This study focuses on a cost-effective tool for optimizing steel cross-sections. It explores different concepts and applications of pre-engineered buildings, detailing structural members and components such as primary, secondary, and sheeting elements. The research concludes that P-EB are more cost-effective than conventional steel buildings for low-rise structures (Bhojkar and Darade 2014). For engineers seeking large-span structures at low costs, steel structures remain the optimal choice due to their inherent advantages in design flexibility. This paper delves into the design aspects of both Pre Engineered Buildings and Conventional Steel Buildings for an industrial building with an 18-m span near New Delhi, India. It explores fully stressed designs of PEB with varying member thickness, CSB with conventional steel members, and CSB with different hollow and compound sections. Through the study of five cases, the objective is to determine the most economical design. A comparison between the designed structures leads to the selection of the most suitable and cost-effective building solution (Gilbile and Mane 2020). This paper emphasizes the significant advantages of pre-engineered steel building systems for single-storey structures, offering a practical and efficient alternative to conventional buildings. By adopting PEB over Conventional Steel Building (CSB) designs, numerous benefits such as economy and simplified fabrication are realized. The study focuses on analyzing and designing an industrial warehouse in accordance with Indian standards IS 800-1984, IS 800-2007, as well as referencing MBMA-96 and AISC-89. Economy is evaluated through weight comparisons between Indian and American codes, highlighting the advantages of PEB in terms of structural efficiency (Thakre and Vairagade 2016). This paper conducts a comparative analysis of tubular structure, conventional steel building (CSB), and pre-engineered building (PEB) designs in accordance with IS 800-2007 standards. IS 875 sections I, II, and III are followed in the calculations for dead load, live load, and wind load, respectively. The focus lies on optimizing sections to meet specific requirements, offering advantages over CSB designs with roof trusses. Design and analysis utilize STAAD Pro V8i Software (Shujat and Desai 2018). This study addresses the need to introduce web-tapered portal frames to Indian construction practices, citing potential financial benefits and local supply challenges with hot-rolled members. The research aims to evaluate the safety and viability of this construction method compared to conventional techniques. It involves analyzing and designing web-tapered portal frames buildings and conventional steel buildings for spans of 20 m using Staad Pro v8i software. The web-tapered steel portal frames are particularly suitable for construction in remote and hilly areas (Lamkhade et al. 2020). This study focuses on compares the steel usage in industrial buildings of various spans using traditional hot rolled sections versus Pre-Engineered Building tapered sections. By analyzing different spans (10–50 m), it aims to establish a relationship between building parameters and steel reduction in primary frames. The analysis is conducted according to Indian Standards (IS 800-2007) using STAAD PRO Connect edition (Kumar and Grover 2015). This paper compares Pre-engineered Building and Conventional Steel Building systems for industrial warehouses. PEB offers economic, eco-friendly, and sustainable structures compared to CSB, which is time-consuming and costly due to uniform hot-rolled sections. PEB utilizes built-up sections and bolted connections for efficiency, especially for large spans and column-free spaces. The study analyzes a G + 3 warehouse using IS 800-2007 standards and STAAD-pro software, focusing on tonnage economy and comparing hot-rolled sections in CSB with cold-formed purlins in PEB (Qureshi et al. 2020).

16.8 Conclusion

Based on the previous research paper study, its shows that the frame structure plays a crucial part in defining the performance and characteristics of the overall building system. Here's a concise summary of the frame-related conclusions.

The comparison amid Pre-engineered Building and Conventional Steel Building structures reveals several advantages of the PEB system. According to IS 800:2007 standards, the PEB structure exhibits greater displacement, lower support reaction, and reduced, bending moment shear force, and axial compared to C-SB. Additionally, PEB structures are significantly lighter, with a weight reduction of 64% compared to CSB structures, leading to better wind resistance. Cold-formed purlins in PEB are also 32.5% lighter than hot-rolled purlins used in CSB.

References

Bhojkar M, Darade M (2014) Comparison of pre engineering building and steel building with cost and time effectiveness. Int J Innov Sci Eng Technol (IJISET) 1(10)

Gilbile MJ, Mane SS (2020) A review on comparative study on the structural analysis and design of pre-engineered building [PEB] with conventional steel building [CSB]. Int Res J Eng Technol (IRJET) 9

Kumar N, Grover KS (2015) A comparative study of conventional steel building and pre-engineered building structure using Staad Pro. Int J Res Appl Sci Eng Technol 25–28

Lamkhade DT, Wayal GS, Sharma MR (2020) Analysis and design of web tapered beams and columns using finite element analysis in portal frame buildings. Int J Eng Res Technol 9(8):757–761

Qureshi HA, Dabhekar KR, Shahakar A, Khedikar IP (2020) Comparative analysis of pre engineered and conventional steel building. JETIR 07:370–377

Sharma H (2017) A comparative study on analysis & design of pre-engineered & conventional industrial building. Int J Innov Res Sci Technol 3(10):91–99

Shujat QS, Desai R (2018) Comparative study of design of industrial warehouse using CSB, PEB and tubular sections. Int J Eng Res Appl. ISSN: 2248-9622

Thakre V, Vairagade L (2016) Analysis and cost comparative study of conventional industrial building with PEB structure. J Inform Knowl Res Civil Eng. ISSN: 0975-6744

Titiksh A (2015) Comparative study of conventional steel building and pre-engineered building to be used as an industrial shed. Int J Eng Res Appl 5(11)

Wankhade S, Pajgade P (2014) Review paper on comparison of conventional steel building & pre-engineering building. Int J Res Advent Technol 2(5):271–276

Chapter 17
Experimental and Numerical Investigation on Expansive Soil Treated with Fly Ash Coupled Red Mud

Ravichandra Honnalli, K. E. Abilash, Adarsh S. Chatra, and Muttana S. Balreddy

Abstract Due to rapid urbanization, the infrastructural developments are taking place at faster pace in developing countries like India. As engineers often face complexities while undertaking construction activities when expansive soils are encountered. In the present study, a novel binder is developed to modify the properties of an expansive soil subjected to grain size analysis, standard proctor compaction and unconfined compressive strength (UCS) tests. An optimum fly ash content of 12% for treated soil is reported as maximum UCS. Further, the soil is treated using optimum fly ash content blended with red mud (RM) varying from 5% to 20% at an increment of 5% along with 6 M calcium hydroxide solution as a pore fluid. To assess the behaviour of a strip footing resting on fly ash-red mud (FR) treated soil is analyzed using Plaxis 2D. The present investigation indicates that the novel binder is found effective in improving the engineering properties as well as settlement characteristics and bearing capacity of soil upon treatment.

R. Honnalli (✉)
Assistant Professor, Department of Civil Engineering, Ballari Institute of Technology and Management, Ballari 583104, Karnataka, India
e-mail: ravibitm9@gmail.com

K. E. Abilash
Assistant Professor, Department of Civil Engineering, Sri Siddhartha Institute of Technology, Tumakuru 572105, Karnataka, India
e-mail: abiman9069@gmail.com

A. S. Chatra
Associate Professor, Department of Civil Engineering, Basaveshwar Engineering College, Bagalkot 587102, Karnataka, India
e-mail: adarshchatra@gmail.com

M. S. Balreddy
Associate Professor, Department of Civil Engineering, Siddaganga Institute of Technology, Tumakuru 572103, Karnataka, India
e-mail: msb@sit.ac.in

B. V. Khode et al. (eds.), *Innovations in Technologies: Pioneering Sustainable Infrastructure for a Resilient Future*, Sustainable Civil Infrastructures,
https://doi.org/10.1007/978-3-031-82133-2_17

Keywords Expansive soil · Compaction characteristics · Unconfined compressive strength · Fly ash · Red mud and Plaxis 2D

17.1 Introduction

Expansive soils popularly known as black cotton soils are subjected to volume change with the moisture variation. Due to their high compressibility, non-uniform stress distribution causing severe damages to the foundations of structures and eventually leads to failure. As they possess high base exchange capacity with large specific area associated with complex problems pose challenges to engineers during the execution of any construction activities like bridges, embankments for roads, canals, foundations for towers, power plants and multi storied buildings etc. Numerous instances of failures have been observed around the globe in these soils and hence they are considered to be problematic that makes undesirable to build any structures unless their properties are modified. In this view point, prior to any construction activities suitable ground improvement techniques have to be adopted to treat such soils to ensure the safety of structures. Extensive researches indicate the effective utilization of industrial byproducts such as GGBS (Kumar and Sivapullaiah 2012), fly ash (Zha et al. 2008), mine tailings (Ramesh et al. 2019), rice husk ash (Phanikumar and Nagaraju 2018), cement kiln dust (Salahudeen et al. 2014) and bagasse ash (James and Pandian 2018) to treat expansive soils. As the direct disposal of such byproducts would pose serious threat resulting in pollution of ground water table; also creates unfavorable environment for human life as well for vegetation growth too. In practical applications, treating the existing soil with industrial by products proves to be cost effective as well as sustainable. At the same time, it provides the feasible option of disposal which avoids environmental pollution further.

Red mud as a byproduct obtained from aluminium industry generated during the extraction of aluminium from bauxite ore is highly alkalinic pose challenge for its safe disposal. In the past, there are instances of environmental pollution caused by its direct disposal. Hence, there is a scope to dispose it safely and one such effective way is bulk utilization for various engineering applications. The bulk utilization of RM includes manufacturing of cement, hybrid composites and cement replacement in concrete (Tsakiridis et al. 2004; Vigneshwaran et al. 2020; Venkatesh et al. 2021). Past researches indicate the red mud is found effective to treat expansive soil with lime (Aswathy et al. 2016), fly ash (Sridevi et al. 2016), as a geopolymerized material with GGBS (Singh 2018; Alam et al. 2019), as a geomaterial for embankments blended with fly ash and gypsum (Sarath Chandra and Krishnaiah 2022) and bio enzyme (Panda et al. 2017). It is also utilized to treat silty subgrade (Ma et al. 2022) and is suitable as a clay liner (Kalkan 2006) as well. Based on the above engineering applications, the red mud RM can be utilized to treat expansive soils effectively.

Fly ash (FA) as a useful byproduct obtained from thermal power plants has been utilized in various geotechnical applications which inturn, found successful in stabilizing the expansive soils. Based on the previous studies, lime blended with fly ash (Renjith et al. 2021), bagasse ash (Pradeep and Mayakrishnan 2023) and cement (Chethan and Ravi Shankar 2021) were much more effective for strength gain in expansive soils. Since, FA of class F possess lesser pozzolanic properties than that of class C and hence it needs additional chemical additives for the activation of alumina and silica. Past studies suggested that FA is also found effective with industrial products like GGBS (Sharma and Sivapullaiah 2017), rice husk ash (Murmu and Patel 2020), phosphogypsum (Degirmenci et al. 2007) and red mud (Mishra et al. 2019) which contain CaO that induces the pozzolanic reaction necessary for strength gain in clayey soils. This is the clear indication of calcium based admixtures blended with RM were found beneficial in enhancing the strength of expansive soils. From considering the aforementioned benefits, the present investigation is focused on utilizing the mixtures of FA and RM along with calcium hydroxide solution as a pore fluid to stabilize expansive soil and the studies on the same are rarely explored.

17.2 Materials and Methodology

In the present study, the expansive soil (ES) is procured from Bagalkot district of Karnataka, fly ash (FA) from locally available supplier and red mud (RM) from HINDALCO Pvt Ltd, Belgaum, Karnataka. The geotechnical properties of soil, red mud and chemical composition of fly ash, red mud is shown in Tables 17.1 and 17.3 respectively. The nomenclature of mix proportion of expansive soil stabilized by fly ash alone and fly ash coupled red mud is mentioned in the Table 17.2.

The soil is amended with fly ash at different % such as 3, 6, 9, 12 and 15 with water as pore fluid. The fly ash content is optimized by UCS test conducted for samples

Table 17.1 Geotechnical properties of expansive soil and red mud

Parameters	Values	
	Soil	Red mud
Gravel	6%	-
Sand	12%	12%
Silt	28%	70%
Clay	54%	18%
Soil classification	CH	MI
Specific gravity	2.68	2.86
Liquid limit	64%	41%
Plastic limit	37%	27%
Plasticity index	27%	15%
Maximum dry unit weight	15 kN/m^3	15.7 kN/m^3
Optimum moisture content	25.7%	33%
Unconfined compressive strength	70 kPa	114 kPa

Table 17.2 Nomenclature for mix proportion

Sl No	Nomenclature	Description of mix
1	SF3	ES + 3% FA
2	SF6	ES + 6% FA
3	SF9	ES + 9% FA
4	SF12	ES + 12% FA
5	SF15	ES + 15% FA
6	FR	Optimum Fly ash + Red mud
7	SFR5	ES + OFA + 5% RM
8	SFR10	ES + OFA + 10% RM
9	SFR15	ES + OFA + 15% RM
10	SFR20	ES + OFA + 20% RM

Table 17.3 Chemical composition of red mud and fly ash

Constituent	Composition in %	
	Red mud	Fly ash
Fe_2O_3	42.6	14.2
Al_2O_3	19.6	20.1
SiO_2	15.2	47.9
TiO_2	10.2	1.1
Na_2O	9.4	1.3
CaO	1.3	8.7
Others	1.7	6.7

cured at 7, 14 and 28 days. The soil is further treated by optimum fly ash content blended with different % of red mud from 5 to 20% at a rate of 5% along with $Ca(OH)_2$ solution of 6 M as pore fluid. Accordingly, consistency limits and compaction tests and UCS tests (samples cured at 7 days, 14 days, 28 days) were carried out. The load-settlement characteristics of a strip footing placed over stabilized soil (UCS values corresponding to 28 days) is analyzed by PLAXIS 2D software.

17.3 Results and Discussions

17.3.1 Optimization of Fly Ash Content

The optimized fly ash content (OFA) is obtained through UCS tests and the samples were pre pared corresponding to maximum dry unit weight (MDU) and optimum moisture content (OMC) as obtained in light compaction test.

Table 17.4 Compaction characteristics

% of fly ash	MDU in kN/m^3	OMC in %
0	15	25.7
3	15.5	25.2
6	15.9	24.9
9	16.2	23.4
12	16.5	23.1
15	16	23.2

17.3.1.1 Compaction Characteristics

The MDU and OMC for expansive soil treated with various % of FA from 3 to 15% at an increment of 3% are shown in Table 17.4. The dry unit weight in FA treated soil is enhanced upto 12% and is decreased thereafter. It is attributed to the improved cation exchange capacity and reduced double layer thickness (Mir and Sridharan 2013). While, the OMC is decreased continuously upto 12% and a little increase is seen for 15% of FA. The OMC is reduced due to the flocculation caused with the addition of FA into the soil and thereby the soil can be compacted at lower water content.

17.3.1.2 Unconfined Compressive Strength

Table 17.5 indicated that UCS of FA treated soil is significantly enhanced for all curing periods. The UCS is continuously improved upto 12% of FA and declined thereafter which is considered as optimum. This suggests the right amount of FA which can enhance shear strength by promoting pozzolanic reaction which is the indication of availability of free lime (Balreddy et al. 2020). However, the strength is reduced beyond optimum; as fly ash behaves like unbounded silt particles due to reduced friction and cohesion.

Table 17.5 UCS of fly ash treated soil

% of fly ash	UCS (kPa)		
	7 days	14 days	28 days
0	70		
3	120	148	179
6	170	199	244
9	210	254	288
12	245	280	324
15	220	254	292

Table 17.6 Plasticity characteristics of fly ash treated soil

% of red mud	Consistency limits (%)			Soil classification
	W_L	W_P	I_P	
0	64	37	27	CH
5	58	35	23	CH
10	52	33	19	CH
15	47	31	16	MI
20	45	30	15	MI

17.3.2 *Influence of $Ca(OH)_2$ on Soil Treated with Fly Ash Blended Red Mud*

To observe the influence of $Ca(OH)_2$ solution as a pore fluid; the soil is further treated with optimum fly ash content and RM varied from 5 to 20% at a rate of 5% to determine the consistency limits, compaction characteristics and UCS for different curing periods.

17.3.2.1 Plasticity Characteristics

The intervention of calcium hydroxide solution as a pore fluid with FR mixtures, the consistency limits of FR treated soil (refer Table 17.6) are reduced to a greater extent which is because of suppressed double layer thickness. At the same time, the continuous reduction in consistency limits is observed resulting in the transformation of soil classification from CH (clay of high plasticity) to MI (silt intermediate plasticity). As the clay particles charged negative at surface are balanced by the cations resulted in decreased repulsive forces in the soil matrix. Subsequently, they tend to move each other just gets attached to impart particle agglomeration (Samantasinghar and Singh 2023).

17.3.2.2 Compaction Characteristics

With reference to the Fig. 17.1, the OMC decreased while MDU increased. The soil containing higher clay content exhibits lesser resistance to compactive efforts which is responsible for higher dry unit weight. When a pozzolanic material like FA is blended with RM, silt size particles in treated soil matrix increases and in turn, the soil can be compacted at lower pore fluid content (Jha and Sivapullaiah 2016).

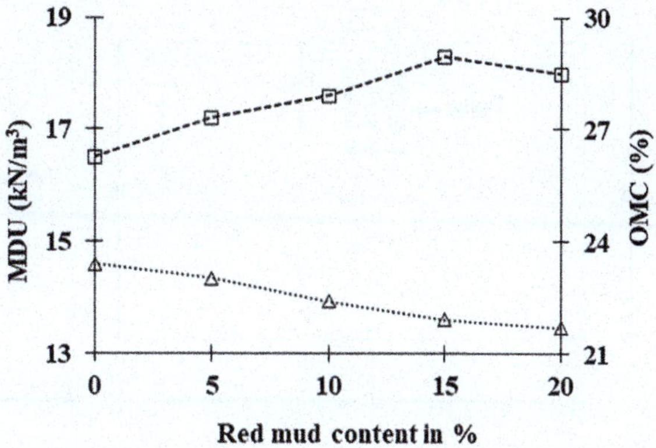

Fig. 17.1 Variation of MDU-OMC of FR treated soil

17.3.2.3 Unconfined Compressive Strength Test

The improvement in UCS of specimens prepared with FR mixtures with calcium hydroxide as a pore fluid is shown in Fig. 17.2. The FR mixtures possesses alkalinic environment with higher pH which is favorable for accelerating the quick pozzolanic reaction (Moghal 2017). With the addition of FR mixtures, the UCS in treated soil is enhanced by 3 to 6 times for different curing periods and maximum UCS is reported for 15% of binder content. Therefore, the optimum red mud content is 15% with an optimum fly ash content of 12%. However, the UCS in soil is gained by 1.5 to 4 times with alone and thus; the strength is improved significantly higher with FR mixtures. It indicates FA blended with RM developed good binding property. While the intervention of $Ca(OH)_2$ solution with FR mixtures improved the shear resistance in treated soil by accelerating the necessary pozzolanic reaction that led to the formation of CASH gel with the affinity of alumina and silica present in FR mixtures (Singh et al. 2018).

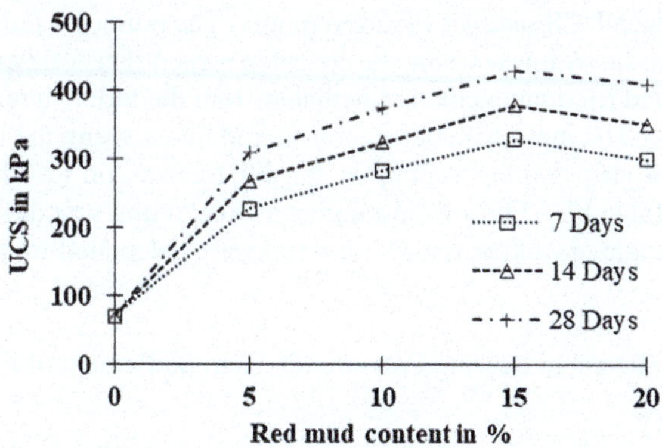

Fig. 17.2 Variation of UCS for FR treated soil

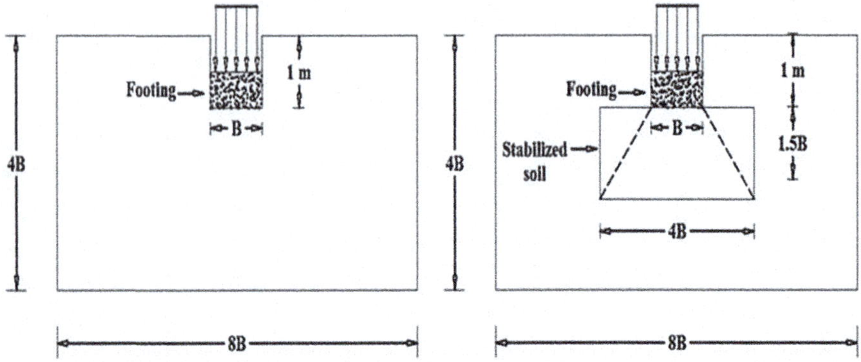

Fig. 17.3 (**Left**) Footing on untreated expansive soil (**Right**) Footing on treated soil with 4B width and 1.5B depth

17.3.2.4 Load-Settlement Characteristics of Strip Footing Resting on Treated Soil

The width of footing is kept as 1, 1.2 and 1.5 m with load intensity varied from 10 to 450 kPa for both the cases are analysed using PLAXIS 2D (Fig. 17.3a, b). The footing is placed at 1 m below ground level whose thickness is taken as 0.5 m and Young's modulus of footing material (concrete) considered as 2.0×10^7 kPa. The thickness of stabilized soil is considered upto significant depth i.e. 1.5B below the footing base and the corresponding width is considered based on the stress distribution which takes place at an angle equal to 45°.

Determination of Input Parameters (E_S, C, Φ and M)

The Young's modulus of soil is obtained as the slope of the stress–strain curve of each UCS samples of different mix. The poisson's ratio is taken as the ratio of lateral strain by linear strain; the cohesion is recorded corresponding to peak stress obtained and frictional angle is calculated from the failure angle obtained from UCS test.

The input parameters considered in assessing the load-settlement characteristics of strip footing resting on the FR treated soil (28 days strength) is shown in the Table 17.7. For soil the parameters like Young's modulus (E_S), cohesion (c), frictional angle (ϕ) and poisson's ratio (μ) were calculated from UCS tests.

Allowable Bearing Pressure (Q_a) and Settlement of Footings

The allowable bearing pressure (Q_a) is obtained from for different L-S curves corresponding to 25 mm settlement; either it can be calculated as ratio of ultimate bearing

Table 17.7 Input parameters of FR treated soil

Mix	MDU in kN/m^3	OMC (%)	E_S (kPa)	C (kPa)	ϕ	μ
ES	16.5	23.4	1640	30	8	0.28
SFR5	17.2	23.9	2240	130	10	0.30
SFR10	17.6	24.2	2890	149	13	0.32
SFR15	18.3	24.8	3560	165	15	0.35

Table 17.8 Allowable bearing pressure of footings

Footing width (m)	Allowable bearing pressure (kPa)			
	ES	SFR5	SFR10	SFR15
1 m	68	79	90	106
1.2 m	50	82	96	115
1.5 m	42	86	102	126

Table 17.9 Settlement of footing

Footing width (m)	Settlement (mm)			
	ES	SFR5	SFR10	SFR15
1 m	300.45	142.22	110.00	78.89
1.2 m	360.82	146.34	113.41	75.61
1.5 m	395.63	155.26	118.42	68.42

pressure (Q_u) to the factor of safety (F). The allowable bearing pressure of footings resting on FR treated soil corresponding to 25 mm settlement is shown in Table 17.8.

The settlement experienced by the footing resting on treated soil is continuously reduced with increased binder content (Table 17.9). Since, the settlement is assessed upto optimum binder content (12% of FA and 15% of RM) as obtained in UCS test conducted on FR treated soils.

The load-settlement (L-S) curves of strip footing with different widths resting on untreated and treated expansive soil are shown in Figs. 17.4 and 17.5. It is also observed that the allowable bearing pressure of footings resting on treated is increased for all footing widths which is evident from the previous study (Mahesh Bhat and Nayak 2021).

17.4 Conclusions

The following conclusions were drawn from the present study

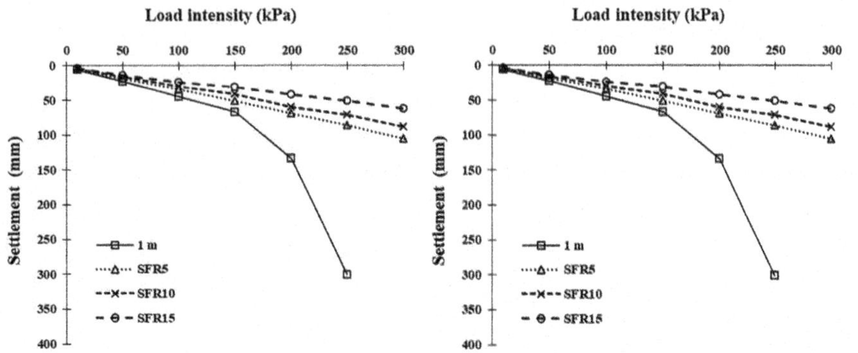

Fig. 17.4 L-S curves of footing on (**a**) untreated soil for different widths (**b**) treated soil for 1 m wide footing

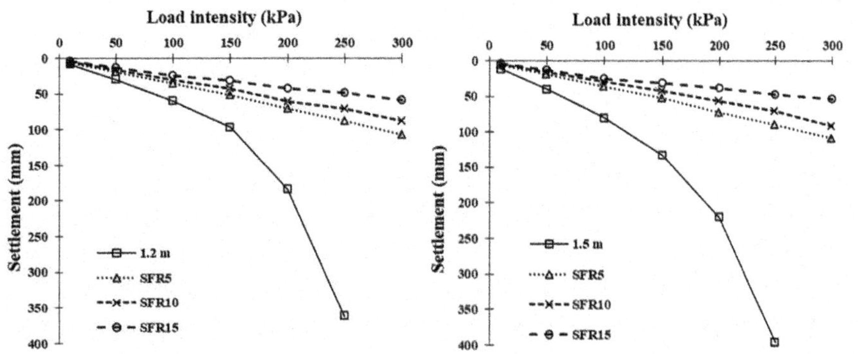

Fig. 17.5 L-S curves of footing on treated soil for (**a**) 1.2 m wide footing (**b**) 1.5 m wide footing

1. The soil treated with FR mixtures and $Ca(OH)_2$ solution has transformed from CH to MI is the clear indication of reduction in plasticity index from 27 to 15%. Such a reduction is attributed to the suppressed diffused double layer thickness.
2. By adding FA upto 12%, the MDU is increased from 15 to 16.5 kN/m^3 and OMC is decreased from 25.7 to 23.1%. While, the MDU is increased (16.5 to 18.3 kN/m^3) with decrease in OMC for soil treated with FR mixtures and $Ca(OH)_2$ solution upto 15% of RM with optimum FA content of 12%.
3. The UCS of soil treated with FA alone enhanced from 70 to 245 kPa, 280 kPa and 324 kPa for samples cured at 7, 14 and 28 days respectively. While the soil samples treated with FR mixtures exhibit higher UCS values in the range from 70 to 329 kPa, 379 kPa and 430 kPa for 7, 14 and 28 days respectively.
4. With the increase in binder content upto 15% of RM and 12% of FA, settlement were reduced (300.45 mm to 78.89 mm, 360.82 mm to 75.61 mm and 395.63 mm to 68.42 mm) for footings of widths 1 m, 1.2 m and 1.5 m respectively.

5. The allowable bearing pressure of footings resting over FR treated soil were enhanced from 68 kPa to 106 kPa, 50 kPa to 115 kPa and 42 kPa to 126 kPa for footings of width 1 m, 1.2 m and 1.5 m respectively.

References

Aswathy M, Salini U, Gayathri VG (2019) Utility of lime and red mud in clay soil stabilization. In: Geotechnical characterisation and geoenvironmental engineering: IGC 2016, vol 1. Springer Singapore, pp 19–26. https://doi.org/10.1007/978-981-13-0899-4_3

Balreddy MS, Abilash KE, Manjunath HK (2021) Experimental investigation on stabilization of silty sand using fly ash. In: Transportation, water and environmental geotechnics: proceedings of Indian geotechnical conference 2020, vol 4. Springer Singapore, pp 281–291. https://doi.org/10.1007/978-981-16-2260-1_27

Chandra KS, Krishnaiah S (2022) Strength and leaching characteristics of red mud (bauxite residue) as a geomaterial in synergy with fly ash and gypsum. Transp Res Interdiscip Perspect 13. https://doi.org/10.1016/j.trip.2022.100566

Chethan BA, Ravi Shankar AU (2021) Strength and durability characteristics of cement and class f fly ash-treated black cotton soil. Indian Geotech J 51(5):1121–1133. https://doi.org/10.1007/s40098-020-00488-2

Degirmenci N, Okucu A, Turabi A (2007) Application of phosphogypsum in soil stabilization. Build Environ 42(9):3393–3398. https://doi.org/10.1016/j.buildenv.2006.08.010

James J, Pandian PK (2018) Bagasse ash as an auxiliary additive to lime stabilization of an expansive soil: strength and microstructural investigation. Adv Civ Eng 2018. https://doi.org/10.1155/2018/9658639

Jha AK, Sivapullaiah PV (2016) Volume change behavior of lime treated gypseous soil—influence of mineralogy and microstructure. Appl Clay Sci 119:202–212. https://doi.org/10.1016/j.clay.2015.09.017

Kalkan E (2006) Utilization of red mud as a stabilization material for the preparation of clay liners. Eng Geol 87(3–4):220–229. https://doi.org/10.1016/j.enggeo.2006.07.002

Ma S, Sun Z, Wei J, Zhang X, Zhang L (2022) Utilization of modified red mud waste from the bayer process as subgrade and its performance assessment in a large-sale application. Coatings 12(4):471. https://doi.org/10.3390/coatings12040471

Mahesh Bhat K, Nayak S (2021) Experimental studies and its application using PLAXIS-2D for lithomargic clay stabilized by GBFS and lime. Geotech Geol Eng 39(7):4901–4915. https://doi.org/10.1007/s10706-021-01802-2

Mir BA, Sridharan A (2013) Physical and compaction behaviour of clay soil-fly ash mixtures. Geotech Geol Eng 31(4):1059–1072. https://doi.org/10.1007/s10706-013-9632-8

Mishra MC, Babu KS, Reddy NG, Dey PP, Rao BH (2019) Performance of lime stabilization on extremely alkaline red mud waste under acidic environment. J Hazard Toxic Radioact Waste 23(4):04019012. https://doi.org/10.1061/(ASCE)HZ.2153-5515.0000448

Moghal AAB (2017) State-of-the-art review on the role of fly ashes in geotechnical and geoenvironmental applications. J Mater Civ Eng 29(8). https://doi.org/10.1061/(asce)mt.19435533.0001897

Murmu AL, Patel A (2020) Studies on the properties of fly ash–rice husk ash-based geopolymer for use in black cotton soils. Int J Geosynth Ground Eng 6(3). https://doi.org/10.1007/s40891-020-00224-z

Panda I, Jain S, Das SK, Jayabalan R (2017) Characterization of red mud as a structural fill and embankment material using bioremediation. Int Biodeterior Biodegrad 119:368–376. https://doi.org/10.1016/j.ibiod.2016.11.026

Phanikumar BR, Nagaraju TV (2018) Effect of fly ash and rice husk ash on index and engineering properties of expansive clays. Geotech Geol Eng 36:3425–3436. https://doi.org/10.1007/s10706-018-0544-5

Pradeep PS, Mayakrishnan M (2023) Understanding the engineering behaviour of expansive soil amended with bagasse ash and lime using microstructural analysis. Int J Geosynth Ground Eng 9(3):26. https://doi.org/10.1007/s40891-023-00445-y

Ramesh HN, Manjunatha BV, Madhavi GRK (2021) Micro-structural and mineralogical studies to evaluate the effectiveness of industrial solid wastes for stabilization of expansive soils. In: Indian geotechnical conference 2019: geotechnics for infrastructure development and urbanisation (GeoINDUS). Springer Singapore, pp 77–93. https://doi.org/10.1007/978-981-33-6590-2_7

Renjith R, Robert D, Setunge S, Costa S, Mohajerani A (2021) Optimization of fly ash based soil stabilization using secondary admixtures for sustainable road construction. J Clean Prod 294. https://doi.org/10.1016/j.jclepro.2021.126264

Salahudeen AB, Eberemu AO, Osinubi KJ (2014) Assessment of cement Kiln dust-treated expansive soil for the construction of flexible pavements. Geotech Geol Eng 32:923–931. https://doi.org/10.1007/s10706-014-9769-0

Samantasinghar S, Singh SP (2023) Red mud-slag blends as a sustainable road construction material. J Constr Build Mater 375:1–12. https://doi.org/10.1016/j.conbuildmat.2023.130926

Sharma AK, Sivapullaiah PV (2012) Improvement of strength of expansive soil with waste granulated blast furnace slag. In: GeoCongress 2012: state of the art and practice in geotechnical engineering, pp 3920–3928. https://doi.org/10.1061/9780784412121.402

Sharma AK, Sivapullaiah PV (2017) Swelling behaviour of expansive soil treated with fly ash–GGBS based binder. Geomech Geoengin 12(3):191–200. https://doi.org/10.1080/17486025.2016.1215548

Shamshad A, Das SK, Rao BH (2019) Strength and durability characteristic of alkali activated GGBS stabilized red mud as geo-material. Constr Build Mater 211:932–942. https://doi.org/10.1016/j.conbuildmat.2019.03.261

Singh NB (2018) Fly ash-based geopolymer binder: a future construction material. Minerals 8(7) https://doi.org/10.3390/min8070299

Singh SP, Samantasinghar S, Sindhuja D (2018) Stabilization of red mud using ground granulated blast furnace slag by geopolymerization technique. In: Contemporary issues in geoenvironmental engineering, GEOMAST 2017, pp 338-350. https://doi.org/10.1007/978-3-319-61612-4_28

Sridevi G, Sanjeet S, Subhrajeet S (2019) Stabilization of expansive soil with red mud and lime. In: Ground improvement techniques and geosynthetics: IGC 2016 vol 2. Springer Singapore, pp 259–268. https://doi.org/10.1007/978-981-13-0559-7_29

Tsakiridis PE, Agatzini-Leonardou S, Oustadakis P (2004) Red mud addition in the raw meal for the production of Portland cement clinker. J Hazard Mater 116(1–2):103–110. https://doi.org/10.1016/j.jhazmat.2004.08.002

Venkatesh C, Nerella R, Chand MSR (2021) Role of red mud as a cementing material in concrete: a comprehensive study on durability behavior. Innov Infrastruct Solut 6(1). https://doi.org/10.1007/s41062-02000371-2

Vigneshwaran S, Uthayakumar M, Arumugaprabu V (2020) Potential use of industrial waste-red mud in developing hybrid composites: a waste management approach. J Clean Prod 276. https://doi.org/10.1016/j.jclepro.2020.124278

Zha F, Liu S, Du Y, Cui K (2008) Behavior of expansive soils stabilized with fly ash. Nat Hazards 47:509–523. https://doi.org/10.1007/s11069-008-9236-4

Chapter 18
Review on Hoop Stress Behaviour for Circular Water Tank Wall with Uniform and Varying Thickness

Eshwari Kotnake, M. P. Bhorkar, and K. R. Dabhekar

Abstract The paper highlights the critical role of water tanks in storing and distributing potable water, essential for human health. It focuses on circular water tanks, emphasizing their design to withstand circumferential hoop tension and vertical bending moments. The review includes various research methods analyzing tank behavior under different conditions using software tools. Key findings underline the importance of calculating hoop stress values for structural performance, guided by standards like IS 3370: PART IV. These standards provide coefficients to determine appropriate wall thicknesses. The study also notes that circumferential expansion induces hoop tension, increasing linearly from top to base, which is crucial for robust tank design. This paper shows the study on various water tank from previous research.

Keywords Ground resting water tank · Analysis using staad-pro · Maximum absolute stress values · Hydrostatic pressure · Design codes

18.1 Introduction

Humans depend on water, which is also the most valuable and necessary resource on Earth for day-to-day existence. The amount of water that is present on Earth's surface is constant despite its continual transformation (Jibhakate et al. 2017). Adequate distribution of water depends on the design of the tank in an area. A raised tank is known for a large water container whose purpose is to store water at a certain height to increase the flow of water. Land storage, soil support, hope, etc. Liquid storage

E. Kotnake (✉) · M. P. Bhorkar · K. R. Dabhekar
Civil Engineering Department, G H Raisoni College of Engineering, Nagpur, Maharashtra 440016, India
e-mail: eshwari.kotnake.mtechstr@ghrce.raisoni.net

K. R. Dabhekar
e-mail: kuldeep.dabhekar@raisoni.net

B. V. Khode et al. (eds.), *Innovations in Technologies: Pioneering Sustainable Infrastructure for a Resilient Future*, Sustainable Civil Infrastructures,
https://doi.org/10.1007/978-3-031-82133-2_18

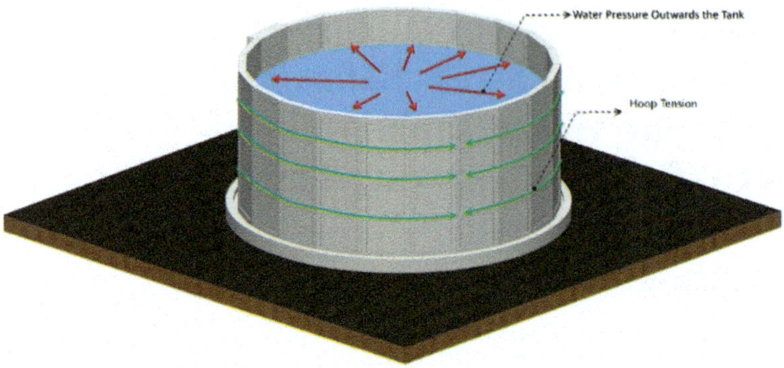

Fig. 18.1 Circular tank resting on ground

tanks are widely used by cities and businesses to store water, liquids and other chemicals (Chavhan et al. 2019). For this reason, the water tank is very important in terms of electricity and work. High-pressure tanks contain large volumes of water at lower levels; this is the most important situation in terms of tank's failure during an earthquake.

A circular water tank, as the name implies, is a tank that has a circular or cylindrical shape, as shown in Fig. 18.1. These tanks are designed to hold water and are often used for various purposes, such as domestic water storage, agricultural needs, industrial applications, or even in firefighting systems. To store large volumes of water, a circular water tank is more preferable (Ghosal 2019). Compared to square or rectangular designs, this design minimizes stress concentrations and is better able to handle the forces involved in storing huge amounts of water. Therefore, a cylindrical water tank will fare better against all external factors than a rectangular tank, such as high winds and heavy rain. Ground Service Reservoirs (GSR), also known as distribution reservoirs. These are the storage reservoirs, which can store the treated water for supply purposes (Chavhan et al. 2019; Kishori and Patel 2015). One kind of stress that the walls of a circular or cylindrical object can encounter is called hoop stress. It functions circumferentially, with its direction perpendicular to the structure's axis. It is an essential mechanical idea that clarifies how cylindrical constructions behave when subjected to internal pressure.

18.1.1 Features and Advantages of Circular Water Tanks

- Circular water tanks have a cylindrical or round-shaped structure. This shape helps to distribute the pressure of the stored water evenly. According to Payal Chavhan, Sneha Dorle et al. water tanks can be constructed from various materials, including plastic (polyethylene), concrete, steel, or fiberglass, each with its own advantages in terms of durability, cost, and maintenance (Chavhan et al. 2019).

- The circular shape inherently provides strength and stability, especially when dealing with large volumes of water. Also, the cylindrical shape allows for efficient use of space, particularly in areas where space is limited.
- Circular tanks can be easily connected to various systems, such as filtration, pumps, or distribution systems, depending on their purpose. The design helps to distribute the pressure of the stored water, minimizing stress points on the structure.
- Regular inspection and maintenance are necessary to ensure the tank remains in good condition and free from leaks or damage. Depending on the material and size, installation might require professional assistance. Costs vary depending on material, size, and installation requirements.
- Circular water tanks are popular due to their structural integrity, space efficiency, and adaptability for different applications. The choice of a specific tank depends on factors like intended use, available space, and budget (Dave et al. 2017; Pole and Khedikar 2017; Aware and Mathada 2015).

18.1.2 Circular Tanks Resting on Ground

In a circular tank resting on ground with a flexible base, the water pressure causes the tank wall to expand circumferentially. This expansion increases linearly from zero at the top of the wall to a maximum at the base. As the water level rises, the pressure exerted on the wall increases, resulting in a gradual expansion that is most pronounced at the base where the pressure is greatest. This flexible base allows the tank to accommodate changes in water volume and pressure, ensuring structural integrity and stability over time.

Because of hydrostatic pressure, the diameter of the tank tends to increase. As seen in Fig. 18.1. This rise in diameter over the tank's height is determined by the kind of joint at the intersection of the slab and wall.

Hydrostatic pressure causes the largest rise in diameter at the base and no increase in diameter at the top when the joints at the base are flexible. This is because hydrostatic pressure has a linear variation from zero at the top to maximum at the base. Figure 18.2 depicts the tank's deformed shape. The base remains stationary when the joint at that location is stiff. Figure 18.2 illustrates the deflection of the vertical wall.

18.1.3 Hoop Stresses

- Hoop stresses are mainly of two types: (a) Tension and (b) compression
- "Hoop stress" is the term used to describe the tension that forms around the perimeter when pressure is applied. The stress that develops around the circumference when pressure is applied is typically referred to as "hoop stress". Because

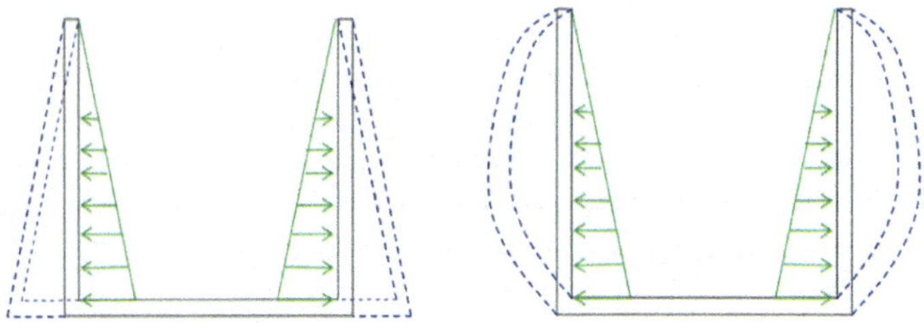

Fig. 18.2 Deformed shape of tank

hoop stresses are tensile, their action is perpendicular to the direction of axiality. The compressive nature generally found for external pressure.

- Tensile tensions are produced in the form of hoops to withstand the bursting action that arises from applying pressure. In the study of Mehul S. Kishori and Chirag N. Patel the hoop tension parameters are provided for different end conditions (Kishori and Patel 2015). A beam-like structure can be seen in the cross-section of circular RCC water storage tanks. In other words, tension occurs after the neutral axis and compression occurs inside it. So, steel is the only material resisting the hoop stress in the water tank. The coefficient of H^2/Dt can be used to compute the hoop tension in the cylindrical tank based on the available design codes (Daftardar et al. 2017).
- In summary, hoop stresses, encompassing both tension and compression, are critical considerations in the design and analysis of structures subjected to pressure. Understanding the behavior of hoop stresses is essential for ensuring the structural integrity and safety of cylindrical tanks in various engineering applications.

18.1.4 Mechanism and Elements

In a circular tank, circumferential expansion results in hoop tension along the wall. This tension increases linearly from the top to the base of the wall. At the top of the wall, where the water depth (h) is zero, the hoop tension is also zero. Conversely, at the base of the wall, where the water depth is at its maximum (D), the hoop tension reaches its maximum value. This tension is proportional to the unit weight of water (γ) and the depth (h) of water within the tank, expressed as (γ * h). This phenomenon highlights the dynamic relationship between water depth and hoop tension along the wall of the tank. Figure 18.3 shows the section of various thickness of wall.

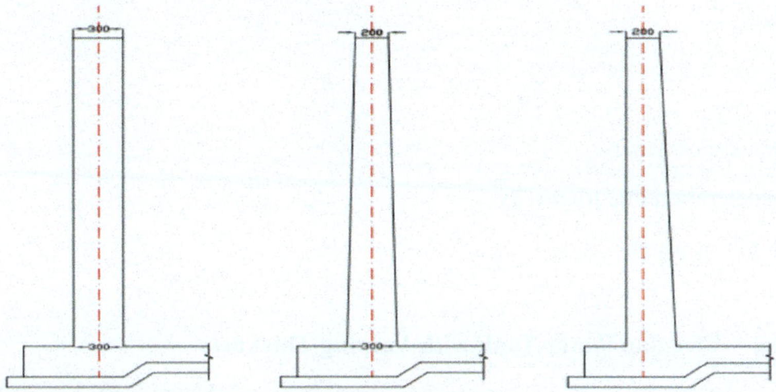

Fig. 18.3 Various thickness of walls

18.1.4.1 Circular Water Tank with Uniform Thickness

Mechanism

For a circular water tank with uniform thickness, the structural analysis can be simplified as the material properties and dimensions are consistent throughout the tank wall. The primary mechanism involves the distribution of hoop (circumferential) and radial stresses due to the internal water pressure.

Elements and Analysis

a) Hoop Stress (Circumferential Stress)

Mechanism: This stress acts along the circumference of the tank and is caused by the internal pressure of the water.

Formula:

$$\sigma h = p \cdot r / t$$

σh = Hoop stress

p = Internal pressure

r = Radius of the tank

t = Thickness of the tank wall

b) Hydrostatic Pressure Distribution

Mechanism: Pressure due to the water column increases linearly with depth.

Formula:

$$p = \rho \cdot g \cdot h$$

ρ = Density of water

g = Acceleration due to gravity

h = Depth of water

18.1.4.2 Circular Water Tank with Varying Thickness

Mechanism

For a tank with varying thickness, the analysis becomes more complex as the material properties and dimensions change along the tank wall. This variation is often designed to optimize material usage and to better withstand varying stress distributions.

Elements and Analysis

c) Hoop Stress (Circumferential Stress)

Mechanism: It is similar to uniform thickness tanks, but the varying thickness affects the stress distribution.

Formula:

$$\sigma h = p \cdot r / t(r)$$

(r) = Thickness as a function of radius or height.

d) Hydrostatic Pressure Distribution

Mechanism: Same as in uniform tanks but applied over varying thickness.

Formula:

$$p = \rho \cdot g \cdot h$$

18.2 Objectives of the Study

- To study how various kinds of water tanks behave.
- To study the hoop stresses for circular water tanks.

18.3 Literature Review

According to **Mangesh L. Jibhakate, M. P. Bhorkar et al.** sand filtration is the method used for most water treatment plant filtration. Sand pores clog as a result of the ongoing filtration process, which reduces efficiency. Backwashing, or the reverse flow of water and air, is a technique used to mitigate such issues. Four percent of the purified water was used for the backwashing procedure, which produced muddy water known as backwash water. After that, the backwash water is released for recirculation into a nearby storage tank or natural stream. A trial for the reuse and recirculation of backwash water is part of the current investigation (Jibhakate et al. 2017).

A case study by **Chavhan et al.** which serves as a foundation for creating a basic, dependable, and cost-effective design for water tanks. This study focuses on the optimization of liquid-retaining structures, particularly those that are grounded and composed of reinforced concrete. **Inspiration for Design**: The case study provides inspiration for developing water tank designs that are both practical and cost-effective. It aims to strike a balance between simplicity and reliability, ensuring that the designs are safe and affordable. **Understanding Safe Tank Optimization**: The study aids in understanding the principles behind optimizing water tank designs to enhance safety. This involves analyzing the structural integrity and support conditions necessary for the tanks to function effectively without compromising on safety. **Design Requirements**: The study outlines specific design requirements that are crucial for the structural strength and stability of water tanks. These requirements are essential for ensuring that the tanks can safely retain liquids without risk of failure. **Focus on Reinforced Concrete Structures**: The case study specifically examines liquid-retaining structures made of reinforced concrete that are situated on the ground. Reinforced concrete is chosen for its durability and strength, making it a suitable material for such applications. **Optimization Techniques**: The study provides ideas and methodologies for optimizing the design of water tanks. This involves precise and practical approaches to ensure that the tanks are not only structurally sound but also cost-effective. **Application to Different Shapes**: The insights from the study are applicable to various shapes of water tanks, emphasizing that the design principles can be adapted to different configurations while maintaining safety and cost-effectiveness (Chavhan et al. 2019). **Mehul S. Kishori and Chirag N. Patel's** paper provides a comprehensive analysis of ground circular concrete tanks under static stress conditions, using IS:3370:2009, PCA standards, and STAAD-Pro. By examining different tank aspect ratios and end conditions, and performing a detailed parametric study, as shown in Figs. 18.4 and 18.5, the paper contributes valuable knowledge for the structural design and optimization of these tanks, ensuring their safety, efficiency, and compliance with established standards (Kishori and Patel 2015).

The paper by **Daftardar et al.** focuses on the tensile stress, specifically hoop tension, experienced around the circumference of a water tank due to the internal pressure from the stored liquid. The study provides an analytical solution for hoop

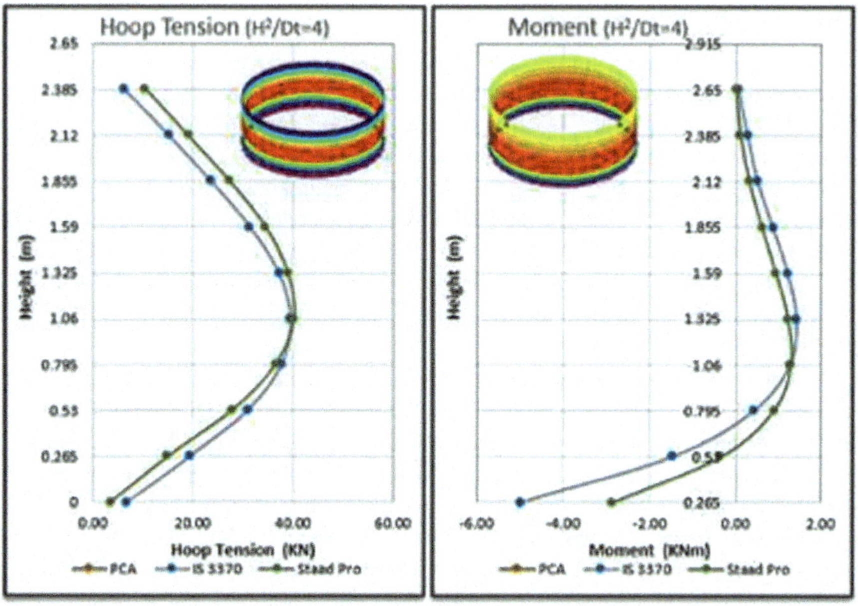

Fig. 18.4 Fixed base and free top (Kishori and Patel 2015)

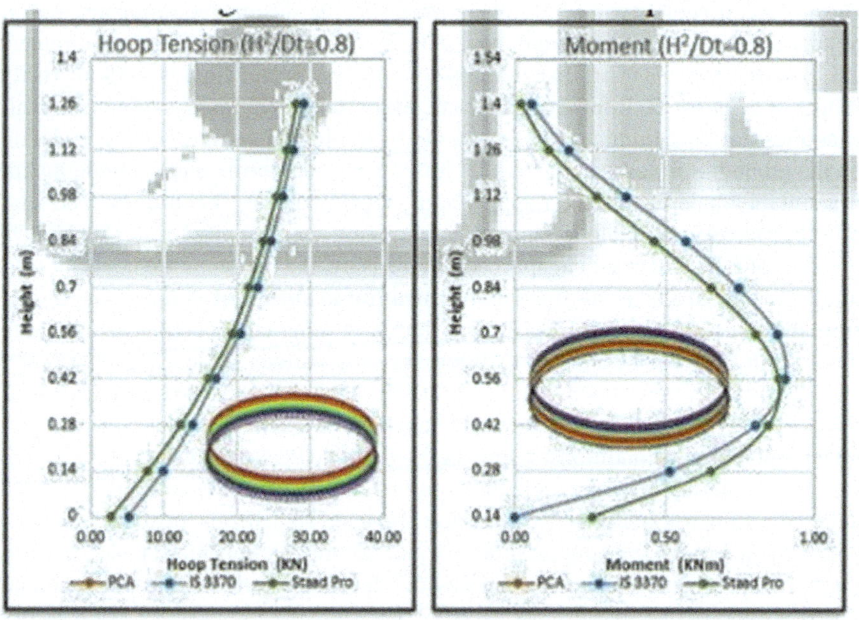

Fig. 18.5 Hinged base and free top (Kishori and Patel 2015)

tension based on the tank's geometry and loading conditions. It highlights a gap in the IS 3370:2009 code, noting that for certain tank dimensions, the necessary coefficients for accurate stress calculations are not provided. This underscores the need for further refinement or supplementary methods to ensure precise stress evaluation for all tank sizes (Daftardar et al. 2017). Hoop tension for the tank wall due to hydrostatic pressure is given by the following formulae

$$\mathrm{T} = pr\frac{D}{2}$$

$$\text{Hoop stress } = pr\frac{D}{2t}$$

(where, pr = water pressure)

$$\text{Hoop strain } = pr\frac{D}{2tE}$$

In the course of the study, it was observed that the coefficients for certain ranges of tank dimensions are not explicitly available in the IS 3370:2009 code. This lack of explicit coefficients presents a challenge for engineers and designers who need precise data to ensure the structural integrity of water tanks. To address this gap, the paper provides comprehensive methodologies for dealing with a variety of tank dimensions and calculating the required coefficients as the results are compared below in the Fig. 18.6. By doing so, it equips practitioners with the necessary tools to derive these coefficients accurately, thereby filling the void left by the existing code. The results obtained from the research study are shown below-

The study by **Ridhdhesh Bhojavia and Dhruti J Dhyani** evaluates the seismic behavior of a ground-supported circular water tank on three soil types: soft, medium, and hard. It examines key parameters, including the masses of the tank wall, water, and base slab, using a spring mass model. The analysis covers time intervals in impulsive and convective modes, base shear, bending moment, overturning moment, vertical excitation pressure, and maximum hydrodynamic pressure. Utilizing IS 1893 Part 2:2014 and IS 1893 Part 1:2016 standards for earthquake analysis, the study compares the diameter-to-height (D/h) ratio with these factors to assess their seismic response, offering insights into how different soil conditions impact tank stability during earthquakes (Bhojavia and Dhyani 2018) (Figs. 18.10 and 18.11).

The 2019 study by **Ghosal M.** utilizes STAAD V8i Pro for the analysis and design of an elevated circular water tank. It combines STAAD structural analysis with manual load calculations, addressing wind load, dead load, self-weight, and hydrostatic load. The study follows the limit state design approach to ensure compliance with IS standards, successfully providing sufficient water supply for families' daily needs while meeting safety and structural requirements. This demonstrates the practical application of advanced structural analysis tools in designing reliable water storage solutions (Ghosal 2019).

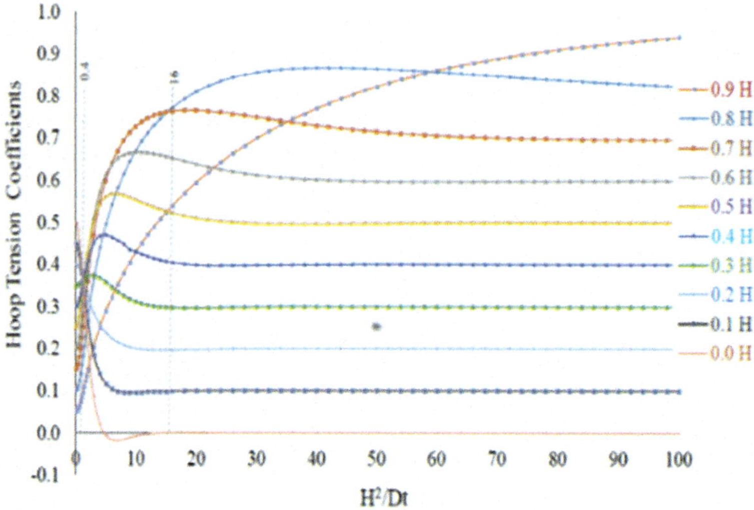

Fig. 18.6 H^2/Dt range from 0.1 to 100 (Daftardar et al. 2017)

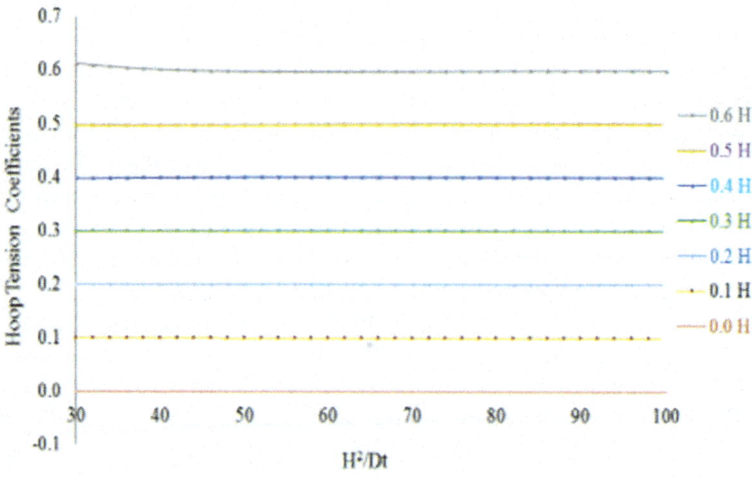

Fig. 18.7 H^2/Dt range from 30 to 100 (Daftardar et al. 2017)

The 2017 paper by **Dave R. Patel R. et al.** discusses the construction of a rectangular, ground-supported water tank using the working stress method. It includes the design and seismic analysis of the tank, with spreadsheet programs employed to derive necessary values. The study establishes relationships between various parameters: tank capacity with moment capacity and reinforcement area, base shear with impulsive height, and overturning moment with convective height. These relationships are valuable for efficiently determining base moment, wall reinforcement, base shear, and overturning moment, thus aiding in the practical and time-saving design of water tanks (Dave et al. 2017).

The research of Prajapati et al. (2021) provides a concise overview of the theory governing the design of structures that retain liquid, including all the essentials in accordance with both the outdated and updated rules. It covers the design and analysis of software for an underground circular tank that is fixed at the base and free at the top. The rectangular tank, which is employing an excel sheet and the IS code approach to rest it on the ground, is also taken into consideration. A comparison of fracture width calculation techniques and codes. According to IS 3370 (Part 2) (2009), IS 456 (2000), and the international code cited in the paper, the smallest permissible crack width for flexure is 0.2 mm. Comparative results are generated in graph form for better understanding. Software calculation little more economical with respect to IS code method. The limit state design by crack width with reference of IS 3370 (Part 2) 2009, IS 456:2000 and the international code of practice found that the crack width limit of 0.2 mm results is more rational and economical design compared to the working stress design (Prajapati et al. 2021). The research of Rathod and Ishwaragol (2018) described that the performance of reinforced concrete with a capacity of one lakh liters and to compare and analyze models of different heights and widths have been shown. This article attempts to examine the behavior of tanks using the STADD.PRO software. Consider the following water levels: full tank, partial tank and empty tank, earthquake zone III. After the analysis was completed, a comparative study was conducted on the changes in bending moment, shear strength and displacement of lines and beams with different tank heights and wide bases. The bending moment value increased slightly depending on the increase in the section height, and increased by 2.9% and 3.1%, respectively, depending on the air load on the tank. The bending moment values were found to increase by 32.0% and 38.8%, respectively, in response to the cylinder wall's height decreasing and the base width increasing. In accordance with the height increase, the base shear value increased by 3.3% and 3.6%, respectively, and increased marginally. The decrease in the base's breadth and the height of the cylindrical wall caused the displacement to increase, increasing it by 46.4% and 39.38%, respectively (Rathod and Ishwaragol 2018). This article of Adiyanto et al. (2019) express the relocation plans, feedback techniques, and pressure analysis to assess the seismic capabilities of an existing high pressure RC tank are shown. The SAP2000 program was used to assess the pressurized RC water tank's seismic capacity. The impact of column size on the seismic capacity of reinforced concrete tanks (RC tanks) is examined using pushover analysis, and all models make use of RC tanks that are currently in use in Malaysia. The 7-story model's floor heights and gap widths are 3.5 and 5.0 m, respectively (Adiyanto et al. 2019). In this paper Vakilaadsarabi et al. (2012) explains the impacts of RCC water tanks are explained in brief. Liquid sloshing causes the majority of damage to liquid tanks during earthquakes, particularly in long-duration ground motions. Long-period ground movements were mostly caused by large earthquakes and effective propagation channels like accretionary prisms. The computational fluid dynamics (CFD) simulation application OpenFOAM (Open Field Operation and Manipulation) is used to look into the effects of ground motions inside a sloshing tank. An improved

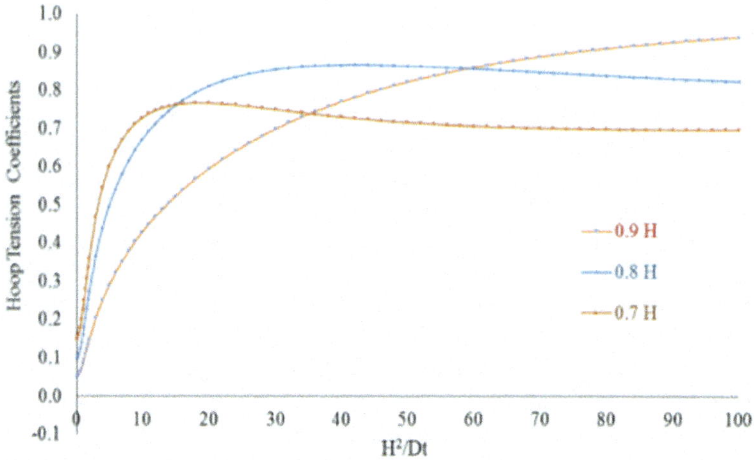

Fig. 18.8 H^2/Dt range from 0.1 to 100 for the height of the tank (0.7 H to 0.9 H) (Daftardar et al. 2017)

VOF technique is used to ensure an accurate description of water displacement. Computational results are confirmed and contrasted to provide some crucial insight into how the length and period of seismic vibrations impact the sloshing phenomena in water tanks (VakilaadSarabi et al. 2012). According to **Chandak, Mayuri, and P. Y. Pawade,** many structural characteristics, such as the structure's stiffness and strength to bear the lateral seismic load caused by the earthquake, are impacted by proper structural modeling. A comparative and parametric analysis is conducted using joint displacement, axial force, maximum bending moment, shear force, time period, etc. (Chandak and Pawade 2021).

18.4 Methodology

Methodology adopted in current study:

The collected data from prior research provides insights into the analysis and design considerations for various water tank types. This information can be used to optimize the design of specific water tank project.

- **Analytical Techniques**: Thin-Walled Cylinder Theory: This is the fundamental approach for calculating hoop tension in a circular tank. It assumes the tank wall is thin compared to its diameter and neglects bending moments.
- **Advanced Analytical Methods**: For complex tank geometries or loadings, more advanced analytical methods like finite element analysis (FEA) can be employed. These methods involve dividing the tank structure into smaller elements and solving complex stress–strain relationships numerically.

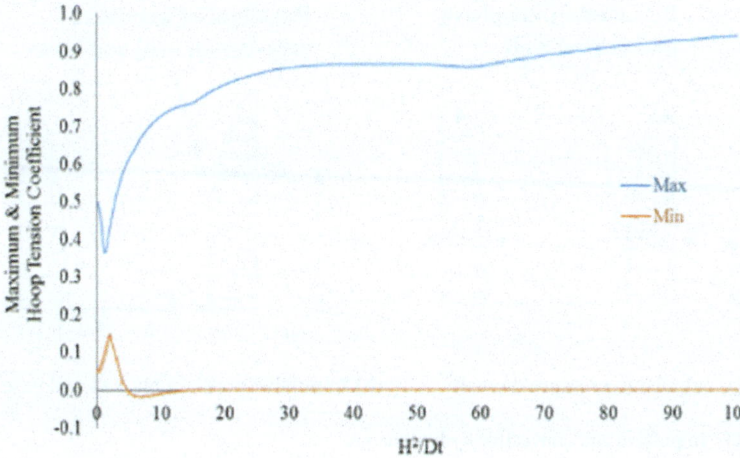

Fig. 18.9 Maximum & minimum hoop tension coefficients for H^2/Dt range from 0.1 to 100 (Daftardar et al. 2017)

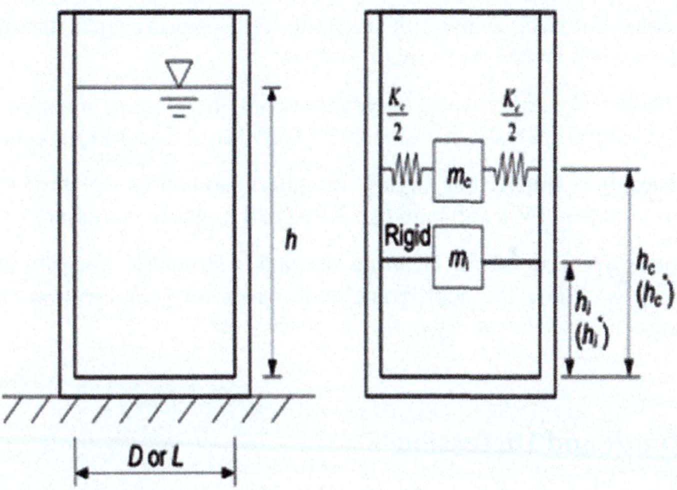

Fig. 18.10 Spring mass modal

- **Software Tools**: **Spreadsheets**: Simple spreadsheets can be used for basic calculations using the thin-walled cylinder theory for uniform pressure distribution. **Structural Analysis Software**: For more complex analysis, including varying water pressure, wind loads, or seismic effects, sophisticated structural analysis software like: **STAAD.Pro**: As mentioned earlier, STAAD Pro allows modeling the tank with shell elements, applying hydrostatic pressure, and analyzing stresses like hoop tension. **SAP2000**: Similar to STAAD.Pro, SAP2000 is another popular structural analysis software that can be used for detailed analysis of circular

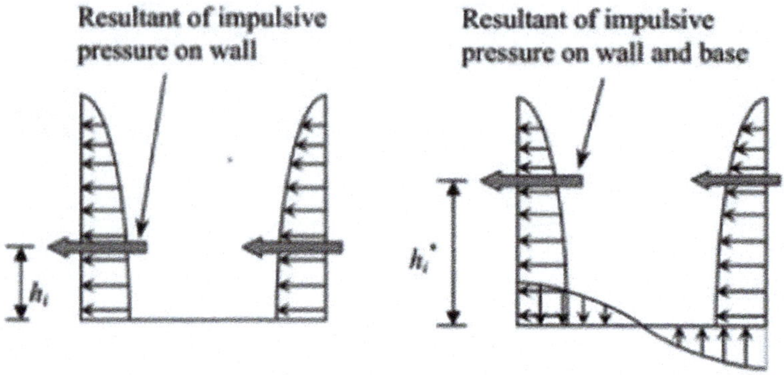

(a) Impulsive pressure on wall (b) Impulsive pressure on wall and base

Fig. 18.11 Impulsive and convective hydrodynamic

tank.The choice of methodology and software depends on the complexity of the tank and the level of detail required.

Simple Tanks: For basic analysis of small, uniform-pressure tanks, the thin-walled cylinder theory and spreadsheets might suffice.

Complex Tanks: For larger tanks, varying pressure distribution, or complex loading scenarios, structural analysis software like STAAD.Pro or SAP2000 is recommended.

Highly Complex Tanks: For highly irregular geometries or intricate loading conditions, advanced FEA software like ABAQUS might be necessary.

Design codes are crucial for ensuring the tank's structural integrity and safety. It's essential to choose the appropriate code based on your location and project requirements.

18.5 Result and Discussion

Comparing results of research paper, it's significantly found that, when it comes to designing, software offers a level of reliability and efficiency that surpasses traditional manual methods. Results produced by software are often more cost-effective and efficient, primarily due to the ability to account for numerous factors under designing conditions that are challenging to manage manually.

Accuracy and Precision: Software enables precise modelling and analysis of complex structures, ensuring accuracy in calculations and predictions. Manual methods may introduce errors or inconsistencies due to human limitations.

Speed and Efficiency: Designing with software significantly reduces the time required to complete tasks compared to manual methods. Automated processes streamline workflows, allowing engineers to generate and iterate designs more quickly.

Complexity Handling: Software can handle intricate design scenarios and factors that may be challenging to address manually. It can simulate various loading conditions, material properties, and geometric configurations with ease, enabling comprehensive analysis.

Cost-Effectiveness: While initial investment in software may be required, the long-term cost savings are significant. Software eliminates the need for physical prototypes or extensive testing, reducing expenses associated with material wastage and labor.

Iterative Design Process: Software facilitates an iterative design process where engineers can quickly modify parameters and evaluate multiple design alternatives. This iterative approach leads to optimized designs that meet performance requirements efficiently.

Automation of Repetitive Tasks: Software automates repetitive tasks, such as calculations and drafting, freeing up engineers to focus on higher-level design decisions and problem-solving.

Integration with Standards and Codes: Many software tools incorporate industry standards and building codes, ensuring compliance and adherence to best practices in design and construction.

Overall, software offers unparalleled advantages in the design process, providing engineers with powerful tools to create innovative, cost-effective, and reliable solutions. By leveraging the capabilities of software, designers can overcome the limitations of manual methods and deliver projects that meet the highest standards of quality and efficiency.

The insights gained from studying water tanks can lead to improved design and materials, enhancing the efficiency and durability of these essential storage units. This knowledge benefits a wide range of applications from ensuring clean water storage in residential areas to maintaining large-scale water supplies in agricultural and industrial sectors. Additionally, understanding tank behavior helps in disaster preparedness, ensuring that water storage systems remain functional during emergencies.

18.6 Conclusion

From the above research study, it is concluded that.

- **Superiority of Software in Design**:

Efficiency and Reliability: Designing with software is highlighted as more reliable and efficient compared to manual methods.

Cost-Effectiveness: Software results tend to be more cost-effective.

Comprehensive Analysis: Software can handle numerous design factors that are challenging to account for manually, improving overall design accuracy and performance.

Literature Review Insights: Existing Research: There is some research on hoop tension analysis, particularly focusing on tank walls with uniform thickness.

Research Gap: A notable gap is identified in the comparison between uniform and varying thickness walls for circular water tanks, suggesting a direction for future studies.

Optimization and Standards: Advanced Design Methodologies: Using advanced design methodologies and software analysis, engineers can optimize the performance, resilience, and longevity of circular water tanks.

Adherence to Standards: By following industry standards and guidelines, such as IS 3370: PART IV, engineers ensure that water tanks meet high safety and quality standards.

Sustainable Water Management: Properly designed water tanks contribute significantly to the sustainable management of water resources.

Contributions

Enhanced Design Processes: Emphasizes the importance and benefits of using software for designing water tanks, advocating for its broader adoption in engineering practices.

Identification of Research Gaps: Highlights the need for further research on the comparative analysis of uniform versus varying thickness walls, paving the way for future studies.

Promotion of Standards Compliance: Reinforces the significance of adhering to established industry standards to ensure the safety, quality, and sustainability of water tank designs.

Focus on Sustainability: Underlines the role of well-designed water tanks in promoting sustainable water resource management, reflecting broader environmental and resource management goals.

Future Scope

- Hoop Stress Behavior for Varying Thickness: Conducting detailed investigations into the hoop stress behavior of circular water tanks with varying thicknesses. Exploring how varying thicknesses impact stress distribution and structural integrity. Utilizing advanced analytical tools, such as finite element analysis (FEA), to model and simulate the behavior of tanks with varying thicknesses under different loading conditions.
- Economical Design of Circular Storage Tanks: Investigating methodologies for achieving economical designs of circular storage tanks. Explore how design

parameters such as material selection, wall thickness, and reinforcement placement can be optimized to minimize costs without compromising structural performance. Implement cost–benefit analyses to compare different design alternatives and identify the most cost-effective solutions.

- Comparison of Stress Coefficients with FEM: Conducting comparative studies between stress coefficients derived from analytical methods and those obtained through finite element analysis (FEA). Evaluate the accuracy and reliability of analytical methods in predicting stress distribution and structural behavior compared to FEA results. Explore potential discrepancies and limitations of analytical methods and identify areas for improvement or refinement.
- Optimization Techniques: Developing optimization techniques for circular water tank design that consider multiple objectives, including structural integrity, cost-effectiveness, and material efficiency. Utilize optimization algorithms and computational tools to iteratively refine design parameters and achieve optimal solutions that balance performance and cost.
- Sustainability and Environmental Impact: Investigating the environmental impact of circular water tank design and construction processes. Explore sustainable materials, construction methods, and maintenance practices that minimize environmental footprint and promote long-term sustainability.
- Dynamic Analysis and Seismic Performance: Extend research to dynamic analysis and seismic performance evaluation circular water tanks. Investigating how dynamic loading, such as seismic events, affects the structural response and integrity of water tanks, and develop mitigation strategies to enhance resilience.
- Field Testing and Validation: Conducting field testing and validation studies to verify the performance of proposed design methodologies and analytical models in real-world conditions.
- Collaborating with industry partners to implement and evaluate new design approaches in practical applications, gathering feedback and refining methods based on field observations.

References

Adiyanto MI, Damit AA, Yaakup SN, Mustapha SA, Chin SC (2019) Effect of column size on the seismic capacity of elevated reinforced concrete water tank. IOP Conf Ser Earth Environ Sci 244:012010

Ali S, Sanghai SS (2021) Seismic vulnerability assessment of reinforced concrete buildings using pushover analysis. AIP Conf Proc 2417

Aware RJ, Mathada VS (2015) Seismic analysis of cylindrical liquid storage tank. IJSR. Index Copernicus Value (2013):6.14 | Impact Factor (2014):5.611

Bhojavia R, Dhyani DJ (2018) Comparative seismic analysis of ground supported circular water tank using codal provision. IJSRSET 4(5)

Bhorkar MP, Shukla J, Pitale NH, Vinchurkar SM (2023) Planning and design of sustainable water harvesting structures using geospatial system, smart and sustainable technologies for disaster resilient infrastructure IOP Conf Ser Earth Environ Sci 1193

Bhurse PS, Sanghai SS, Kumari NL (2021) Seismic response of rectangular RC building with and without infill walls considering soil–structure interaction. In: Advances in civil engineering and infrastructural development: select proceedings of ICRACEID 2019. Springer Singapore

Chandak M, Pawade PY (2021) Influence of metakaolin and steel fibres on stress strain behaviour of concrete. J Phys Conf Ser 1913(1). IOP Publishing

Chavhan P, Dorle S, Bendwar A, Jangere AK, Singh R, Raza R (2019) A case study on water tank resting on ground. IJRESM 2(3)

Chawla S, Jamle S, Meshram K (2020) To study the economical design of Intz water tank. IJAERS. 10.22161.71.19

Daftardar A, Vichare S, Vashi J (2017) An analytical solution for hoop tension in liquid storage cylindrical tanks. IJEAS 4(7). ISSN:2394–3661

Dave R, Patel R, Vyas P (2017) Design and seismic analysis of ground supported water tank. ICRISET 1:1–6. Kalpa Publications in Civil Engineering

Ghosal M (2019) Water tank analysis using STAAD PRO. Int Trans Eng Sci 1(2)

Haqbin SH, Patel CD, Patel BR (2017) Seismic analysis of RCC elevated rectangular water tank using is 1893 (Part2):2006 draft code (2016). Eng Int J Approx Reason

IS: 1893–1984 Criteria for earthquake resistance design of structures. a. Part 1-1962 Recommendations for earthquake resistant design of structures. b. Part 2-2014 design base shear of elevated water tanks

IS: 11682–1985 Criteria for design of RCC staging for overhead water tank

IS 456: 2000 Code of practice for plain and reinforced concrete

IS: 875 Code of practice for design loads. a. Part 1-dead load, b. Part 2-live load, and c. Part 3-wind load

IS: 3370-Design tables for liquid retaining structure. a. Part-1 general requirements. b. Part-2 reinforced concrete structures. c. Part-4 design tables

Jibhakate ML, Bhorkar MP, Bhole AG, Baitule PK (2017) Reuse and recirculation of filter backwash water of water treatment. J Eng Res Appl 7(4), (Part -1)

Kishori MS, Patel CN (2015) Parametric study of circular water tank using different methodology. IJSRD 3(05)

Maheshwari U, Sravani B (2016) Performance of elevated circular water tank in different seismic zones. Int J Technol Res Eng 3(5)

Maidankar S (2015) Seismic analysis of elevated circular water tank using various bracing systems. IJAERS 2(1)

Mohammed HJ (2011) Economical design of water concrete tanks. Eur J Sci Res 49(4):510–520. Euro Journals Publishing Inc. ISSN:1450–216X

Pole SM, Khedikar AR (2017) Seismic investigation of RC elevated water tank for different types of staging systems. IJIRSET 6(7)

Prajapati KD, Mehta NS, Modi RN (2021) Comparative study of water tank by WSM and LSM. IJCRT 9(7). ISSN:2320–2882

Rathod S, Ishwaragol MB (2018) Analysis of overhead water tank with different staging height and base width. IRJET 05(06)

Vakilaadsarabi A, Miyajima M, Murata K (2012) Study of the sloshing of water reservoirs and tanks due to long period and long duration seismic motions. Eng Environ Sci Geol

Chapter 19
Review on "Seismic Performance of Buildings with Fixed Base and Lead Rubber Bearing Base Isolator"

Gaurav Gupta, Kuldeep Dabhekar, and Dilip Mase

Abstract In recent times, there has been a noticeable rise in the frequency of seismic events, leading to the collapse or damage of numerous structures. The primary goal of this project is to enhance building resilience through earthquake-resistant structures. To mitigate structural damage during earthquakes, the base isolation system is widely adopted and employed globally. In this literature survey, experienced shows that regular structures perform really good as compare to irregular structures. Braces and shear walls, however these practices made the structure heavier. The Lead Rubber Bearing represents an innovative form of earthquake-resistant rubber bearing. It is created by incorporating a lead core into a standard laminated rubber bearing, combining vertical support, horizontal displacement, and hysteric damping within a unified unit. Lead rubber bearing such base isolation mechanisms are designed to separate the superstructure from the foundation, thereby enhancing the structural response to dynamic forces. This approach involves used specialized bearings that are installed between the building's foundation and its lower portion. This review consolidates existing literature and theoretical perspectives on base isolation systems. Some papers explore the efficacy of various isolators and their suitability. The primary objective of this paper is to delve into the foundations of isolators and offer practical guidelines for making informed choices among the numerous available options.

Keywords Lead rubber bearing · Tremor defensive framework · Base isolation

G. Gupta (✉) · K. Dabhekar · D. Mase
G H Raisoni College of Engineering, Nagpur, India
e-mail: gaurav.gupta.mtechstr@raisoni.net

K. Dabhekar
e-mail: kuldeep.dabhekar@raisoni.net

B. V. Khode et al. (eds.), *Innovations in Technologies: Pioneering Sustainable Infrastructure for a Resilient Future*, Sustainable Civil Infrastructures,
https://doi.org/10.1007/978-3-031-82133-2_19

19.1 Introduction

19.1.1 General

The conventional seismic design of buildings relies on structural ductility and redundancy. The impact of severe earthquakes is mitigated through the energy dissipation capacity of structural components and their connections.

Base isolation systems serve to partially disconnect a structure from seismic ground motion through specially designed, replaceable devices positioned between the structure and its foundation. This is exemplified by comparing a standard G + 10 storey building with one protected by base isolation. The conventional building reduces dynamic forces through the plasticity of both structural and non-structural components. To comprehend a building's performance under seismic effects and lateral loads, evaluating its time period is crucial. Understanding the significance of the time period precedes its assessment, as it plays a pivotal role in estimating lateral loads and contributes to seismic evaluations. The time period is contingent on mass and stiffness, making it a key factor in assessing a building's behavior under lateral loads. Despite the challenge of determining the exact time period, it is indispensable for designing earthquake-resistant structures that are both safe and cost-effective. This project aims to explore the various parameters influencing a building's time period through a parametric study aligned with IS code (IS 1893(part I):2016). The study involves modeling and analyzing the structure using ETABS software, considering diverse parameters. A comparison is then made between the time period values obtained from the code and the software.

19.1.2 Base Isolation System

Seismic isolation technology for earthquakes, the structure is either separated or sideways movements of the structure are minimized from ground or foundation motion. Base isolation bearings are engineered to be vertically stiff and horizontally flexible, accommodating the disparity in lateral movement while providing support to the superstructure. This decouples the building from seismic waves, significantly reducing structural response compared to a fixed base. By lowering the fundamental frequency relative to fixed-base buildings, base isolation helps maintain the building's elasticity during an earthquake. This concept is applicable not only to buildings but also extends to other structures like bridges, nuclear power plants, and liquid storage tanks.

19.1.3 Lead Rubber Bearing System

Lead Rubber Bearing (LRB) was firstly invented by Williamson in 1975. It was found in New Zealand. The application of Lead Rubber Bearings in building and bridge constructions is a practical and cost-effective solution for seismic isolation. Consisting of a laminated elastomeric bearing pad, top and bottom sealing, and connecting plates with an inserted lead plug, it offers an effective approach.

19.1.4 Components of Lead Rubber Bearing System

Lead rubber bearings (LRBs) consist of several key components that work together to provide seismic isolation for structures. The main components include:

1. Lead Core: The central component of an LRB is a cylindrical or spherical core made of lead, which is a dense and malleable metal. This lead core is designed to deform under seismic forces, absorbing and dissipating energy during an earthquake.
2. Rubber Layers: The lead core is surrounded by layers of natural or synthetic rubber. These rubber layers provide flexibility and elasticity, allowing the LRB to move and deform while maintaining stability. They also act as isolators, reducing the transmission of seismic forces and vibrations to the structure.
3. Steel Plates: Steel plates are bonded to the rubber layers on both sides of the lead core. These plates help distribute the forces and provide structural support, ensuring that the LRB can transport the vertical load of building or structure.
4. Top and Bottom Bearings: LRBs typically have top and bottom bearings, which allow for controlled movement. These bearings can be designed as sliding bearings or spherical bearings, depending on the specific application and design requirements.
5. Anchor Bolts: Anchor bolts secure the LRB to the structure's foundation and the superstructure. They prevent horizontal movement of the bearing, ensuring proper functioning during an earthquake.

These components work in concert to allow controlled movement of the building during an earthquake, absorbing and dissipating the seismic energy and protecting the structure from significant damage. LRBs are a crucial technology for earthquake-resistant design in regions prone to seismic activity (Fig. 19.1).

19.1.5 Future Scope of Base Isolation System

1. Nations such as the United States, New Zealand, Japan, China, and various European countries have incorporated these methods as standard practice in constructing both public and residential buildings.

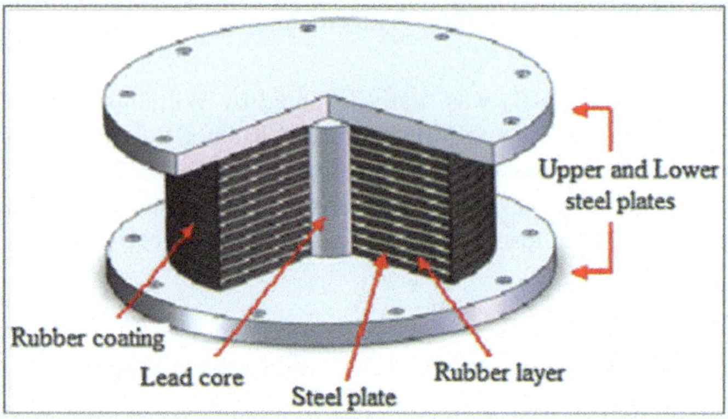

Fig. 19.1 Cross sectional area of lead rubber bearing

2. However, in India, the integration of base isolation techniques in public and residential structures is still in its early stages. Apart from a few exceptions like the hospital building in Bhuj and the experimental structure at IIT Guwahati, most buildings are constructed without utilizing these techniques.
3. Currently, there is a lack of national-level guidelines and codes available for engineers and builders to reference.
4. Buildings equipped with isolation technology remain secure even during powerful earthquakes, with their response typically reduced to 1/2 to 1/8 compared to conventional structures.
5. As isolated structures are subjected to reduced seismic forces, their construction costs are generally lower than those of traditional buildings designed for the same earthquake conditions.
6. Seismic isolation can be implemented in both new and existing structures, allowing for flexibility in design, whether the buildings have regular or irregular layouts.
7. In addition to reducing the overall movement of the structure during earthquakes, base isolation also safeguards non-structural elements and instruments, providing comprehensive protection beyond structural reinforcement alone (Fig. 19.2).

19.2 Literature Review

In a study conducted by Parhi and Sahoo (2018), the researchers compared impact of earthquake force on RC framed buildings with fixed bases and seismic base isolated. Through the ETAB software, they compared the GF + 10, GF + 15 stories RC Building in seismic zone II on various parameters including storey shear, storey drift, moment, deformed shape, time period, displacement, and rigidity which generated because of the earthquake on diverse plots. The study revealed that the damage to base isolated structures was significantly smaller comparison in fixed base structures.

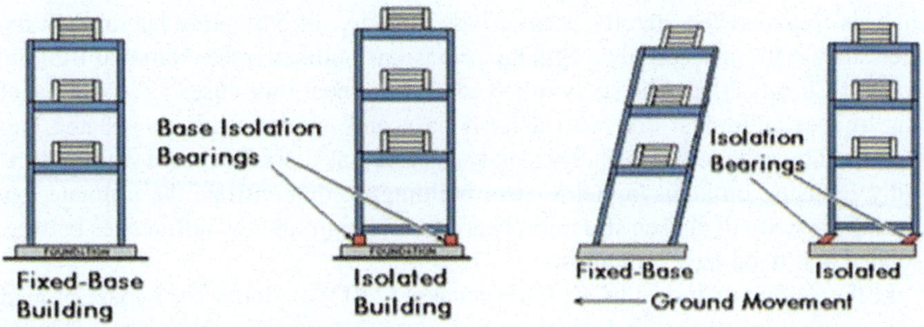

Fig. 19.2 Base isolation system control

As a result, base isolated structures can be occupied immediately after an actual earthquake.

In a study conducted by Lin and Shanto (1993), the researchers compare the structural integrity of fixed base and isolated base coordinated braced and special moment resistant steel frames. They used various design codes for fixed base and base isolation and referred to the 1990 Structural Engineering Association of California (SEAOC) suggested design base shear. The fixed base frames were designed based on the SEAOC recommendations, while the base isolated building was designed for 100, 50, and 25 percent of the suggested horizontal forces. Non-linear time history method was conducted to evaluate various performance measures, including collapsed frames, roof displacement elements, yielded frames, and total relative roof displacement. The results showed that the base isolated building designed for 50 percent of SEAOC of the recommended horizontal force performed superior than the alternative combinations. A comparative study was also conducted to evaluate the peak response of fixed and isolated moment resulting braced steel frames.

In the study conducted by Anwar et al. (2016), they focused on investigating the seismic response of a base isolated system's soil-structure interaction in various soil having different types, including stiff clay, limestone, and loose sand. The study involved 18 and 14-story reinforced concrete buildings with both symmetric and asymmetric floor plans. Their research consisted of nonlinear analyses of 18 different models, categorized into three groups: models with a fixed base, models considering soil-structure interaction, and models with a base isolation system. The aim was to compare key seismic parameters such as base shear, storey displacement, and torsion using SAP2000 software. One of the notable findings was that structures situated on medium and soft soils exhibited more significant seismic effects compared to those on hard soil. The study also highlighted that a base isolation system was particularly effective in hard soil conditions, leading to reduced storey displacement, base shear, and torsion. Specifically, the base-isolated structures with soil-structure interaction showed a substantial reduction in torsion: 81% on limestone, 76% on compact clay, and 75% on unstable sand.

In a study conducted by Mahmoud et al. (2012), investigating behavior of steel buildings incorporating structural rubber bearings. They conducted a comprehensive analysis to assess the effectiveness of base isolation in mitigating lateral drift and inter-story drift, in accordance with the provisions outlined in the National Building Code of Canada 2010. The study involved the nonlinear time-history response analysis of two buildings: one with a fixed base and another with an isolated base supported by rubber bearings. This analysis served as a practical case study, specifically focusing on an asymmetric steel building, to demonstrate the ultimate load bearing capacity of chosen structural bearings and highlight the differences between isolated and fixed-base buildings.

In the study conducted by C. P. Providakis (2007) explored the impact of lead-rubber isolation systems and supplemental viscous dampers on base-isolated buildings exposed to near-fault earthquakes. Providakis examined two plan-asymmetrical buildings, one 6-storey and the other 5-storey, both made of 3D reinforced concrete. These buildings were equipped with lead-rubber isolation systems and supplemental viscous dampers placed beneath each column. The analysis involved nonlinear time history simulations using ETABS software. The findings of this research indicated that the utilization of lead-rubber isolation systems and supplemental viscous dampers is crucial in regions prone to near-fault earthquakes, as they effectively mitigate the destructive effects of seismic activity. However, it was also noted that in far-fault earthquakes, the high damping characteristics of these isolation systems can lead to an increase in the building's stiffness. This increase in stiffness controls structural displacement but may transfer energy to higher modes, potentially contradicting the primary objective of reducing superstructure drift through the use of isolation systems.

In a study conducted by Sekar and Kadappan (2015), an examination was carried out to compare the performance of base-isolated bearing and rigid base reinforced concrete buildings under varying soil conditions, seismic zones, and ground slopes. The research involved the analysis of a multi-story RC building situated on both sloping ground and normal surfaces, designed both with and without isolators. Different story heights, ranging from 1st to 10th stories, were considered across different seismic zones. The building's layout was rectangular, measuring 12×16 m and situated in a seismic zone. The study employed dynamic linear analysis utilizing the response spectrum method, yielding diverse results contingent on the terrain and seismic zone conditions. The introduction of isolators was observed to extend the fundamental natural time period of the structure while concurrently reducing base shear. However, it was noted that the building's story drift increased. These outcomes indicated that the use of base isolators had the effect of rendering the building more rigid, resulting in an extended natural time period. It's important to emphasize that the findings were influenced by the specific terrain conditions and seismic zones under consideration.

In a study conducted by Tian Xue and Ming in 2008, they introduced the concept of seismic isolation using laminated rubber bearing isolators within a frame-shear wall structure. Their research involved a comparative analysis between an isolated structure and a non-isolated one. The findings demonstrated that employing a base isolation method can effectively reduce both displacement and base shear in a structure. Moreover, it was noted that the connection of the isolation bearing without the need for a top fixing plate is a safe design choice, which also contributes to a significant reduction in the overall cost of the isolation bearing system.

In 2008, Kalantari et al. conducted research on the impact of utilizing two distinct seismic isolators: LRB isolators and friction pendulum isolators. They generated structural framework with varying numbers of stories (two, five, eight and twelve) for three scenarios: lead rubber isolator, and friction pendulum isolator, each with divergent rigidity levels. The findings indicated that friction pendulum isolators effectively reduced structural displacement compared to both fixed-base and lead rubber isolators.

In 2014, Dhawade conducted a research study to investigating impact of employing base isolation in a 14-story frame structure subjected to seismic forces, comparing it to a fixed-base structure. The GF + 14 framework structure was design with base isolation utilizing ETAB software. HDRB was chosen as the isolator due to its exceptional flexibility and energy absorption properties. The application of base isolation resulted in a notable reduction in both storey overturning moments and storey shear forces when compared to the non-isolated, fixed-base building.

In 2015, Jong Wan Hu conducted a comprehensive examination of the benefits associated with the implementation of Lead Rubber Bearing as a isolation systems. The study involved non-linear dynamic time-history method utilizing near-fault ground motions. It focused on assessing seismic responses concerning base shears and inter-storey drifts, comparing structures with and without LRB isolation systems. The primary purpose of the LRB isolator at the base was to extend the structural vibration period by enhancing lateral flexibility within the frame structure. Consequently, this led to a significant reduction in ground accelerations transmitted to the superstructure. The LRB isolator devices were found to effectively prolong the fundamental vibration period throughout the entire frame structure.

In 2014, Ashish R. Akhare conducted a comparative study on impact of High-Density Rubber Bearings and Friction Pendulum Systems on a standard GF + 12 reinforced concrete hospital building. The analysis involved non-linear time history simulations using SAP2000. The results revealed a significant reduction in base shear for buildings equipped with Friction Pendulum Systems, indicating their superior performance compared to those with High-Density Rubber Bearings. Interestingly, the effects on storey displacement were minimal and similar for both isolator types. Additionally, it was observed that, across different storey heights, Friction Pendulum Bearings exhibited less variation in storey displacement, story drift, and story accelerations compared to buildings equipped with HDRB, although these priciples converged at the upper storey of the structure.

P. B. Mrunal and Sameer S. Shaikh explored concept of base isolation at various levels within a three-story building. They conducted experiments on a building, both with and without base isolation, positioned at different levels within the structure. Their investigation revealed that attaching isolators to upper floors led to a reduction in acceleration, resulting in a lower risk of damage to those upper floors when compared to isolators placed at the building's foundation. For their study, they employed laminate rubber bearings (LBR) as the isolating elements due to LBR's composition, which consists of different layers of rubber and steel. The researchers subjected both building configurations to time history motion analysis, recognized as the most suitable method for replicating earthquake effects. The results of their analysis demonstrated a significant reduction in acceleration when base isolators were positioned at the plinth level and the first-floor level.

In 2021, a research team led by N. Lingeshwaran, Tummuru Manish Kumar Reddy, Satrasala Koushik, P. Preethi, and others conducted a study involving both symmetric and asymmetric structures with various shapes, including T-shapes, L-shapes, H-shapes, and rectangular shapes. Their models consisted of G + 9 story buildings, and they utilized the ETABS software for their analysis. The primary focus of their research was to compare story displacement and story drift across these different structural configurations. The study's findings indicated that symmetric structures exhibited lower displacement compared to asymmetric ones. Notably, T-shaped buildings were identified as more vulnerable to earthquake loads than symmetric structures. Additionally, the research showed similar displacement behavior in both L-shaped and H-shaped structures. Overall, symmetric structures demonstrated superior seismic performance in comparison to their asymmetric counterparts.

In 2013, Chandak N. R. conducted a comparative analysis of RC framed buildings with fixed bases and those with isolated bases employing rubber bearings and friction isolators. The study employed both the response spectrum method and the finite element method for assessment. In the case of friction-type isolators, there was a noticeable reduction in base shear and relative drift. It was observed that the relative drift values were inversely related to the number of stories in the building. Interestingly, the Indian Standard (IS) code provided the maximum displacement value, while the Euro code offered the less displacement value for buildings with fixed bases.

In the study conducted by Sushma G. Sawadatkar in 2017, various methods of base isolation were investigated to mitigate potential damage to structures, particularly in the field of civil engineering. The fundamental idea of base isolation is exemplified through an illustration of a building settling on frictionless rollers. When the ground experiences seismic activity, these rollers allow free movement, while the building itself remains stationary, effectively preventing the transfer of seismic forces to the structure. In essence, base isolation involves introducing a structural discontinuity at the foundation, ensuring that the isolators bear the full brunt of the earthquake forces. The study's conclusion highlights the effectiveness of the lead rubber bearing method when compared to alternatives like high-density rubber bearing and the friction pendulum method.

In their 2013, Donato Concellara et al. discussed on distinctions among lead rubber isolators and friction sliders. They devised an isolator that combined a lead rubber bearing with a friction slider (FS), naming it a high damping hybrid seismic isolator (HDHSI). The seismic response of the HDHSI was contrasted with that of a traditional lead rubber isolator under varying seismic conditions, including differences in frequency and intensity. The study focused primarily on HDHSI, comparing it to the conventional lead rubber isolator. A range of seismic scenarios were considered, and the results were analyzed in terms of parameters such as base shear, displacement, and shear force at the superstructure's base. The comparative analysis indicated that HDHSI provides superior protecting during several seismic events compared to other isolators.

In 2015, Hosseini and Rao conducted a decision that made seismic behavior tall buildings equipped with shear wall positioned of both the central core and the center of various external perimeter, complete openings. They developed a model of a 40-story building and carried out an analysis using E-TABS software, employing various methods including the response spectrum method, equivalent static lateral force method, and time history analysis in accordance with IS codes. The study's findings revealed that placing shear wall at central core and along the external perimeters is openings led to improved performance of maximum storey displacement, storey drift, and base shear, and exhibited higher stiffness characteristics. Additionally, it was observed that the more stress induced by openings in shear wall deviated by approximately 5% compared to shear walls without openings.

In 2019, Dar and Singh conducted a dynamic performance of a base isolator tubular high building system utilizing Lead Rubber Bearings (LRB) within the E-TABS software. They developed models of rigidly connected RCC frame buildings with 29 stories (GF + 29) in various configurations, including simple tube, simple frame, tube-in-tube, and bundled tube, both with and without LRB-based base isolation. Relational assessments were carried out for each model based on IS 1893:2002 (Part 1) criteria. The study revealed a significant decrease in the model period, with reductions of 24, 13, 10 and 26% when base isolation was implemented in simple RCC frame, simple tube, tube-in-tube models, and bundled tube respectively. Moreover, the application of base isolation led to a reduction in base shear ranging from 10 to 30% across all models, along with a decrease in overturning moments within the buildings.

19.3 Conclusion

After analyzing the above mentioned study, it is evident that the performance of RC and Steel frame structures varies between fixed base and base isolation systems. Base isolation techniques prove to be more effective in enhancing structural performance during seismic events. The role of base isolators is crucial in earthquake resilience. In areas with hard soil strata, the structural response is generally satisfactory. However, in soft soil conditions, acceleration increases, leading to decreased energy dissipation and higher frequencies. The placement of base isolators at the foundation is preferable as it yields consistent results. Notably, base-isolated buildings exhibit superior bending capacity compared to fixed base and shear wall systems, even in regions with settled density near fault lines. Isolation systems demonstrate good efficiency in low to medium-height buildings.

While implementing base isolation may initially increase construction costs, it offers significant benefits for general use. This technique, primarily essential for critical facilities like hospitals and police stations, etc. ensures safety and minimizes structural damage. Notably, buildings with base isolation can achieve higher performance levels, such as Intermediate Occupancy and Life Safety, which are often unattainable in structures lacking this technology.

References

Akhare AR, Wankhade TR (2014) DzSeismic performance of RC structure using different base isolator. IJESRT 3(5):724–729. ISSN:2277–9655

Anwar SF, Hasmath M, Khan MA (2016) Share, "study on soil-structure interaction and base isolated system for seismic performance of structures resting on different types of soils." Glob J Eng Sci Res 3(10):61–67

Cancellara D, De Angelis F, Pasquino M (2013) A novel seismic base isolation system consisting of a lead rubber bearing in series with a friction slider. J Appl Mech Mater 256–259:2174–2184

Chandak NR (2013) DzEffect of base isolation on the response of reinforced concrete building. J Civ Eng Res 3(4):135–142. https://doi.org/10.5923/j.jce.20130304.02

Dhawade SM (2014) Comparative study for seismic performance of base isolated & fixed based RC frame structure. Int J Civ Eng Res 5

Dhiraj NS, Pravat KP (2018), Base isolation of residential building using lead rubber bearing technique. Int J Eng Res Technol (IJERT) 07(05) Bhubneswar, Odisha, India

Hosseini M, Ramana NV (2015) Earthquake analysis of high rise buildings with shear walls at the center and center of each side of the external perimeter with openings. Int J Sci Res 5(12):59–71

Kalantari SM, Naderpour H, Hoseini Vaez SR (2008) Investigation of base-isolator type selection on seismic behavior of structures including story drifts and plastic hinge formation. The 14th world conference on earthquake engineering, China

Lingeshwaran N, Koushik S, Reddy TMK, Preethi P (2021) Comparative analysis on asymmetrical and symmetrical structures subjected to seismic load. ELSEVIER. https://doi.org/10.1016/j.matpr.2020.11.340

Mahmoud S, Mahmoud A (2012) Building with base isolation techniques. Al-Azhar Univ Eng J 7

Nagpure A, Sanghai SS (2019) Effect of floor diaphragms on seismic response of RCC framed building—a review. Smart technologies for energy, environment and sustainable development: select proceedings of ICSTEESD 2018, pp 323–333

Providakis CP (2007) Effect of LRB isolators and supplemental viscous dampers on seismic isolated buildings under near-fault excitations. Eng Struct 30:1187–1198

Sekar T, Kadappan PL (2015) Seismic analysis of multistory buildings resting on normal and sloping grounds in different seismic zones with and without base isolator. Imanager J Struct Eng 4(1)

Shenton HW, Lin AN (1993) Relative performance of fixed based and base isolated concrete frame. ASCE J Struct Eng 119(10):2952–2968

Chapter 20
A Review on Cable Stay Bridge

Pradnya Nimsarkar and P. Y. Pawade

Abstract Cable-stayed bridge designs exemplify the harmony of innovation, elegance, and practicality that characterizes modern civil engineering. The purpose of this review is to investigate and assess general facts on pipelines, providing a complete analysis of their historical development, design, modification, construction, and usage in diverse contexts. This article delves into the history of cable connections, highlighting the technological advances that led to their invention, tracking their evolution to current designs, and assessing practical applications. An examination of the link's structure and performance indicates that it is inefficient in terms of transportation and the environment. Case studies of well-known cable-stayed bridges provide insight into the effective use of novel concepts, architectural characteristics, and inventive solutions. The assessment also discusses sustainability, future prospects, and innovations that have the potential to radically transform the area of transmission line electrical engineering. This study combines knowledge from numerous domains to provide relevant information to scientists, engineers, and anybody else interested in the development and advancement of cable connections.

Keywords Cable stay bridge · Analysis · Seismic behaviour · CRPF

20.1 Introduction

The bridge is an example of architecture that expertly combines architectural beauty with technological functionality. The cable-stayed bridge, with its distinctive profile and innovative design, has come to symbolize modern engineering philosophy. These bridges' massive spans and gorgeous skylines, which decorate skylines all over the world, are made possible by the interconnection of cables that distribute stresses

P. Nimsarkar (✉) · P. Y. Pawade
G H Raisoni College of Engineering, Nagpur, India
e-mail: pradnya.nimsarkar.mtechstr@ghrce.raisoni.net

P. Y. Pawade
e-mail: prashant.pawade@raisoni.net

B. V. Khode et al. (eds.), *Innovations in Technologies: Pioneering Sustainable Infrastructure for a Resilient Future*, Sustainable Civil Infrastructures,
https://doi.org/10.1007/978-3-031-82133-2_20

exceedingly well and provide structural stability. Steel bridges are becoming the most preferred alternative for spanning varied terrains and meeting public transit demands due to its ability to expand over long periods of time while maintaining their visual appeal. This booklet serves as the foundation for a complete examination of the many aspects of cable-stayed bridges, including their historical evolution, design, modifications, construction techniques, and functional characteristics. Furthermore, the goal of this research is to bring together knowledge from a variety of academic publications and real-world applications to throw light on the complex web of innovations and advances that are transforming the area of cable-stayed bridge engineering.

Advantages of cable Stay Bridge:

Because of its numerous advantages, cable-stayed bridges are the favored choice in civil engineering. The following are some major benefits:

1. **Spanning Long Distances**:

Cable-stayed bridges are ideal for bridging huge bodies of water or steep valleys because they can span longer distances with fewer supports than other types of bridges.

Design Flexibility: Their design is adaptable to a variety of environmental and geographical conditions, allowing for changes in tower heights and span lengths.

2. **Aesthetic Appeal**:

Architectural elegance: Their lovely features, phone connections, and faces all add to their appeal, and they usually serve as key metropolitan icons.

3. **Structural Efficiency**:

Cable-stayed bridges are more structurally efficient than other types of bridges while remaining sturdy and stable.

Better Load Distribution: The cable system's effective load distribution over the bridge reduces stress concentrations and allows for more uniform loading conditions.

4. **Construction and Maintenance**:

Ease of Construction: Because cable-stayed bridges are built with simpler procedures than other types of bridges, they may be erected faster.

Maintenance: Because cable arrangements are often easier to inspect and maintain, they often require less maintenance than suspension bridges.

5. **Suitable for Various Environments**:

Cable-stayed bridges may be designed to withstand a broad range of environmental variables, including severe winds, seismic activity, and temperature variations, making them suitable for usage in a number of locales.

6. **Cost Considerations**:

Cost-Effectiveness: Given that they require less maintenance and have a longer lifespan than some other types of bridges, cable-stayed bridges may prove to be more cost-effective in the long run, even though the initial construction costs may vary.

20.2 Literature Review

1. **"Damage tolerance of prestressing steel in a bridge stay-cable after30 years of service in a harsh environment." Andrés Valiente, M. Iordachescu, A. Aznar, et al.**

This essay's primary focus is on the degradation of prestressing steel in bridge cables after 30 years of usage in harsh environments. The authors tested the tensile strength of corroded steel wire and looked at the effects of corrosion damage on steel products. The report presents the general results and conclusions of the study, without delving into the specific limitations or criteria that were used. This article outlines tests, such as tensile testing and sample scanning electron microscopy (SEM), that are performed on damaged cables and wires. Tensile testing was performed on samples from stationary electrical equipment, and the results were used to evaluate the tensile properties and damage of the prestressing steel. This essay's primary focus is on the degradation of prestressing steel in bridge cables after 30 years of usage in harsh environments. The authors tested the tensile strength of corroded steel wire and looked at the effects of corrosion damage on steel products. The report presents the general results and conclusions of the study, without delving into the specific limitations or criteria that were used. This article does not clearly state how important the research findings are for bridge cable design and maintenance. Advice or instructions for improving prestressed steel's performance and durability in similar publications would be beneficial, given the findings of the study. This report does not highlight any limitations or unanswered questions associated with this experiment. It's important to discuss how limits impact the outcome while testing or interpreting data.

2. **"Application study on the first cable stayed bridge with CFRP cable in China." Kuihua Mei, Yajuan Li, Zhitao Lu.**

The overall material composition of the bridge, the integration of CFRP cables, the tension and static performance of the cables, their quality features, and the reaction of the bridge are all examined. This work analyzes simple and complex models alike. The basic model, which was used for frequency calculations, showed good agreement with the complex model, except for the torsional frequency values. These models, which include the three-girder truss and shell girder models, are used to evaluate the performance of the bridge. The finite element method was used for the study. Testing was done in addition to the bridge's construction. Micrometers, oil pressure gauges, gauges, and gauges are used to measure, extract, and filter the cable tension. When compared to the genuine values, the measured values are determined to be consistent. Overall, this study makes use of models, statistics, and experiments to evaluate several aspects of bridge performance.

3. **"Seismic performance of cable stayed bridges with different geometric conditions." Ahmed M Fawzy, Khaled F El-Kashif & Hany A Abdalla.**

This study investigates the effects of various factors on the static shape and static response of the cable extension. A numerical model developed with the help of the ANSYS program was employed in this investigation. This work uses a three-dimensional structural model to examine connected chains. The analysis was carried out using finite element analysis (FEA) and Midas software. The inverted Y-shape, the H-shape, and the A-shape are among the forms that the model takes into account. This study's examination also took cable sag into account. Overall, this work uses mathematical models and finite element analysis to investigate how tower modifications affect cable extension operations. Research has shown that a variety of factors, such as cable inclination and tower height, influence a bridge's performance. Studies show that the geometry of bridges, which comprises This study also demonstrated that the width of the bridge deck has a major impact on the exterior shear force brought on by seismic forces. This study presents a method for developing a linear model to anticipate deck length. The results of the study indicate that the proposed method may be used in the design of steel pipes affected by seismic reinforcement.

4. **"An experimental study on bond-type anchorages for carbon fiberreinforced polymer cables." Kuihua Mei, Rudolf Seracino, Zhitao Lv.**

The experiment looked into the type of contract, how long it lasted, and bad conduct. Examples from experiments are used in this study. This experiment looked at five circular multi-row matrix bonded anchors that are meant to be used in CFRP cable systems conduct, an anchoring sealing, etc. To evaluate the mechanical properties and weight carrying capability of CFRP bars, conduct out experimental testing. This study evaluated five muscle matrix-bonded fastening techniques for CFRP cable systems, evaluating their ability to sustain loads, maintain tensile strength, and handle load shear. The results showed that the failure modes in each sample were caused by ruptured tendons, proving the efficacy of the sticky fastening method. As the bond length increases, the average stress on the bond decreases, and the distribution of tension on the bond varies with bond length. The load shear curves of the resin-based anchors show non-linear variations. A rise in measurement is thought to suggest that steady system operation may be possible when the load is less than the 2 mm limit. It was found that the muscle tension levels in each cable model vary, which enhances handling. This research provides designers with information on how to create connection systems for cable-stayed bridges using CFRP cables.

5. **"Force based & displacemet based reliability assessment approaches for highway bridges under multiple hazard actions." Chao Huang, Jeery J Shen, Mi Zhou & George C Lee.**

This article does not address the reliability of bridges in atypical conditions, especially when dead, live, and seismic loads are involved. The model used in this investigation was mathematical in nature. The two methods were compared using nine simple bridge models proposed in previous research. This sample was selected as a representative extension for further analysis. Column heights, box beam heights, spans, and seismic design for columns are among the details of these models. The force-based technique and the perception-based method were the two efficient approaches that Lee et al. employed to evaluate the bridge's dependability. The migration path offers more integration and confidence in the chosen model. Calculating the percentage and overall failure rates as well as the results' normal distribution was part of the reliability tests used in the study. This research, in general, uses mathematical models to evaluate bridge dependability and analyzes the results of different approaches. This article describes two methods for assessing a bridge's dependability in the event of severe weather that consider continuous, live, and seismic loads. The theory-based technique can provide more integration and reliability than the force method when it comes to bridge building. The results showed that the variable's test dependability was steady and reliable. However, more calibration and validation are required to ensure that these methods are accurate. This study highlights how important it is to consider a range of hazards to ensure the efficacy and security of bridge construction.

6. **"Geometry Control Simulation for Long-Span Steel Cable-Stayed Bridges." Jie Wu, Dan M. Frangopol, Mohamed Soliman.**

This paper presents a numerical model for the geometric control simulation of long-span steel cable-stayed bridges. In addition to the geometric and physical properties of the bridge, the model takes into consideration the sag caused by lengthy cables. This simulation uses the Sutong Bridge in China as an example to show how beneficial methods are for computing geometric nonlinear effects. This content is unrelated to any clinical or experimental research. This paper presents the geometric control simulation technique and software design for a long-span steel cable bridge that takes geometric nonlinear effects into consideration. BGCS design software may be used to test the building process and reinforce the bridge connection. The method uses several short points to simulate the drooping effect of lengthy cables and adjusts the length of the unstretched wire to simulate the stretching phase. In geometric control, three forms of actuation are taken into consideration: fixed position, parallel, and tangent types. The results show that neglecting geometric nonlinearity may lead to large errors in the final geometric arrangement. The program developed within the AutoCAD environment has outstanding computational performance and user interface.

7. **"Influence of earthquake input angle on seismic response of curved girder bridge" Yongjun Ni, Jiayu Chen, Hualiang Teng, Hui Jiang. Operating parameters:**

This study looks into how the seismic penetration angle affects the seismic response of curved girder bridges. The seismic study of the bridge identifies changes in the radius of curvature and the arrangement of the supports. This research looks at the function of curved bridges as well as strut yielding and deformation, expansion joint sliding and collapse, curvature, and pier-beam connection type. This study also looked at the effects of factors such as insulation measurement, foundation position, foundation table form, and beam curvature on the seismic response of the curved beam. The maximum seismic response of a curved girder bridge is significantly influenced by the prescribed seismic input angle. Because of the structure's non-uniformity, moments and torsional forces combine to create communication. The most negative seismic input angle that produced the maximum seismic response of the curved bridge was found by carefully examining and contrasting the computing results from various methods. Bridge piers have a triangular seismic field fluctuation in relation to the entry angle, with one pier displaying a maximum field that is significantly higher than the other piers. The theoretical model makes it easy to evaluate the worst aspects and requires less data to be investigated. For the seismic design of curved bridges, the SRSS combination provides a conservative measure, and the response values of different combinations are often comparable. This paper provides a single formulation for the seismic penetration angle and illuminates the seismic field of curved bridges.

8. **"Optimization of cable pre-tension forces in long-span cable-stayed bridges considering the counterweight" Chaolin Song, Rucheng Xiao, Sun Bin**

Verifying parameters: The cable pre-tension determines the counterweights of long-span cable-stayed bridges dispersion of the beam's bending time. Details of the Model:

Create a finite element (FE) model in order to determine geometric nonlinearity. The continuous beam method is used to determine the distribution of cable pre-tension. The revised structure is meant to reduce the total bending strength. Optimization algorithms do a variable-step search for each design in order to address optimization challenges. The optimal model was created using ANSYS, taking geometric nonlinearity into consideration. ANSYS input files (APDL) are created using the ANSYS Parametric Design Language. The optimization algorithm software is built in MATLAB. Accuracy analysis is used in ANSYS to determine target values. This paper proposes an optimization technique that, by taking counterweights into account while calculating the strength of long steel pipes, lowers the maximum bending moment of beams and towers in the building of asymmetric bridges. The recommended method is precise and effective even while processing a wide range of designs, including ones with heavy elements. The results emphasize the importance of weight assessment since it increases the bending moment of the beam in the design of asymmetric cable-stayed bridges. The proposed approach is more exact and accurate than actual models and load models because it computes the cable pre-tension while accounting for weight using ANSYS and MATLAB. To calculate the step size—which is necessary to determine the optimal cable pre-tension—the cross section is set to 10 kN.

9. **"Design and behaviour of super-long span cable-stayed bridge with CFRP cables and UHPC members" Liang Ren, Zhi Fang, Kai Wang.**

The geometric length of the steel girder cable tower, normal stone tower, and cable bridge is 1088 m. The stiffness, shear strength, local stability, and puncture shear resistance of the ultra-high performance concrete (UHPC) component of the carbon fiber reinforced polymer (CFRP) cable determine the design and operation of cable junctions combining UHPC and CFRP cables. The strength characteristics of two belts—static performance, overall stability, and aerostatic stability—are examined using the finite element method. According to the energy balancing concept, the UHPC tower's physical characteristics meet the same requirements. Performance concerning seismicity and modality. This study compares and displays a cable-stayed bridge with carbon fiber reinforced polymer (CFRP) cables and ultra-high performance concrete (UHPC) components to steel structures. Because of their exceptional attributes and careful design, the results show that using CFRP cables and UHPC pieces results in robust, high-quality bridges. Because of the properties of the CFRP and UHPC materials, the bridge's main span may extend up to 1000 m. UHPC beams have several advantages, including as stability, anti-slip properties, and drilling capabilities, and are customized to meet local specifications. Sag effects are reduced and voltage and temperature uniformity are raised by employing CFRP cables. UHPC towers and CFRP cables improve the bridge's overall stability, aerostatic stability, dynamic mode, and seismic performance. Compared to the prior steel cable-stayed bridge, the new system has a greater seismic deformation resistance, critical frequency, and stability coefficient. This research shows that the recommended connection employing UHPC parts and carbon fiber cables gives fair value in terms of capacity and performance, despite the article's high cost.

10. **Transverse seismic response of diamond-shaped pylon in cable-stayed bridge: Experiment and analysis. Chao Zhang, Matthew Brewer, Jianian Wen, Qiang Han**

The effects of the shape, loading methods, and beam stiffness of the Diamond Tower. Damage modes and hysteretic response to cyclic quasi-static loads of a rhombic tower. The many failure modes and forms of damage that can happen to diamond-shaped towers are determined by the loading methods, mid- and low-beam stresses, and tower design. The exterior performance of the diamond tower was examined through experimentation and computational analysis. When the length reduction of the diamond-shaped wire tower was examined under cyclic quasi-static stress, it was found that plastic hinges formed around the center junction of the two tower legs. When the rhombus tower numerical model was developed and evaluated in compliance with the experiment, similar outcomes were obtained. A parametric study was conducted to find out how the Diamond Tower's geometric layout, loading method, and beam stiffness influenced its seismic performance. The damage and its location in the diamond-shaped tower are determined by the load plan, beam stiffness, and tower design, according to the study's findings. Specific designs that are derived from fiber content are valid and beneficial for further study. Unlike other towers, the lower beam of the diamond-shaped tower has the lowest possible beam, which reduces the stiffness of the beams and can reduce earthquake damage by just tickling the legs of two persons. The damage certificate (psn) is a tool used in engineering applications to find damage to diamond towers.

11. **"Deformation-based pushover analysis method for transverse seismic assessment of inverted Y-shaped pylons in kilometer-span cable-stayed bridges: Formulation and application to a case study" Jingcheng Wang, Xiaowei Wang, Aijun Ye, Zhongguo Guan**

This work focuses on evaluating the external performance of the Y-shaped tower of the mile-span cable-stayed bridge using the deformation-based pushover analysis

(DPA) technique. Details of the model: A numerical model was utilized for analysis in this study. The DPA concept was used in the study paper to analyze the performance of the inverted Y-shaped tower of Sutong Bridge. We examined the computational expenses associated with several analyses, including additive analysis (IDA), competitive analysis (CPA), and multivariate analysis (MPA). The deformation-based pushover analysis (DPA) approach is a reliable and effective method for seismic study of the inverted Y-shaped tower of a kilometers-long cable-stayed bridge. It is shown that the DPA method, which utilizes a new variable load model and uses the equivalent displacement rule, is 2.3 times more efficient for tall skyscrapers. Through a comparative comparison of studies employing modal pushover analysis (MPA), incremental dynamic analysis (IDA), and conventional pushover analysis (CPA), it was shown that DPA provides a reasonably accurate estimation of the elastic-inelastic seismic response of the landmarks, with error rates typically around 30%. In contrast to impulse collision analysis (CPA), which may provide an inappropriate seismic response, motion analysis (MPA) might produce accurate predictions in the opposite direction. Because DPA is effective and only needs one test using the original model, it is appropriate for the initial construction of the Y-shaped bridge. This article does not go into great detail on the results or findings of the Sutong Bridge case study or how well DPA performs in different settings.

12. **"Seismic performance assessment of a sea- crossing cable-stayed bridge system considering soil spatial variability" Chao Li, Yucheng Diao, Hong-Nan Li, Haiyang Pan, Ruisheng Ma, Qiang Han, Yihan Xing**.

This work aims to evaluate the collapse performance of coastal cable-stayed bridges (SCCS) by taking spatial ground movement into consideration. The impact of varying ground elevations on the seismic dynamic response at several locations along the SCCS bridge was evaluated. To ascertain the relative importance of various soil components in the seismic reaction, sensitivity analysis was employed. Stability is the primary aspect affecting how SCCS bridges respond seismically. The aim of this research is to ascertain the impact of ground spatial variability on the seismic performance of marine cross-cable bridges (SCCS). With the use of random field theory and the Latin hypercube sampling technique, it generates an uncertain field that represents different soil conditions. An enhanced 3D finite element model of a sample SCCS bridge was developed for use in soil interaction (SSI) and to stochastically simulate spatially variable seafloor seismic motions (SVSSM) with soil uncertainty. Using the storm test method, the relative sensitivity of various soils to the dynamic response of SCCS bridges was examined. The results show that different ground conditions affect SVSSM and SSI, which results in significant differences in the seismic response of the bridge. Ignoring fluctuations in soil conditions devalues the lifespan of SCCS bridge systems and components. Soil density and shear modulus have the greatest effects on the seismic response of SCCS.

13. **"Surrogate-based fragility analysis and probabilistic optimisation of cable-stayed bridges subject to seismic loads" Andrea Franchini, Wendel Sebastian, Dina Ayala**

This essay's primary subjects are fragility analysis and steel pipe optimization under seismic pressures. Column and tower sections were two of the main design parameters that were examined. This study also considers the bridge's dynamic response, and the dynamic response of the continuous bridge is used to characterize the mathematical model. This article discusses the best way to improve water supply in areas that are prone to earthquakes. Parametric fragility functions (PFF) are used in the design of cable-stayed bridges to define the choice variables for learning and optimization. Four optimization strategies are proposed to reduce earthquake damage, building charges, and repair costs. This research modifies the test model and provides a case study of the structure in order to evaluate the quality aspects of the 542 m three-span cable-stayed bridge. Numerical optimization techniques aim to manage inconsistent assumptions and diverse interactions. When optimization targets are satisfied independently, there is a considerable improvement in damage, repair costs, and construction and repair expenses. A set of Pareto-optimal design variables is created to simultaneously optimize the mean maximum ground velocity (PGV) and construction and maintenance expenses. This concept serves as the basis for further developments in multi-hazard design.

14. **"Cable loss performance investigation of cable-stayed bridge equipped with Roll-N-Cage isolator" Mohammad Omran, Abbas Karani**

The wire loss performance of connectors with separate wires was investigated. We looked at the circumstances of increasing collapse during seismic ground movements. Details of the model: The connection structure of RNC insulators was investigated using mathematical modeling in line with the three characteristics of RNC insulators. Create three-dimensional finite element models with Opensees software for nonlinear dynamic analysis. The bridge model was examined using three different earthquake scenarios: the Imperial Valley, Northridge, and San Fernando. The bridge model in its entirety and the bridge model without the wires were the two scenarios that were studied. The RNC isolation model includes parameters for power dissipation, design modification, return stiffness, hysteresis, and efficiency. This study looks at how an isolated Roll-N-Cage (RNC) cable bridge performs under difficult conditions and how it is affected more when there are earthquakes. RNC insulators with mechanical damping can reduce axial stresses and the ongoing movement of adjacent lines, provide structural stability, and halt cracking. Losses in the cables connecting the deck to the towers may cause increased axial stresses in neighboring cables and eventually cause permanent displacement. RNC isolation effectively reduces power and variance, especially while the machine is idle. Time course dynamic investigation indicates that the permanent area, maximum lay displacement, and stress history of surrounding cables are significantly reduced when RNC insulation is used in conjunction with active buffering devices, hence minimizing the hazard of development. The study's findings demonstrate how RNC insulation might improve cable extensions' earthquake performance by lowering pulse amplification.

15. **"Estimating tension of a prestressed concrete cable-stayed bridge under construction and traffic use conditions using a vision-based system". Sung-Wan Kim, Dong-Uk Park, Jin-Soo Kim, and Jae-Bong Park**.

This research focuses on the use of visual-based stress prediction for prestressed concrete cable-stayed bridges in construction and traffic. The main analytical metric in this article is cable tension, which is important in assessing the integrity and stability of cable-stayed bridges. This paper also covers the dynamic responsiveness of cables as measured by image-based sensors. The accuracy of stress estimation utilizing the vision system was evaluated by comparing the stresses measured by the vision-based system with the stresses seen during testing and the stress measurements performed using the vibration approach. Furthermore, the duration was evaluated and the tension between the wires in the current cable-stayed bridge and the tension recorded during construction were compared. The research focused on assessing stress in prestressed concrete cable-stayed bridges using an image-based method and an indirect measurement of cable tension. Vision-based systems assess the dynamic response of cables using sensors and image processing techniques to improve identification and accuracy. The less than 3% discrepancy between the voltage calculated by the vibration technique and the voltage measured during disassembly shows the accuracy of visual measurement. The fact that the cable tension is still good after 10 years of use demonstrates the integrity and safety of the bridge. Vision-centered vibration systems have proven to be successful and economically feasible because they provide a non-contact means of controlling stiffness throughout the construction and operation of vehicles.

20.3 Conclusion

Cable-stayed bridges represent an innovative and efficient engineering solution for spanning long distances with elegance and structural integrity. Through this review, we have examined various aspects of cable-stayed bridges, including their design principles, structural components, construction techniques, and notable examples worldwide. These bridges offer several advantages over traditional bridge designs, such as enhanced aesthetic appeal, reduced material usage, and greater flexibility in span length. Their unique cable-stay system efficiently distributes loads, allowing for longer spans without the need for excessive support structures.

Furthermore, cable-stayed bridges have demonstrated exceptional performance in withstanding environmental factors such as wind, earthquakes, and dynamic traffic loads. Their adaptability to various site conditions and architectural requirements makes them a preferred choice for many modern infrastructure projects. However, it's important to acknowledge that the construction and maintenance of cable-stayed bridges entail significant engineering challenges and costs. Proper planning, meticulous design, and rigorous inspection protocols are essential to ensure the safety and longevity of these structures.

Overall, the review underscores the importance of cable-stayed bridges in contemporary civil engineering, offering a balance of functionality, efficiency, and aesthetic appeal. As technology continues to advance, we can expect further innovations in the design and construction of these iconic structures, contributing to the sustainable development of transportation infrastructure globally.

Bibliography

Athira VV, Sruthy S (2018) Investigation on the use of cold formed perforated steel sections as columns and purlins. Int J Eng Technol 7(4.5)

Chao H, Shen JJ, Zhou M, Lee GC (2015) Force based & displacemet based reliability assessment approaches for highway bridges under multiple hazard actions. J Traffic Transp Eng 4:223–232

Deepla N, Venkat R (2018) Comparitive study on cold formed light gauge steel angle sections and channel sections of compression members. Res Rev J Appl Sci Innov RRJASI 2(2)

Indian Standards IS: 800: 2007 "Code of practice - General Construction Steel" New Delhi: BIS; 2007

EN 1993-1-3 Design of steel structures–General rules–Supplementary rules for cold-formed members and sheeting (Euro code)

Fawzy AM, El-Kashif KF, Abdalla HA (2022) Seismic performance of cable stayed bridges with different geometric conditions. J Eng Appl Sci 69:82

Franchini A, Sebastian W, Ayala D (2022) Surrogate-based fragility analysis and probabilistic optimisation of cable-stayed bridges subject to seismic loads. Eng Struct 256

IS 811(1987): Cold formed light gauge structural steel sections

IS 801 (1975): Code of practice for use of cold formed light gauge steel structural members in general building construction

IS 2062-2011: Hot Rolled Medium & High Tensile Structural Steel

Jie W, Frangopol DM, Soliman M (2015) Geometry control simulation for long-span steel cable-stayed bridges. J Eng Struct 90:71–82

Journal-A Beginner's Guide to Cold-Formed Steel Framing-Build steel USA

Journal-Rapid Building System (Light Gauge Steel Buildings)-Volta Green Structures Hyderabad, Telangana

Kakade SA, Bhandarkar BA, Sonar SK, Samare AD (2014) Study of various design methods for cold formed light gauge steel sections for compresive strength. IJRET Int J Res Eng Technol 03(09). NCETCE2014

Kalbhor SR, Shelke N, Kadlag V (2020) Optimization of Industrial Structure using Light Gauge Steel Section. Int Res J Eng Technol (IRJET) 07(08)

Katkar DD, Phadtare NP (2015) Comparative Study of an Industrial Pre—Engineered Building with Conventional Steel Building. Int Res J Eng Technol (IRJET) 05(10)

Kim SW, Park DU, Kim JS, Park JB (2022) Estimating tension of a prestressed concrete cable-stayed bridge under construction and traffic use conditions using a vision-based system. Structures

Konkong N (2017) An Investigation on the ultimate strength of cold-formed steel bolted connections. Eng, Technol Appl Sci Res 7(4)

Kumbhar SS, Jadhav HS (2021) Current research on cold-formed steel structures: a review. Int Res J Eng Technol (IRJET) 08(01)

Landolfoa R, Mammana O, Portioli F, Di Lorenzo G, Guerrieri MR (2009) Experimental investigation on laser welded connections for built-up cold-formed steel beams. J Constr Steel Res 65

Li C, Diao Y, Li HN, Pan H, Ma R, Han Q, Xing Y (2023) Seismic performance assessment of a sea- crossing cable-stayed bridge system considering soil spatial variability. Reliab Eng Syst Saf 235

Liang H, Roy K, Fang Z, Lim JBP (2022) A Critical review on optimization of cold-formed steel members for better structural and thermal performances. Buildings

Makebo GM, Paulmakesh A (2021) Analysis of light gauge steel by using fem subjected to tension load. J Univ Shanghai Sci Technol 23(9)

Meher S, Nar R, Jagadale S, Kalal G, Chandanshive V (2018) Design of industrial warehouse. Int J Eng Res Technol (IJERT) 7(02)

Mei K, Li Y, Lu Z (2015) Application study on the first cable stayed bridge with CFRP cable in china. J Traffic Transp Eng 2:242–248

Mei K, Seracino R, Lv Z (2016) An experimental study on bond-type anchorages for carbon fiber reinforced polymer cables. J Constr Build Mater 106:584–591

Navyashree PM, Sowjanya GV, Mallesh TV (2020) Analysis of cold formed steel sections: review. Pap Int J Eng Res Technol (IJERT) 8(14)

Ni Y, Chen J, Teng H, Jiang H (2015) Influence of earthquake input angle on seismic response of curved girder bridge. J Traffic Transp Eng 2:233–241

Omran M, Karani A (2022) Cable loss performance investigation of cable-stayed bridge equipped with roll-N-cage isolator. Structures 41:1329–1344

Prabhakaran S (2020) Study on load carrying capacity of cold formed steel built-up column. J Xidian Univ 14(5)

Ren L, Fang Z, Wang K (2019) Design and behavior of super-long span cable-stayed bridge with CFRP cables and UHPC members. Compos B 164:72–81

Song C, Xiao R, Bin S (2018) Optimization of cable pre-tension forces in long-span cable-stayed bridges considering the counterweight. Eng Struct 172:919–928

Thorat AR, Patil SK (2017) A study of performance of pre-engineered building of an industrial warehouse for dynamic load. Int Res J Eng Technol (IRJET) 04(06)

Valiente A, Iordachescu M, Aznar A et al (2021) Damage tolerance of prestressing steel in a bridge stay-cable after30 years of service in a harsh environment. J Constr Build Mater 285:122945

Wang J, Wang X, Ye A, Guan Z (2023) Deformation-based pushover analysis method for transverse seismic assessment of inverted Y-shaped pylons in kilometer-span cable-stayed bridges: Formulation and application to a case study. Soil Dyn Earthq Eng 169:107874

Chapter 21
An SEM Model for Identification of Important Delay Parameters Affecting Precast Construction

Rohan Sawant, Shruti Wadalkar, Deepa A. Joshi, and Radhika Menon

Abstract Precast construction has acquired significant recognition as an alternative to conventional cast-in-situ construction due to its improved performance. Precast concrete construction projects, on the other hand, are regularly delayed. As a result, it is vital to identify the major variables causing delays in precast concrete construction projects and research ways to reduce them. The delay factors for precast construction projects were chosen and classified using literature research and an expert interview. Data is then collected using questionnaires. Data analysis has been done using structural equation model. Ranking of factors has been done using Relative Importance Index. According to the results, inadequate structural connections for precast components are the most important cause and are most readily influenced by other delay factors. This report also recommends that large delays be prioritized.

Keywords Construction delay · Precast construction · Structural equation model · SPSS

21.1 Introduction

The COVID-19 epidemic has posed considerable obstacles to the building sector, which is an important pillar of economic, cultural, and political growth. The pandemic's effects on building projects have been diverse, resulting in delays, interruptions, and financial pressure. In response, academics have investigated techniques to reduce these negative consequences. One interesting strategy is to utilize structural equation modeling (SEM) to find key characteristics influencing building results. SEM offers a robust framework for describing and measuring the relationships

R. Sawant (✉) · S. Wadalkar · D. A. Joshi
Department of Civil Engineering, Dr. D. Y. Patil Institute of Technology Pimpri, Pune, India
e-mail: sawantrohan883@gmail.com

R. Menon
Department of Mathematics, Dr. D. Y, Patil Institute of Technology Pimpri, Pune, India

B. V. Khode et al. (eds.), *Innovations in Technologies: Pioneering Sustainable Infrastructure for a Resilient Future*, Sustainable Civil Infrastructures,
https://doi.org/10.1007/978-3-031-82133-2_21

between diverse construction elements. Researchers created a theoretical SEM model after examining data from questionnaire surveys and personal interviews. This model accounts for the impacts of four latent components on the construction industry's response to COVID-19. These components have a significant impact on project delays and industry performance. As we navigate the complexity of the epidemic, SEM is a useful tool for identifying and addressing the problems that construction professionals face. Precast concrete (PC) is a popular type of construction technique in which the structural components are cast away from the building site and transported there. PC has been shown to lead to shorter project times, better structural performance, less onsite construction labor, and less construction waste (Tam et al. 2007), fewer greenhouse gas emissions (Mao et al. 2013; Ji et al. 2324), better quality and faster construction (Tam et al. 2015a; Hong et al. 2016), and improved worker health and safety (Jaillon and Poon 2014; Lu et al. 2018). Due to all the advantages, the PC construction industry has been rapidly expanding in recent years. China, Hong Kong, and Singapore are only a few examples of places that have recently begun to use PC-based construction for tall structures (Hong et al. 2016; Jaillon and Poon 2014), and this trend is expected to continue. Even though one of the best things about PC-based construction is that it saves time, PC-based projects can still get delayed. According to Lu et al. (2018); Hatmoko et al. 2019), the average PC project of medium size takes 22% more time than expected.

Precast construction is becoming more popular since it allows builders to design and create building components in factories, transport them to construction sites, and assemble them with a greater degree of finish than is feasible with typical piecemeal construction (Li et al. 2018; Smit 2014). Off-site construction, prefabrication, manufacture, construction, and production are all terms used to describe the same process, which is often referred to as "prefabrication." Precast concrete construction has many advantages for investors and contractors over traditional cast-in-place concrete construction, including faster construction, higher-quality outputs, greater tolerances, lower costs, fewer on-site labor re-works, and safer construction environments (Wang and Hu 2018; Tam et al. 2015b; Li et al. 2014).

Precast concrete construction is good for the environment because it reduces waste (Blismas et al. 2006), lets some parts be used again (Matic et al. 2016), and uses less water (Minunno et al. 2008). Due to these many advantages, the building sector has expanded greatly in both industrialized and developing nations (Pons 2014). In particular, precast concrete buildings are becoming more dependent on meticulous planning across the board. This is due to the increased number of stakeholders, the greater complexity of building methods, and the longer duration of the construction process. This makes it harder for construction management tasks like tracking progress and making plans. Before starting construction (Salavatian et al. 2019), researchers and engineers should pay special attention to the scheduling risks of prefabrication projects, which are not present in traditional building methods.

21.2 Factors Causing the Delays in Precast Construction

Delays in numerous sectors, such as construction, transportation, and manufacturing, have a considerable impact on efficiency, costs, and overall project performance. Understanding the variables that cause delays is critical for good project management and decision-making. This literature review looks at the methodology and criteria employed by researchers to identify elements that influence delay in their investigations. Some researchers work with domain experts or stakeholders to identify and prioritize aspects based on their practical experience and expertise. Expert perspectives may give useful insights into variables that are not immediately evident in the literature. Many investigations begin with a thorough literature analysis to identify common variables influencing delay. This technique assures that the chosen parameters are based on established theoretical frameworks and empirical facts. We often select factors based on their frequency of occurrence and projected influence on project timeframes and results. We prioritize the inclusion of factors that have a high probability of occurrence and serious consequences. Key stakeholders includes customers, contractors, and regulatory authorities, rank factors in order of importance. This guarantees that the chosen elements are relevant to the project's stakeholders' interests and concerns. The selection of elements influencing delay in research projects is an important phase that determines the breadth, emphasis, and application of the results. Researchers use a range of approaches and criteria to find elements that are theoretically founded, experimentally substantiated, and practically relevant to their field of study. Understanding how criteria are picked may help guide future research and lead to more effective project management and delay mitigation measures.

(1) Construction Techniques:

While off-site building decreases on-site construction strain greatly, dealing with numerous precast components is tough. The components are also rather big and heavy, necessitating enough site space as well as suitable scheduling management for lifting and installation. Moreover, high-precision installation procedures, particularly for difficult connection operations, are critical for precast construction (Li et al. 2016).

(2) Workforce:

As compared to conventional on-site building, precast construction requires less labor. Unfortunately, the existing precast construction method is not technologically sophisticated and depends heavily on labor, which necessitates machine-oriented abilities on site and throughout the production method. Poor performance and insufficient worker experience have a substantial influence on plan development.

(3) Resources:

A material deficiency might cause a building delay. The primary reason for material scarcity is when demand exceeds supply. The necessity for on-site construction varies depending on the size of the precast components used in off-site precast systems. Damage to stored components might also be a source of material scarcity.

(4) Machinery:

Equipment shortages or unavailability are key factors that lead to delays. Some contractors may encounter a shortage of machines to produce work during the building stage due to excessive costs and equipment breakdowns.

(5) Contractors:

Estimated times are made to aid in project administration and planning. Inadequate planning and scheduling on the part of contractors is another common cause of project delays. One other reason projects go behind schedule is poor site management and oversight. Heavy-duty cranes and precise measurements may be required for moving and installing massive precast sections. It's obvious that rehandling components affects scheduling effectiveness. Parts are sometimes returned to the manufacturer because they failed or were poorly designed.

(6) Clients:

Construction delays may arise if customers make modifications to the design or offer new needs throughout the construction term. In such circumstances, contractors are unable to begin work until architects produce new designs.

(7) External Conditions:

Climate is an issue that might cause delays during the project since it interferes with scheduled operations like on-site concreting chores. Delays are sometimes attributed to ineffective regulation and unstable policies. Customs delays at border crossings, for example, may arise when moving components overseas.

(8) Location

1. Project location affects precast material availability. Precast materials may be harder to deliver to distant projects, delaying development.
2. Weather: Weather may greatly affect precast construction projects. Severe temperatures, winds, rain, and snow may slow precast material delivery and installation.
3. Infrastructure for transportation: The quality and availability of transportation infrastructure could affect how precast materials get to the project site. Inadequate roads, highway access, and bridges might delay deliveries.
4. Labor availability: Skilled labor availability might delay precast construction projects. If competent personnel are scarce and training is needed, the project may take longer.
5. Permits and approvals: Local permit and approval requirements might delay development. Before building, allow time for permits and approvals.

Finally, a precast building project's location might affect its timeframe and cost. To avoid delays and finish the project on time and within budget, consider these aspects during planning. (Fenner et al. 2017) Found that in dense urban areas like Hong Kong, traffic and public transportation might be a hindrance to the widespread use of precast buildings.

(9) Project Type:

Several different types of construction projects make use of computers, and precast-based building has progressed in terms of precast component size, robustness, and construction strategy. As a result, depending on the nature of the project, precast construction building projects may experience schedule slips (Choi et al. 2019).

(10) Contractor Experience:

Contractors working on precast construction projects should have solid experience and skills in this area. The contractor has to be aware of and prepared for any unfavorable project circumstances, such as design faults, defects in the construction plan, and damaged precast construction, early enough in the project to eliminate schedule delay risk factors (Tserng et al. 2006; Ji et al. 2018).

(11) Handling of adverse weather conditions:

Environmental factors like rain, snow, and temperature may have an impact on the assembly of precast construction components, just like any other building procedure. Crane operation, for example, becomes difficult in snow and/or severe wind, and PC supply might suffer as a result (Wuni and Shen 2019; Gong et al. 2019). As a result, how construction managers and project policies foresee, monitor, and handle severe weather conditions may have a substantial influence on schedule performance in precast concrete building projects.

(12) Multitude of Precast Construction types:

Several other types of precast construction are used in construction, as was previously mentioned. These include load-bearing PC components like columns, beams, and slabs, as well as non-load-bearing PC components like walls, balconies, and so on. Using a wide variety of precast construction may increase the complexity of wiring, crane operating weights, and site logistics, all of which can have an effect on the timeline and performance of the project. As several different precast construction models are likely to be used on a given project, this might create significant delays in the overall schedule (Tserng et al. 2006).

(13) Supply of Precast Construction:

Since PCs are the primary building material in precast construction-based construction projects, any problems with PC supply might cause significant delays (Wuni and Shen 2019; Gong et al. 2019). If all precast construction for the building project is provided by the same PC vendor, there is a danger of a timetable delay. In such cases, the capacity of the PC manufacturer may have a significant impact on the project's timeline and performance.

21.3 Methodology

An online survey was conducted to better understand precast construction delays. A comprehensive but easy survey was designed. The questionnaire is divided into two sections. The first section looks at survey answers, while the second looks at precast construction delays. To rate anything use a 1-to-5 Likert scale. Owners, architects, site engineers, contractors, and quantity surveyors were among those who answered. The questionnaire was sent to commercial businesses, governmental and private organizations, academic institutions, and others. Some people didn't know how to react, while others responded poorly. There were 178 completed survey forms assessed, with 18 being eliminated due to incompleteness. All replies were used to determine the results. The survey received 178 responses. All of the data was analyzed using SEM. The PLS method employs evaluation and structural equations. Internal models provide component relationships. SPSS factor analysis is performed next. A standardized questionnaire was used to collect information on the variables that contribute to project delays. The poll addressed a wide spectrum of stakeholders from several sectors, including construction, transportation, and manufacturing. Participants submitted personal information, such as their project job and years of experience, and assessed the relevance and frequency of several delay reasons using a Likert scale. Furthermore, open-ended questions enabled respondents to share qualitative insights into their experiences with project delays. Data was analyzed using both quantitative approaches, such as descriptive statistics and factor analysis, and qualitative techniques. This holistic approach provided nuanced knowledge of the multidimensional nature of project delays, allowing for the creation of focused mitigation and management methods.

21.4 Data Analysis

Figure 21.1 gives information regarding data collection through questionnaire survey. PLS-SEM was shown to be better for examining precast-related development-affecting factors in this study. PLS-SEM was chosen over other relapse approaches because it takes dependent and free-parts multi-collinearity into account. The multivariate PLS-SEM model employs straight relapses and factor reduction to investigate many interdependent subordinate connections. The structural model (Kolo et al. 2014) can look at the route diagram and observe all the associated components, both dependent and independent, to estimate the relationship. When regression and path analysis are combined, you obtain causal modeling. There are three major advantages to employing the equation model rather than the existing PLS-based models (Wang and Hu 2018; Tam et al. 2015b). Those are the folks. Covariance-based approaches, on the other hand, assess four or more constructs, while PLS methods only evaluate one. The PLS approach is particularly effective in analyzing survey data that does not have a normal distribution. PLS considers interaction effects and corrects measurement errors (Figs. 21.2 and 21.3).

Following the reliability analysis, the internal consistency of the study is evaluated using a correctness or fineness evaluation. Cronbach's alpha was found to be 0.816. This indicates it is greater than the standard criterion of 0.7. After that, analyzing

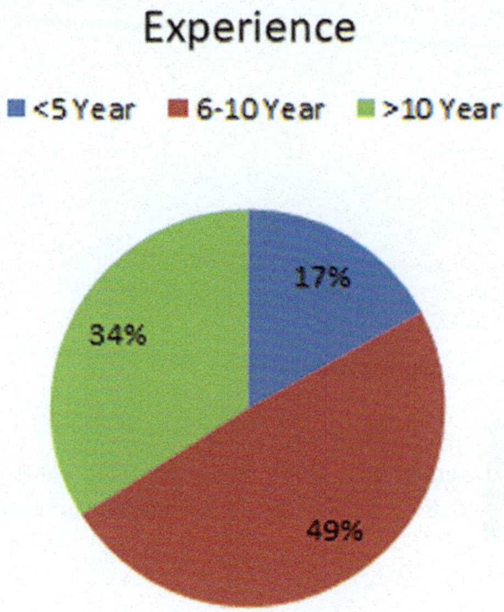

Fig. 21.1 Experience Information

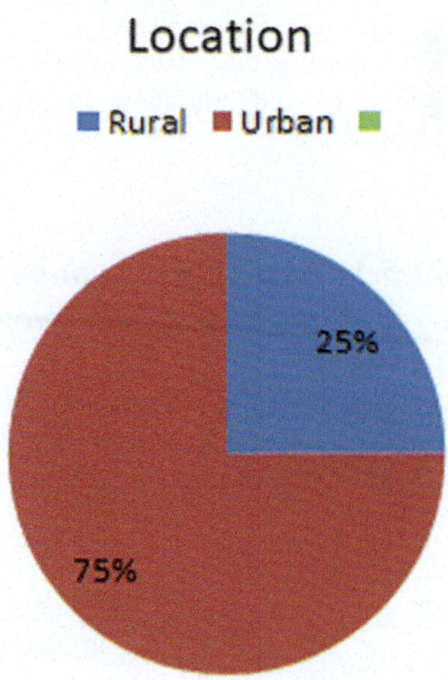

Fig. 21.2 Location Information

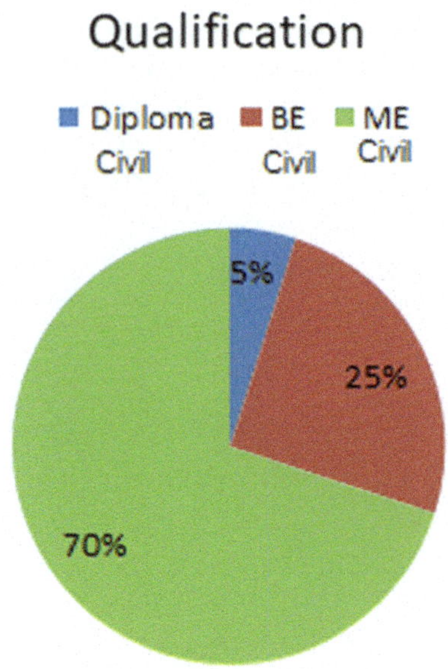

Fig. 21.3 Qualification Information

dependability is allowed since it is valid. Table 21.1 contains every single effect statistic. The conclusion is that the survey findings may be trusted.

21.5 SEM Model for Construction

Consequently, the created SEM model provides a structure for examining the interactions of study variables. Figure 21.4 shows factors affecting delay in precast construction.

21.5.1 Structural Equation Modeling for Causes of Delay in Precast Construction Analysis

A. Structural Equation Modeling (SEM) for causes of delay in precast construction analysis is shown in Figs. 21.5, 21.6, 21.7, 21.8, 21.9, 21.10, 21.11.

Table 21.2 represents coefficient of latent variables for SEM Model A.

Table 21.1 Causes of delay in precast construction with factor loadings and Cronbach's alpha

Code	Causes	Factor loading	α
A1	Low-precision installation techniques	0.7079	0.8331
A2	Improper lifting operations	0.8214	0.8614
A3	Insufficient site space to layout	0.7772	0.8306
A4	Location	0.585	0.9122
A5	Inadequate worker experience	0.5901	0.9128
A6	Low productivity	0.775	0.9238
A7	Insufficient proficiency	0.5712	0.8222
A8	Multitude of Pc types	0.6769	0.9203
A9	Project type	0.7549	0.92
A10	Shortage and delay in material supply	0.8514	0.82
A11	Damage to stored components	0.7809	0.9205
A12	Equipment failure	0.6985	0.8338
A13	Vehicle and equipment unavailability	0.812	0.8621
A14	Inefficient decision-making	0.7678	0.8313
A15	Additional requirements	0.5756	0.9129
A16	Contractor experience	0.5807	0.9135
A17	Variations and changes	0.7656	0.9245
A18	Slow decision-making	0.5618	0.8229
A19	Improper construction planning and schedule design	0.6675	0.921
A20	Wrong delivery requirements and routes	0.7455	0.9207
A21	Poor site layout	0.842	0.8207
A22	Lack of communication among participants	0.7715	0.9212
A23	Information gaps among companies	0.6977	0.8328
A24	Rehandling components	0.8112	0.8611
A25	Inefficient structural connections	0.767	0.8303
A26	Return to manufacturers owing to design errors	0.5748	0.9119
A27	Re-manufacturing owing to component damage	0.5799	0.9125
A28	Policy uncertainty	0.7648	0.9235
A29	Energy and water supply failure, transportation disruption	0.6667	0.92
A30	Handling of adverse weather conditions	0.7447	0.9197

21.5.2 Model Fit and Quality Indices

APC is 0.949 on average. An acceptable and good match has a route coefficient of more than 0.90. $R^2 = 0.924$. R2, a statistical measure, is always between zero and one hundred percent. A response variable value of 0% indicates that the model does

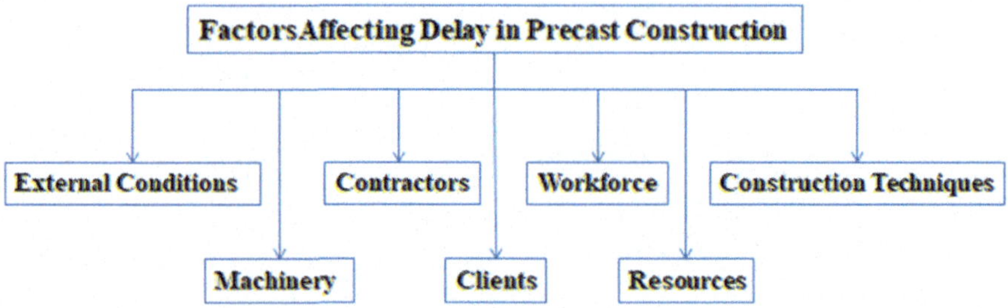

Fig. 21.4 Factors affecting delay in precast construction

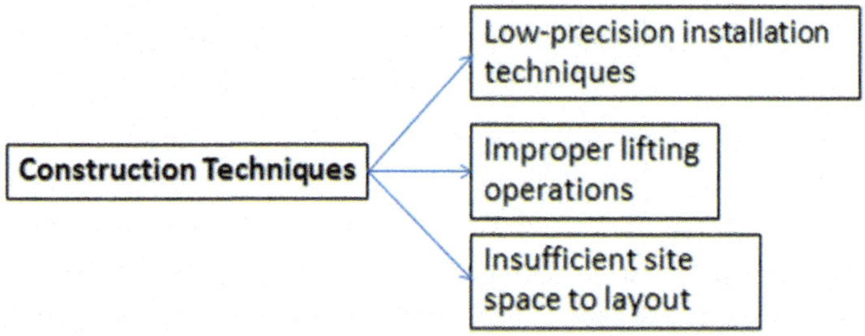

Fig. 21.5 SEM Model A- causes of delay in precast construction

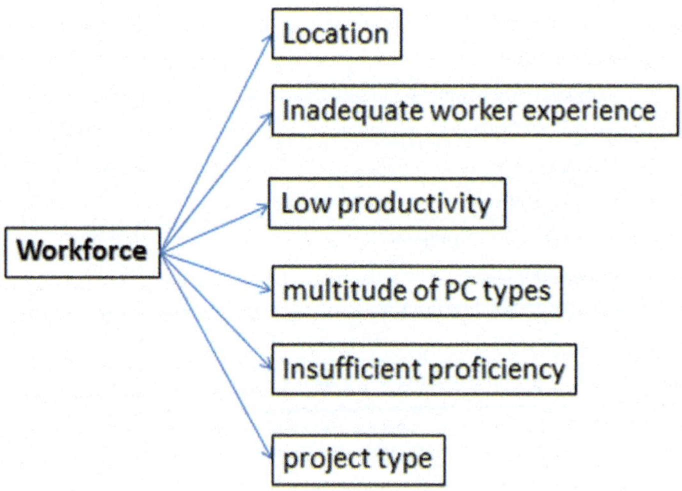

Fig. 21.6 SEM Model B- causes of delay in precast construction

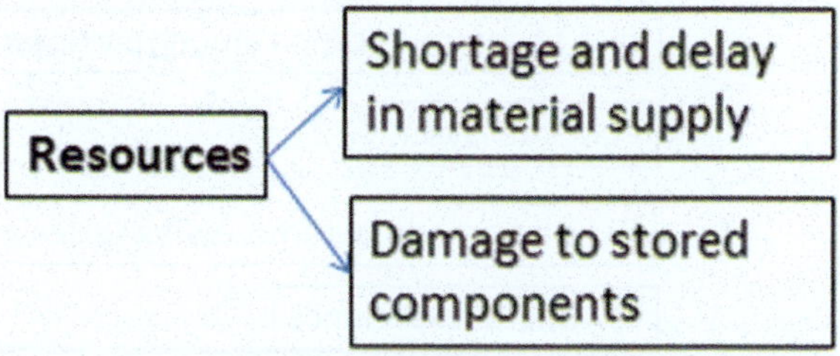

Fig. 21.7 SEM Model C- causes of delay in precast construction

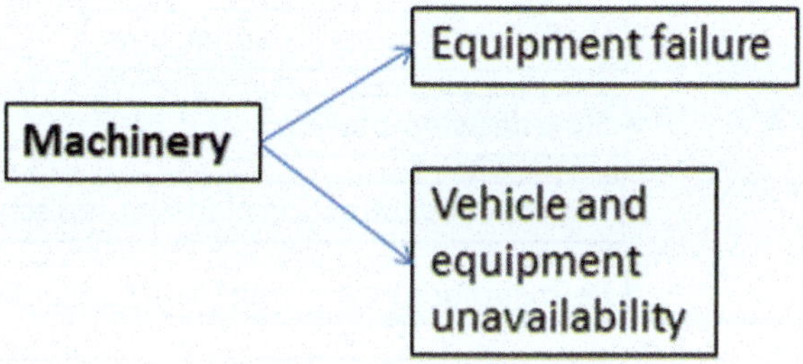

Fig. 21.8 SEM Model D- causes of delay in precast construction

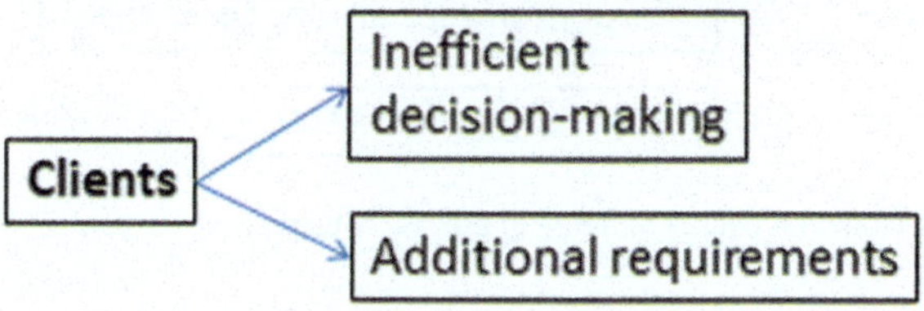

Fig. 21.9 SEM Model E- causes of delay in precast construction

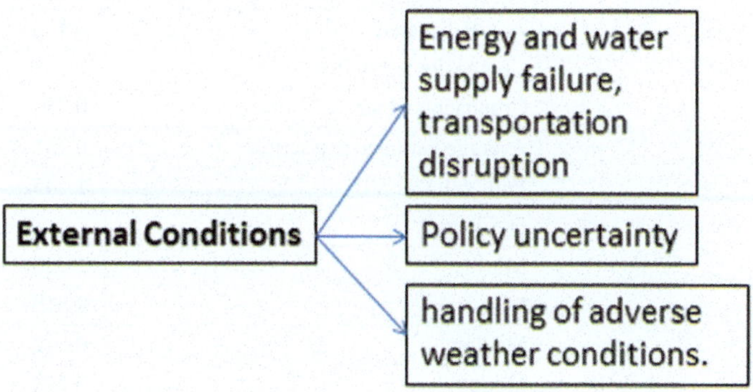

Fig. 21.10 SEM Model F- causes of delay in precast construction

Fig. 21.11 SEM Model G- causes of delay in precast construction

Table 21.2 Coefficient of latent variable SEM for model A

Model A	
R-squared	0.335
Adj. R-squared	0.126
Composite reliability	0.843
Cronbach's alpha	0.75
Average variance extraction	0.751
Full collinear VIF	1.245
Q-Squared	
Min	-3.758
Max	2.613
Median	0.542
Mode	0.348
Skewness	0.678
Exc. Kurtosis	1.259

not account for dispersion around the mean. The dependent variable mean predicts the regression model and dependent variables. A model fits perfectly if it describes all variance in response variables around the mean. A higher R2 value indicates that the regression model fits your data better. With a GoF of 0.135, the model fits perfectly. The goodness-of-fit (GOF) metric verifies that the model corresponds to the experimental data. The GOF values range from 0 to 1, with 0.10 being low, 0.25 medium, and 0.36 denoting large. A good model fit demonstrates a simple but effective model. Kurtosis as well as Skewness, using descriptive statistics, compares Skewness and kurtosis. SEM accepts Skewness and kurtosis between -3 and -3.

21.5.3 Relative Importance Index

The questionnaire survey replies were thoroughly examined using SPSS to get further information. Table 21.3 shows ranks components by expected criticality based on condition (2), using the Relative Importance Index (RII). RII = W/(A*N), where 0 = 1. W stands for "weight," and respondents assigned it a number between 1 and 5, with 1 indicating "strongly disagree" and 5 indicating "strongly agree." There were N replies, with the highest number being 5.

Thus, the eight most affected variables by the Precast Construction Delays have been identified by the factor analysis. Below are several ways they may be categorized.

1. Inefficient Structural Connections (0.30 Coefficient)
2. Location (0.29 Coefficient)
3. Improper Construction Planning and Schedule Design (0.29 Coefficient)
4. Contractor Experience (0.28 Coefficient)
5. Insufficient Site Space to Layout (0.26 Coefficient)
6. Low-Precision Installation Techniques (0.26 Coefficient)
7. Re-Manufacturing Owing to Component Damage (0.25 Coefficient)
8. Inadequate Worker Experience (0.22 Coefficient).

21.6 Result and Discussion SPSS Analysis

21.6.1 Reliability Statistics

The reliability of the sample size was checked using SPSS. Table 21.4 displays the value (as calculated by SPSS).

Possibly has a value between 0 and 1. The fact that the value of "" in this example is 0.842, which is greater than 0.6 and very close to 1, demonstrates that the sample size is sufficient and the questionnaire is reliable to draw conclusions from.

Table 21.3 Relative importance Index for causes

Code	Causes	RII	Rank
A4	Location	0.945	1
A8	Multitude of Pc types	0.938	2
A25	Inefficient structural connections	0.929	3
A19	Improper construction planning and schedule design	0.927	4
A5	Inadequate worker experience	0.918	5
A27	Re-manufacturing owing to component damage	0.915	6
A22	Lack of communication among participants	0.912	7
A2	Improper lifting operations	0.907	8
A16	Contractor experience	0.903	9
A10	Shortage and delay in material supply	0.9	10
A3	Insufficient site space to layout	0.89	11
A6	Low productivity	0.884	12
A21	Poor site layout	0.878	13
A9	Project type	0.871	14
A1	Low-precision installation techniques	0.86	15
A24	Rehandling components	0.842	16
A30	Handling of adverse weather conditions	0.828	17
A23	Information gaps among companies	0.812	18
A20	Wrong delivery requirements and routes	0.789	19
A13	Vehicle and equipment unavailability	0.775	20
A11	Damage to stored components	0.76	21
A12	Equipment failure	0.755	22
A15	Additional requirements	0.743	23
A28	Policy uncertainty	0.712	24
A18	Slow decision-making	0.702	25
A7	Insufficient proficiency	0.685	26
A14	Inefficient decision-making	0.674	27
A29	Energy and water supply failure, transportation disruption	0.666	28
A26	Return to manufacturers owing to design errors	0.645	29
A17	Variations and changes	0.632	30

Table 21.4 Reliability statistics

Cronbach's alpha	Based on standardized items, Cronbach's alpha	No. of items
0.842	0.684	30

Table 21.5 KMO and the bartlett test (SPSS output)

Kaiser–Meyer–Olkin measure of sampling adequacy	0.715
Bartlett's test of Sphericity approx. Chi-Square	1735.514
df	247
Sig	0.000

21.6.2 Feasibility of Factor Analysis Data

The suitability of the factor analysis results was checked using sample adequacy metrics such as the Kaiser–Meyer–Olkin (KMO) and the Bartlett's Test. If the survey questionnaire passes Bartlett's Test of Sphericity, it may be used for factor analysis (Table 21.5).

21.6.3 Mitigation Measures for Reduce Delays in Precast Construction

Reducing delays in precast construction involves implementing effective planning, communication, and management strategies throughout the project. Here are some recommendations that can help:

1. Robust project planning: Prior to starting the project, it is important to create a detailed project plan that takes into consideration factors such as lead times, site preparation, transportation, installation, and quality control. The plan should be given to everyone who is working on the project so that everyone knows what is expected of them and what their roles and responsibilities.
2. Efficient procurement: To avoid delays, it is important to ensure that all materials and components required for the precast construction are procured in a timely manner. This can be done by setting up a reliable supply chain, placing orders far in advance, and keeping an eye on when the materials will arrive.
3. Effective communication: Communication is essential for keeping everyone on the same page and working towards the same objectives. This can be done by holding regular meetings, reporting on progress, and changing the project plan from time to time.
4. Quality control: Quality control is critical in precast construction to ensure that the components meet the required standards and are fit for use. At each step of the building process, quality control measures should be put in place so that problems can be found and fixed as soon as possible.
5. Skilled workforce: Having a qualified staff is critical to completing the precast construction on schedule and at a high level. This may be accomplished by providing employees with proper training and assistance, as well as ensuring that they have the tools and equipment they need to do their duties.

By implementing these strategies, delays in precast construction can be reduced, leading to a smoother and more efficient construction process.

21.7 Conclusion

Overall, precast construction has been largely acknowledged as an alternative to traditional cast-in-situ concrete construction since it may provide multiple advantages to investors and contractors. Despite significant attempts to enhance project schedule estimates and management, delays occur regularly in precast concrete construction projects. To enhance the scheduling performance of precast construction, the delay reasons should be recognized and analyzed. This article uses the combination approach to analyze the interplay between the many delay variables and their causes and to rank the elements that contribute to delays according to their significance in the building industry. Thirty risk factors were found after sifting through the available research. Questionnaires were used to gather information that was used to aid in the strategy's practical implementation in precast construction. Inefficient structural connections, location, improper construction planning and schedule design, contractor experience, insufficient site space to layout, low-precision installation techniques, re-manufacturing due to component damage, and insufficient worker experience are the factors most easily affected by delay in the construction industry, according to the results of structural equation modeling. Overall, this study adds an evaluation framework for decision-making in prefabrication construction schedule management.

References

Arashpour M, Bai Y, Aranda-mena G, Bab-Hadiashar A, Hosseini R, Kalutara P (2017) Optimizing decisions in advanced manufacturing of prefabricated products: theorizing supply chain configurations in off-site construction. Autom Constr 84:146–153

Blismas N, Pasquire C, Gibb A (2006) Benefit evaluation for off-site production in construction. Constr Manag Econ 24:121–130

Choi JO, Chen XB, Kim TW (2019) Opportunities and challenges of modular methods in dense urban environment. Int J Constr Manage 19(2):93–105

Fenner AE, Razkenari M, Hakim H, Kibert CJ (2017) A review of prefabrication benefits for sustainable and resilient coastal areas: In: Proceedings of the 6th international network of tropical architecture conference, tropical storms as a setting for adaptive development and architecture, Gainesville, FL, USA

Gong P, Teng Y, Li X, Luo L (2019) Modeling constraints for the on-site assembly process of prefabrication housing production: a social network analysis. Sustainability (Switzerland) 11(5):1387

Hatmoko JUD, Wibowo MA, Astuty MD, Arthaningtyas DR, Sholeh MN (2019) Managing risks of precast concrete supply chain: A case study. In: Proceeding of the, MATEC Web of Conference, 05004. Les Ulis, France: EDP Sciences vol 270

Hong J, Shen GQ, Mao C, Li Z, Li K (2016) Life-cycle energy analysis of prefabricated building components: an input–output-based hybrid model. J Cleaner Prod 112 (Jan):2198–2207
Jaillon L, Poon CS (2014) Life cycle design and prefabrication in buildings: a review and case studies in Hong Kong. Autom Constr 39(Apr):195–202
Ji Y, Qi L, Liu Y, Liu X, Li HX, Li Y (2018) Assessing and prioritising delay factors of prefabricated concrete building projects in China. Appl Sci (Switzerland) 8(11):2324
Ji Y, Qi L, Liu Y, Liu X, Li HX, Li Y (2018) Assessing and prioritising delay factors of prefabricated concrete building projects in China. Appl Sci (Switzerland) 8 (11):2324
Kolo SJ, Rahimian FP, Goulding JS (2014) Offsite manufacturing construction: a big opportunity for housing delivery. Nigeria. Procedia Eng 85:319–327
Li Z, Shen GQ, Xue X (2014) Critical review of the research on the management of prefabricated construction. Habitat Int 43:240–249
Li CZ, Hong J, Xue F, Shen GQ, Xu X, Mok MK (2016) Schedule risks in prefabrication housing production in Hong Kong: a social network analysis. J Clean Prod 134:482–494
Li CZ, Xu X, Shen GQ, Fan C, Li X, Hong J (2018) A model for simulating schedule risks in prefabrication housing production: a case study of six-day cycle assembly activities in Hong Kong. J Cleaner Prod 185 (Jun):366–381
Lu W, Chen K, Xue F, Pan W (2018) Searching for an optimal level of prefabrication in construction: an analytical framework. J Cleaner Prod 201 (Nov):236–245
Mao C, Shen Q, Shen L, Tang L (2013) Comparative study of greenhouse gas emissions between off-site prefabrication and conventional construction methods: two case studies of residential projects. Energy Build 66(Nov):165–176
Matic D, Calzada JR, Todorovic MS, Erić M, Babin M (2016) Cost-effective energy refurbishment of prefabricated buildings in Serbia. In: Pacheco-Torgal F, Granqvist C-G, Jelle, BP, Vanoli GP, Bianco N, Kurnitski J (eds) Cost-Effective Energy Efficient Building Retrofitting, Woodhead Publishing: Cambridge, UK. Chapter 16, pp 455–487
Minunno R, O'Grady T, Morrison G, Gruner R, Colling M (2008) Strategies for applying the circular economy to prefabricated buildings. Buildings 8:125
Mostafa S, Chileshe N, Abdelhamid T (2016) Lean and agile integration within offsite construction using discrete event simulation: a systematic literature review. Constr Innov 483–525
Pons O (2014) Assessing the sustainability of prefabricated buildings. In: Pacheco-Torgal F, Cabeza LF, Labrincha J, de Magalhães A (eds) Eco-efficient construction and building materials, Woodhead Publishing: Cambridge, UK pp 434–456
Salavatian S, D'Orazio M, Di Perna C, Di Giuseppe E (2019) Assessment of cardboard as an environment-friendly wall thermal insulation for low-energy prefabricated buildings. In: Sayigh A (ed) Sustainable building for a cleaner environment: selected papers from the world renewable energy network's med green forum 2017. Springer International Publishing: Cham, Switzerland, pp 463–470
Shen GQ, Xu G, Liu Y, Wang Y (2019) Stakeholder associated supply chain risks and their interactions in a prefabricated building project in Hong Kong. J Manage Eng 35(2):05018015
Smit RE (2014) Off-site construction implementation resource: off-site and modular construction explained. In: Off-site construction council; national institute of building sciences: Washington, DC, USA
Tam VWY, Tam CM, Zeng SX, Ng WCY (2007) Towards adoption of prefabrication in construction. Build Environ 42(10):3642–3654
Tam VWY, Fung IWH, Sing MCP, Ogunlana SO (2015) Best practice of prefabrication implementation in the Hong Kong public and private sectors. J Cleaner Prod 109(Dec):216–231
Tam VWY, Fung IWH, Sing MCP, Ogunlana SO (2015) Best practice of prefabrication implementation in the Hong Kong public and private sectors. J Clean Prod 109;216–231
Tserng HP, Yin SY, Li S (2006) Developing a resource supply chain planning system for construction projects. J Constr Eng Manage 132(4):393–407

Wang Z, Hu H, Gong J (2018) Framework for modeling operational uncertainty to optimize offsite production scheduling of precast components. Autom Constr 86:69–80

Wuni IY, Shen GQ (2019) Critical success factors for modular integrated construction projects: a review. Build Res Inf 48(7):763–784

Chapter 22
Performance Analysis of Indian Metro System

Abdul Sami Ansari and Sujesh D. Ghodmare

Abstract The architecture and administration of India's present metro train systems provide essential lessons for future system planning and funding. The 2017 formulation of the Metro Rail Policy has resulted in the expansion of a variety of rail-based public transportation networks across the country. The Government of India has continued to provide funding support over the last 20 years. There are presently subway, commuter, and train networks in use in India. Probably by the year 2025, metro rail networks in major cities of India will be about 697 km which was recorded as 222 km in three cities in 2011. The authorities have also claimed that an additional 1,032 km metro rail network will service around 27 municipalities. This means that the permitted metro rail projects are expected to cost INR 3 trillion over the next five years. In this study, we attempt to assess the operational and financial efficiency and efficacy of India's present metro train networks. The performance of Indian metro rail systems has been compared to the best practices from across the world. The outcomes can be used to plan out existing and future metro networks in any developing region.

Keywords Lighter urban rail systems · Operational performance · Financial performance

22.1 Introduction

The authorities have announced that an additional 1,032 km of metro rail network will service around 27 municipalities. This shows that the permitted metro rail projects are expected to cost INR 3 trillion over the next five years. This article attempts to examine the operational and budgetary efficiency and efficacy of India's present

---

A. S. Ansari (✉) · S. D. Ghodmare
G. H. Raisoni College of Engineering, Nagpur, India
e-mail: abdul.ansari.mtechts@ghrce.raisoni.net

S. D. Ghodmare
e-mail: sujesh.ghodmare@raisoni.net

B. V. Khode et al. (eds.), *Innovations in Technologies: Pioneering Sustainable Infrastructure for a Resilient Future*, Sustainable Civil Infrastructures,
https://doi.org/10.1007/978-3-031-82133-2_22

metro train networks. This has been accomplished by taking into account the four major cities with the longest approved metro networks, namely Delhi, Mumbai, Bengaluru, and Chennai, with a focus on different performance factors of Delhi Metro. The performance of Indian metro rail systems has been compared to best practices from across the world (Molan and Hummer 2019; Bared 2005). The dense and high-rise development (retail, workplace, and housing) above and around metro stops, which provides for more room, value, and ease in crowded areas, contributes significantly to the exceptional standard of the urban environment and life (Sharma 2011; Wengler and Urbanik 1995). Metros are long lasting infrastructure demonstrating the seriousness related to sustainable transit (Maji et al. 2013; Mohommed Balley Madhi 2015).

Aside from the numerous advantages, it is worth mentioning that developing subway systems is difficult and complicated (Ghodmare , 2022). These are the key challenges that the metro system faces as it grows (D.D., Khode and Ghodmare 2022). Proactive asset management is used to maintain the value and usability of Greenfield activities after the initial development costs have been incurred (Sannikon and Timonovetz 2018; Ghodmare et al. 2022).

Furthermore, one of the most major impediments to system development is a lengthy planning process (Rodke et al. 2022). It is usual for 10–20 years to pass between the introduction of a new proposal and the initiation of a route in established, built-up communities (Lee et al. 2004). Given the lengthy time to market, parties must achieve a high level of unity and agreement, which necessitates government protection and continuous efforts (Ghodmare et al. 2019; Yang et al. 2014). To overcome the drawbacks and maximize the benefits, appropriate system execution is essential. Many Indian subway systems have been found to be lacking in this area, suggesting the necessity for more expansion to attain success (Mujammel et al. 2020).

22.2 Performance Analysis

The performance of a few Indian metro networks has been compared to that of world champions like London, Singapore, and Hong Kong on a number of metrics in order to pinpoint possible areas for development (Nelson et al. 2000; Yadava et al. 2021). Figure 22.1 shows the different performance indicators taken into consideration.

22.3 Methodology Adopted

Two alternative situations were investigated in order to determine financial success. It was well known that a substantial sum of money was needed to break even monetarily (Lina 2020). Prior to the COVID-19 pandemic, a three-year performance study was

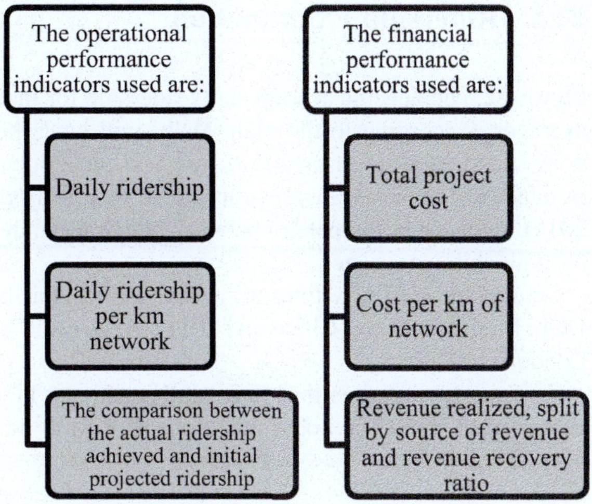

Fig. 22.1 Performance indicators under consideration

performed by few of the researchers in a situation with full-scale economic activity, support, and transit requirements (Chlewicki 2011; Lina 2020).

22.4 Results and Discussion

In addition to that the study of network lengths and average daily ridership was also performed for the towns in question for the three fiscal years (FY) from 2017 to 2020, as shown in Fig. 22.2. India's towns have a higher population density than cities in other nations (Nelson et al. 2000; Yadava et al. 2021).

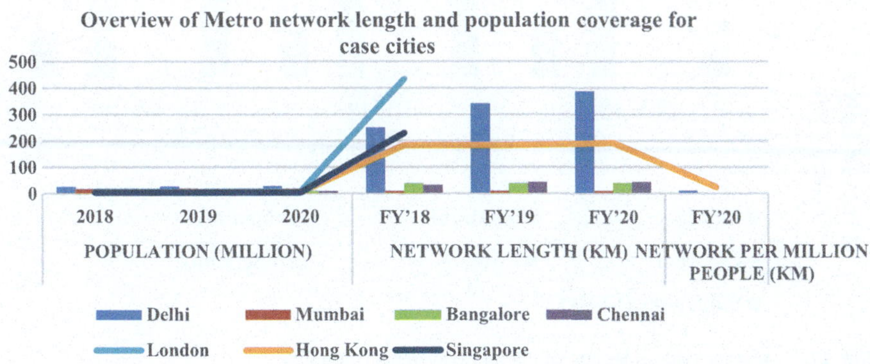

Fig. 22.2 Overview of metro network length and population coverage for case cities

22.5 Ridership Examination

Figure 22.3 depicts the average daily ridership for Indian cities as well as worldwide averages. According to the plan, Delhi will have the most daily traffic among its domestic and foreign competitors by FY 20; in comparison, its circulation increased by a factor of two over the previous three years (Maji et al. 2013; Molan and Hummer 2019). Because of the patchy network penetration, the other Indian metro networks have much lower usage.

The addition of new lines at present are responsible for some of the increase in Delhi Metro ridership (Chlewicki 2011). As a result, Delhi has 55% more riders than London.

Figure 22.4 examines the typical daily traffic per km for the different Metro System networks. It should be noted that Hong Kong has the highest traffic per km of any overseas city.

22.5.1 Ridership: Projections Verses Actuals

Because of alterations in counting techniques, the Delhi metro only achieved 38% of its anticipated ridership in FY18, but this increased to 79% by FY20. Projected utilization by Mumbai Metro, Bangalore and Chennai was 74%, 27% and 13% of their respective targets. Figure 22.5 contrasts expected and actual ridership for India's four metro networks whereas Fig. 22.6 represents it for Delhi metro.

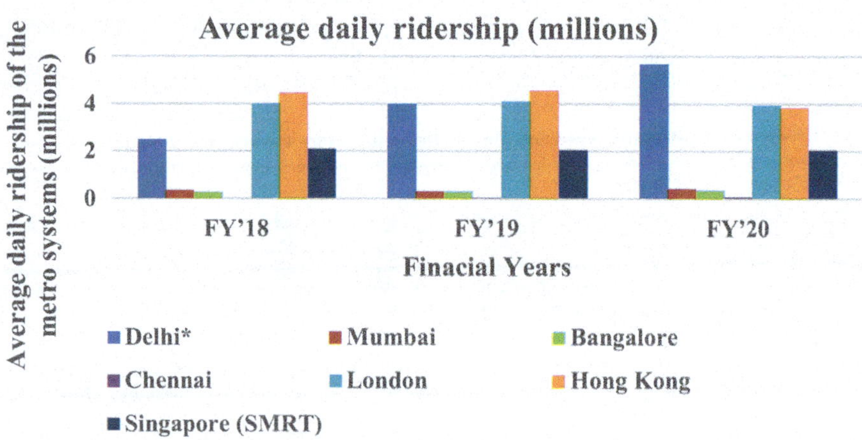

Fig. 22.3 Average daily ridership of the metro systems (millions)

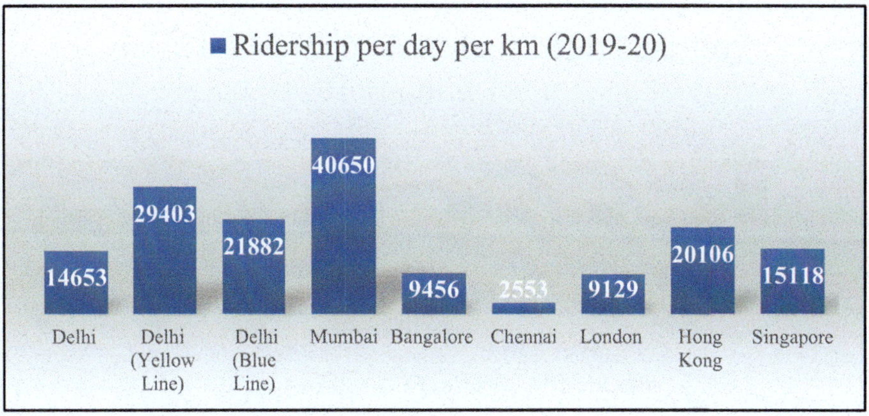

Fig. 22.4 Ridership per day per km (2019–2020)

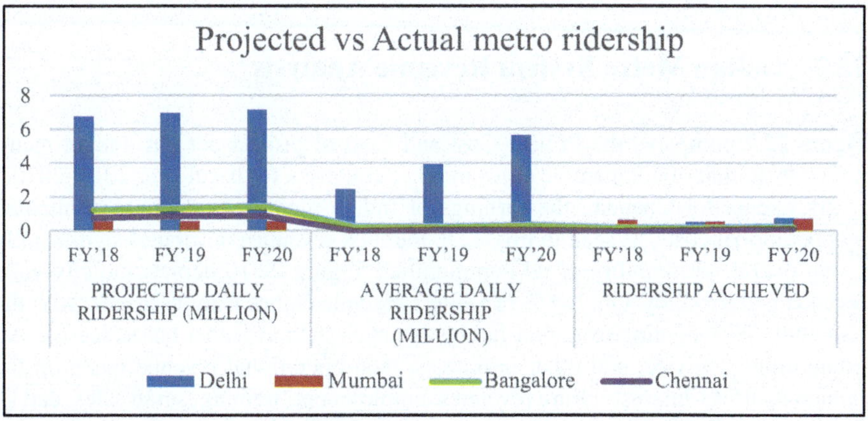

Fig. 22.5 Projected versus actual metro ridership

22.6 Budgetary Analysis

Here the overall financial performance of metro system is compared to that of their foreign rivals for the purpose of analyzing expenses and revenues separately. Figure 22.8 Illustrate the sort of network share, as well as the overall cost and cost per km, of constructing the four metro networks. When compared to Chennai, Bangalore can have cheaper expenses despite having a bigger under-ground network share.

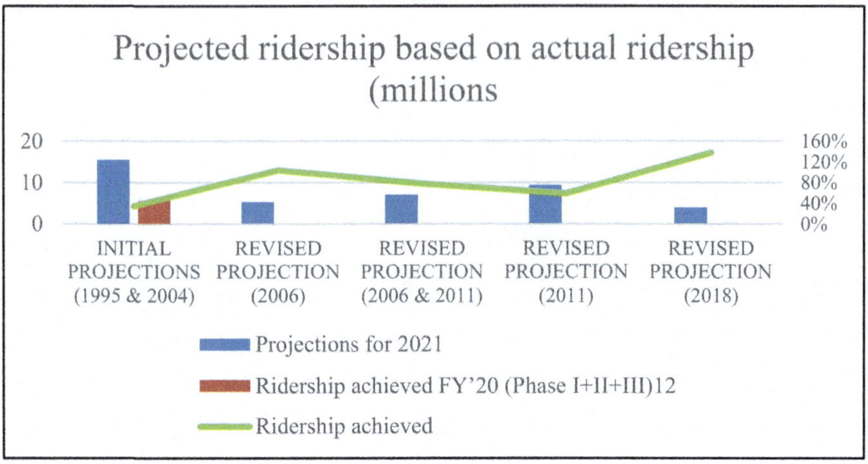

Fig. 22.6 Projected ridership based on actual ridership (millions)

22.7 Indian Metro System Revenue Analysis

Figure 22.7 compares the Project Cost and Cost of project per Km Indian metro networks to that of comparable cities in other countries for fiscal year 2019–2020.

As Except for Chennai, the earnings of Indian metro networks are dominated by operational costs, as seen in Fig. 22.8. Farebox revenue contributes significantly to the overall profitability of all communities. Figure 22.10 depicts the city-wide revenue pattern study. Only 48% of Delhi's revenue comes from user fees, with the remaining 37% coming from non-farebox sources such as feeder buses, leases, real estate money, advice, and other initiatives. Non-operational revenue made up the remaining 15%. Interest, future revenues, employment, training, scrap sales, and so forth. Unlike Delhi, Mumbai Metro One earns 89% of its revenue from passenger

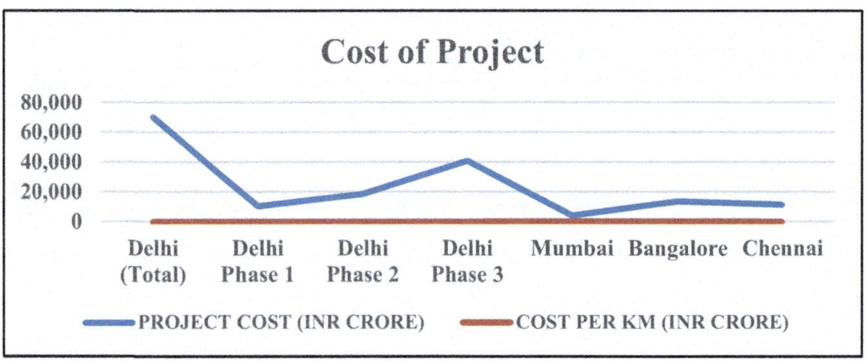

Fig. 22.7 Project cost and cost of project per Km

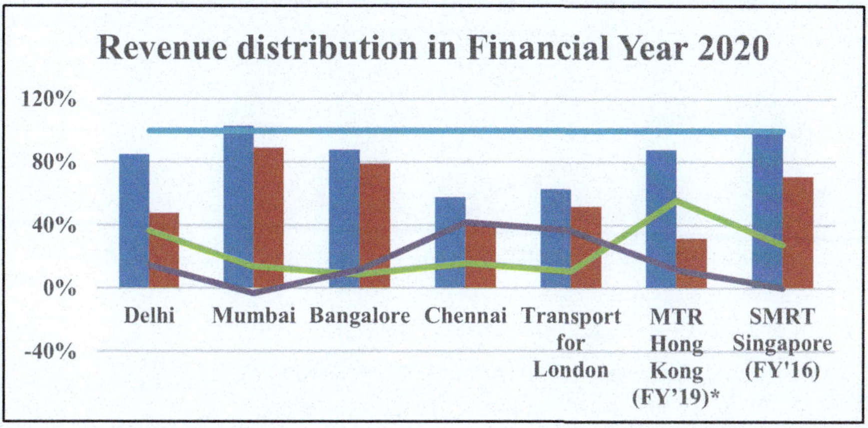

Fig. 22.8 Revenue distribution in financial year 2020

passes. The remaining 11% is made up of leasing revenue, marketing, and other costs14.

22.8 Overall Financial Performance of Indian Metro Systems and Their International Counterparts

The revenue recovery ratio compares Indian metros' financial performance to that of their international peers by assessing the system's ability to recuperate expenditures from its various sources of income. The three revenue recovery rates listed have also been examined.

Figure 22.9 depicts the financial prosperity of three worldwide standard cities. This organisation manages metro services as part of this infrastructure. The figures take into account the performance of all public transportation choices in these cities, not just the metro. All three towns saw operational and general advantages, allowing them to use the money for additional initiatives.

1. The farebox income recovery ratio is the percentage of fare money to operating expenses.
2. The operational revenue recovery ratio is the percentage of total operational revenue to total operational expense.3-The ratio of total income to total expenditure is known as the total revenue recovery ratio.

Figure 22.10 depicts the revenue recovery rates for Indian metro networks and their international counterparts. All metros have total recovery ratios of less than one; Mumbai and Bangalore, with recovery ratios of 1.9 and 1.0, respectively, only recoup their running costs from firebox income.

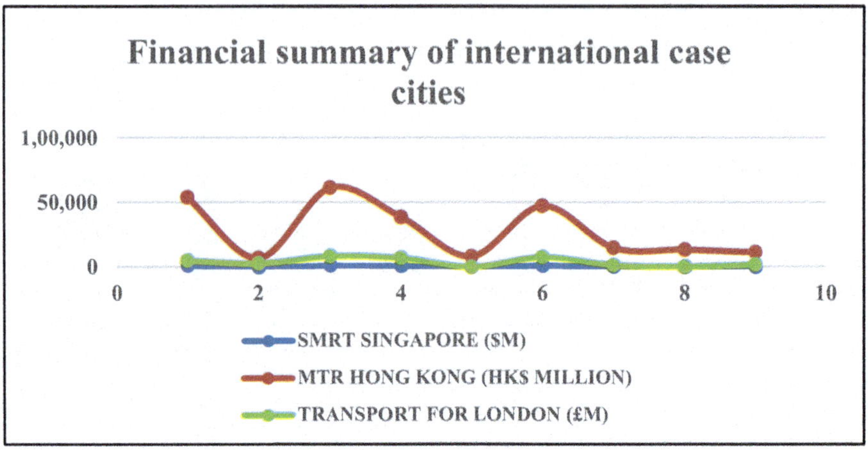

Fig. 22.9 Financial summary of international case cities

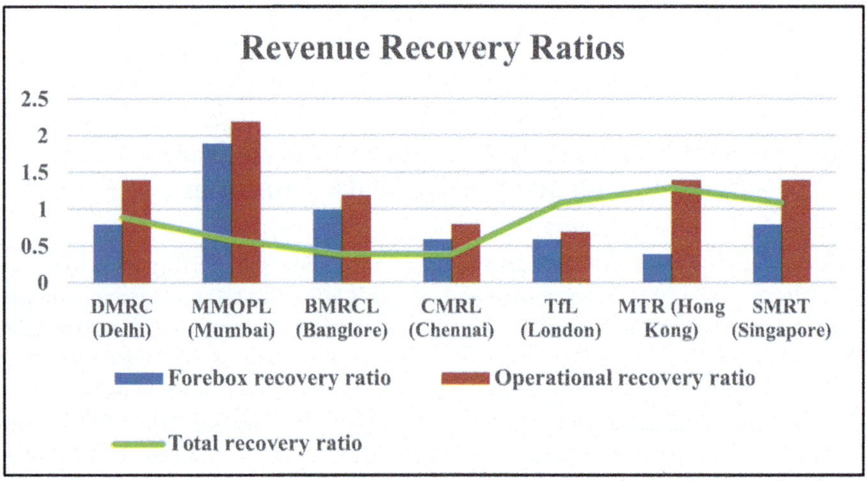

Fig. 22.10 Revenue recovery ratios of Indian metros and international peers

22.9 Recommendations for Upcoming Rail Projects Performance of Metro Ridership

a. **Additional steps for increased ridership:** Additional Connecting metros with other modes of public transportation can help increase traffic by providing proper first- and last-mile connection. Operators and authorities are both equally liable for this, and collaboration from all parties is required.

b. **Comprehensive ridership estimates:** Accurate ridership projections are essential for a practical evaluation of the project's contribution to the city's mobility demands and funding system planning. When compared to overseas rivals, the present metro systems have traffic per kilometer equivalent to some of the world's largest metro systems.

22.10 Conclusion

This article recommends changes to India's current and future public transportation networks. Indian metro railway ventures surpass global best practices, yet there is still an opportunity for financial performance to increase. To ensure the long-term survival of present metro rail systems, governments must acknowledge that they will continue to lose money and adopt appropriate measures to manage both regular and capital finance deficiencies. When assessing possible metro and other foreign rail efforts, the advantages and disadvantages of existing systems must be considered. Metropolitan regions bring significant economic, social, and environmental benefits. The advantages are unparalleled and ongoing, spanning generations rather than a few years or decades. Despite having significantly lower total traffic than anticipated during the planning process, Indian metro systems have usage per km similar to their foreign counterparts.

References

Bared J (2005) Crash comparison of single point and tight diamond interchange. J Transp Eng 131(5):379–381

Chlewicki G (2011) Should diverging diamond interchange is consider has diamond interchange. Transp Res Board 2223:88–95

Ghodmare SD, Gulghane A (2022) An attempt for sustainability evaluation considering implementation of rapid transit system. Materials Today: Proceedings. ISSN 2214-7853, https://doi.org/10.1016/j.matpr.2022.01.262. https://www.sciencedirect.com/science/article/pii/S2214785322002942

Ghodmare SD, Bajaj P, Khode BV (2019) Comparative analysis of BRTS and MRTS—an approach required for selection of system. In: Kolhe M, Labhasetwar P, Suryawanshi H (eds) Smart technologies for energy, environment and sustainable development. Lecture notes on multidisciplinary industrial engineering. Springer, Singapore. https://doi.org/10.1007/978-981-13-6148-7_46

Sustainable transportation using automated highway system for intercity connectivity Mendhe (2022) In: Ghodmare PD, Titarmare SD, Wadhai AS, PJ. International conference on emerging trends in engineering and technology, ICETET, 2022

Metro Follows an Optimized Speed Trajectory to Save Energy Yadav (2022) In: Khode DD, Ghodmare BVSD. International conference on emerging trends in engineering and technology, ICETET, 2022

Lee S, Messer CJ, Youngtae Oh, Lee C (2004) Assessments of three simulation models for diamond interchange analysis. J Transp Eng 130(3):304–312

Madhi MB (2015) Microscopic simulation on the design and operational performance of diverging diamond interchange. J Transp Eng
Maji A, Mishra S, Jha MK (2013) Diverging diamond interchange analysis: planning tool. J Transp Eng 139:1201–1210
Molan AM, Hummer JE (2019) Modeling safety performance of new super DDI design in terms of vehicular traffic and pedestrian. Accid Anal Prev 198–209
Nelson EJ, Bullock D, Urbanik T (2000) Implementing actuated control of diamond interchanges. J Transp Eng 126(5):390–395
Critical analysis of saturation flow rate and affecting parameters (2022) In: Rodke SJ, Ghodmare SD, Deshmukh A. International conference on emerging trends in engineering and technology, ICETET, 2022
Sannikon S, Timonovetz V (2018) A procedure for an individual design of an interchange. Transp Res Procedia 682–688
Shaikh MZ, et al (2020) Diverging diamond interchange: a comprehensive review. J Emerg Technol Innov Res 7(5)
Sharma S (2011) Performance evaluation and diverging diamond interchange in comparison with conventional diamond. J Transp Eng
Wengler S, Urbanik T (1995) Real-time optimization at diamond interchanges. IEEE
Wu L, Ci Y (2020) Fuel consumption at the oversaturated signalized intersection considering queue effect: a case study in Harbin, China Energy 192:116654
Mitigation of blackspots on highways by the application of safe system approach (2021) Yadava DK, Ghodmare SD, Kumar N. Materials Today: Proceedings, Elsevier. https://doi.org/10.1016/j.matpr.2021.11.042, online 19 November 2021
Yang XF, Chang GL (2014) Development of a signal optimization model for diverging diamond interchange. J Transp Eng 04011010–1

Chapter 23
Comparison of Performance of Various Bus Rapid Transit Systems in India

Waseem Bari and Kapil Jajulwar

Abstract One way to conceptualize the BRTS is as an integrated system that encompasses parking, land use planning, affordable transportation, bike and pedestrian infrastructure, housing policy, road design and space utilization, and public transportation prices. Since the debut of the Bus Rapid Transit (BRT) idea, around ten Indian cities have indicated interest in it, with some of these towns now in various stages of the planning or system implementation process. The Bus Rapid Transit System (BRTS) is the primary tactic used in this article to satisfy the needs of equity and sustainability. In addition, it has attempted to investigate the BRTS's technology and functionality, as well as the system's architecture. This article seeks to close the gap by doing a thorough analysis of BRTS in the Indian context and focusing Ahmedabad BRTS, with largest network in India. Here, an attempt was made to illustrate the BRTS assessment procedure by assessing the performance cities and local authorities in implementing the bus rapid transit system planned.

Keywords Affordable transport · Bus rapid transit system · Sustainability

23.1 Introduction

23.1.1 BRTS as Per Indian Situation

The population in six largest metropolises in India grew by 1.9 times between 1981 and 2001, but the number of automobiles climbed by about 7.75 times during that same period, according to a 2007 report from the Ministry of Urban Development (Agrawal et al. 2010). Approximately 142 million motor vehicles were registered

W. Bari (✉) · K. Jajulwar
G. H. Raisoni College of Engineering, Nagpur, India
e-mail: Waseem.bari.mtechts@ghrce.raisoni.net

K. Jajulwar
e-mail: Kapil.jajulwar@raisoni.net

B. V. Khode et al. (eds.), *Innovations in Technologies: Pioneering Sustainable Infrastructure for a Resilient Future*, Sustainable Civil Infrastructures,
https://doi.org/10.1007/978-3-031-82133-2_23

Fig. 23.1 Different types of buses using the BRTS corridor

as of the end of March 2011, an increase from approximately 0.3 million in March 1951. The country's registered car count rose at a 9.9% Compound Annual Growth Rate (CAGR) between 2001 and 2011. Over three times as many motor vehicles were registered in India during the last ten years as there were roads (Ghodmare and Yadav 2021; Gannon and Liu. 1997). The percentage of buses, including omnibuses decreased to just 1.1% in 2011, according to the Transport Research Wing (2012) (Ahmedabad Municipal Corporation 2011, 2010). With more and more motor vehicles on Indian roadways, the overall situation presents a serious challenge to the country's transportation industry and infrastructure.

23.1.1.1 Public Transport in India

India's public transportation system is comparable to that of many other developing countries globally. According to a study done by the Urban Development Unit in 2008, public transportation systems in various cities in India are unable to keep up with the rapid and large development in needs (AMC, Auda and CEPT University 2008; Gakenheimer 2008).

Bus services, in particular, have deteriorated, and their modal share has declined even more as a result of customers transferring to more customized modes of transportation and intermediate public transit (such as taxis and three-wheelers) (Anand and Tiwari 2006; Tiwari 2001), which has a detrimental influence on bus operations. Furthermore, building flyovers and highways was considered as the only method to enhance urban transportation, which did nothing to expand everyone's mobility and instead packed the streets with a growing number of private motorized transport vehicles (Astrop 1996; Chen 2009).

23.1.1.2 The National Policy Context for Low-Carbon Mobility

The Energy and Resources Institute estimates that road transportation accounts for 80% of all emissions in India's transportation sector, which consumes around 16.9%

Fig. 23.2 Corridors at a junction and Bus pile up at green signal while changing lanes

of the country's total energy. According to a separate study, the road transportation industry accounted for 94.5 percent of the transportation sector's total emissions of CO2 and CO (Badami 2009; Clifton 2003). It is critical to note that investments in India's transportation industry need an understanding of low-carbon transportation initiatives. Since 2005, the Indian government has made considerable investments in India's capital cities and metropolises under the JnNURM, a seven-year mission to reform governance, construct infrastructure, and provide housing and basic services for the urban poor (Badami et al. 2007; Ghodmare et al. 2022).

23.2 BRT Projects in India

In 2011, JnNURM supported BRTS projects in nine Indian cities (Fig. 23.3). There are presently BRT lines in Ahmedabad, Jaipur, Vizag, Pune, and Delhi, varied in size and kind (Baker et al. 2005). Ahmedabad now has the biggest BRTS route in service, and it is increasing. Because of its early success, other cities now see it as a model for imitation (Mujammel and Ghodmare 2020). The Delhi government, rather than the JnNURM, funds the 14.6-km BRT line in the National Capital Territory. India has used a variety of configurations for its BRT lanes, including the Bogotá model, which is a dedicated BRT system with a bus route that accommodates all types of buses.

In terms of adapting and evolving to their surroundings, BRTSs in India still have a long way to go. As a result, an assessment of existing barriers is required (Barter 1999; Cheng and Yang 2018). In general, many cities that desired to build BRTs already had bus networks in place, such as Ahmedabad's Municipal Transport System (AMTS) and Pune's Mahanagar Parivahan Mahamandal Limited.

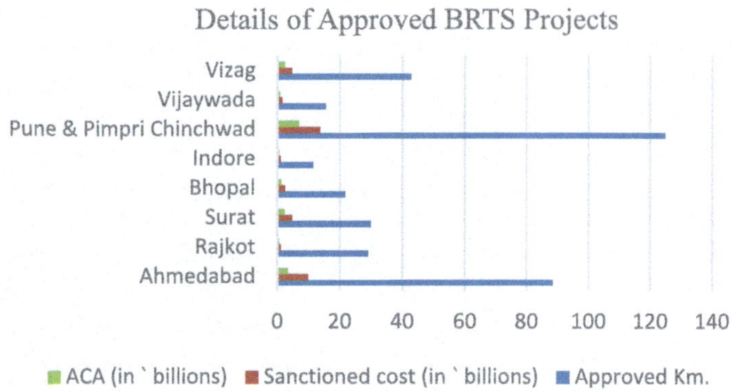

Fig. 23.3 Approved BRTS projects under JNNMURM

23.3 Research Framework and Approach

The assessment framework for diverse BRT initiatives must take into account the overarching principles of equity, accessibility, and sustainability (Maji et al. 2013; Yang and Chang 2014). Determining the BRT's stakeholders, service prices, and inclusion are critical from the standpoints of equity and accessibility (Songa 2006; Yang and Chang 2014). Understanding the transition from private motor transport to BRTS, as well as the quality of NMT infrastructure constructed as part of the BRT system, is critical for sustainability (Molan and Hummer 2019; Ghodmare et al. 2019b).

The study begins by contextualizing the introduction of Bus Rapid Transit (BRTS) in Indian cities within the global conversation about the advantages of BRTS over traditional bus-based public transportation systems, as well as its affordability in comparison to capital-intensive and expensive metro systems (Gupta et al. 2021; Madhi 2015). One interpretation of the system's talks at various levels is that they led to the development of Indian public policy and an examination of the 2006 National Urban Transport Policy (NUTP) (Claros and Edara 2017; Metre et al. 2021). Thanks in large part to the JnNURM, cities around the country have begun to de-develop their BRTS funding plans. BRTS have been planned in several cities and are in various stages of development (Chlewicki 2011; Pai and Hidalgo 2009).

This takes us to the context of Indian public policy and the 2006 National Urban Transport Policy (NUTP). Cities around the nation have begun to submit BRTS bids for funding through the JnNURM, due in large part to the JnNURM. Numerous cities have BRTS plans at various phases of construction.

The second key part being investigated is a detailed examination of the Ahmedabad BRT system, which is currently India's biggest network (Yadava et al. 2021; Lee et al. 2004). The BRTS are being evaluated in terms of inclusion and a shift away from private motor vehicles and toward public transportation. The BRT system's inclusiveness is described in this context by how well-suited it is to a variety of income brackets, what the impoverisher's mobility preferences and demand patterns are, and how well-maintained the infrastructure is for bikers and pedestrians near to the BRT lines.

As stated in the objectives, the major focus of the assessment is on the system's social inclusion and sustainability. The review would include both the potential for BRTS design and the technology employed, since both would affect accessibility and BRTS trip costs (Shaikh et al. 2020; Wu and Ci 2020). The following sections have addressed these general concerns about BRTS in India. Table 23.1 presents an analysis of policy hurdles and system bottlenecks for BRTS in India.

23.3.1 Delhi BRTS

Delhi's morphology results in short commutes by dispersing commercial concentrations and residential regions around the city. The Delhi Transport Service (DTS) was Delhi's first public transportation system, founded in 1948 by the Ministry of Transport, Government of India, and transferred to the Municipal Corporation of Delhi in 1958. After the federal government assumed control in 1971, the Delhi Road Transport Acts Amendment Act established the Delhi Transport Corporation (DTC) with the goal of increasing efficiency.

23.3.2 Pune HCBS

Pune had a population of 5.05 million in 201,112. During 2001–11, the city's population grew by 3.0% year. It is located on the main railway route connecting Mumbai to the southern states.

23.3.3 Jaipur BRTS

The city's population was 3.07 million in 201,115, with a 2.8% yearly growth rate between 2001 and 2011. Jaipur, like many other Indian towns, has both a walled city and a new town. The overall road network is 1,500 km long, with 34% two-lane and 54% four-lane roads. The city's vehicles have an average yearly growth rate of 13%.

23.3.4 Indore BRTS

The city's population is rapidly growing, and roads are limited. Car traffic on city streets is also rapidly increasing. Private businesses also offer long-distance buses. Six private enterprises around the city manage the system's bus fleet. In December

Table 23.1 System bottlenecks and policy barriers for BRTS systems in India

BRTS System at	System bottlenecks/Policy barriers	Impact	Result
Delhi	Traffic signaling system. The lack of institutional coordination	Created long waiting times for the BRTS buses at traffic junctions	Increased the trip time. The mismanagement of the BRTS corridor
	The short length of the bus corridor. DIMTS is acting as the corridor manager of the BRTS, while the Transport Department regulates the bus operations	Fails to make any impact either on commuters	Less bus friendly System. DIMTS does not have a say in the junction design or signal cycling or bus operations
	The navigation facilities at the junctions	The buses experience a pile-up	Instead of stopping at a designated place, they stop at other stops and creates operational problems
Pune	The lack of institutional coordination as PMA comprises two municipal corporations and three cantonment boards	The PMC and PCMC continue with the old routes with different BRT systems	Non coordination. PMPML does not have financial autonomy
	The lack of administrative or political will to build the BRT. System have been reluctant to enforce the new traffic rules that would make the system run smoothly	Road widening—By PMC using JnNURM funds	Nonfunctioning of BRTS
	The complete lack of social marketing of the BRT system. The bus drivers of only the new (exclusive BRT) buses were trained	The social media were not taken into confidence to promote the system	The public was unaware of the benefits of the system
Jaipur	The project stalled midway after plans of the metro project were floated. (Very late)	Low BRT ridership	The corridor had to be re-aligned at many places. Land acquisition for achieving the ROWs
Indore	Land acquisition- Major problem	Obstructing the ROWs	Incomplete construction
	The lack of underground sewage and water lines on the sanctioned corridor	Discontinued footpaths and bicycle tracks	Improper services

of 2011, the fleet consisted of 107 buses. Currently, the Atal Indor City Transport Service operates 25 routes around the city.

23.4 The Status of BRT Projects in Selected Cities

Aside from Ahmedabad, which will be extensively covered, there are just a few operational BRT lines in Delhi, Pune, and Jaipur; they are insufficiently long to quantify the project's effects or evaluate using any performance indicators. There are differences in the systems; Delhi and Pune have an open BRT system that allows various bus types to operate, whilst Jaipur and Indore have yet to achieve this level.

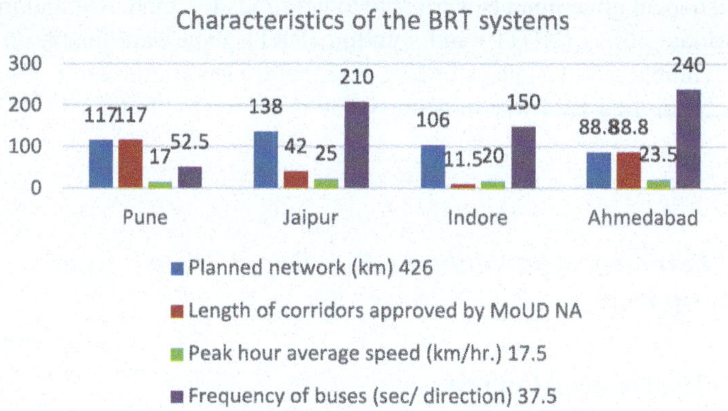

Fig. 23.4 Characteristics of the BRT systems of the selected cities

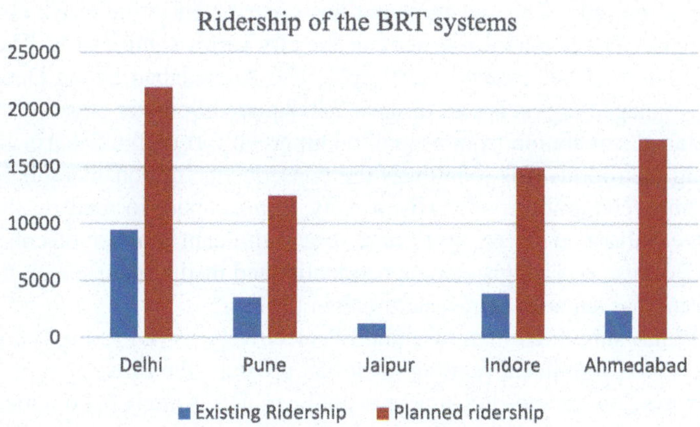

Fig. 23.5 Ridership of the BRT systems of the selected cities

The results show that insufficient comprehensive planning, subpar operational performance, and implementation errors are the main reasons why all BRTS initiatives are failing to solve issues. The degree to which the BRT system should be prioritized as the main mass rapid transit system in each city was not clearly defined or consistent (Ghodmare et al. 2019a; Mahadevia and Chand 2010). It has also been noted that the Delhi BRTS first drew criticism before becoming widely accepted.

It is difficult to develop BRT networks in Indian towns (Sannikon and Timonovetz 2018; Ghodmare et al. 2019). The local media and others are putting tremendous pressure on political and bureaucratic elites to keep the present proportion of private automobiles, which renders the systems ineffective. Even after covering implementation costs, the municipalities of all four cities were unable to construct more than one or two isolated corridors. Put another way, there has been more interest in the underlying technology of BRT systems than in their software.

Although local governments were able to get funds for JnNURM and implement the appropriate policy (NUTP) and solution (BRT), they were unable to take use of these favorable circumstances to alter regional transportation arrangements. The almost 44.5-km-long BRT line in Ahmedabad is known as "India's most successful BRTS."

23.4.1 Janmarg: Ahmedabad's Bus Rapid Transit System (BRTS)

23.4.1.1 The Planning Context

The Ahmedabad Municipal Corporation (AMC) expanded its size from 198 to 450 square kilometers in 2006 by combining seven municipalities and districts that encircled the old city. This development has contributed to the city's rapid population increase. The population density of the city varies significantly. The average density in the old AMC region is 189 ppH. The Ahmedabad Urban Development Authority-managed region is part of the 1,330 square kilometer city.

Ahmedabad's evolution from a small commercial fort on the Sabarmati River to the thriving metropolis it is today was the consequence of considerable historical, political, and economic growth. Historically, Ahmedabad, located on the eastern bank of the Sabarmati River, was home to a significant number of cotton textile factories. Because it is a high-income residential and institutional sector, the west is wealthier and has superior infrastructure than the east.

The city has submitted many Detailed Project Reports (DPRs) to the JnNURM and has received significant funding for urban projects like as water supply, sanitation, highways, low-income housing, and public transit. Among major cities, the city received the second-highest per capita financing for JnNURM projects. The residential component is being used to rehabilitate project-affected individuals. Transportation projects are one of the civic activities designed to present the image of a "World Class" metropolis.

The city has yet to produce a comprehensive mobility strategy that takes into account both future metro rail projects and existing public transportation infrastructure. Ahmedabad was one of the first cities to build a public transit system, and it also spearheaded other transportation initiatives, including the establishment of a BRT system and the conversion of public transportation to run on compressed natural gas.

The city's BRT system, dubbed "Janmarg," has attracted national and worldwide attention as an illustration of best practices in public transportation since it was introduced in 2009. Janmarg started two years ago and has since grown to a 44.5-km route, carrying over 0.15 million passengers daily and earning over $0.75 million in revenue. Municipal Corporation of Ahmedabad (2011). The number of Indian colonies is thus expected to increase dramatically. Finding out if the title "People's Pathway," or Janmarg, is inclusive and well-known as a municipal public transit route is the aim of this study.

23.5 Conclusion

It is vital to determine if the BRTS offers low-cost and reasonable transportation options to low-income groups in order to ensure a modal shift away from private two- and four-wheelers and toward public transit. This study on BRTSs in India addressed these two issues through a thorough case study of Ahmedabad City, which has the largest BRTS network among those already in service in Indian cities. Any mechanism put in place to bridge a gap in public transportation availability usually works. The main concern, however, is whether any new public transportation system can overcome the two key challenges listed above. When the difficulties of BRTS planning and execution in varied cities are addressed, this will be investigated further. Furthermore, capital- and energy-intensive transportation options often exclude and move low-income inhabitants from cities. Sustainability and justice, in this view, are interwoven transportation goals rather than mutually exclusive concepts. BRT, if correctly implemented, may be regarded a major system capable of dealing with carbon footprints.

References

Agrawal A, Agrawal AK, Chakravarti D (2010) Bus rapid transit Delhi: mobility for all. Retrieved 21 May, 2011, from Social Research in Transport (SORT) Clearinghouse: http://www.sortclearinghouse.info/cgi/viewcontent.cgi?article=1718&context=research

Ahmedabad Municipal Corporation (2010) AMTS budget 2010–11. Retrieved 20 December 2011, from http://www.amts.co.in/SitePage.aspx?id=45

Ahmedabad Municipal Corporation (2011) Janmarg-Ahmedabad bus rapid transit system: accessible Ahmedabad. Ahmedabad: booklet issued by Ahmedabad municipal corporation at the sub-regional EST Workshop for Policy Dialogue in South Asia on 26–28 August 2011

AMC, AUDA and CEPT University (2008) Bus rapid transit system plan phase II detailed project report. Retrieved 12 September 2011, from http://www.ahmedabadbrts.com/images/Ahmedabad%20BRTS%20Phase-2%20DPR_April%202008.pdf

Anand A, Tiwari G (2006) A gendered perspective of shelter-transport-livelihood link: the case of poor women in Delhi. Transp Rev 26(1):63–80

Astrop A (1996) The urban travel behaviour and constraints of low income households and females in Pune, India. Retrieved 11 May 2011, from Women's Travel Issues, Proceedings from the Second National Conference at Baltimore MD: http://www.fhwa.dot.gov/ohim/womens/chap12.pdf

Badami M (2009) Urban transport policy as if people and the environment mattered: pedestrian accessibility the first step. Econ Pol Wkly 44(33):43–51

Badami M, Tiwari G, Mohan D (2007) Access and mobility for the urban poor in India. In: Laquian A (ed) The inclusive city: infrastructure and public services for the urban poor in Asia. New Delhi: Woodrow Wilson Centre Press

Baker J, Basu R, Cropper M, Lall S, Takeuchi A (2005) Urban poverty and transport: the case of Mumbai. Retrieved 18 November 2011, from Open Knowledge Repository, World Bank, Working Paper- 3693: https://openknowledge.worldbank.org/handle/10986/8602

Barter PA (1999) Transport and urban poverty in Asia: a brief introduction to the key issues. Regional Development Dialogue 20(1):143–163

Chen H (2009) Evaluating the safety impact of the number and arrangement of lanes on freeway exit ramps. Accid Anal Prev 543–551

Cheng Y, Yang X (2018) Concurrent optimization of signal progression and crossover spacing for diverging diamond interchange. J Transp Eng 144(3)

Chlewicki G (2011) Should diverging diamond interchange is consider has diamond interchange. Transp Res Board 2223:88–95

Claros B, Edara P (2017) When driving on the left side is safe: safety of the diverging diamond interchange ramp terminals. Accid Anal Prev 133–142

Clifton KJ (2003) Examining travel choices of low income populations: issues, methods and approaches. Retrieved 11 May 2011, from Moving through Nets: The Physical and Social Dimensions of Travel, 10th International Conference on Travel Behaviour Research, Lucerne 10–15 August 2003: http://www.ivt.ethz.ch/news/archive/20030810_IATBR/clifton.pdf

Gakenheimer R (2008) Landuse and environment in transport planning- as an option among others in rapidly growing and transforming cities. Retrieved May 21, 2011, from http://www.internationaltransportforum.org/Topics/Workshops/WS4Gakenheimer.pdf

Gannon C, Liu Z (1997) Poverty and transport, discussion paper TWU30. Washington D.C.: World Bank Transport Division

Ghodmare SD, Yadav G (2021) Transportation planning using conventional four stage modeling. Turk J Comput Math Educ 12(12):2891–2897. Published online: 23 May 2021

Ghodmare SD, Bajaj P, Khode BV (2019) Comparative analysis of BRTS and MRTS—an approach required for selection of system. In: Kolhe M, Labhasetwar P, Suryawanshi H (eds) Smart technologies for energy, environment and sustainable development. Lecture Notes on Multidisciplinary Industrial Engineering. Springer, Singapore. https://doi.org/10.1007/978-981-13-6148-7_46

Ghodmare SD, Gulghane A (2022) An attempt for sustainability evaluation considering implementation of rapid transit system. Materials Today: Proceedings. ISSN 2214-7853, https://doi.org/10.1016/j.matpr.2022.01.262. https://www.sciencedirect.com/science/article/pii/S2214785322002942

Ghodmare SD, et al (2019) Application of the multi attribute utility technique with it's for sustainability evaluation of emerging metropolitan city of Nagpur. Int J Civ Eng Technol. EID: 2-s2.0–85060924119

Ghodmare SD, et al (2019) Sustainable transportation system with intelligent transportation system: a framework for assessment and application for emerging metropolitan city©. JETIR February 2019, 6(2)

Gupta MP, Ghodmare SD, Khode BV (2021) Feasibility study on widening of roads as per traffic survey. Des Eng 2021(04):210–217

Lee S, Messer CJ, Youngtae Oh, Lee C (2004) Assessments of three simulation models for diamond interchange analysis. J Transp Eng 130(3):304–312

Madhi MB (2015) Microscopic simulation on the design and operational performance of diverging diamond interchange. J Transp Eng

Mahadevia D (2010) Urban Reforms in Three Cities – Bangalore, Ahmedabad & Patna. In: Chand VK (ed) Public Service Delivery in India: Understanding the Reforms Process. Oxford University Press, New Delhi, pp 226–295

Maji A, Mishra S, Jha MK (2013) Diverging diamond interchange analysis: planning tool. J Transp Eng 139:1201–1210

Metre K, Baghel H, Suman G, Batra M, Ghodmare SD (2021) Compact city and related impact on sustainable development in urban areas. In: Gupta LM, Ray MR, Labhasetwar PK (eds) Advances in civil engineering and infrastructural development. Lecture Notes in Civil Engineering, vol 87. Springer, Singapore. https://doi.org/10.1007/978-981-15-6463-5_50

Molan AM, Hummer JE (2019) Modeling safety performance of new super DDI design in terms of vehicular traffic and pedestrian. Accid Anal Prev 198–209

Pai M, Hidalgo D (2009) Indian bus rapid transit systems funded by the jawaharlal Nehru national urban renewal mission. Transp Res Rec: J Transp Res Board 2114:10–18

Sannikon S, Timonovetz V (2018) A procedure for an individual design of an interchange. Transp Res Procedia 682–688

Shaikh MZ, Ghodmare S, Islam AU, Urade S (20220) Diverging diamond interchange: a comprehensive review. J Emerg Technol Innov Res (JETIR)© 7(5). ISSN-2349–5162

Shaikh MZ, et al (2020) Diverging diamond interchange: a comprehensive review. J Emerg Technol Innov Res 7(5)

Songa H (2006) Comparison of Operation Performance of Diamond Interchanges between China and U.S.A. J Transp Eng

Tiwari G (2001) Traffic segregation: a case for bus priority lanes with segregated cycle tracks-case study Delhi, workshop on transportation Landuse & the environment, December 3–4, Pune Retrieved June 2010 from http://tripp.iitd.ernet.in/publications/paper/planning/gtpune.pdf

Wu L, Ci Y (2020) Fuel consumption at the oversaturated signalized intersection considering queue effect: a case study in Harbin, China. Energy 192(2020):116654

Yadava DK, Ghodmare SD, Kumar N (2021) Mitigation of blackspots on highways by the application of safe system approach. Materials Today: Proceedings, Elsevier. https://doi.org/10.1016/j.matpr.2021.11.042, online 19 November 2021

Yang XF, Chang GL (2014) Development of a signal optimization model for diverging diamond interchange. J Transp Eng 04011010–1

Chapter 24
Impact Assessment of Replacement of Waste Bituminous Concrete with Virgin Bituminous Mixes

Bhalchandra V. Khode, Prafull J. Wadhai, Sujesh D. Ghodmare, and Nalini Vaidya

Abstract It is common knowledge that building a highway requires a significant financial outlay, and when designing flexible pavements, there are important factors to take into account while designing the pavement's mix. The development of high volume pavements in India has significantly improved recently compared to prior years. However, this has brought about a need for concentration in the provision of essential and rare aggregates of natural resources. A section of the current pavements has been repaired, involving the removal of the top layer of asphalt to make room for new layers beneath. Natural deterioration has been triggered by the removal of the old solid surfacing layer and asphalt layer in the open spaces. Low volume streets have been left in a deplorable shape due to a lack of suitable assets. Determining whether a mixture of recovered asphalt concrete, new aggregates, and a cationic emulsion makes sense as a surfacing medium for low volume pavement construction was the main aim of the experiment. The impact of using recycled asphalt pavement (RAP) to replace bituminous concrete, either partially or completely, is being evaluated in India. This study examined the differences in mechanical properties of recycled asphalt pavement (waste BC) and virgin HMA mix, ranging from 10 to 40%.

Keywords RAP · Flow value · Bituminous concrete · Marshall stability · Density · Virgin HMA mix

24.1 Introduction

Use of road network is growing daily. It is the demand of time that something new is to be introduced in order to preserve the network efficiencies. This is needed in order to accommodate the additional volume of traffic safely. New methods should be

B. V. Khode (✉) · P. J. Wadhai · S. D. Ghodmare · N. Vaidya
G. H. Raisoni College of Engineering, Nagpur, India
e-mail: bhalchandra.khode@raisoni.net

B. V. Khode et al. (eds.), *Innovations in Technologies: Pioneering Sustainable Infrastructure for a Resilient Future*, Sustainable Civil Infrastructures,
https://doi.org/10.1007/978-3-031-82133-2_24

affordable and ecologically. These are to be well disposed, regardless of the consistent increase in costs of bitumen. The intense lacking of good aggregates that is vital for a creation of asphalt concrete is also the major issue to be handled. A few novel techniques, like as recycling—that is, repurposing previous asphalt mixtures after grinding them into new mixes with varying ratios—can be used. It is feasible to design a laboratory programme that produces asphalt mixtures with ratios similar to those of standard mixes. This would reduce the economical aspect of RAP in hot mix asphalt. Lot of greenhouse gases and dangerous pollutants are discharged due to boiling of bituminous binder aggregates of Hot Mix Asphalt (HMA). Also, there is an issue of the shortage of aggregates. It also demands the transportation of materials from significant distance which causes combustion of diesel in transport resulting in emanation of pollutants. In spite of that recycling of black-top asphalts is a noble and specialized approach which is also affordable and favorable for protection of regular assets and ecological issues. Use of RAP in asphalt restoration is very important because of the regular rising expenses of scarcity of aggregates, asphalt and the saving of environment. It is predicted to be the common resources reproducing asphalt. It may also fell under the category of the most recycled product in the world. The main idea behind recycling asphalt is to cut cost for new asphalt pavement. It is also utilized and represented as an innovation. It is the need of time that specific instructions and regulations should be available for RAP in order to confine the amount that it can be used for each mix. Many of the studies have indicated an averages of RAP utilized in new blends are around 12% to 15% at national level. The National Asphalt Pavement Association (NAPA) has defined objectives to build the normal RAP content all through the world. Reuse and Conservation of non-sustainable power sources, Energy preservation and improved asphalt perfection are few of the advantages of RAP. It is generally achieved through improved asphalt physical properties by alteration of existing grade of aggregates and binding properties of asphalt. To determine the dynamic modulus and flow quantity, Copeland et al. evaluated the effectiveness of two preparation mixes: an elevated RAP–hot-mix asphalt (HMA) control mix with a high RAP–WMA. (Lastname et al. 2007). Results of flow number and Dynamic modulus experiment reflected that high RAP-WMA mix is softer than the high RAP-HMA control mix showing lower flow number and slightly lower E* values. Mallick et al. (2008) found that the use of high quantity RAP offers a very economic method of rehabilitation. Its use is very important to save natural resources, money along with environmental. Higher RAP substance would require the utilization of aberrant warmth procedures. Mohammad and Cooper (2011) described characterize HMA mixtures containing High Reclaimed Asphalt Pavement content with Crumb Rubber additives. This was examined based on the lab assessment of traditional HMA mixes and those containing high RAP, crumbed rubber and additives content. It was observed that the added rubber crumbs remolds the asphalt content into less plastic material for the HMA mix with high RAP content. Resistance to moisture is also considered. Apeagyei et al. (2013) developed on the production of high-RAP mixes to assess the qualities of asphalt mix containing distinctive RAP. The work was carried out to accomplish a superior comprehension of how high RAP (>20%) influences the mix execution properties significant for increasingly tough and savvy black-top. Hugo et al. (2012) described reused HMAs can be a decent option for paving of roads, especially if specialized agents are used to lower the temperature at the time of production and to improve their sustainability. Arshad Hussain et al. (2012) found the Results of HMA with RAP and without RAP showed satisfactory performance over 16 years of the service period. The results reflected that pavement prepared with RAP content as 35%, can be sustainable if designed properly during its life span. Due to oxidation the penetration and softening point show variation respectively in terms of increment or decrement. Al-Qadi et al. (2014) suggested HMA asphalt gave a saving range from 14 to 34% for a RAP content changing between 20 to 50%. Mixes were planned by Marshall -Method at a wide scope of 0 to 100% RAP mixes to deal with low, medium and high traffic loads. The impact of adding chemical as well as mineral admixtures to enhance concrete with high performance. The outcomes show that low RAP content up to 30% can be utilized in wearing courses infrastructure (Khode and Ghodmare 2019; Sanjay et al. 2022; Sonkusare et al. 2021; Chandak and Pawade 2021a, 2021b, 2020; Solanke et al. 2022; Sonkusare and Pawade 2021; Solanke and Pawade 2021). It was tried to represent the result for mix utilizing these tests arranged at various RAP substances and tried to decide the ideal RAP content. Same materials mix was utilized by RAP level of 10%, 20%, 30% and 40% and tests arranged of Bituminous Concrete. Marshall Stability, thickness, firmness, flow value, VA, MAV, VFB, s by setting them up at various RAP rate.

24.2 Reclaimed Asphalt Pavement (RAP)

RAP basically represents the elimination and repurposing of top layer (asphalt) of prevailing pavement. Its necessary degree is accomplished by pummeling the material in a crusher and the characteristics are shown in Table 24.1. It can also be obtained from processing of HMA layer as replacement of pavement at full depth removal and HMA generated wastes at plants. Processing of HMA layer also known as milling which is the process removal or scraping of tattered uppermost sections of any existing pavement to a certain depth. It is usually done using shovels, cranes or scrapers at a certain specified location up to calculated depths. Then it is thereafter transported for recycling.

24.3 Bituminous Mix Design

Asphalt/Bituminous solid comprises of a blend of materials. It is ceaselessly received from greatest size, ordinarily under 25 mm, through the considerably fine filler that is usually less than 0.075 mm. An ample quantity of bitumen is incorporated into the mixture with the aim of achieving a compacted blend that is effectively impermeable. This ensures the resulting mixture possesses commendable dissociative and flexible

Table 24.1 Reclaimed asphalt pavement—typical properties

S. No	RAP-parameters	Range/Values
01	Unit-Weight of material (Kg/m^3)	1900 to 2250
02	Water/Moisture Content	Maximum between 3%–5%
03	Contents of asphalt	5%–6%
04	Asphalt penetration in % at 25 °C	10–80
05	Compacted-unit-weight of material (Kg/m^3)	1500–1950
06	California bearing ratio of material (CBR)	100% RAP: 20%–25%

properties. The mix design process of bituminous involves determining the optimal proportions of bitumen, fine aggregates, filler and coarse aggregates. This is helpful to create a blend which is serviceable, solid, sturdy and affordable. The target of the blend configuration is to deliver a bituminous blend by ratiocinating different segments in order to have.

1. Adequate bitumen to guarantee a strong pavement
2. Adequate strength in order to the resist shearing even in high-temperature traffic situations.
3. Conferring considerable air voids volume in the compacted bitumen considering more traffic-related compaction.
4. Workability so as to allow ease of laying pavement without segregating it
5. Reducing cracking caused due to repetitive traffic loading
6. Preventing cracks caused due to shrinkage at low temperatures

24.4 Theoretical Considerations for Rap

For arrangement of Bituminous Concrete (BC) materials according to Marshall Test Data for Virgin Bituminous Concrete (As Per ASSHTO T245) is given as maximum for 5.8% bitumen by weight of total mix. Whereas Ministry of Road transport & Highway (MORTH) specified the values for stability (kg) 900 min, flow (mm) 2 to 4, Va (%) 3 to 5, VMA (%) 13 min and Marshal Quotient as 2 to 5 and are also shown in Table 24.2.

24.5 Aggregates and Minerals

For arrangement of Bituminous blends (BC) materials according to Ministry of Road Transport &Highway (MORTH) are evaluating and shown in Table 24.3 and a specific kind of binder in necessary amounts were obligatory according to Marshall Procedure and Properties of mineral aggregates for Elongation Index and FI is Flakiness Index obtained by Experiment are shown in Table 24.4.

Table 24.2 Statistics from the marshall test involving virgin bituminous concrete

Physical & volumetric characteristics of virgin bituminous concrete (VG-30) as per ASSHTO T245						As Per MORTH
Bitumen percent	5.4	5.6	5.8	6.0	6.2	–
Density in gm/cc	2.388	2.421	2.443	2.432	2.424	–
Stability in Kg	902.12	1071.74	1232.70	1008.38	905.33	900
Flow(mm)	2.33	2.63	3.27	4.33	5.30	2–4
VA (%)	6.75	5.17	3.99	3.85	4.14	3–5
VMA (%)	15.77	14.79	14.19	14.52	15.23	13
VFB (%)	57.22	65.08	71.89	73.49	72.80	65–75

Table 24.3 Aggregate gradation for BC was implemented (MORTH)

Sieve size (mm)	Percentage passing of material	Required limit (MORTH) Table No: 500–17
19	100	100
13.2	99.66	79–100
9.5	84.24	70–88
4.75	63.41	53–71
2.36	52.95	42–58
1.18	38.37	34–48
600μ	30.08	26–38
300μ	20.71	18–28
150μ	14.72	12–20
75 μ	05.68	04–10

Table 24.4 Important mineral aggregates properties

Property	Result obtained by experiment	Requirements as per MORTH, Table No: 500–16	Test method
Aggregate impact value in %	21.6	30 max	IS 2386 (Part IV)
Water absorption in %	0.7	2 max	IS 2386 (Part III)
Specific Gravity,G	2.66	2.5–3.0	IS 2386 (Part II)
Combined (EI + FI) Index, %	25.2	35 max	IS 2386 (Part I)
Stripping, %	98	Min retained coating 95	IS 6241
Water sensitivity	92	Retained tensile strength 80%	AASTHO 283

Table 24.5 Characteristics of bitumen (VG-30)

Characteristics	Value	Requirements as per IS 73: 2006	Test method
Penetration, (25 °C, 100 g, 5 s), 0.1 mm	64	50–70	IS 1203–1978
Softening point (Ring and Ball), °C	48	≥ 47	IS 1205–1978
Ductility at 27 °C (5 cm/min)	78	–	IS 1208–1978
Specific gravity	1.01	–	IS 1202–1978
Viscosity at 60 °C, Poise	2570	≥ 2400	IS 1206–1978
Viscosity at 135 °C	725	≥ 350	IS 1206–1978

24.6 Binder

The Viscosity Grade (VG 30) infiltration grade bitumen is utilized as binder for readiness of Mix, Specific Gravity G = 1.01. Its significant characteristics is demonstrated in Table 24.5.

24.7 Methodology and Testing Procedure

The experimental analysis is split under two sections. First part illustrates the tests performed on the materials (Aggregates, Bitumen, and RAP), and another part addresses the testing done on bituminous mixtures. Showing various tests led on the bituminous mix in with varying binders and their amount, and RAP concentrates in the mix. The aggregates were taken from a neighboring mixing plant. A mix utilizing these aggregates was arranged and tests at various binder contents and finding its optimum amount. Similar aggregates mix was utilized and some portion of it was supplanted by RAP level of 10%, 20%, 30% and 40% was determined at Bitumen 5.8% were used for preparing Concrete of Bitumen content. Characteristics like Marshall Stability, Flow value and thickness were seen at rates of bitumen 5.8% three samples by preparing at each RAP rate are shown in Table 24.6 and Figs. 24.1, 24.2, 24.3, 24.4, 24.5, 24.6.

24.8 Validation of Data

The comparison between the MORTH Specification (section 500) mix design specifications of bituminous mix Obtained Value of mix @ 5.8% by Virgin Aggregate carried out with Obtained Value of mix of RAP @ 30% of Experimental value of Density(gm/cc), Stability (kg), Flow(mm), Voids in Aggregates (VA) (%), Voids

Table 24.6 Marshall test data for BC with RAP 10%, 20%, 30% & 40%

Reclaimed asphalt pavement (RAP) (%)	Density (gm/cc)	Stability (Kg)	Flow (mm)	MQ	Voids in aggregates (VA) (%)	Voids filling bitumen (VFB) (%)	Mineral aggregate voids (MAV) (%)
RAP (10%)	2.359	882.12	2.23	3.96	7.01	56.84	16.24
RAP (20%)	2.402	1043.35	2.9	3.6	5.31	63.92	14.71
RAP (30%)	2.434	1148.85	3.2	3.59	4.05	70.17	13.57
RAP (40%)	2.416	979.03	4.43	2.21	4.75	66.56	14.2

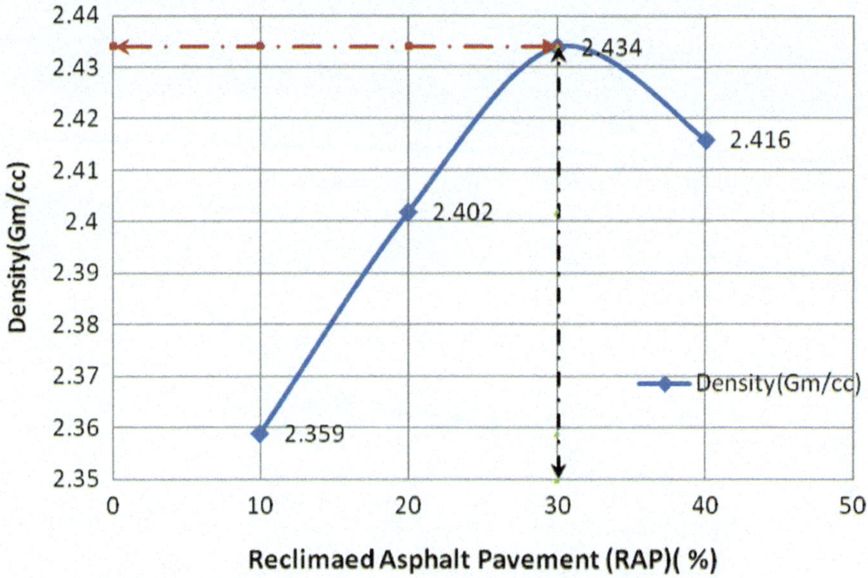

Fig. 24.1 Reclaimed asphalt pavement (RAP) (%) with density (gm/cc)

Filling Bitumen (VFB) (%) and Mineral Aggregate Voids (MAV) (%). The comparison of MORTH as well as value obtained from this study shows a close agreement as Shown in Table 24.7

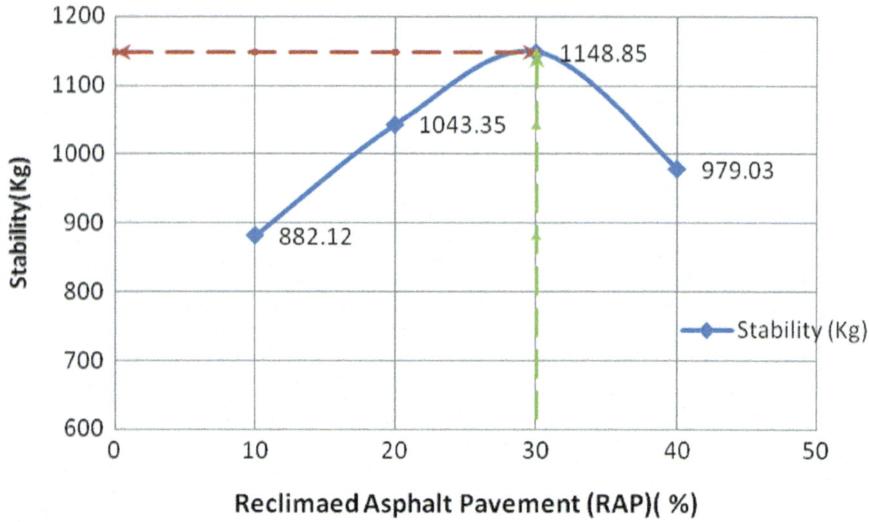

Fig. 24.2 Reclaimed asphalt pavement (RAP) (%) with stability (kg)

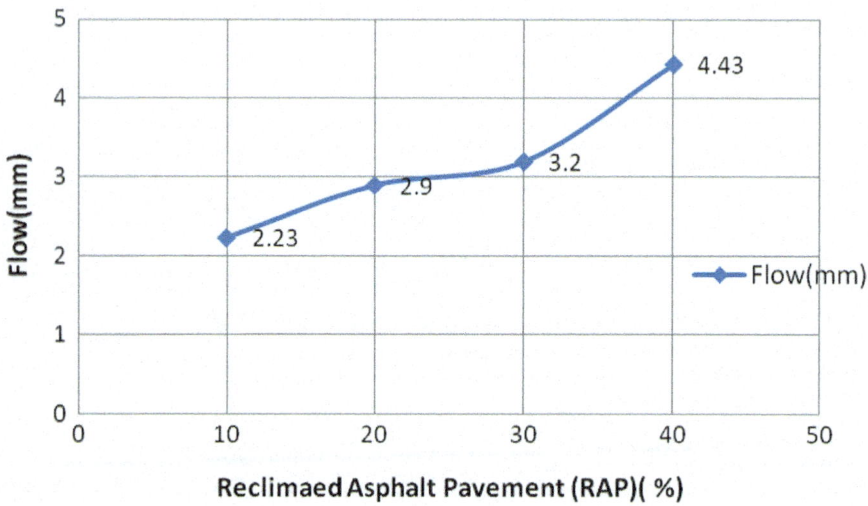

Fig. 24.3 Reclaimed asphalt pavement (RAP) (%) with flow (mm)

24.9 Results and Conclusions

The results allow for the drawing of the relevant conclusions and observations of tests conducted on many different parameters for the analysis, computation of BC's Optimum Binder Content (OBC), and Marshall characteristics of concrete with bituminous or without adding RAP:

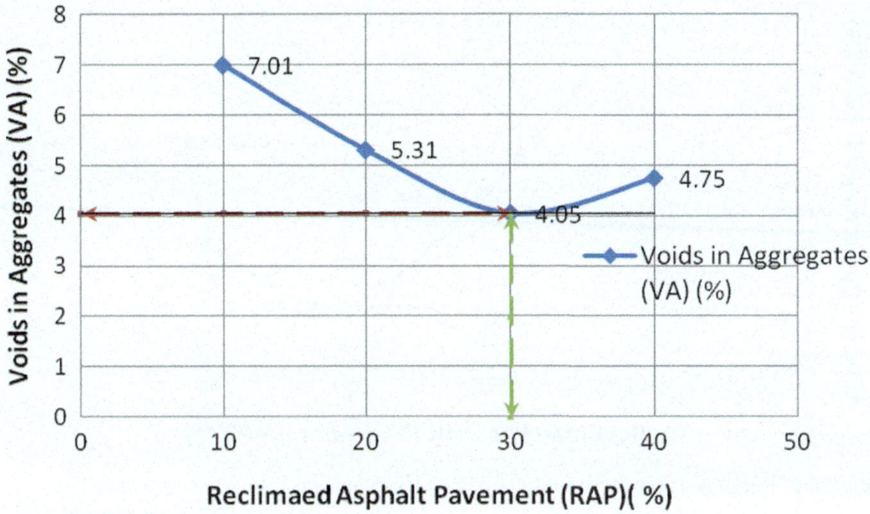

Fig. 24.4 Reclaimed asphalt pavement (RAP) (%) with voids in aggregates (VA) (%)

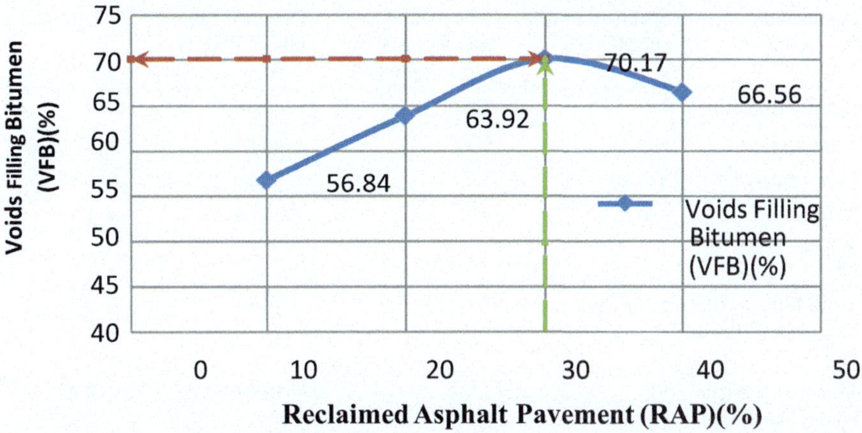

Fig. 24.5 Reclaimed asphalt pavement (RAP) (%) with voids filling bitumen (VFB) (%)

(1) The greatest Marshall stability measured at RAP 27.65% is 900 kg, whereas Marshall Stability at RAP 30% is 1148.85 kg.
(2) The bituminous mix has a density of 2.443 gm/cc and a density of 2.434 at 30% RAP; the difference between these two densities is minimal.
(3) The difference between the flow values at OBC of virgin mix (3.27 mm) and RAP content (30 mm) is small but acceptable.
(4) Voids in the total VA at the virgin mix's OBC are 4.0%, and at the 30% RAP content, they are 4.05%. The disparity is tolerable and insignificant.

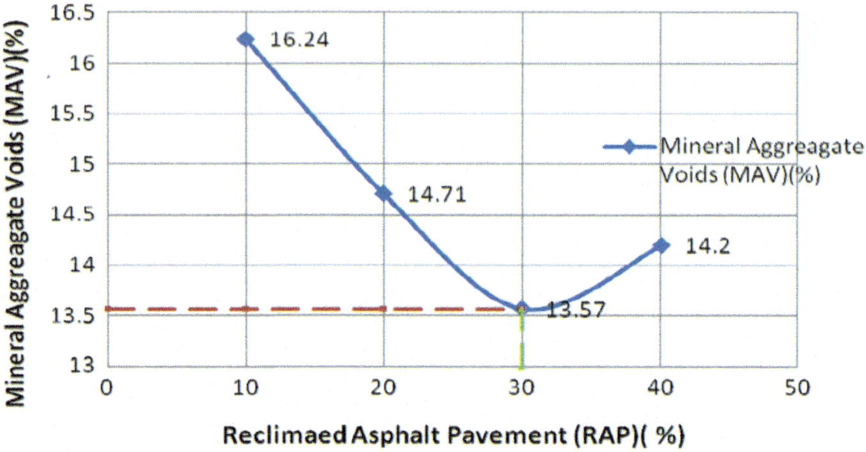

Fig. 24.6 Reclaimed asphalt pavement (RAP) (%) with mineral aggregate voids (MAV) (%)

Table 24.7 Comparison between the MORTH requirements (section 500) mix design and obtained value of mix of RAP @ 30%

Parameters	Specified limit as per MORTH	Obtained value of mix @ 5.8% by virgin aggregate	Obtained value of mix of RAP @ 30%	% Difference
Density (gm/cc)	–	2.443	2.434	0.37
Stability (kg)	900	900	1148.85	+ 27.65
Flow (mm)	2 to 4	3	3.20	+ 6.67
Voids in aggregates (VA) (%)	3 to 5	4	4.05	+ 1.25
Voids filling bitumen (VFB) (%)	65 to 75	70	70.17	+ 0.24
Mineral aggregate voids (MAV) (%)	13	13	13.57	+ 4.38

(5) The amount of voids in the aggregate of mineral MAV at the virgin mix's OBC is 13.57%; the required amount is 13%, and the difference is insignificant. Minimal.

(6) Bitumen-filled voids: the difference between VFB at OBC of virgin mix (71.79%) and amount at RAP content (30%) is 70.17%, which is well within the range (65–75%).

References

Al-Qadi IL, et al (2014) Reclaimed asphalt pavement–a literature review. Illinois Center for Transportation Series No. 07–001:1–19

Apeagyei AK, Clark TM, Rorrer TM (2013) Stiffness of high-RAP asphalt mixtures: Virginia's experience. ASCE, J Mater Civ Eng

Arshad Hussain and QiuYanjun (2012) Evaluation of asphalt mixes containing reclaimed asphalt pavement for wearing courses. IPCSIT 26:43–48

Chandak MA, Pawade PY (2020) Compressive strength and ultrasonic pulse velocity of concrete with metakaolin. Civ Eng Arch

Chandak M, Pawade PY (2021a) Influence of metakaolin and steel fibers on stress strain behavior of concrete. J Phys: Conf Ser. https://doi.org/10.1088/1742-6596/1913/1/012070

Chandak M, Pawade PY (2021b) Experimental study on the effect of aspect ratio and volume fraction of steel fibre on the mechanical properties of metakaolin SFR concrete with mathematical model. J Phys: Conf Ser. https://doi.org/10.1088/1742-6596/1913/1/012070

Khode BV, Ghodmare SD (2019) Effect of mineral and chemical admixture for improvement of high- performance concrete. In: Smart technologies for energy, environment and sustainable development. Lecture Notes on Multidisciplinary Industrial Engineering. Springer, Singapore. https://doi.org/10.1007/978-981-13-6148-7_41

Copeland et al (2007) Field evaluation of high reclaimed asphalt pavement- warm mix asphalt project. J Transp Res Board. No. 2179, Florida, USA, PP 93–101

Mallick RB, et al (2008) Using-warm asphalt technology to incorporate high percentage of reclaimed asphalt pavement material in asphalt mixture. Transportation Research Record 2051, Washington, D.C, pp 71–79

Mohammad LN, Cooper SB (2011) Characterization of HMA mixtures containing High Reclaimed Asphalt Pavement content with Crumb Rubber additives. ASCE J NOV

Sanjay SS, Pal A, Chakraborty SK, Khan HA (2022) Reduction of shrinkage of self-compacting concrete using polycarboxylate ether as shrinkage reducing admixture. Mater Today: Proc 60(1):448–451. https://doi.org/10.1016/j.matpr.2022.01.316

Silva HMRD, Oliveira JRM, Jesus CMG (2012) Are totally recycled hot mix asphalts a sustainable alternative for road Paving? Resources, Conservation and Recycling 60:3848, Elsevier

Solanke SS, Pawade PY, Khan HA (2022) An experimental study on tensile as well as flexural strength of concrete by using sugarcane baggase ash & steel fiber. Mater Today: Proc 60(1):627–637. https://doi.org/10.1016/j.matpr.2022.02.129

Solanke SS, Pawade PY (2021) An investigation of mechanical properties of concrete by addition of sugarcane baggase ash and steel fiber. J Phys: Conf Ser 1913(1):012069. https://doi.org/10.1088/1742-6596/1913/1/012069

Sonkusare HG, Pawade PY, Pujara HP (2021) Experimental mix design approach of reactive powder concrete to recognize compressive strength through non-destructive test. J Phys: Conf Ser. https://doi.org/10.1088/1742-6596/1913/1/012149

Sonkusare HG, Pawade PY (2021) Investigate the temperature effects on curing of reactive powder concrete containing silica. Lect Notes Civ Eng

Chapter 25
Challenges and Recommendations in Urban Transportation

Vinod Kumar Singh and Amol Pote

Abstract Urban mobility and accessibility are critical to Indian cities' long-term economic success. The consolidation of built forms and spatial extension are inextricably linked to urban stock and flows. Urban mobility has several ramifications for public policy and practice. Therefore, cross-process communication and policy consistency are critical. A systems approach, which provides a comprehensive knowledge of the obstacles and their root causes, is the most effective way to address complicated urban transportation concerns. Significant policy adjustments will be difficult to undertake without a thorough understanding of the interactions between the system elements that drive the symptoms. Bus journeys, NMT, and personal autos are all forms of transportation. Cities and transportation networks are viewed as tools for achieving actual socioeconomic goals, with mobility serving as a facilitator. The goal of this research is to discover and suggest solutions to urban transportation issues. Furthermore, the significance of government involvement in achieving this aim has been assessed. It has been underlined that tackling all of these urban transportation challenges would need a combination of planning reforms, capacity building, knowledge management, and the transferability of best practices.

Keywords Multi-dimensional · Reduce travel distances · Public transport options

25.1 Introduction

Reliable and efficient urban transportation networks are critical for India's rapid economic expansion. Urban transportation is important in India because it reduces poverty by increasing access to labor markets, resulting in greater wages in undeveloped regions (Road Transport Yearbook (2009-10 and 2010-11), (Mohanty et al.

V. K. Singh (✉) · A. Pote
Depatment of Civil Engineering, G H Raisoni College of Engineering, Nagpur, India
e-mail: vinod.singh.mtechts@ghrce.raisoni.net

A. Pote
e-mail: sujesh.ghodmare@raisoni.net

B. V. Khode et al. (eds.), *Innovations in Technologies: Pioneering Sustainable Infrastructure for a Resilient Future*, Sustainable Civil Infrastructures,
https://doi.org/10.1007/978-3-031-82133-2_25

2009). Physical connectedness is required for both urban and rural communities to succeed economically (Ghodmare et al. 2021). Growing population patterns, mispricing, and urban planning for machines than people occur in localities constructed for vehicles than people, resulting in insufficient public transit networks (Metre et al. 2021; Kamyotra et al. 2012). Because services and industrial enterprises are concentrated in and around major cities, efficient and reliable urban transportation networks are required to transfer people and connect manufacturing sites to the supply chain (Ghodmare et al. 2021). At its current levels, urban mobility is already contributing to a crisis scenario characterized by high levels of congestion, pollution, road deaths, and injustice, which will eventually lead to a poor accessibility situation (Bhatt et al. 2013). As poverty affects mobility problems of poor public are particularly severe (Foster and Kumar 2007). Despite significant differences in metropolitan size, structure, and growth trends across India's 468 cities, several common features contribute to the severity of urban transportation challenges (Baindur 2011; Nimburkar et al. 2022).

25.1.1 Urban Mobility Influencing Trends (India)

It includes rapid urbanization, growing motorization, and a diminishing modal share of non-motorized transport (NMT) (Baindur 2011).

25.1.1.1 Rapid Urbanization

Metropolitan transport networks are encountering new problems as a result of urbanization trends and patterns. Only five Indian towns had one million or more people, and 42 cities had 0.1 million or more; the great majority of India was rural (Tiwari 2011). In 2011, there were 53 towns with populations larger than one million, as illustrated in Fig. 25.1, and 468 districts with populations greater than 0.1 million (Tiwari 2007).

Almost half of them (>0.5 million) live in small cities (Ghodmare and Deshmukh 2022; WHO 2013). Surat, Nashik, and Faridabad saw the fastest decadal growth, whereas Mumbai, Delhi, Kolkata, Chennai, Hyderabad, and Bengaluru experienced slow peripheral growth as outlying cities integrated with the greater metropolitan zone (Baindur 2011; Siddhaye 2011).

25.1.2 Rapid Motorization

Since 2001, there have been considerably more cars per 1,000 people in Indian urban areas. Between 1991 and 2009, the number of registered automobiles in the nation

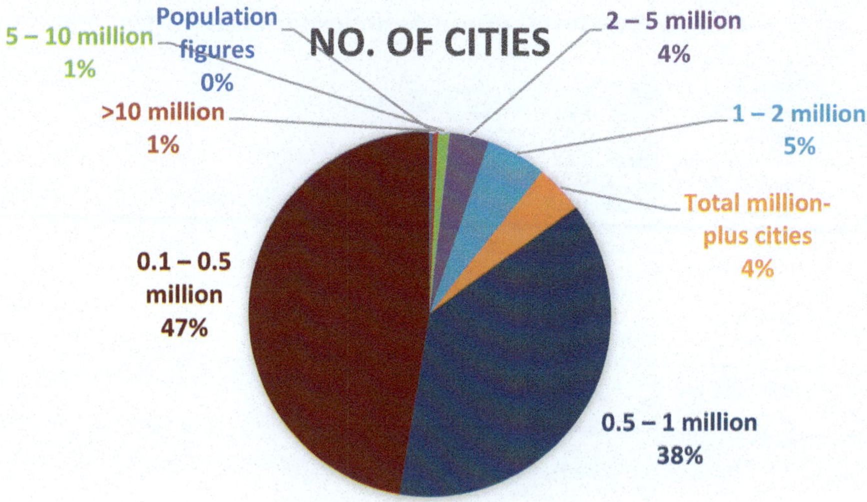

Fig. 25.1 Number of cities as per population; population figures as per 2011 census

increased at a CAGR (Compound Annual Growth Rate) of 9.8%. Customized automobile and two-wheeler accommodations saw annual CAGRs of 9.6% and 10.3%, respectively (Foster and Kumar 2007; Metre et al. 2021). The increase of registered autos is substantially greater in cities with populations above one million than in the rest of India (Ghodmare et al. 2019; Singh et al. 2008). Meanwhile, automotive sales in metropolitan areas increased at almost twice the rate of those in cities with populations of one million or more. In 2011, total automobile entries in 22 municipalities climbed at an 8.7% CAGR, accounting for around 28% (39.7 million) of the country's total automobiles (141.5 million) (NUTP 2008; Sarma et al. 2011).

Since 2001, there have been substantially more cars per Five urban regions had registration rates of more over 500 per 1,000 persons, accounting for 54% of all automotive registrations in 2011 (Sarma et al. 2011). As illustrated in Fig. 25.2, Delhi has more than 6.3 million autos per citizen. According to Ministry of Road Transport and Highways figures, India's automobile population has grown by 10% each year over the last decade. Approximately 71.8% of the 141.5 million registered automobiles on Indian roads in 2011 were two-wheelers. Figure 25.2 demonstrates this.

25.1.3 Dwindling Share of Non-Motorised Transportation

Non-motorized transportation includes walking, cycling, push scooters, hand trolleys, and other small-wheeled mobility devices. (NMT). Walking to work is more common in Indian cities than driving a personal car (Sriraman 2012; Yadav et al.

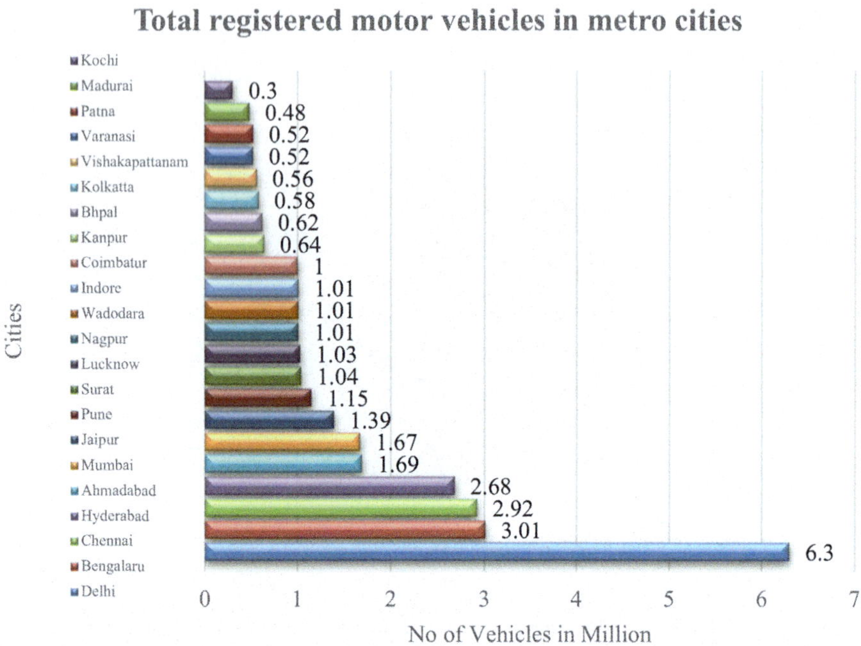

Fig. 25.2 Total registered motor vehicles (Sarma et al. 2011)

2022a).The percentage of NMT falls significantly as towns grow in size, increasing reliance on private transportation (Yadav et al. 2022b) (Thornton 2011). Urban design that promotes cycling and walking is under fire as sprawl-based development becomes increasingly common in major cities. Large roadways and low-density, split land use are increasingly being included into new neighborhood and project layouts (Jha and Chandiramani 2012; Ghodmare and Gulghane 2022).

25.2 Methodology and Scope of Work

The purpose of this article is to provide information about the growth of urban transport networks in India following economic liberalization in the early 1990s through a thorough literature analysis. The reviews revealed recurring themes, issues, and barriers. Following that, informal meetings with important players in the metropolitan transit business were conducted. To collect a comprehensive variety of viewpoints on urban mobility in India, efforts were made to solicit participation from a wide range of stakeholders, including politicians, think tanks, industry representatives, and academics. Finally, the perspectives on the challenges and potential solutions are presented based on five years of work experience in various firms in Indian cities. The responses have been combined to provide an overview of the current situation,

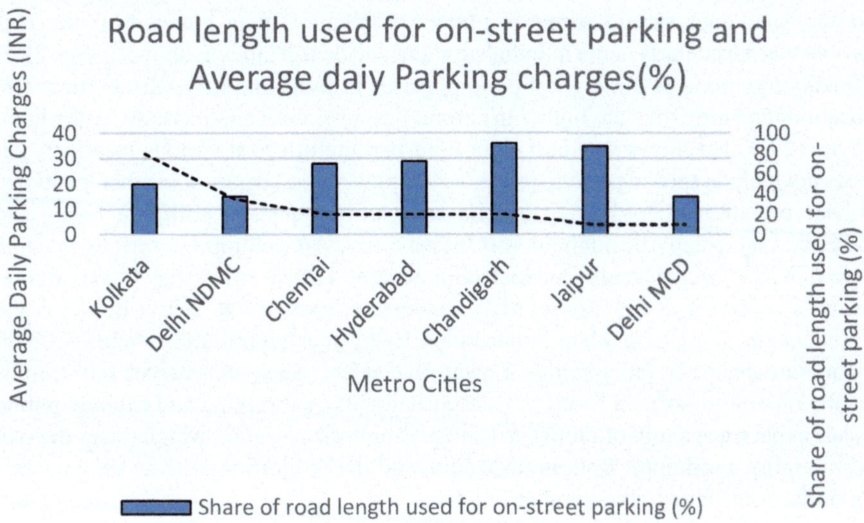

Fig. 25.3 Road length used for on-street parking and average daiy parking charges (%)

which is presented in the sections below, as well as key issues and areas for strategic action.

25.2.1 Urban Transport Problems

25.2.2 Road Congestion

Given the apparent relationship between the two variables, we expect the average trip time and intensity to increase as the population expands. Refer to Fig. 25.3. While most metropolitan areas have an average mileage of less than 6 km, Bangalore has an average mileage of around 8 km (Sriraman 2012). Rising salary levels, migration, women's involvement, a service-oriented economy, and longer and more frequent travel are all expected to contribute to these trends' continuation. There is little question that traffic will worsen as more people commute great distances for work or school (Singh et al. 2008).

25.2.3 Parking Problems

In Indian cities, the scarcity of on-street and off-street parking spaces increases the time it takes to find a parking space and causes traffic congestion. According to published data, street parking is a major problem on many streets in India. This problem is particularly acute in India's smaller and denser cities as shown in Fig. 25.3. Kolkata has some of the highest parking prices in India and they vary by time and location, meaning they are higher in certain business areas and increase by the hour. Figure 25.3 also shows standard eight-hour parking prices at various locations, but does not include special parking prices at shopping malls, airports and other locations.

Air pollution-According to official air quality statistics, pollution levels and severity vary greatly throughout 180 Indian cities. Air pollution is serious in some areas. According to study conducted in several Indian cities, the transportation sector's influence grows when finer particles are investigated. Air pollution is the sixth leading cause of death in Indian cities. Every year, air pollution causes 620,000 premature deaths in Indian cities. Lower road safety. Because refugees rely significantly on non-motorized forms of transportation like as walking and cycling, public roads experience a mix of motorized and non-motorized traffic, which raises the risk of mortality and injury. Both motorization and the proportion of slow-moving cars in traffic contribute to the number of deaths.

25.2.4 Challenges

Although each city has its own history, characteristics, and challenges, urban Indian cities commonly share characteristics and issues with other growing cities. According to expert lectures, literature assessments, and direct information, the following are some important concerns and obstacles that contribute to or exacerbate urban mobility issues in Indian cities:

25.2.5 Gaps in Laws and Regulations

Currently, no federal, state, or local law exists to satisfy Indian communities' transit needs. To address this mode of transportation in a systematic manner, three laws related to the Indian metro system were enacted: the Tramway Act of 1886, the Metro Railway (Civil Works Construction) Act of 1978, and the Delhi.

25.2.6 Fragmented Institutional Frameworks

Unfortunately, they are implemented by multiple organizations run by federal, state, and local governments that may or may not cooperate. In some states, the transportation agency is responsible for urban transportation planning, in others it is done by the urban development or municipal government.

25.2.6.1 Distorted Land Markets Affecting Transport Infrastructure Development

The high cost of real estate acquisition and the long and tedious procedures involved greatly hinder the growth of an integrated metropolitan transportation system. A significant amount of public property keeps large blocks of strategically located farmland away from the market. Regulations that are complicated and time-consuming to understand add to the cost of purchase. Many properties are vulnerable to litigation due to legislation such as ULCRA (Urban Land Heading and Regulation Act), which prevents them from being included in the supply of building land.

25.2.6.2 Comprehensive Design Standards for Transport Infrastructure Lacking

In 2008, land acquisition concerns caused around 70% of all building project delays. One in India, there are few standardized criteria for building, repairing, and operating rolling stock and transportation infrastructure. There are no criteria for the building, operation, or maintenance of mass transport technologies such as metro, light rail, train, or bus rapid transit systems, with the exception of road and traditional railway infrastructure.

25.2.7 Human Resource Challenges

There are no guidelines for the construction, operation, or maintenance of such public transportation technologies (e.g., subway, light rail, train, or bus rapid transit systems) for infrastructure systems other than road and conventional rail.

25.2.8 Absence of Reliable Transport Data

Without a database for scientific administration and analysis of urban traffic data, it is difficult to develop effective urban traffic plans and accurately examine the consequences of various municipal initiatives. Even readily available data raises questions about its quality and trustworthiness, since the majority of the material obtained is either part of a specific research or was gathered with a specific objective in mind. Another difficulty is that information is scattered across several businesses and might be difficult to get.

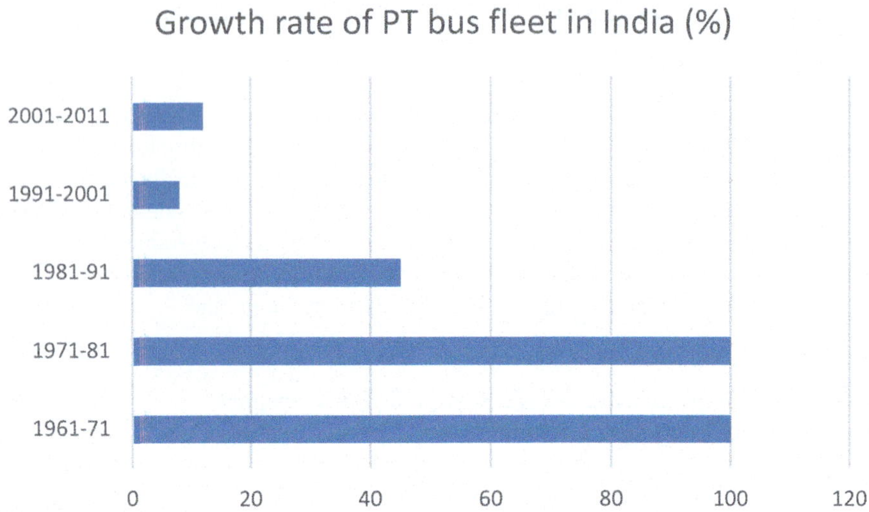

Fig. 25.4 Decadal growth rate of PT bus fleet in India (%) (Roychowdhury 2013)

25.2.9 Inefficiencies in Bus Based PT Services

According to SRTU time series data, the number of bus permits has steadily decreased since 1961 (Fig. 25.4). In locations where there are no SRTUs, private buses or intermediate public transport (IPT) are used to meet public transportation demand.

25.2.10 Energy Security

Between 2000–2001 and 2010–2011, oil output increased by 1.6 percent per year, but consumption of hydrocarbon products increased by more than 4 percent per year. India's oil reserves to production ratio (R/P) indicates that reserves will last 30 years, however the worldwide R/P ratio suggests that oil will last 46 years. India has and will continue to rely heavily on foreign oil. Currently, exports cover around 80% of India's oil needs. If present hydrocarbon consumption trends continue, this percentage will reach 90% by 2032. The present trend of rising use of fossil fuels in transportation and increased reliance on oil imports jeopardizes India's energy security and local economic development, making the country vulnerable to swings in global oil prices. Instead of using the funds to support economic growth, it sends significant sums of rupees abroad to buy fuel.

25.3 Policy Responses to Address Urban Transport Issues

The federal government continues to provide just a minor part to urban transit. Because rural areas housed the majority of India's population, connecting them to metropolitan centers remained the primary focus of investment and transportation policy guidelines until the mid-1990s.

25.3.1 Planning Commission

To fulfill demand, planning papers from the First Five-Year Plan to the Twelfth Five-Year Plan prioritized intercity travel via various forms of transportation. The Twelfth Five-Year Plan has a more extensive section on Urban Management, which for the first time addresses a 20-year horizon. The emphasis was on operating metro rail systems and examining land value capture strategies, as well as corporate financing solutions for such large projects.

25.3.2 National Urban Transport Policy (NUTP)

To address these issues, the NUTP proposed four strategies aimed at increasing road space efficiency by prioritizing public transportation, improving traffic management tools, limiting the growth of private vehicle traffic, and reducing vehicle emissions through vehicle technological development. The NUTP regarded states as the primary facilitators of policy implementation, with the central government's role confined to providing financial help and technical knowhow.

25.3.3 Jawaharlal Nehru National Urban Renewal Mission (JNNURM)

Launched by the state in December 2005, the Jawaharlal Nehru National Urban Renewal Mission (JNNURM) chose 63 municipalities as candidates for national financing in its first phase. In order to meet these objectives, JNNURM needed several state and local level modifications. Funding for urban transportation made about 11% of JNNURM. Except for Delhi, most urban transport networks and initiatives are owned and funded by state governments. The state's engagement in this industry is new, particularly after the National Urban Transport Policy and JNNURM were introduced in 2006. Through subsidizing public transportation, developing detailed planning for urban transportation, and facilitating urban transportation, the JNNURM program sought to enhance the networks of public transportation in large cities.

25.3.4 Proposed Policy Reforms

The extreme objective of the urban transportation technique is to supply naturally neighborly urban travel. The self-evident challenge is getting the different offices locked in in urban transportation to distinguish, execute, and assess arrangement measures that are effective in handling specific issues in a synchronized and composed way. Following sections describes the involvement of the different agencies in policy reformation process.

25.3.4.1 Re-Aligning Legal and Regulatory Instruments

Each state should enact comprehensive urban transportation law that defines the roles of various local and state-level agencies in public transportation, land use and integration, safety, NMT, IPT, and other issues. Finally, regional and municipal administrations should be in charge of urban traffic.

25.3.4.2 Institutional Restructuring

The planning teams' talents and capacities to create appealing ideas of the future of cities are also evaluated as part of the overall process. Furthermore, collaborative strategies must be implemented to ensure that planning and finance decisions are socially equitable and include individuals in a common vision of their city and neighborhoods. As part of institutional reform, financial resources must be dedicated to train and strengthen the capacity of important personnel so that they are equipped to face the complex difficulties of urban transportation.

25.3.5 Mobilization of Resources

The availability of resources is critical for the UMTA or DPC to successfully manage and carry out the urban transport coordination function. Resources include public money, a competent and highly educated personnel, and current and relevant information for planning and decision-making. The next sections go over these crucial tools in further depth.

25.3.6 Public Financing

If the current imbalance in funding and investment in public transport and car-centric road development projects is to be corrected, it must be addressed gradually. More funding is needed to create non-motorized and efficient public transport infrastructure to offset the current preference for roads and highways.

25.3.7 Capacity Building

It entails retraining municipal employees in management and higher education abilities. The primary goal of educating current municipal officials should be to increase their knowledge, practical abilities, and overall awareness of urban traffic issues. States should be required to establish specialist urban transportation agencies at the municipal and state levels as soon as feasible, choose officers to work in these agencies, train them, and then transfer them to those agencies or departments. All stakeholders, including the general public, require a trustworthy source of information in order to establish and evaluate sustainable municipal transportation policies and initiatives. Planning reforms Multimodal mobility must be organically integrated into a holistic land use system that promotes synergies and connections as part of a comprehensive approach to land use and transit management. In terms of the Indian situation, a few cities have implemented distinct planning reforms for various kinds of transportation.

25.3.8 Transferability of Best Practices

Mumbai has India's largest commuter train network, transporting over five million people each day and accounting for 80% of the region's total passenger-kilometers, or 58% of all regional public transportation users. To build these networks, suburban train operations must be entrusted to a state or municipal railway operator, which is governed by the state or local government. The city's first watercraft system connects the Howrah rail station to the city centre. The docks, a historic structure, are near multiple public transportation hubs, including a metro station, bus stop, and train terminal.

25.4 Conclusion

Sustainable mobility can help to eliminate poverty and promote shared wealth; it is an important component of urban development in India. A thorough integration of urban transportation and land use planning systems is required to reap the benefits of maximizing benefits, promoting linkages, and increasing utility through multimodal mobility solutions. Traditional efforts to prioritize public funding and investment for private transportation over public transit must address current urban transportation challenges such as traffic congestion, road accidents, pollution, and so on. Infrastructure development for NMT and high-capacity public transit need more government funding. The urban transportation industry must be considered a coherent unit in terms of system funding and pricing. Urban transportation networks are known to have an impact on the built form and spatial morphology of cities. It might also be claimed that strong, integrated, and participatory organizations must be developed and encouraged in order to successfully improve city accessibility. It has also been stated that transportation organizations must have well defined responsibilities, legislative authority, financial independence, and professional experience. The most essential issue is to ensure that such organizations have systems in place for monitoring, accountability, and transparency with stakeholders. Strong political will and a relentless public desire for change are other crucial factors (Ghodmare et al. 2019; Mendhe et al. 2022).

References

Baindur D (2011) Planning sustainable urban transport system—case study of Bengalurumetropolitan region. J Econ Policy Res 7(1)

Bhatt C et al (2013) Sustainable urban transport—principles and implementation guidelines for Indian cities

Foster A, Kumar N (2007) Have CNG regulations in Delhi done their job? XLII(51). http://www.epw.in/special-articles/have-cng-regulations-delhi-done-their-Gautheir,A.,2012.SanFrancisco knowsHowtoparkIt

Nimburkar AS, Ghodmare SD, Tenpe A (2022) Traffic signal synchronization system for highway corridor in urban area. In: international conference on emerging trends in engineering and technology, ICETET

Ghodmare SD, Gulghane A (2022) An attempt for sustainability evaluation considering implementation of rapid transit system. Mater Today Proc. https://doi.org/10.1016/j.matpr.2022.01.262; https://www.sciencedirect.com/science/article/pii/S2214785322002942. ISSN 2214-7853

Rodke SJ, Ghodmare SD, Deshmukh A (2022) Critical analysis of saturation flow rate and affecting parameters. In: International conference on emerging trends in engineering and technology, ICETET

Ghodmare SD, Bajaj P, Khode BV (2019) Comparative analysis of BRTS and MRTS—an approach required for selection of system. In: Kolhe M, Labhasetwar P, Suryawanshi H (eds) Smart technologies for energy, environment and sustainable development. Lecture notes on multidisciplinary industrial engineering. Springer, Singapore. https://doi.org/10.1007/978-981-13-6148-7_46

Ghodmare SD, Khode BV, Ladekar SM (2021) The role of artificial intelligence in industry 4.0 and smart city development. Lecture notes in civil engineering 87, pp 591–604

Ghodmare SD et al (2019) Application of the multi attribute utility technique with it's for sustainability evaluation of emerging metropolitan city of Nagpur. Int J Civ Eng Technol EID: 2-s2.0-85060924119

Jha R, Chandiramani J (eds) (2012) Perspectives in urban development : issues in infrastructure planning and governance. Capital Publishing Company

IRC 92(2007) Manual of specification and standards, Indian road congress; Syed U, Ghodmare SD, Khode BV (2016) Performance impact of various bus priority strategies using vissim 7.0. Int Res J Eng Technol 3(4):2479–2483

Kamyotra JS, Basu DD, Agrawal S (2012) National ambient air quality status and trends in India—2010. Central pollution control board (CPCB) ministry of environment & forests. http://www.cpcb.nic.in/upload/NewItems/NewItem_192_NAAQSTI.pdf

Mendhe PD, Ghodmare SD, Titarmare AS, Wadhai PJ (2022) Sustainable transportation using automated highway system for intercity connectivity. In: international conference on emerging trends in engineering and technology, ICETET

Metre K, Baghel H, Suman G, Batra M, Ghodmare SD (2021) Compact city and related impact on sustainable development in urban areas. In: Gupta LM, Ray MR, Labhasetwar PK (eds) Advances in civil engineering and infrastructural development. Lecture notes in civil engineering. vol 87. Springer, Singapore. https://doi.org/10.1007/978-981-15-6463-5_50

Mohanty N, Sarkar R, Pandey A (eds) (2009) Land—a critical resource for Infrastructure. 3i Network

NUTP (2008) National urban transport policy. Published by the ministry of urban development. Government of India. http://urbanindia.nic.in/policies/TransportPolicy.pdf

Road Transport Yearbook (2009–10 and 2010–11). MoRT&H, Government of India

Sen SK et al (2011) Road transport year book (2007–2009), vol 1

Siddhaye N (2011) The skywalk: a road above hell. DNA news. http://www.dnaindia.com/mumbai/report-the-skywalk-a-road-above-hell-1603873

Singh V et al (2008) Study on traffic and transportation policies and strategies in urban areas in India. http://urbanindia.nic.in/programme/ut/final_Report.pdf

Sriraman S (2012) The need for a new orientation to the national urban transport policy

Thornton G (2011) Final report volume 1: appraisal of Jawaharlal Nehru national urban renewal mission (JNNURM). http://JNNURM.nic.in/wp-content/uploads/2012/06/Appraisal-of-JNNURM-Final-Report-Volume-I-pdf

Tiwari G (2007) Urban transport in Indian cities. Articles|LSE Cities. http://lsecities.net/media/objects/articles/urban-transport-in-indian-cities/en-gb/

Tiwari G (2011) Key mobility challenges in Indian cities. http://www.internationaltransportforum.org/jtrc/DiscussionPapers/DP201118.pdf

WHO (2013) Global status report on road safety 2013—supporting a decade of action

Yadav DD, Khode BV, Ghodmare SD (2022a) Metro follows an optimized speed trajectory to save energy. In: international conference on emerging trends in engineering and technology, ICETET

Yadava DK, Ghodmare SD, Kumar N (2022b) Mitigation of blackspots on highways by the application of safe system approach. Mater Today Proc Elsevier. https://doi.org/10.1016/j.matpr.2021.11.042

Chapter 26
Effect of Seasonal Variation on Inactivation EC (*Escherichia Coli*), FC (Feacal Coliform) and TC (Total Coliform) by Solar Disinfection

Isha P. Khedikar, A.R.Tembhurkar, and Kuldeep R. Dabhekar

Abstract Solar radiations are natural disinfectant (UV-A) which is abundantly available in India which is tropical country. Hence there is need to use the solar energy for disinfection of grey water. Hence this study is aimed at to do investigation of effect of seasonal variation, three sets of experimentations were considered depending up on the solar intensity 400–500 W/m^2 as Rainy, 500–600 W/m^2 as Winter and 600–700 W/m^2 as Summer with the optimized condition of turbidity (30 NTU), angle (45°), flowrate (135 ml/min) and reflective rear surface. Performing parameter is bacteriological inactivation of EC *(Escherichia Coli),* FC (Feacal Coliform) and TC (Total Coliform). It was observed that with increase in intensity of solar light behavior of inactivation changed.

Keywords Disinfection · Escherichia coli (EC) · Feacal coliform (FC) · Grey water · Solar intensity · Total coliform (TC) · SODIS · Solar disinfection

26.1 Introduction

Increasing population one of the leading cause of water scarcity is the India, there will be water scarcity may happened in due course of time (Amin et al. 2013; Edwin et al. 2014). This issue of water scarcity can be solved up to certain extent by reuse of water (Khatun and Amin 2011). Now a days, to reduce fresh water demand grey water is extensively used for various purposes except drinking (Edwin et al. 2014; Liu et al.

I. P. Khedikar (✉)
Department of Civil Engineering and Centre of Excellence Smart City and ITS, G H Raisoni College of Engineering, Nagpur, MS 440016, India
e-mail: isha.khedikar@raisoni.net

A.R.Tembhurkar · K. R. Dabhekar
Department of Civil Engineering, Visvesvaraya National Institute of Technology, Nagpur, MS 440010, India

B. V. Khode et al. (eds.), *Innovations in Technologies: Pioneering Sustainable Infrastructure for a Resilient Future*, Sustainable Civil Infrastructures,
https://doi.org/10.1007/978-3-031-82133-2_26

2010) (USEPA2004). Researcher reported that demand of fresh water is reduced by about thirty percent by reuse of grey water for non-potable purposes (Eriksson et al. 2002). Researcher investigated characterization of grey water and observed that it is necessary to treat grey water before reuse (Khatun and Amin 2011).

26.1.1 Need of Reuse of Grey Water and Disinfection

Researchers reported that grey water comprises between 50 and 80% that of the total wastewater generated from residential area (Birben et al. 2016). Further, grey water physio-chemical and microbiological characteristics varies from region to region with regards to consumption, applications and development (Birben et al. 2016). Applications of grey water for non-potable purposes were found to be significant (Winward et al. 2008; Kundu et al. 2015). However it was observed that pathogenic microorganisms present in the untreated grey water cause diseases (Eriksson et al. 2002). Grey water contains lesser organic contaminants than domestic wastewater (Winward et al. 2008). Hence, more preferable to reuse for non-potable purposes (Winward et al. 2008). There is limited data available on grey water characteristics (Eriksson et al. 2002). Depending upon the source of generation of raw grey water fluctuation in organic loading (low, medium and high strength) was observed by various researcher which leads to variation in the characteristics of grey water (Edwin et al. 2014; Ghunmi et al. 2011; Albalawneh and Kuochang 2015). In grey water faecal bacteria has been reported by researchers (Eriksson et al. 2002; Casanova et al. 2001; Ottoson and Stenstrom 2003). Hence, there is a need to study the faecally transmitted pathogens which are potentially present in grey water (Gerba 2000). Prior to reuse grey water should be free from pathogenic microorganisms to reduce risk of dispersal diseases (Eriksson et al. 2002).

26.1.1.1 Chlorine as Disinfectant

Chlorine is chemical disinfection conventionally used worldwide (Burch and Thomas 1998). Organic matter present in wastewater is higher than water from other sources, therefore higher residual chlorine is required (Rabbani and Hooshyar 2011). Chlorine disinfection produces byproducts like trihalomethanes (THMS), halo-acetic acids. Disinfection By-products (DBPs) which are resultant from the reaction of organic compounds and chlorine in wastewater are carcinogens in nature (Amin et al. 2013; Bitton 2005). However, it is difficult to notice all DBPs formed within wastewater disinfected with chlorine.

26.1.1.2 Solar Disinfection (SODIS)

Solar disinfection is one of the alternatives which is naturally available in plenty (Wegelin et al. 1994; Kehoe et al. 2001). The Indian sub-continent communities were using this method as early as 2000 BC to 4000 B.C, when water was filtered through charcoal and exposed to sunlight in open trays so that it could be "blessed" by the Sun without knowledge about the actual process (Wegelin et al. 1994; McGuigan et al. 2012). Downes and Blunt first reported the bactericidal effect of sunlight in 1877 (Gill and Price 2010; Khedikar and Tembhurkar 2016a).

Wegelin et al. (1994) supported Arca's findings that inactivation of bacteria such as *Escherichia coli,, Enterococci* and *Str.feacalis* viruses, *encephalomyocarditis*, rotavirus and bacteriophage f2 were done by UV-A radiation using the germicidal properties of natural sunlight at 555 Wh/m^2 for *E.coli* for 3-log inactivation for midday summer sun (5 h) was reported (Wegelin et al. 1994). They reported that bacterial inactivation rate considerably increases when water temperature is above 50°C whereas, virus inactivation rate steadily increased when water temperatures reached 20–50°C. Solar radiation along with thermal water treatment gave rise to synergistic effect as temperature from 12 to 40°C was not important for solar disinfection. However for highly turbid water UV dose was ineffective hence either needed higher penetration or decreased turbidity (Khedikar and Tembhurkar 2016a) as microorganisms get shielded and survived longer. Contamination should not exceed 1000 faecal coliform/100 ml for effective disinfection using SODIS. Bacteria and total coliform inactivation were 99.9%, in contrast, the tested virus showed 3–4 times, bacterial spores showed 9 times, and amoebic cysts showed 15 times higher doses to have same effect when compared with required dose for bacterial inactivation. Then research was conducted using laboratory isolation to assess effectiveness of SODIS against various bacteria, by use of laboratory isolates (Wegelin et al. 1994; Joyce et al. 1996) and naturally contaminated waters (Sommer et al. 1997). Research after early 1990's on SODIS has become multi-dimensional and several researches studies are going on. Earlier research was done predominantly using PET bottles for disinfecting contaminated water to control water borne diseases (McGuigan et al. 1998; Conroy et al. 2002; Meierhofer 2006). Reflection systems and concentration enhance the irradiance in the sample (Mani et al. 2006) and enhance its effect for disinfecting other bacteria, protozoa and virus as which are also of interest to researchers (Gomez-Couso et al. 2009). Since 1991, development and application of the SODIS was supported by Swiss Federal Institute of Aquatic Science and Technology (EAWAG), one million people used this treated drinking water EAWAG 2008 worldwide (Meierhofer and Wegelin 2002).

Solar Radiation Wavelengths: The UV wavelengths range from 400 and to 100 nm (Acra et al. 1990). Solar UV radiation consists of UV-A which have wavelength 400–315 nm, UV-B have wavelength 315–280 these wavelength are able to reach earth surface (Chau et al. 2008), and UV-C have wavelength 280–100 nm (McGuigan et al. 2012; Paidalwar and Khedikar 2016; Khan et al. 2015), unable to reach to earth surface as it is absorbed by ozone (Wegelin et al. 1994). UV-C wavelength showed maximum absorption by DNA hence greatest efficiency in inactivation (Burch and Thomas 1998). As wavelength decreases the inactivation of microorganism increases (Wegelin et al. 1994).

The SODIS manual suggests that most favorable regions for implementing solar disinfection lie between 35°N and 35°S, where throughout the year 300 days clear sky is available to use SODIS where the majority of the developing countries are located (McGuigan et al. 1999).

Many researchers have reported that as the solar intensity increased, the percentage inactivation also increased (Wegelin et al. 1994; Kehoe et al. 2001; McGuigan et al. 1998). Solar pasteurization was effective in both batch and flow-through devices with less maintenance but yielded high treatment costs (Saitoh and EL-Ghetany 2002). Pasteurization temperature (above 65 °C) was achieved with reflectors or reactors using (Saitoh and EL-Ghetany 2002). Thermo-tolerant coliforms are indicator microorganism to assess efficiency of the disinfection process (Hachich et al. 2012).

However, near UV-A (320–400 nm), sunlight radiations are majorly responsible for microorganisms inactivation (Wegelin et al. 1994; Sommer et al. 1997; Meierhofer and Wegelin 2002). When temperature elevated above 45 °C it was observed there is synergistic effect occurred between temperature and UV-A (Wegelin et al. 1994). Presence of solar radiation of 555 Wh/m^2 is required for *E. Coli* to have 3 log reduction (Wegelin et al. 1994).

Therefore, in this study investigation of effect of the seasonal variation i.e. solar intensity 400–500 W/m^2 (Rainy), 500–600 W/m^2 (Winter) and 600–700 W/m^2 (Summer) with the optimized condition of turbidity (30 NTU), angle (45°), flowrate (135 ml/min) and reflective rear surface. Performing parameter is bacteriological inactivation of EC (Escherichia Coli), FC (Feacal Coliform) and TC (Total Coliform).

26.2 Methodology

26.2.1 Influent Sample Collection

Influent samples of grey water were collected in the morning at 9.45 am from residential building. Influent samples collected were always passed though pre-treatment unit i.e. designed roughing filter for experimental work so that foam and turbidity of influent can be maintained at 30 NTU. Influent turbidity was high which was maintained at 30 NTU by dilution method.

26.2.2 Structure of Solar Collector

Basic shape of solar collector was rectangular with Length 70.5 cm; Width 45 cm and Height of 55.25 cm and having semicircle of diameter 50.5 cm within rectangular section. This solar collector was made up of material steel. There were 5 borosilicate tubes placed 20 cm apart from each other in the semicircular section. Total capacity was 2000 ml. On the stand of size 70.5 × 1000 cm solar collector was mounted. Glass wool was filled in the space of solar collector which act as insulation (Khedikar and Tembhurkar 2016b).

26.2.3 Experimental Procedure of Solar Collector

Influent samples collected from residential building was fed to solar collector at 9.00 am so that it should be properly exposed to solar radiation. On the solar collector having 4000 ml capacity (2000 ml in solar collector + 2000 ml in bottle reservoir) experimentations were conducted with solar intensity as rainy (400–500 W/m^2), winter (500–600 W/m^2) and summer (600–700 W/m^2), with the optimized condition of turbidity (30 NTU), angle (45°), flowrate (135 ml/min) and reflective rear surface. Solar collector was exposure to solar radiation for 8 h of sunshine. Temperature (°C) was recorded for atmosphere, bottle reactor and reflective rear surface. All readings were taken at 09:00AM, Noon, 14:00 PM, 16:00 PM and 17:00 PM for the every set of reading of experiment. Simultaneously, Pyranometer (First class; Make: Sivara Systems and Solutions) was used for measurement of UV intensity.

26.2.4 Analytical Procedure

As soon as samples were collected glass bottle (250 ml) it was brought for microbial analysis in laboratory. As it is microbial analysis each glass bottle was properly sterile. For microbial study method adapted was Most Probable Number (MPN) to determine EC, FC and TC as per Standard Methods for the Examination of Water and Wastewater (1999) (Khedikar et al. 2021; Standard Methods for the Examination of Water and Wastewater 1999; Khedikar and Tembhurkar 2016b).

26.3 Results and Discussion

26.3.1 Seasonal Variation Effect When Mean Solar Intensity Varies from 400 to 500 W/m²

In the present study when the experimentation was done with the mean solar intensity varies from 400 to 500 W/m^2 that is in the rainy season. It was observed that maximum percentage inactivation occurred at 5.00 PM for EC and TC as 60.00% and 85.37% respectively as shown in (Fig. 26.1). However, FC inactivation in terms of percentage

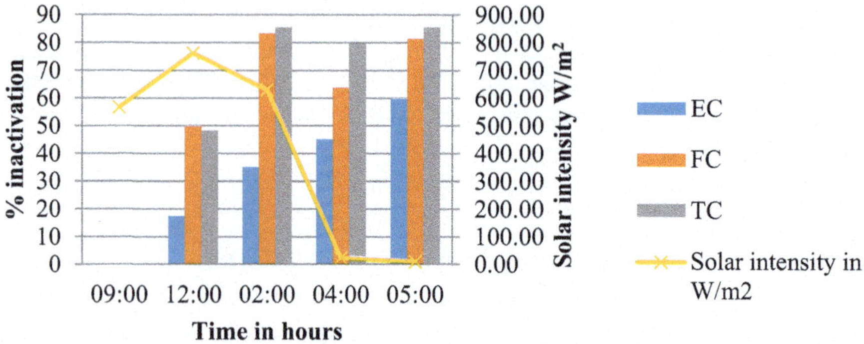

Fig. 26.1 Variation in time of exposure with mean solar intensity range 400 to 500 W/m^2 (Rainy Season)

was slightly less at 5.00 PM than at 2.00 PM but about 1% only. Though the intensity at 2.00 PM was observed to be 632.40W/m^2 the inactivation of EC was observed as 35.29%. It was observed that inactivation in terms of percentage was enhanced which may be due to prolonged exposure period of 8 h. Furthermore, time of exposure leads to increase bottle reservoir temperature through the reflective rear surface similar finding was observed by Kehoe et al. (2001).

Ambient air temperature also played an important role in the inactivation. It was noted that when ambient air was less percentage inactivation was also decreased. Wegelin et al. (1994) reported that no synergistic effect was observed with temperature between 12 and 40°C.

When the solar intensity was less the microorganisms were exposed to sub-lethal doses of solar intensity was also reported by McGuigan et al. (2012). However, this can be overcome by prolonged exposure. Therefore, it seems that when the solar intensity is between 400 and 500 W/m^2, 8 h exposure time may be required.

26.3.2 Seasonal Variation Effect When Mean Solar Intensity Varies from 500 to 600 W/m^2

With the solar intensity varies from 500 to 600 W/m^2 that is in winter season experimentation was done with the same optimized conditions. It was observed that inactivation in terms of percentage in EC, FC and TC at 5.00 PM was93.82%, 99.17%and 98.08% respectively (Fig. 26.2). However, inactivation in terms of percentage of EC, FC and TC at 2.00 PM was observed to be 95.09%, 99.42% and 99.66% respectively (Fig. 26.2). It was observed that in the 9.00 AM the solar intensity was 645.5W/m^2. Solar intensity increased from 9.00 AM to 2.00 PM constantly up to 712.4 W/m^2, which lead to increase in the temperature of bottle reservoir through reflective rear surface. With the help of reflectors optical component of solar radiation was increased

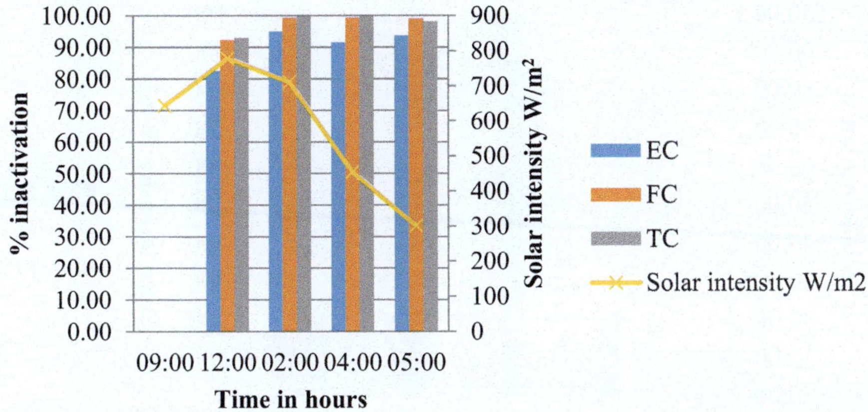

Fig. 26.2 Variation in time of exposure with mean solar intensity range 500 to 600 W/m^2

in flow reactors (Saitoh and EL-Ghetany 2002). Basic reason for to use reflectors to concentrated radiation and to have synergistic effects of temperature under strong and moderate weather conditions (Amin et al., 2013), present findings were in agreement with researchers (Amin and Han 2009; Wegelin et al. 1994, Sommer et al. 1997). Due to the synergistic effect, the inactivation rate was three time slightly as reported by Wegelin et al. (1994). As the solar intensity increased, the percentage inactivation also increased which was similar to the findings by McGuigan et al (2012). It was observed that small reduction in the percentage inactivation at the end of the 8 h of exposure time for EC, FC and TC as 1.27%, 0.25% and 1.58% respectively. This may be due to the flow reactor system, in which bacteria were exposed to sub-lethal dose of UV similar to McGuigan et al. (2012). Another, factors to be were taken in to account were viability to microorganisms to the solar intensity (Wegelin et al. 1994). Different microorganisms showed different viability to the solar radiation (Birben et al. 2016). It was observed that the EC were resistant to the solar intensity similar to Hachich et al. (2012). It was also reported that inactivation was dependent on concentration of bacteria in the influent. Furthermore, in influent if high concentration of bacteria was present and low UV intensity which causes lower inactivation in terms of percentage (Sommer et al. 1997). When intensity ranged from 400 to 600 W/m^2 prolonged exposures were needed.

26.3.3 Effect of Seasonal Variation When Mean Solar Intensity Range Between 600 to 700 W/m^2

In the present study when experimentation with the solar intensity of 600 to 700 W/m^2 was conducted with same optimized conditions in the summer season, it was observed that at 2.00 PM the percentage inactivation in EC, FC and TC was 99.99% for each

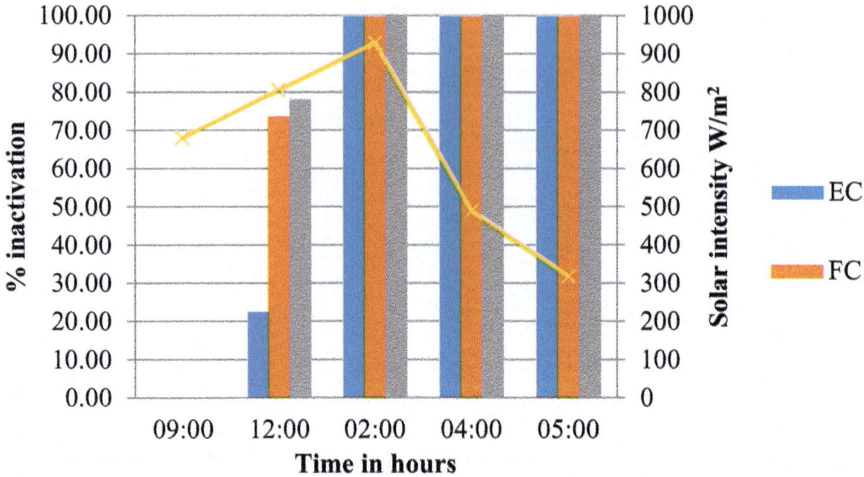

Fig. 26.3 Variation in time of exposure with mean solar intensity range 600 to 700 W/M^2

of them respectively (Fig. 26.3). Solar intensity at 9.00 AM was observed to be 682.6 W/m^2 which was much higher for the initial solar intensity and it further increased from 9.00 AM to 12.00 PM and 12.00 PM to 2.00 PM at 810.7 and 931.2 W/m^2 respectively. Inactivation rates EC may be greatly dependent on wavelength (Wegelin et al. 1994). At the mean intensity range 600 to 700 W/m^2, 99.99% inactivation was possible for EC, FC and TC respectively.

In all sets of experiments it was observed that inactivation rate of the coliforms were as follows: TC was more susceptible to solar radiation with the solar intensity varying from 400 to 600 W/m^2 than FC than EC.

TC > FC > EC.

Furthermore, with the solar intensity between 600 and 700 W/m^{2}equal susceptibility was exhibited.

TC = FC = EC.

The destruction of bacteria was depended on the cell wall complexity and thickness (Birben et al. 2016) and it was reported that bacterial inactivation rates were as TC > FC > fecal streptococci, for low organic loading grey water (Birben et al. 2016).

26.4 Conclusion

More inactivation of bacteria either minimum lethal UV-A doses for short period time or prolonged exposure is required. Further, bacterial inactivation efficiency was dependent on the initial concentration of bacteria. Higher the solar intensity higher will be the inactivation as demonstrated. As a result, during summer season the time of exposure of shortened to 5 h and in winner season it was extended to 8 h with good inactivation efficiency. In rainy season inactivation of bacteria was decreases due to reduce solar intensity which may be overcome by prolonged exposure of 2 days. Furthermore, inactivation was also dependent on the type of the bacteria which varies with solar radiation. When the solar intensity 400–600 W/m^2 relationship is TC > FC > EC and solar intensity 600–700 W/m^2 relationship TC = FC = EC.

References

Acra A, Jurdi M, Muallem H, Karahagopian Y, Raffoul Z (1990) Water disinfection by solar radiation: assessment and application. International Development Research Centre IDRC-TS66e, Ottawa, Ontario, Canada

Albalawneh A, Kuochang T (2015) J Res—Granthaalayah 3(12):6–35

Amin MT, Han MY (2009) Roof-Harvested Rainwater for Potable Purposes: Application of Solar Collector Disinfection (SOCO-DIS). Water Res 43:5225–5235

Amin M, Hashemi H, Bovini M, Hung Y (2013) J Environ Health Eng 2:1–9

Birben NC, Uyguner-Demirel CS, Bekbolet M (2016) J Catal 6(91):1–14

Bitton G (2005) Wastewater microbiology, 3rd edn. Wiley

Burch JD, Thomas KE (1998) J. Solar Energy 64(1–3):87–97

Casanova LM, Gerba CP, Karpiscak M (2001) J Environ Sci Health Part A-Toxic/Hazard Subst Environ Eng 36(4):395–401

Chau A, Giang K, Leung M, Tam N (2008) J Exp Microbiol Immunol 12:39–44

Conroy RM, Meegan ME, Joyce T, McGuigan K, Barnes (2002) J Arch Dis Child 85:293–295

Edwin G, Gopalsamy P, Muthu N (2014) J Appl Water Sci 4:39–49

Eriksson E, Auffarth K, Henze M, Ledin A (2002) J Urban Water 4:85–104

Gerba CP (2000) J Quantative Micrbiol 2:55–68

Ghunmi LA, Zeeman G, Fayyad M, van Lier JB (2011) J Crit Rev Environ Sci Tech 41:657–698

Gill LW, Price C (2010) J Energy 35:4607–4611

Gomez-Couso H, Fontan-Sainz M, McGuigan KG, Ares-Mazas E (2009) J Acta Trop 112:43–48

Hachich EM, Bari MD, Christ AG, Lamparelli CC, Ramos SS, Sato MZ (2012) J Braz J Microbiol 43(2):675–681

Joyce T, Mcguigan K, Meegan M, Conroy R (1996) J Appl Env Micro 62:399–402

Kehoe SC, Joyce TM, Ibrahim P, Gillespie JB, Shahar RA, McGuigan KG (2001) Effect of agitation, turbidity, aluminum foil reflectors and container volume on the inactivation efficiency of batch-process solar disinfectors. J Water Res 35(4):1061–1065

Khan A, Khedikar I, Shrivastava R (2015) J Mech Civ Eng 12:35–39

Khatun A, Amin M (2011) Greywater reuse: a sustainable solution for water crisis in Dhaka, Bangladesh. In: 4th annual paper meet and 1st civil engineering congress, 22–24, pp 427–443

Khedikar IP, Tembhurkar AR (2016) Am J Eng Res (AJER) 5(12):187–192

Khedikar IP, Tembhurkar AR (2016) Int J Sci Technol Eng 2(12):573–578

Khedikar IP, Tembhurkar AR, Dabhekar KR, Godbole BJ (2021) J Phys Conf Ser

Kundu S, Khedikar IP, Sudame A (2015) J Mech Civ Eng 12:1–8

Liu S, Butler D, Memon FA, Makropoulos C, Avery L, Jefferson B (2010) J Water Res 44:267–277

Mani SK, Kanjur R, Singh ISB, Reed R (2006) J Water Res 40:721–727

McGuigan K, Joyce T, Conroy R, Gillespie J, Elmore-Meegan M (1998) J Appl Microbiol 84:1138–1148

McGuigan KG, Joyce TM, Conroy RM (1999) J Med Microbiol 48:785–787

McGuigan KG, Conroy RM, Mosler HJ, Preez MD, Jaswa EU, Ibanez PF (2012) J Hazard Mater 235–236:29–46

Meierhofer R (2006) J Madagscacons Conserv Dev 1(1):25–30

Meierhofer R, Wegelin M (2002) Swiss federal institute of environmental science and technology (EAWAG/SANDEC), Switzerland

Ottoson J, Stenstrom TA (2003) J Water Res 37(3):645–655

Paidalwar AA, Khedikar IP (2016) Int J Sci Technol Eng 2(9):213–219

Rabbani D, Hooshyar H (2011) Iran J Environ Health Sci Eng 8(2):117–122

Saitoh TS, EL-Ghetany HH (2002) J Sol Energy 72(3):261–269

Sommer B, Marino A, Solarte Y, Salas ML, Dierolf C, Valiente C, Mora D, Rechsteiner R, Setter P, Wirojanagud W, Ajarmeh H, AlHassan A, Wegelin M (1997) J Water SRT—Aqua 46:127–137

Standard methods for the examination of water and wastewater (1999) 23rd edn

USEPA2004 Guidelines for water reuse report EPA/625/r-04/108, Washington, DC

Wegelin M, Canonica S, Mechsner K, Fleischmann T, Pesaro F, Metzler A (1994) J Water SRT Aqua 43:154–169

Winward GP, Avery LM, Stephenson T, Jefferson B (2008) J Water Res 42:483–491

Chapter 27
Enhancement in Seismic Isolation Using U-Shape Metallic Isolator

Kalyani Chavane, A. A. Manchalwar, and M. V. Waghmare

Abstract Passive Control System for enhancing seismic performance of structure is most popular way. Isolators significantly reduced damage factors of structure. Preventing these damages, seismic isolation techniques is essential. In the present paper, seismic performance of base isolated structure is observed with Lead Rubber Bearing (LRB) and U-Shaped Metallic Isolator as well as investigate the obtained results. Non-linear dynamic analysis is performed on a 5-story lumped parameter model, with each story having a single degree of freedom (SDOF), using SAP2000 software. The analysis considers four different earthquake ground motions. The observed results indicate a significant reduction in response quantities, thereby mitigating structural damages in both isolation systems.

Keywords Seismic performance · Base isolation · Lead rubber bearing · U-shape metallic isolator · Non-linear dynamic analysis · SAP2000

27.1 Introduction

An earthquake occurs when energy is suddenly released within the Earth's crust and lithosphere, leading to vibrations on the Earth's surface through seismic waves. These events primarily stem from the shifting of tectonic plates beneath the Earth's surface, causing strain energy to accumulate within rocks at plate boundaries and areas of weakness. When the strain reaches its limit, sudden slips occur at faults or plate boundaries, releasing stored energy and generating seismic waves. Most

K. Chavane (✉)
Civil Engineering Department, AISSMS COE, Pune, India
e-mail: kalyanichavane@gmail.com

A. A. Manchalwar · M. V. Waghmare
AISSMS College of Engineering, Pune, India
e-mail: aamanchalwar@aissmscoe.com

M. V. Waghmare
e-mail: mvwaghmare@aissmscoe.com

B. V. Khode et al. (eds.), *Innovations in Technologies: Pioneering Sustainable Infrastructure for a Resilient Future*, Sustainable Civil Infrastructures,
https://doi.org/10.1007/978-3-031-82133-2_27

earthquakes are caused by these slips, propagating elastic waves through the Earth's crust (Datta 2010). Alternatively, the Earth's lithosphere is segmented into numerous large and small fragments called tectonic plates. The interactions among these plates are responsible for triggering earthquakes.

During earthquakes, the collapse of residences, educational institutions, healthcare facilities, and other public structures results in significant loss of life and property damage. Additionally, public infrastructure including roads, bridges, dams, as well as essential utilities such as power and water supply, are also severely affected or destroyed. Over 95% of earthquake-related deaths result from building collapses not meeting earthquake-resistant standards. Despite regulations mandating compliance, new constructions in earthquake-prone areas often neglect these codes, especially in India. Buildings deteriorate over time and are subject to various hazards, necessitating continual evaluation and retrofitting. The dynamic nature of earthquake codes further underscores the ongoing need for adapting construction practices to evolving standards.

Base isolation is an innovative earthquake resistance technology designed to minimize the impact of seismic forces on structures. The system diminishes the earthquake force transmitted to the structure by incorporating flexible bearings or isolators at its base. These isolators prevent the direct transfer of seismic loads to the superstructure, shielding it from the full brunt of the earthquake's impact. By isolating the structure from ground motion, base isolation enhances the building's ability to withstand earthquakes, reducing the risk of structural damage. Base isolation represents a recent advancement in structural engineering, offering a dynamic approach to seismic resilience by incorporating flexible and isolating elements at the foundation level. Base Isolation serves as a seismic retrofitting method utilizing active control devices, passive control devices, and semi-active control devices to improve building resilience. Recently, maximum research is concentrated on analyzing the effectiveness and evaluating the structural integrity of buildings in civil industry that are outfitted with Base Isolation systems. This exploration aims to optimize earthquake response and bolster the effectiveness of this innovative technology in mitigating seismic risks.

Kang and Kang (2009) investigates the seismic isolation of buildings through the use of fiber-reinforced elastomeric structures during earthquake events. Utilizing lumped mass system models incorporating hysteretic and nonlinear-elastic properties, the research underscores the effectiveness of isolating structures in managing building movements, especially during periods of increased earthquake acceleration. This study delves into the dynamic attributes of structures equipped with fiber-reinforced elastomeric isolators, focusing on design methodologies and numerical simulations. Results indicate that higher building frequencies reduce interstory deformation, with flexible fibers yielding lower acceleration responses and stiffer fibers leading to higher accelerations. Smoother building motion is observed with isolators, and the study recommends high damping isolators for robust protection against strong earthquakes, contributing valuable insights for advancing fiber-reinforced elastomeric isolators.

Gomase and Bakre (2011) explores the seismic behavior of a multi-story building with base isolation located in proximity to an active fault, using real earthquake records. Modelled with bi-linear hysteretic behavior, the high damping rubber bearing system adheres to Uniform Building Code (UBC-97) provisions for near-fault effects. Non-linear time history analysis reveals successful UBC-97-predicted isolator displacements, showcasing substantial reductions in floor acceleration and inter-story drifts compared to a fixed base configuration during significant earthquakes.

Islam et al. (2012) examines the integration of Lead Rubber Bearings and High Damping Rubber Bearings as base isolators, studying alterations in structural parameters to enhance the isolation effects. The study entails the development of nonlinear models for these bearings, addressing design aspects and structural viability. Linear static, free vibration, and nonlinear dynamic time domain analyses are carried out on buildings. The findings indicate that base isolation notably diminishes seismic response, particularly in medium-rise structures built on soft to medium-stiff soil. The study underscores the stability improvement and cost-effectiveness of base isolation, demonstrating its efficacy in mitigating seismic instability and reducing structural responses in vulnerable areas.

Kubo et al. (2014) investigated seismic design strategies for nuclear reactor buildings, employing seismic isolation systems. They particularly assessed the rubber bearing with a lead plug (LRB) isolation system through seismic response analyses. The results obtained from both isolated buildings and LRBs comply with design criteria, with LRBs reaching ultimate conditions at 2.0–2.5 times the design-basis intensity.

Han et al. (2014), studied the pre-implementation and post-implementation of base isolation using Lead Rubber Bearing by performing earthquake analysis on an aged reinforced concrete frame building which is non-ductile in nature. Corresponding analysis consist of structural, ground motion, and modelling uncertainties, conducting a sensitivity study that identifies key parameters affecting seismic demands, including concrete compressive strength, viscous damping, steel yield stress, and the integrity of beam- column joints. The findings revealed that, the capacity of base isolation to mitigate seismic risks, particularly at elevated damage thresholds.

Cancellara and De Angelis (2016a) compares the seismic response of a multi-story reinforced concrete building employing Lead Rubber Bearings (LRBs) for base isolation with that of a novel high damping hybrid seismic isolator (HDHSI). Nonlinear dynamic analyses are employed to assess both systems. As a result, the HDHSI system showcases its potential for enhanced earthquake resistance, emphasizing inventive passive structural control and seismic base isolation methods compared to Lead Rubber Bearings.

Cancellara and De Angelis (2016b) investigates three hybrid base isolation systems aimed at safeguarding reinforced concrete structures against bi-directional ground motions. These systems comprise combinations of Elastomeric Spring Dampers in parallel with Friction Sliders, Lead Rubber Bearings in parallel with Friction Sliders, and High Damping Rubber Bearings in parallel with Friction Sliders. The design adheres to European seismic codes EC8. The highly nonlinear behavior of the composite sliding isolation systems is scrutinized, considering recorded bi-directional ground motions. Through a comparative analysis, the study examines the time history of base acceleration, shear, displacements, inter-story drifts, and peak shear values. The outcomes of this study, underscore the efficacy of these hybrid base isolation systems in enhancing seismic and dynamic responses when compared to fixed base structures.

Kangda and Bakre (2017) investigates how various isolation parameters, such as damping ratio, yield strength, post-yield stiffness ratio, and yield displacement, affect the structural responses of a base-isolated building. Newmark's step-by-step integration method is employed for the analysis. The results demonstrate the effectiveness of base isolation, especially when employing low bearing displacement, albeit with a trade-off resulting in reduced top-floor acceleration.

In their study, Faal and Poursha (2017) extended the use of modal analysis (MPA) and extended the N2 (EN2) method to mid-rise building frames with isolated bases. They consider the influence of higher modes on seismic demands, particularly focusing on Lead Rubber Bearings (LRBs). Their findings suggest that while the MPA and EN2 methods demonstrate effectiveness for medium-rise frames equipped with stiff isolators, their accuracy diminishes when applied to structures with softer isolators.

Mazza (2017) examines the loss of mechanical properties in High Damping Lead Rubber Bearings (HDLRBs) and Lead Rubber Bearings (LRBs) by predicting their temperature profiles during exposure to fire. Their approach involves three-dimensional finite element modeling, utilizing Eurocode 1 time–temperature fire curves, and comparing them with steady responses. Property reduction in fire-damaged isolators is evaluated using the 200°C isotherm method. Through numerical investigations on six-storey reinforced concrete base-isolated buildings subjected to fire scenarios, the study uncovers notable reductions in mechanical and geometrical properties of isolators, resulting in increased structural responses. This research underscores the necessity of accounting for fire-induced damage when assessing the seismic performance of base-isolated structures.

Antony Vimal et al. (2023) explore the efficacy of base isolation systems in mitigating lateral forces during seismic events. They examine five five-storey buildings with varying Fundamental Natural Time Periods (FNTPs) ranging from 0.4 to 2.0 s. By keeping mass and damping matrices constant while adjusting stiffness matrices, the study compares structural responses with and without base isolation. The results demonstrate that buildings with shorter Fundamental Natural Time Periods exhibit improved performance with base isolation, highlighting the significant impact of these systems on the seismic behavior of five-story structures.

From above literature, the efficacy of Lead Rubber Bearings (LRBs) and U-Shape Metallic Isolators (2020) as base isolation systems for a five-story building model treat as a lumped mass system (Zhang and Phillips 2016) gets examines, subjected to four different earthquakes, with and without isolators. Linear elastic behavior is assumed for the model throughout all seismic events. Various control strategies are evaluated using SAP2000 software through non-linear dynamic analysis. Earthquake data utilized are sourced from FEMA P965. Previously, the effectiveness of U-Shape metallic dampers was experimentally validated. The current study extends this research by investigating their potential as isolators, leading to a detailed investigation of their performance.

27.1.1 U- Shaped Metallic Damper

Following the 1995 Kobe Earthquake, U-shaped metallic dampers have become pivotal hysteretic devices for absorbing earthquake-induced energy in base-isolated structures, enjoying widespread adoption in Japan. Commonly applied in hospitals, residential buildings, and various isolated structures, these dampers underwent detailed testing after the M9.0 Great East Japan Earthquake. The testing aimed to assess existing fatigue life models. During earthquakes, buildings experience multidirectional vibrations and significant horizontal deformation, necessitating homogeneous damping. Nippon Steel's seismic isolation steel damper, designed for plasticity, actively absorbs seismic energy through 360° plasticization in the horizontal force direction, contributing to effective earthquake response.

Suzuki et al. (2005) provides an overview of the properties and test outcomes concerning a U-Shaped Steel Damper, focusing on its horizontal characteristics, velocity dependency, and temperature sensitivity.

Konishi et al. (2012) provides an overview of fatigue life testing conducted before and after earthquakes on U-shaped dampers. It presents post-earthquake testing results and draws conclusions on residual fatigue.

Oh et al. (2012) suggests a seismic mitigation system suitable for low- or medium-rise structures, integrating laminated-rubber bearings and U-shaped hysteretic dampers crafted from high-toughness steel featuring slotted holes to improve deformation capabilities in their study. The experimental investigation of seismic response is conducted through shake-table tests.

Jiao et al. (2014) prepare report on dynamic loading tests of U-dampers, finding that loading speed minimally affects plastic deformation capacity. It establishes a method for evaluating ultimate plastic deformation capacities based on a Manson–Coffin relation. Moreover, there search demonstrates that the U-damper's hysteretic behavior is scarcely affected by the loading rate and initial temperature, suggesting the adoption of a bilinear model to characterize force–deformation relationships.

Ene et al. (2017) underscores the significance of evaluating base-isolated structures equipped with U-shaped steel dampers across diverse intensities of bidirectional ground motion in the corresponding work. An analytical model simulates bidirectional displacement responses, and Incremental Dynamic Analysis (IDA) categorizes damper performance into five levels. Fragility curves are generated, indicating that the use of an in-plane model, as commonly practiced in Japanese design, overestimates safety margins concerning damper fracture under high seismic intensities (Fig. 27.1).

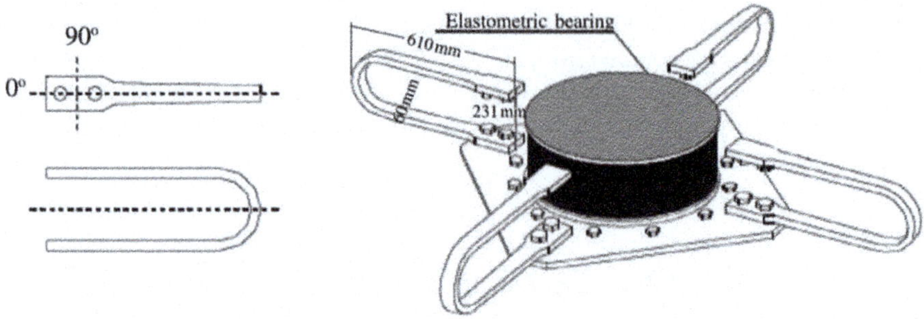

Fig. 27.1 U-Shaped steel damper (Ene et al. 2017)

27.1.2 Lead Rubber Bearing

Seismic isolation for civil structures originated in New Zealand in the late 1960s, led by R. Ivan Skinner. Lead rubber bearings (LRBs), a successful seismic base isolator since the 1970s, operate as laminated rubber bearings with a central lead core enhancing system rigidity (Robinson and Tucker 1977; Robinson 1982). The lead core increases lateral stiffness, enabling greater energy absorption. LRBs are modelled as nonlinear elements in structural analysis.

Makris (2018) documents the origins and initial advancements of seismic isolation, covering developments up to the early 1990s.

Chen et al. (2023) suggests a generalized Bouc–Wen model tailored for lead rubber bearings (LRBs), providing an accurate representation of rubber characteristics. Validated with diverse LRB designs, the model demonstrates precise simulation of large cyclic behavior (Fig. 27.2).

The Lead Rubber Bearing (LRB) is described as a nonlinear element defined by parameters including post-yield stiffness, initial stiffness, yield strength, and post-yield stiffness ratio. Together, these parameters demonstrate the nonlinear behavior of the LRB (Naeim and Kelly 1999). The post-elastic stiffness (k_P) is determined as follows:

$$k_P = \frac{GA_b}{T} \tag{27.1}$$

where,

A_b is an area of rubber.

T is rubber thickness.

G is Shear modulus of rubber.

The effective stiffness k_{eff} of LRB, expressed in terms of k_P, k_u, and Q is provided by

$$k_{eff} = k_P + \frac{Q}{D} \tag{27.2}$$

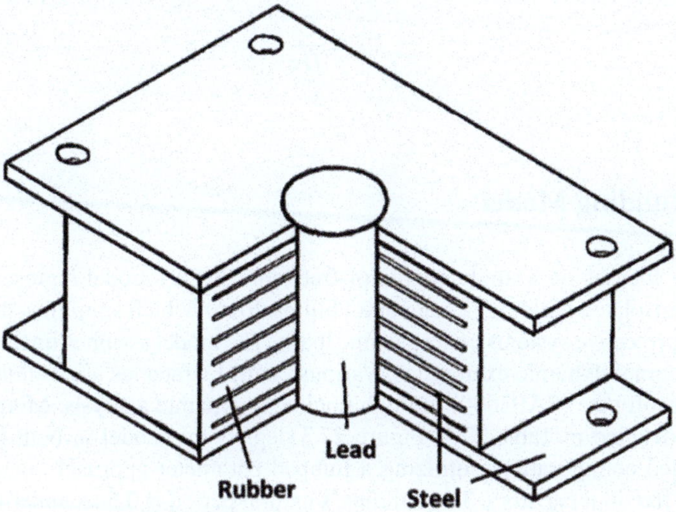

Fig. 27.2 A schematic representation of a Lead Rubber Bearing constructed with multiple layers of steel laminate. (Robinson and Tucker 1977; Robinson 1982)

where,

D is designed displacement.

The design of the effective stiffness of isolation is primarily aimed at achieving the desired value of the isolator period (*T*), as illustrated by:

$$T = 2\pi\sqrt{\frac{M}{k_P}} \tag{27.3}$$

where,

M is total mass.

Here,

Yield strength *Q* in which β_{eff} is assumed as 0.05 is given by equation,

$$Q = \frac{\pi \beta_{eff} k_p D^2}{(2 - \Pi\beta_e ff)D - 2Dy} \tag{27.4}$$

where,

Yield displacement *Dy*

Based on experimental investigations conducted by several researchers, it is recommended that *Dy* is approximately 0.05 to 0.1 times the thickness of the rubber (Naeim and Kelly 1999).

Yield force of bearing can be obtained as,

$$F_y = Q + K_P Dy \tag{27.5}$$

$$k_u = \frac{F_y}{Dy} \tag{27.6}$$

27.2 Building Model

The study focuses on a single-degree-of-freedom (SDOF) model representing a 5-story base-isolated building (Zhang and Phillips 2016), which simplifies the investigation of protective systems under seismic loads. The model assumes linear elasticity during external dynamic excitations. Various control strategies are evaluated using SAP2000 software (SAP2000) through nonlinear dynamic analysis. Model parameters are detailed in Table 27.1. Figure 27.3 depicts the model in both fixed base and isolated configurations, utilizing a lumped parameter approach featuring one degree of freedom per story. The building's natural period is 0.54 s under fixed base conditions, while it shifts to 2.5 s with base isolation, maximizing energy dissipation during seismic events.

Table 27.1 Characteristics of models for fixed-base and base-isolated structures with five stories (Zhang and Phillips 2016)

Floor	Floor mass (kg)	Story stiffness (kN/m)
Base	$m_0 = 61{,}200$	$K_0 = 2{,}129.8$
1	$m_1 = 53{,}073$	$k_1 = 101{,}196$
2	$m_2 = 53{,}073$	$k_2 = 87{,}279$
3	$m_3 = 53{,}073$	$k_3 = 85{,}863$
4	$m_4 = 53{,}073$	$k_4 = 74{,}862$
5	$m_5 = 53{,}073$	$k_5 = 57{,}177$

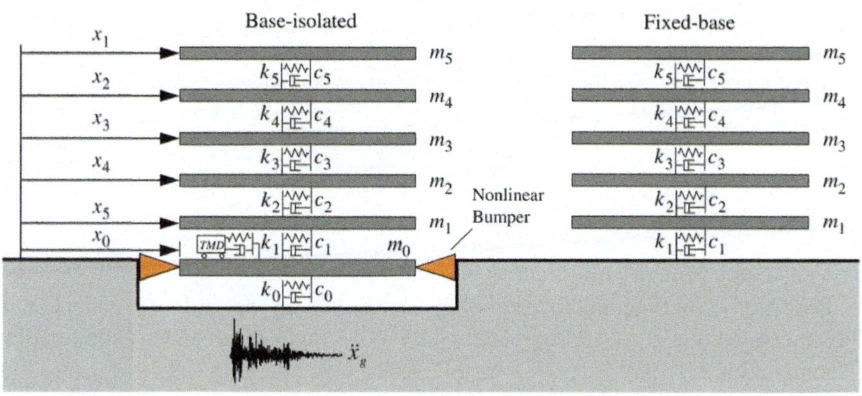

Fig. 27.3 Structural models of a five-story building in both traditional fixed-base and base-isolated configurations (Zhang and Phillips 2016)

Table 27.2 Ground motion data

Sr. No	Earthquake	Year	PGA (g)
1	Imperial valley	1940	0.05
2	Kern county	1952	0.1508
3	San Fernando	1971	0.1319
4	Loma prieta	1989	0.275

27.3 Dynamic Response of Earthquake Ground Motions

The study utilizes four diverse field ground motion records from large earthquakes, varying in magnitude and frequency content. Table 27.2 offers detailed information about each station, including the associated peak ground acceleration (PGA).

Figure 27.4 illustrates the time histories of all earthquake records utilized in the study, offering a comprehensive portrayal of the input excitations.

27.4 Dynamic Responses of Superstructure

The utilization of Lead Rubber Bearing Isolators results in a notable reduction in structural acceleration during seismic events. This reduction is evident across various ground motion scenarios, as shown in Fig. 27.5 through non-linear time history analysis. The deformability of the base isolators absorbs a significant portion of seismic energy, thereby limiting its transmission to the superstructure. Consequently, the application of LRBs effectively decreases joint accelerations compared to fixed base conditions. Moreover, LRBs contribute to lowering the base shear values, consequently enhancing overall structural performance. Graphical representations in Fig. 27.5 illustrate these effects for different earthquakes, both with and without isolator application.

Figures 27.4 and 27.5 provide a comprehensive comparison of top floor acceleration and base shear in a structure during four distinct earthquakes, facilitating a detailed analysis between fixed base conditions and those utilizing Lead Rubber Bearing (LRB) isolation. Both LRB and U-shaped isolator systems, illustrated in Figs. 27.6 and 27.7, demonstrate a notable decrease in top floor acceleration and base shear, indicating their effectiveness in mitigating seismic forces. Particularly, the U-shaped metallic isolator stands out for its efficacy in reducing top floor acceleration and base shear, contributing significantly to the structure's overall resilience against earthquake-induced forces. This collective evidence highlights the considerable benefits of both LRB and U-shaped isolator systems in improving structural stability and reducing the impact of seismic events on the superstructure.

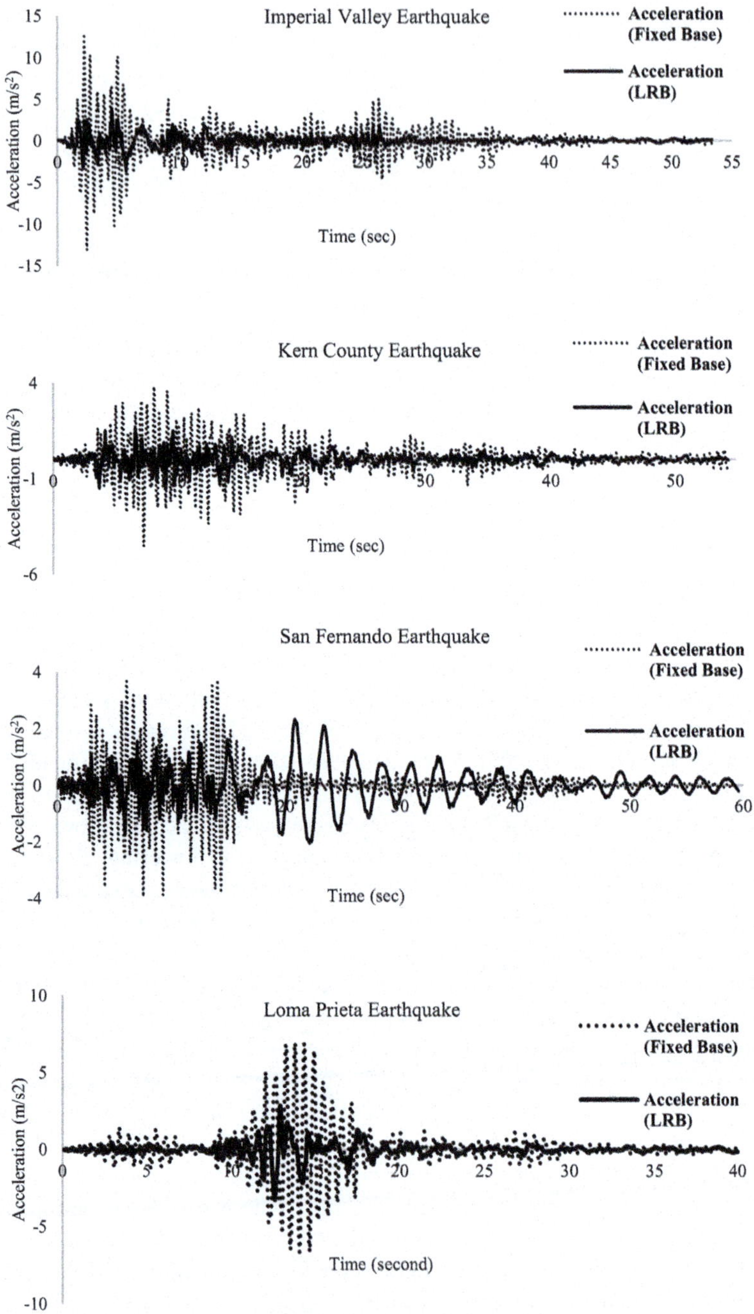

Fig. 27.4 Acceleration data of a five-story building recorded during four distinct earthquake events: (1) Fixed base; (2) LRB

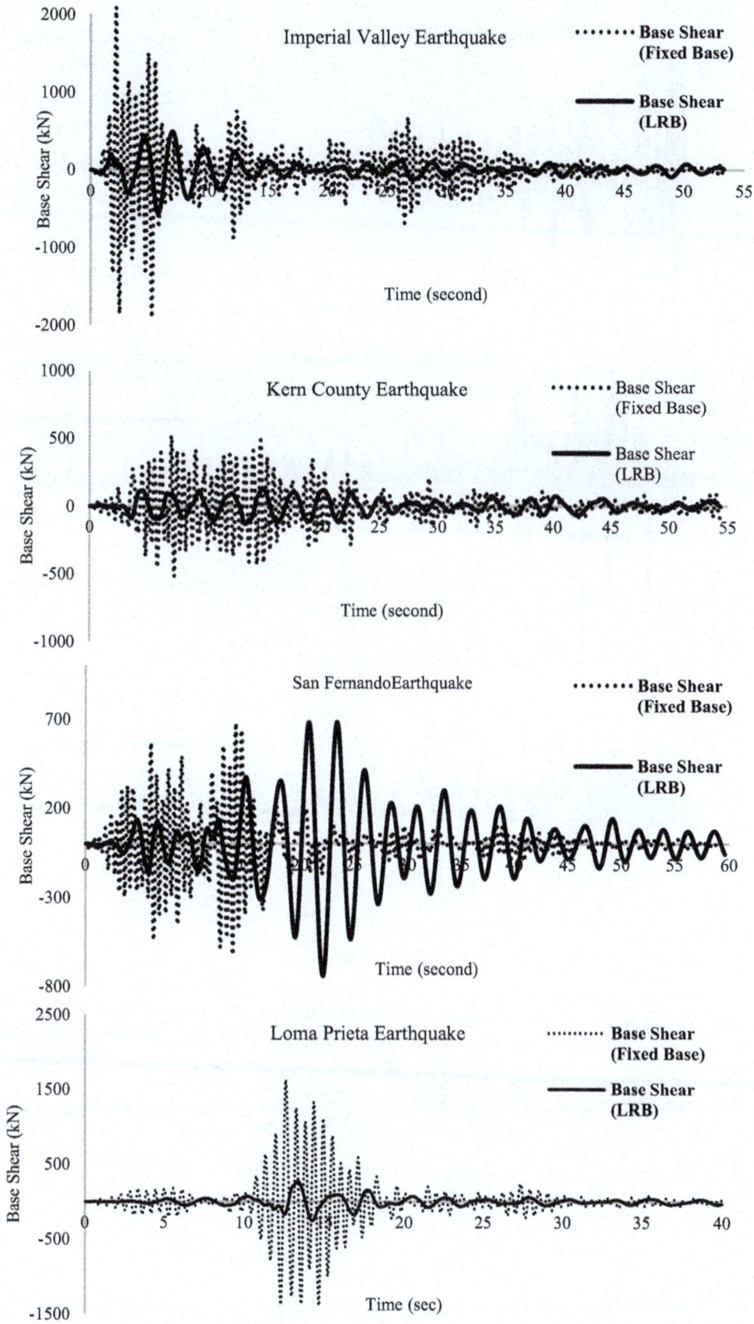

Fig. 27.5 Base Shear data of a five-story building recorded during four distinct earthquake events: (1) Fixd base; (2) LRB

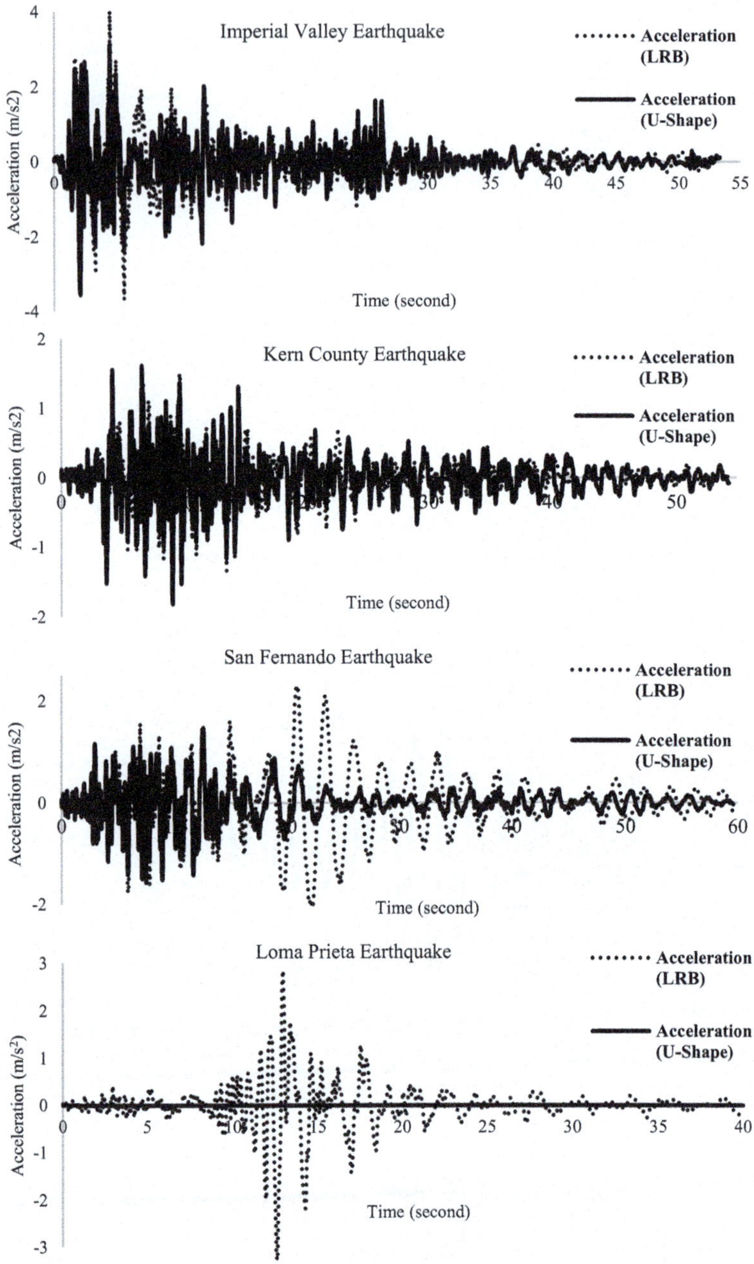

Fig. 27.6 Acceleration data of a five-story building recorded during four distinct earthquake events: (1) LRB; (2) U-Shape Metallic Isolator

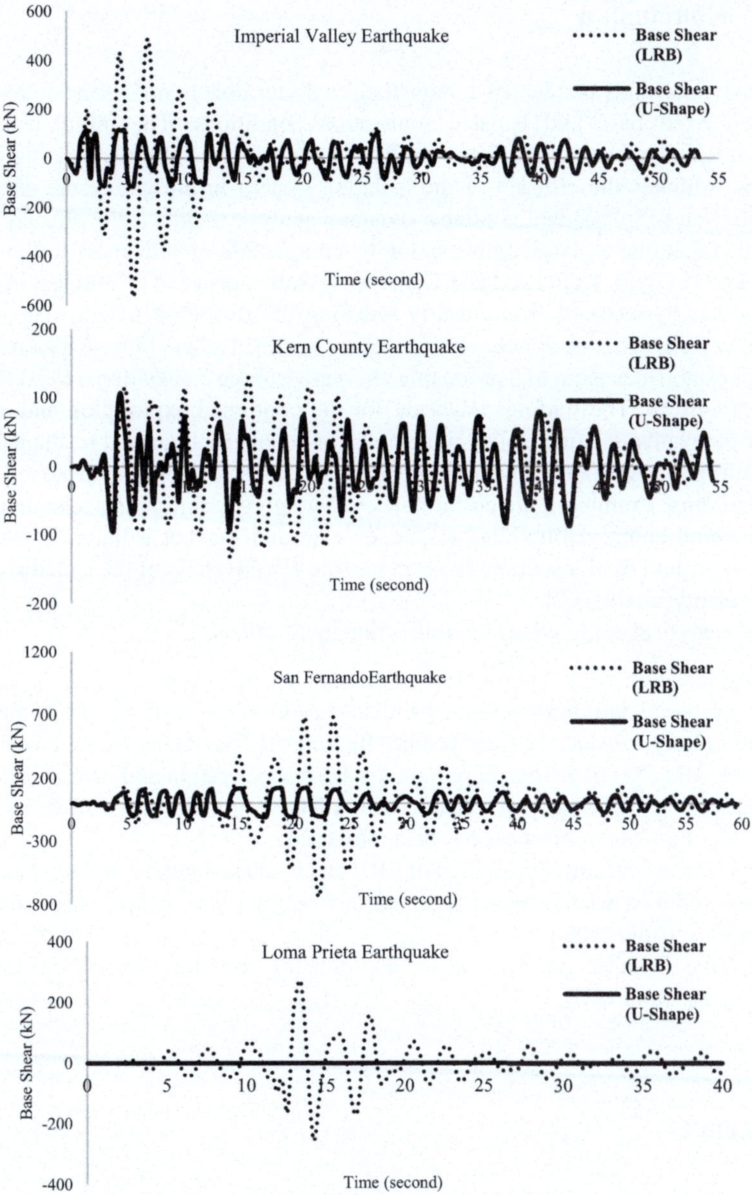

Fig. 27.7 Base Shear data of a five-story building recorded during four distinct earthquake events: (1) LRB; (2) U-Shape Metallic Isolator

27.5 Conclusion

The analytical study conducted demonstrates a discernible trend in seismic response between fixed base and isolated foundation conditions. The initial Nonlinear Dynamic Analysis conducted between fixed base and Lead Rubber Bearing (LRB) isolator validates the efficacy of the isolation system in mitigating the effects of seismic events. Subsequent nonlinear dynamic analysis validates the efficacy of the U-shaped metallic isolator, emphasizing its comparable effectiveness to the established LRB system. Both LRB and U-shaped isolators prove to be efficient methods for isolated foundations, significantly reducing the overall structural response to seismic activity. This study not only highlights the effectiveness of U-shaped isolators but also establishes them as a promising alternative to the conventional Lead Rubber Bearing system. The findings advocate for the continued exploration and utilization of innovative seismic isolation methods for enhanced structural resilience. This study utilized SAP2000 software to conduct nonlinear dynamic analysis on a 5—story building, examining the performance of isolators in mitigating seismic forces. The research compared building responses with and without isolators across four different ground motions, aiming to evaluate the effectiveness of the isolator system under seismic conditions.

The analytical study yields the following observations:

1. Base isolation system plays vital role during seismic events.
2. In comparing buildings equipped with isolators to those without during analysis, particularly considering Lead Rubber Bearings (LRBs) as isolators, it was noted that LRBs effectively decrease both top floor acceleration and base shear.
3. Comparing the U-shaped isolator with LRB, a notable decrease in top floor acceleration and base shear has been observed.
4. Hence, it proves implementation of LRB and U-shape metallic isolator has effectively reduced base shear and acceleration in comparison to fixed base conditions during seismic events.
5. U-shape metallic isolator proves a promising tool through effectiveness in comparison of LRB.

References

Vimal PA, Suresh SL, Vinu M (2003) Effects of fundamental natural time periods on the seismic performance of base isolated multistorey buildings. Asian J Civ Eng 24(6):1801–1815

Cancellara D, De Angelis F (2016a) A base isolation system for structures subject to extreme seismic events characterized by anomalous values of intensity and frequency content. Compos Struct 157:285–302

Cancellara D, De Angelis F (2016b) Nonlinear dynamic analysis for multi-storey RC structures with hybrid base isolation systems in presence of bi-directional ground motions. Compos Struct 154:464–492

Chen P, Wang B, Zhang Z, Li T, Dai K (2023) A generalized model of lead rubber bearing considering large strain stiffening and degradation. Eng Struct 275

Datta TK (2010) Seismic analysis of structures. Indian Institute of Technology Delhi, India

Ene D, Yamada S, Jiao Y, Kishiki S, Konishi Y (2017) Reliability of U-shaped steel dampers used in base isolated structures subjected to biaxial excitation. Earthq Eng Struct Dyn 46(4):621–639

Faal HN, Poursha M (2017) Applicability of the N2, extended N2 and modal pushover analysis methods for the seismic evaluation of base-isolated building frames with lead rubber bearings (LRBs). Soil Dyn Earthq Eng 98:84–100

FEMA P695 (2010) Evaluation of the FEMA P-695 methodology for quantification of building seismic performance factors, NEHRP consultants joint venture a partnership of the applied technology council and the consortium of universities for research in earthquake engineering

Gomase OP, Bakre SV (2011) Performance of non-linear elastomeric base-isolated building structure. Int J Civ Struct Eng 2(1):280–291

Kang GJ, Kang BS (2009) Dynamic analysis of fiber-reinforced elastomeric isolation structures. J Mech Sci Technol 23:1132–1141

Han R, Li Y, van de Lindt J (2014) Seismic risk of base isolated non-ductile reinforced concrete buildings considering uncertainties and mainshock–aftershock sequences. Struct Saf 50:39–56

Islam ABMS, Hussain RR, Jameel M, Jumaat MZ (2012) Non-linear time domain analysis of base isolated multi-storey building under site specific bi-directional seismic loading. Autom Constr 22:554–566

Jiao Y, Kishiki S, Yamada S, Ene D, Konishi Y, Hoashi Y, Terashima M (2014) Low cyclic fatigue and hysteretic behavior of U-shaped steel dampers for seismically isolated buildings under dynamic cyclic loadings. Earthq Eng Struct Dyn 44(10):1523–1538

Kangda MZ, Bakre S (2017) The effect of LRB parameters on structural responses for blast and seismic loads. Arab J Sci Eng 43(4):1761–1776

Konishi Y, Kawamura N, Terashima M, Kishiki S, Yamada S, Aiken I, Black C, Murakami K, Someya T (2012) Evaluation of the fatigue life and behaviour characteristics of U-shaped steel dampers after extreme earthquake loading, In: 15th world conference on earthquake engineering, Lisbon, Portugal

Kubo T, Yamamoto T, Sato K, Jimbo M, Imaoka T, Umeki Y (2014) A seismic design of nuclear reactor building structures applying seismic isolation system in a high seismicity region—a feasibility case study in Japan. Nucl Eng Technol 46(5):581–594

Makris N (2018) Seismic isolation: early history. Earthq Eng Struct Dyn 48(2):269–283

Manchalwar A, Bakre SV (2020) Seismic performance of structure with isolated foundation using U-shape steel damper as an isolator. Soil Mech Found Eng 57(2):170–177

Mazza F (2017) Residual seismic load capacity of fire-damaged rubber bearings of r.c. base-isolated buildings. Eng Fail Anal 79:951–970

Naeim F, Kelly JM (1999) Design of seismic isolated structures: from theory and practice. Wiley, Toronto

Oh SH, Song SH, Lee SH, Kim HJ (2012) Seismic response of base isolating systems with U-shaped hysteretic dampers. Int J Steel Struct 12(2):285–298

Robinson WH (1982) Lead-rubber hysteretic bearings suitable for protecting structures during earthquakes. Earthq Eng Struct Dyn 1(4):593–608

Robinson WH, Tucker AG (1977) A lead rubber shear damper. Bul NZ Soc Earthq Eng 10(3):151–153

SAP2000 Integrated software for structural analysis & design, technical reference manual. Computers & Structures, Inc

Suzuki K, Watanabe A, Saeki E (2005) Development of U-shaped steel damper for seismic isolation system, Nippon steel technical Report No. 9

Zhang R, Phillips BM (2016) Performance and protection of base-isolated structures under blast loading J Eng Mech 142(1)

Chapter 28
Study for the Development of Fly Ash Based Concrete

Anupam Saha, Bidhan Ghosh, and Hasim Ali Khan

Abstract Concrete is extensively used sustainable structural material. Control concrete consists of cement, sand and stone-chips. But 1 kg CO_2 is generated for producing 1 kg of cement, resulting pollution. The use of industrial byproducts such as fly ash can be beneficial to the production of friendly concrete. Present study is an attempt to identify the characteristics of Portland Pozzolana Cement (PPC) based fly ash mixed concrete. The study has been taken up for the designed mix M40 grade and M60. The mechanical nature of concrete were experimented at seven days and twenty eight days in which cement is partly substituted with fly ash ranging from 15 to 35%. To maintain the slump and water cement ratio Sikament 600 (SD) super plasticizer was used in the study. The optimal percentage of replacement of fly ash was found 30% in cement concrete from the study.

Keywords Sustainable · Eco friendly · Fly ash · Mechanical properties · Super plasticizer

28.1 Introduction

Concretes are changing day by day according to the rheological and mechanical performance intended to be used in a particular structure, its environment, construction method and to enhance its performance criteria known as High Performance concretes (HPC). Now-a-days, cement industry is major manufacturers of carbon dioxide. Coal is used as fuel in factory and burning of coal turn out huge fly ash daily (MORTH 2013; ASTM D 2006). Due to pozzolana nature of fly ash, it can be of use in concrete. The utilisation of fly ash is also a step to reduce the use of natural resources as one of the main challenges now confronting the concrete industry in the countries that are developing countries like India. For the study, fly ash was obtained from Bandel thermal power plant, Tribeni, West Bengal. Various literature on physical

A. Saha · B. Ghosh · H. A. Khan (✉)
Adamas University, Barasat, India
e-mail: hiihasim@gmail.com; hasimali.khan@adamasuniversity.ac.in

B. V. Khode et al. (eds.), *Innovations in Technologies: Pioneering Sustainable Infrastructure for a Resilient Future*, Sustainable Civil Infrastructures,
https://doi.org/10.1007/978-3-031-82133-2_28

and mechanical property of fly ash and concrete have been reviewed (Indian Standard 2002, 2013). For sustainable industrial development, industrial ecology must be practiced (Indian Standard 1978; Copeland et al. 2007). It indicates that waste products of one industry are recycled as substitutes of other industries, thereby reducing the environmental hazards. Use of mineral admixtures, especially having pozzolanic characteristics that can replace cement is one of the ways to do so (Mallick et al. 2008; Mohammad and Cooper 2011).

Concretes are changing day by day according to the rheological and mechanical performance intended to be used in a particular structure, its environment, construction method and to enhance its performance criteria known as High Performance concretes (HPC). Now-a-days, cement industry is major manufacturers of carbon dioxide. Coal is used as fuel in factory and burning of coal turn out huge fly ash daily. Due to pozzolana nature of fly ash, it can be of use in concrete (Hussain and Yanjun 2012). The utilisation of fly ash is also a step to reduce the use of natural resources as one of the main challenges now confronting the concrete industry in the countries that are developing like India. For the study, fly ash was obtained from Bandel thermal power plant, Tribeni, West Bengal.

28.2 Background of the Study

Various literature on physical and mechanical property of fly ash and concrete have been reviewed. For sustainable industrial development, industrial ecology must be practiced (Silva et al. 2012). It indicates that waste products of one industry are recycled as substitutes of other industries, thereby reducing the environmental hazards. Use of mineral admixtures, especially having pozzolanic characteristics that can replace cement is one of the ways to do so (Alex et al. 2013; Al-Qadi et al. 2014).

Though High Performance Concrete (HPC) developed gradually over many years by strives to generate concrete with more strengths, nonetheless, it differs from high-strength concrete. The focus has shifted from extremely high strength to other qualities that are desirable in certain situations (Khode and Ghodmare 2019; Sanjay et al. 2022). High Performance concretes (HPC) are concretes with properties which satisfy the performance criteria (Sonkusare et al. 2021; Chandak and Pawade 2021a, b). Fly ash concretes process lower early strength due to the slow pozzolanic reaction of fly ash. This strength reduction is based on the characteristics of the fly ash used. produced enormous amounts of fly ash concrete in the 1980s, with 55–60% of the cement is restore by fly ash and such concrete ultimately confirm outstanding mechanical and durability properties (Sanjay et al. 2022). Research was done for the restitution of 30–60% cement by fly ash to spot the outcomes of fly ash replacement ratio on the characteristics of Self compacting concrete (Solanke et al. 2022; Sonkusare and Pawade 2021; Solanke and Pawade 2021).

Table 28.1 Material characteristics of Portland pozzolana cement (PPC)

Sl. No.	Characteristics	Experimental value
1	Uniformity	34
2	Relative density	3.12
3	Beginning time	140 min
4	Ultimate time	490 min
6	Soundness (By Le-chatelier apparatus)	6 mm
7	Compression strength	
	i. 7 days	30
	ii. 28 days	45

28.3 Aim of Work

A. Now-a-days, many of the construction industries are using Portland pozzolana cement based on fly ash which is friendly. Therefore, in present research Portland pozzolana cement is used as a binder material in concrete.
B. The use of byproducts will further reduce the quantity of environmental hazards with addition to control of pollution happened due to production of cement. So, fly ash is a good unconventional material.
C. The mechanical characteristics of fly ash concrete will provide an idea about the incorporation of such materials in construction work.

28.4 Experimental Programme

Mechanical and Chemical characteristics of all the ingredients are set after testing.

28.4.1 Cement

Portland pozzolana cement (fly ash based) of ACC Branded was used in this research. Its physical properties were identified and shown in Table 28.1. The chemical characteristics are provided by the Manufacturing unit and shown in Table 28.2.

28.4.2 Fly Ash

By burning of coal in thermal power plant, fly ash is obtained as a byproduct and it is collected from the chimney of a manufacturing unit. For the study of fly ash

Table 28.2 Chemical percentage of Portland pozzolana cement

Sl. No.	Characteristics (%)	Value
1	Magnesia	1.2
2	Sulphur	2.4
3	Loss due to ignition	2
4	Chloride	0.023

Table 28.3 Material characteristics of fly ash

Sl. No.	Characteristics	Experimental value
1	Relative density	3.1
2	Modulus of fine	1.94
3	Normal consistency	44%

has collected from Bandel Power plant, Tribeni in West Bengal. Its physical characteristics were identified and shown in Table 28.3 and Fig. 28.1. The chemical characteristics are provided by the Manufacturing unit and shown in Table 28.4.

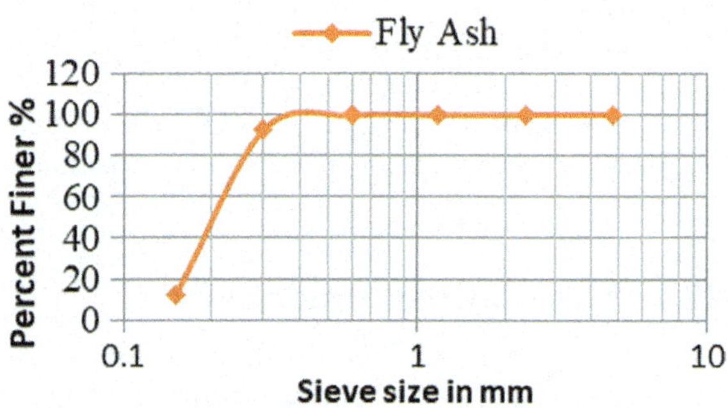

Fig. 28.1 Grading arc for the fly ash

Table 28.4 Chemical percentage of fly ash

Sl. No.	Elements	Mass percent (%)
1	SiO_2	55.11
2	Al_2O_3	21.11
3	Fe_2O_3	5.21
4	CaO	6.72
5	K_2O	2.95
6	MgO	2.60
7	SO_3	0.70

Table 28.5 Material characteristics of fine aggregate

Sl. No.	Characteristics	Experimental value
1	Specific gravity	2.46
2	Modulus of fine	2.42 (Zone IV)
3	Absorption of water	2.48%

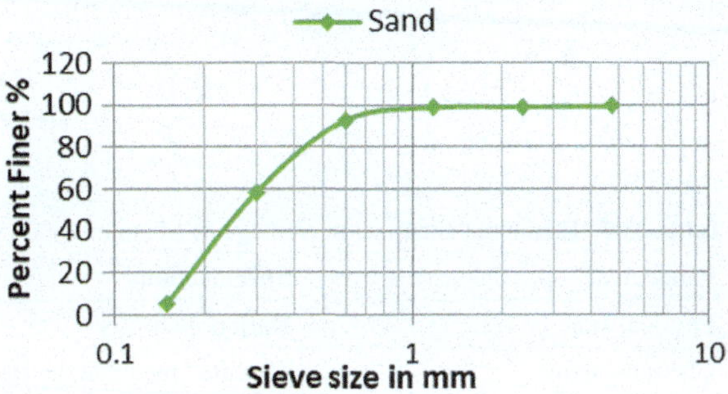

Fig. 28.2 Grading arc for the fine aggregate

Table 28.6 Material characteristics of coarse aggregate

Sl. No.	Characteristics	Experimental value
1	Specific gravity	2.76
2	Modulus of fine	7.8
3	Absorption of water	0.48%
4	Maximum aggregate size	20 mm

28.4.3 Fine Aggregate

Fine aggregate is made from river sand which is easily available in this project. Its physical properties were identified and illustrated in Table 28.5 and Fig. 28.2.

28.4.4 Coarse Aggregate

Fresh dry stones were taken as coarse aggregate in this research. The 20 mm size aggregate is maximum size. Which is obtained from the test. Its physical properties were identified and demonstrated in Table 28.6 and Fig. 28.3.

28.4.5 Water

Portable water is easily available for mixing the concrete mix.

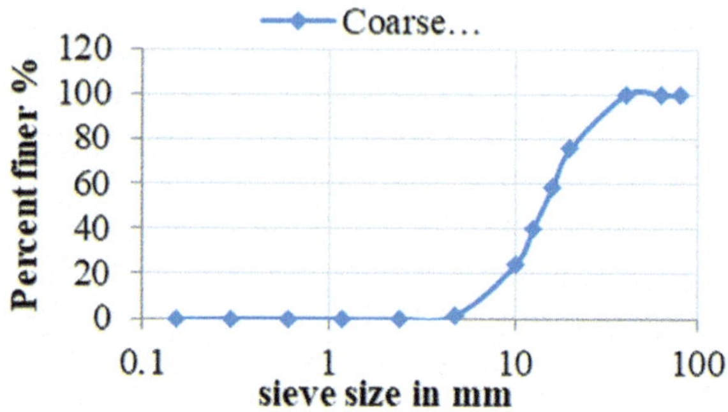

Fig. 28.3 Particle size allocation arc of coarse aggregate

Table 28.7 Elemental chemical percentage of Sikament 600 (SD)

Parameter	Specifications
Physical state	Dark brown
Chemical name	Modified naphthalene formaldehyde sulphonate
Relative density at 25 C	1.21 kg per l
pH	Minimum 6

28.4.6 Super Plasticizer

Sikament 600 (SD) was used as a superplasticizer in this project. It has base material as Modified Naphthalene Formaldehyde Sulphonate. Its chemical characteristics were given by the Manufacturing unit and shown in Table 28.7.

28.4.7 Mix Proportion

Concrete of grade M40 and M60 were investigated in this study. For production M60 concrete with Portland pozzolona cement, 3 mixes are prepared, 1st mix is prepared with only Portland Pozzolona Cement, 2nd mix is prepared using the cementitious components containing 45% fly ash, and 3rd is prepared with cementitious using the cementitious components containing 60% fly ash. Besides this M40 concrete is made with Portland pozzolana cement, the mixtures are made with cementitious material containing 65% fly ash. The compressive strength for each mixture is determined for seven and twenty eight days and the workability slump of all casted specimens were checked. As usual assumption is made as PPC consist of 30% fly ash.

Mixture quantities and characteristics of concrete of grade M40 concrete prepared with PPC and different proportions of fly ash are given in Tables 28.8 and 28.9 and Mixture quantities and characteristics of concrete of grade M60 concrete prepared with PPC and different proportions of fly ash are given in Tables 28.10 and 28.11.

Compressive strength of M40 and M60 grade of concrete for different mix are demonstrated in Tables 28.12 and 28.13.

Table 28.8 Mixture quantities of grade M40 concrete

Mix no.	W/CM	Water Kg/m^3	Cement		Fly ash		Fly ash in total CM, %
			%	Kg/m^3	%	Kg/m^3	
1	0.40	150	100	380	0	0	30
2	0.36	140	85	325	15	50	45
3	0.33	133	70	270	30	100	60
4	0.29	120	65	250	35	138	65

Table 28.9 Mixture quantities and workability of fresh concrete of grade M40

Min no.	W/CM	Total wt. of CM, Kg/m^3	Coarse agg., Kg/m^3	Fine agg., Kg/m^3	Super plasticizer (l/m^3)	Slump, mm
1	0.40	375	1130	760	3.85	87
2	0.36	388.88	1132	762	3.85	101
3	0.33	403.03	1130	760	3.85	96
4	0.29	413.79	1160	785	3.85	93

Table 28.10 Mixture quantities of grade M60 concrete

Mix no.	W/CM	Water Kg/m^3	Cement		Fly ash		Fly ash in total CM, %
			%	Kg/m^3	%	Kg/m^3	
5	0.35	156	100	460	0	0	30
6	0.33	149	85	395	15	62	45
7	0.30	140	70	325	30	138	60

Table 28.11 Mixture quantities and workability of fresh concrete of grade M60

Min no.	W/CM	Total wt. of CM, Kg/m^3	Coarse agg., Kg/m^3	Fine agg., Kg/m^3	Super plasticizer (l/m^3)	Slump, mm
5	0.35	445.71	1145	685	3.85	82
6	0.33	451.51	1142	681	3.85	98
7	0.30	466.66	1155	691	3.85	90

Table 28.12 Compression strength of grade M40 concrete

Mix no.	Cement, %	Fly ash %	Fly ash in total CM, %	W/CM	Compressive strength, MPa	
					7 days	28 days
1	100	0	30	0.40	43	47
2	85	15	45	0.36	50	55
3	70	30	60	0.33	52	56
4	65	35	65	0.29	45	50

Table 28.13 Compression strength of grade M60 concrete

Mix no.	Cement, %	Fly ash %	Fly ash in total CM, %	W/CM	Compressive strength, MPa	
					7 days	28 days
5	100	0	30	0.35	44	60
6	85	15	45	0.33	47	64
7	70	30	60	0.30	45	61

28.5 Conclusions

1. For same grade of concrete, the proportion of water to cementitious components decreased with the increment of fly ash amount.
2. As the total quantity of cementitious ingredients increased, the ratio of water to cementitious constituents dropped for the same slump.
3. It is possible to produce concrete that satisfies the strength requirements of grades M40 and M60 concrete, correspondingly, using up to 35 and 30% fly ash, etc. This type of concrete will outperform control concretes in terms of strength development at later ages and is going to grow to a desirable early age.
4. Compressive Strength is increased up to 30% fly ash for M40 concrete and 15% for M60 concrete but the compressive strength will decrease above the mentioned % of fly ash.

References

Apeagyei AK, Clark TM, Rorrer TM (2013) Stiffness of high-RAP asphalt mixtures: Virginia's experience. ASCE J Mater Civ Eng

Al-Qadi IL et al (2014) Reclaimed asphalt pavement—a literature review. Illinois Center for Transportation Series No. 07-001, pp 1–19

ASTM D 6927 (1997) Test method for Marshall stability and flow of bituminous mixtures. Am Soc Test Mater

Chandak M, Pawade PY (2021a) Influence of Metakaolin and steel fibers on stress strain behavior of concrete. J Phys Conf Ser. https://doi.org/10.1088/1742-6596/1913/1/012070
Chandak M, Pawade PY (2021b) Experimental study on the effect of aspect ratio and volume fraction of steel fibre on the mechanical properties of Metakaolin SFR concrete with mathematical model. J Phys Conf Ser. https://doi.org/10.1088/1742-6596/1913/1/012070
Copeland et al (2007) Field evaluation of high reclaimed asphalt pavement-warm mix asphalt project. J Transp Res Board, Florida, USA 2179:93–101
Hussain A, Yanjun Q (2012) Evaluation of asphalt mixes containing reclaimed asphalt pavement for wearing courses IPCSIT 26:43–48
Indian Standard (1978) Method for testing tar & bituminous materials, IS 1203-1978. Bureau of Indian Standard, New Delhi
Indian Standard (2002) Method for test for aggregates for concrete, IS 2386 (Part I & Part II). Bureau of Indian Standard, New Delhi
Indian Standard (2013) Paving bitumen specification (4th revision), IS 73:2013. Bureau of Indian Standard, New Delhi
Khode BV, Ghodmare SD (2019) Effect of mineral and chemical admixture for improvement of high-performance concrete. In: Kolhe M, Labhasetwar P, Suryawanshi H (eds) Smart technologies for energy, environment and sustainable development. Lecture notes on multidisciplinary industrial engineering. Springer, Singapore. https://doi.org/10.1007/978-981-13-6148-7_41
Mallick RB et al (2008) Using-warm asphalt technology to incorporate high percentage of reclaimed asphalt pavement material in asphalt mixture. Transp Res Rec 2051 Washington, DC, pp 71–79
Mohammad LN, Cooper SB (2011) Characterization of HMA mixtures containing high reclaimed asphalt pavement content with crumb rubber additives. ASCE J Nov
MORTH (2013) Specifications for road and bridge works (fifth revision). Indian Roads Congress
Sanjay SS, Pal A, Chakraborty SK, Khan HA (2022) Reduction of shrinkage of self compacting concrete using polycarboxylate ether as shrinkage reducing admixture. Mater Today Proc 60(1):448–451. https://doi.org/10.1016/j.matpr.2022.01.316
Silva HMRD, Oliveira JRM, Jesus CMG (2012) Are totally recycled hot mix asphalts a sustainable alternative for road paving? Resour Conserv Recycl 60:3848, Elsevier
Solanke SS, Pawade PY, Khan HA (2022) An experimental study on tensile as well as flexural strength of concrete by using sugarcane baggase ash & steel fiber. Mater Today Proc 60(1):627–637. https://doi.org/10.1016/j.matpr.2022.02.129
Solanke SS, Pawade PY (2021) An investigation of mechanical properties of concrete by addition of sugarcane baggase ash and steel fiber. J Phys Conf Ser 1913(1):012069. https://doi.org/10.1088/1742-6596/1913/1/012069
Sonkusare HG, Pawade PY (2021) Investigate the temperature effects on curing of reactive powder concrete containing silica. In: Lecture notes in civil engineering
Sonkusare HG, Pawade PY, Pujara HP (2021) Experimental mix design approach of reactive powder concrete to recognize compressive strength through non-destructive test. J Phys Conf Ser. https://doi.org/10.1088/1742-6596/1913/1/012149

Chapter 29
Seismic Performance of Regular and Irregular Buildings with Fixed Base and Lead Rubber Bearing Base Isolator

Gaurav Gupta, Kuldeep Dabhekar, and Dilip Mase

Abstract A significant portion of the global population resides in areas prone to seismic activity, exposing them to the risk of earthquakes with varying magnitudes and frequencies. Earthquakes consistently result in substantial loss of life and property damage. To alleviate the impact of earthquakes on buildings, the base isolation technique emerges as an effective solution. Base isolation stands out as a powerful tool in earthquake-resistant design. The fundamental concept involves minimizing the transmission of energy to the superstructure during seismic events, gaining increased acceptance over the past two decades. Previous experiences highlight the heightened vulnerability of irregular buildings compared to their regular counterparts. The application of Lead Rubber Bearings proves to be a promising technology for safeguarding diverse structures such as buildings, bridges, and airport terminals. This paper conducts a response spectrum analysis on two plans with equal area using ETAB software. The comparison involves a square-shaped (regular) and an L-shaped (irregular) building, both with G + 10 storied RC frame structures. The seismic effects of a fixed base structure are contrasted with those of a structure employing Lead Rubber Bearings as a base isolator. The results reveal that incorporating lead rubber bearings at the base of a structure leads to less damage compared to a fixed base structure.

Keywords Tremor Defensive Framework · Fixed Base · Base Isolation · Lead Rubber Bearing

G. Gupta (✉) · K. Dabhekar
G. H. Raisoni College of Engineering, Nagpur, India
e-mail: gaurav.gupta.mtechstr@raisoni.net

K. Dabhekar
e-mail: kuldeep.dabhekar@raisoni.net

D. Mase
P T Mase & Associates, Nagpur, India

B. V. Khode et al. (eds.), *Innovations in Technologies: Pioneering Sustainable Infrastructure for a Resilient Future*, Sustainable Civil Infrastructures,
https://doi.org/10.1007/978-3-031-82133-2_29

29.1 Introduction

29.1.1 General

The initial Base Isolation technique was suggested by Dr. Johannes Calanterients, an English Medical Doctor, in 1909, and it was presented by scientists and engineers in premature 1923 in New Zealand. In India, the base isolation skill was initially illustrated after the 1993 Killari (Maharashtra) Earthquake (Mahmoud and Mahmoud 2012). Two single-storey buildingsand, a shopping complex building, and a school building, in the freshly transplanted Killari town, were constructed with rubber base isolators on hard ground. Both were brick masonry buildings with concrete roofs (Rajput et al. 2023). The four-storey Bhuj Hospital building incorporated base isolation after the 2001 Bhuj (Gujarat) earthquake (Fabio De Angelis 2016; Khedikar et al. 2021).

Notable milestones in Base Isolation adoption include:

- 1st implementation in New Zealand in 1974
- 1st US implementation in 1984
- 1st Japanese implementation in 1985
- 1st Indian implementation in 2001

Base isolation involves providing seismic isolation in the foundation. It is a popular method for protecting structures from earthquake forces (Lingeshwaran et al. 2021; Mahakalker et al. , 2023). Base isolation decouples a superstructure from its substructure on shaky ground, preserving a building or non-building structure's integrity (Anwar et al. Oct. 2016; Shenton and Lin 1993). It provides flexibility and energy dissipation by inserting isolators between the foundation and the building structure (Akhare and Wankhade 2014). In structures with Base Isolation Systems, vibrations during an earthquake are mainly experienced by the ground alone (Kalantari et al. 2008). This technique reduces the impact of an earthquake by isolating the structure from hazardous ground motions, increasing the fundamental vibration period, and subjecting the structure to lower earthquake forces (Dhawade 2014).

Lead Rubber Bearing. The second classification of elastomeric bearing is Lead Rubber Bearing (LRB). These bearings provide a blend of vertical bearing, restoring force, horizontal adaptability and suppressing within a single unit (Rodke et al. 2022; Jain and Sanghai 2020). In this system, lead plug is forcefitted into a pre-formed hole in the elastomeric bearing, providing stiffness under typical loads and dissipating energy under substantial lateral loads (Somwanshi and Rina 2015; Sahoo and Parhi 2018). The lead-rubber bearing exhibits rigid in both lateral and vertical directions under low lateral loads, such as minor earthquakes (Nimbekar et al. 2021; Khedikar et al. 2021). Notably, it integrate the functionalities of flexibility at earthquake load levels, rigidity at service load level, and suppressing in a concise unit (Sanghai and Gurmule 2024; Kukwas et al. , 2023).

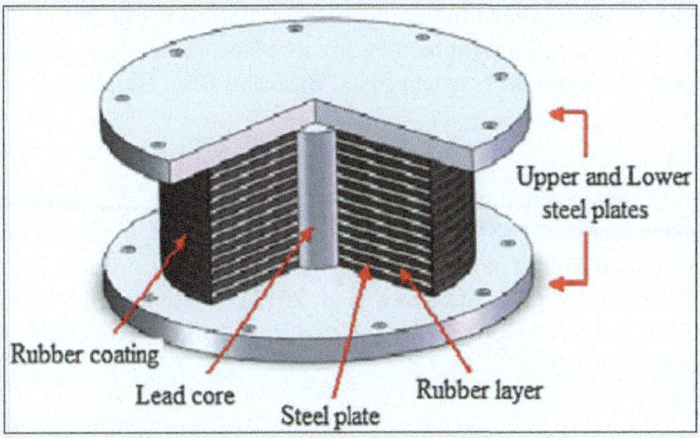

Fig. 29.1 Cross sectional of lead rubber bearing

LRBs possess energy absorption capacity through supplementary hysteretic suppression during the yielding of lead core, reducing lateral displacements, especially under ambient vibrations (Sanghai and Pawade 2022). The production of LRB systems takes into account the lead material's properties, including its low yield point, elasto-plastic behaviour, resistance to repeated loads and self-renewal over time following distortion (Bhurse et al. 2019). The system demonstrates elevated initial rigidity against very less lateral forces caused by mild or wind earthquakes (Chandak 2013; Chhatani et al. 2021). However, a significant drawback is the difficulty in determining damage to lead core in centre when exposed to strong ground motion (Sekar and Kadappan 2015; Providakis 2007) (Fig. 29.1).

Components of Lead Rubber Bearing.

- Upper and Lower plates: Facilitate the shift of loading and constrain distortion of the lead core (Sanghai and Pawade 2022; Mahakalkar et al. 2747).
  - Lead Core: Functions to reduce displacement and dissipate energy.
  - Steel plates: Enhance vertical rigidity and constrain distortion of the lead core.
  - Intrinsic Rubber Layers: Backing the weight of the structure, displacement and accommodate rotation, and restore the moving bearing to its original position.
  - Rubber coating: Serves to guard the rubber layers and reinforced steel plates.

Objective.

This study involves examining G + 10 reinforced concrete buildings, both regular and irregular, based on IS1893-2016 and UBC provisions. One structure has a fixed base, while the other utilizes base isolation (Ali and Sanghai 2417; Nagpure and Sanghai 2018).

The objectives are as follows:-

- To analyze and design lead rubber bearing for high rise structure.
- To study the effect of lead rubber bearing as a base isolator on the structure.
- To compare the response between Fixed Base and Base Isolator.

Methodology.

Regular and Irregular structures are modelled in ETAB software. Shape of the structure are square shape and L-Shape. Modelled consist of G + 10 storey RC frame structure. Modelled consist of response spectrum analysis according to IS 1893:2016 (part 1) & UBC provision (Cancellara et al. 2013). These structures are modelled with fixed base and Lead Rubber Bearing as a Base Isolator (Chandak and Pawade 1913; Mahore et al. 1193).

29.2 Material Properties

Grade of Concrete: M30.

Grade of Steel: HYSD 500.

Floor to Floor Height: 3mt.

Dead Load: Gravity Load (beam, column, slab).

Live Load: 3KN/m^2

Floor Finish: 1.5KN/m^2

Internal Wall: 115 mm.

External Wall: 150 mm.

Column Size: 450 mm × 450 mm.

Beam Size: 300 mm × 500 mm.

Slab Thickness: 125 mm.

Seismic Zone: III.

Response Reduction Factor: 5

Structure: RCC (SMRF).

Importance Factor: 1.2

Site Type: II.

Soil Profile Type: Medium.

29.3 Characteristics of Lead Rubber Bearing

Isolators are provided base of buildings.

Properties of LRB are mentioned below (Tables 29.1 and 29.2):

Table 29.1 Regular building (square shape)

Property type	RSM
Rotational inertia 1	0.009715681
For U1 efficient rigidity	3,048,409.923
For U2 & U3 efficient rigidity	3048.409923
For U2 & U3 efficient attenuation	0.05
For U2 & U3 distance from end-J	0.00254158
For U2 & U3 rigidity	28,089.88367
For U2 & U3 tensile strength	71.39269306

Table 29.2 Irregular building (L-shape)

Property Type	RSM
Rotational inertia 1	0.009715681
For U1 efficient rigidity	3,110,786.628
For U2 & U3 efficient rigidity	3110.786628
For U2 & U3 efficient attenuation	0.05
For U2 & U3 extended from end-j	0.00254158
For U2 & U3 rigidity	28,664.66017
For U2 & U3 tensile strength	72.85353365

29.4 Modelling

G + 10 Stories RC Structure are modelled in ETAB software and the plan having same floor area as shown below (Hosseini and Ramana 2015). These structure are modelled with fixed base and with base isolation. Lead Rubber Bearing (LRB) as a Base Isolator are used in base of the structure because it employing heavy damping. Response Spectrum Method is used for analysis (Rajput et al. 2023; Mahakalker et al. 2023).

- Square Shape (Regular) Building is used for modelling in ETAB Software with Fixed Base & Lead Rubber Bearing (LRB) as shown below.
- L-Shape (Irregular) Building is used for modelling in ETAB Software with Fixed Base & Lead Rubber Bearing (LRB) as shown below (Figs. 29.2, 29.3, 29.4, 29.5).

29.5 Result

Storey Displacement for Models with fixed base (Tables 29.3 and 29.4).

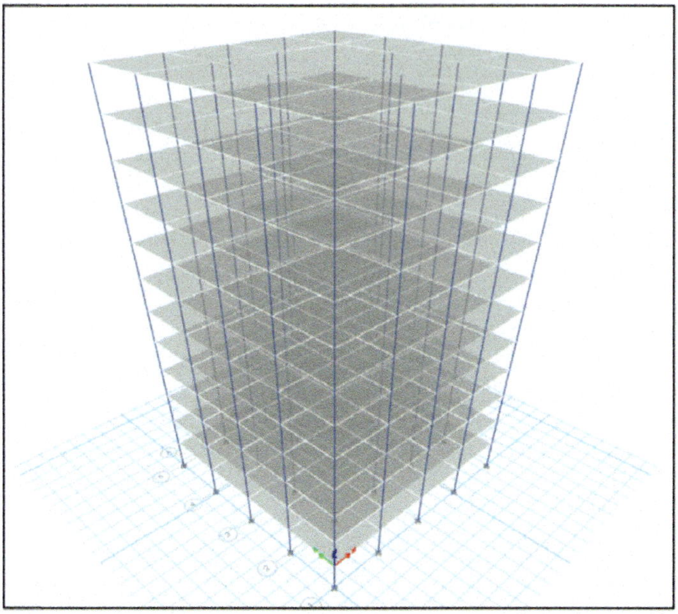

Fig. 29.2 Square shape model with fixed base

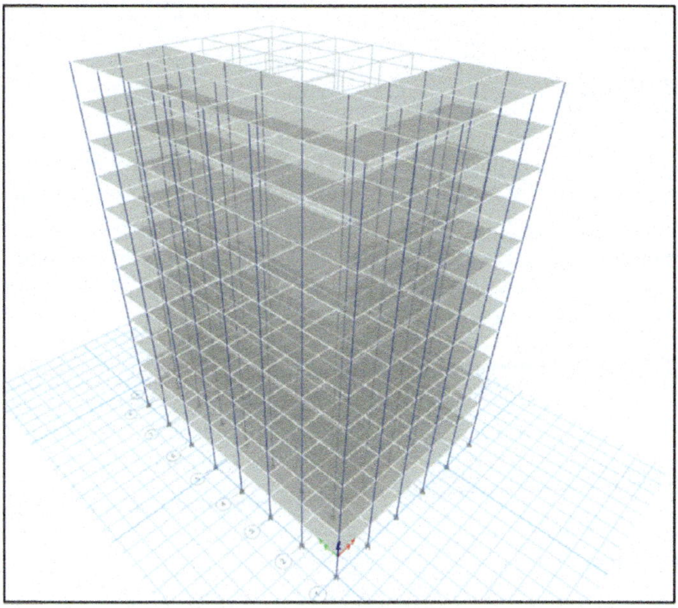

Fig. 29.3 L-shape model with fixed base

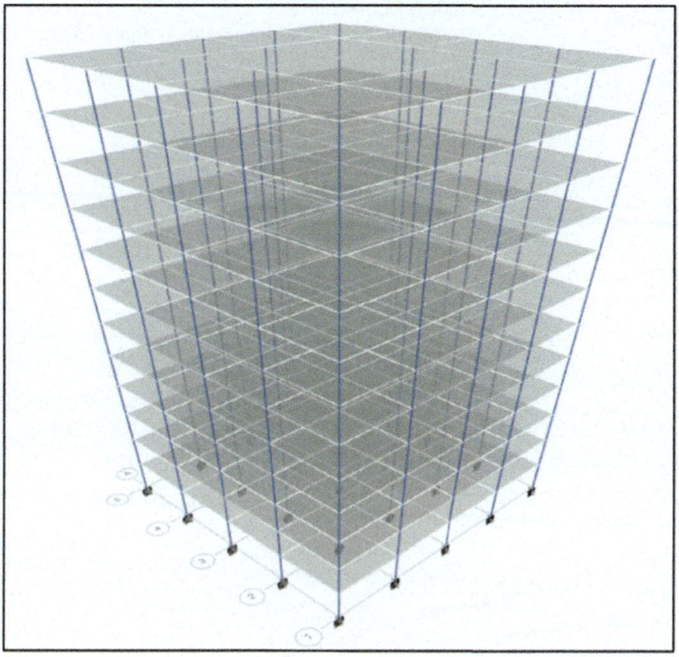

Fig. 29.4 Square shape model with lead rubber bearing (LRB)

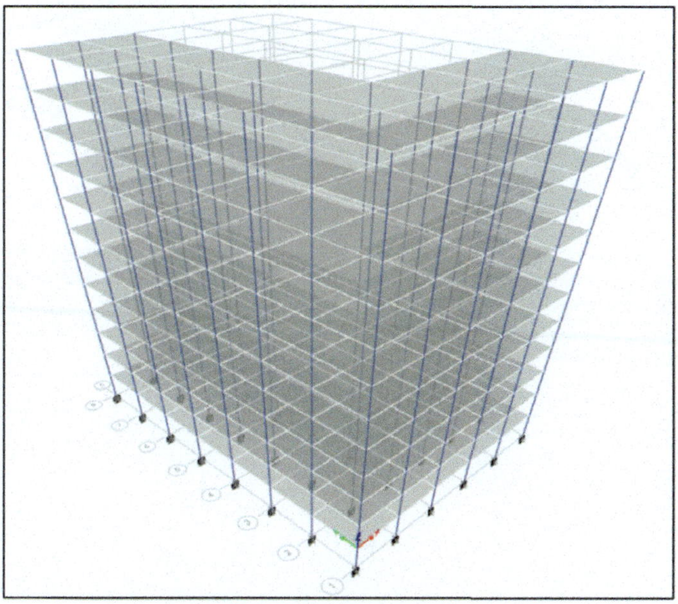

Fig. 29.5 L-shape building with lead rubber bearing (LRB)

Table 29.3 Storey displacement in X-direction in mm

Storey number	Storey height	Square shape	L-shape
Terrace	33	17.156	21.756
Nineth F	30	16.702	21.038
Eighth F	27	16	20.032
Seventh F	24	15.051	18.728
Sixth F	21	13.871	17.15
Fifth F	18	12.481	15.328
Fourth F	15	10.905	13.297
Third F	12	9.17	11.094
Second F	9	7.304	8.76
First F	6	5.338	6.34
Ground F	3	3.315	3.89
Base	0	1.335	1.539

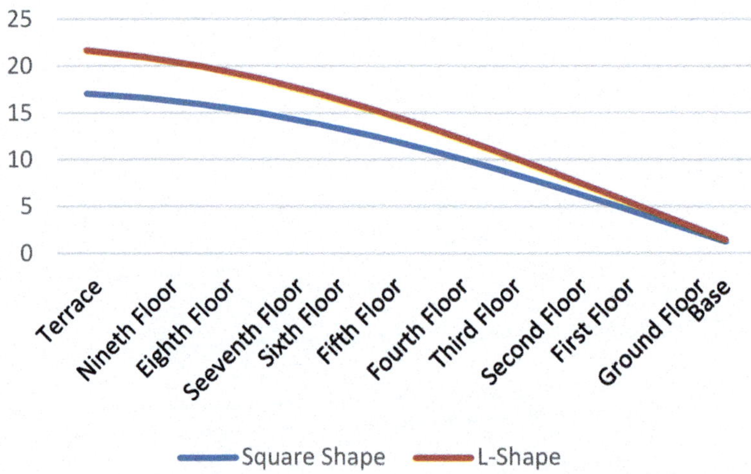

Table 29.4 Story displacement in Y-direction in mm

Storey number	Storey height	Square shape	L-shape
Terrace	33	17.156	17.323
Nineth F	30	16.702	16.849
Eighth F	27	16	16.133
Seventh F	24	15.051	15.163
Sixth F	21	13.871	13.959
Fifth F	18	12.481	12.542
Fourth F	15	10.905	10.94
Third F	12	9.17	9.181
Second F	9	7.304	7.296
First F	6	5.338	5.32
Ground F	3	3.315	3.297
Base	0	1.335	1.327

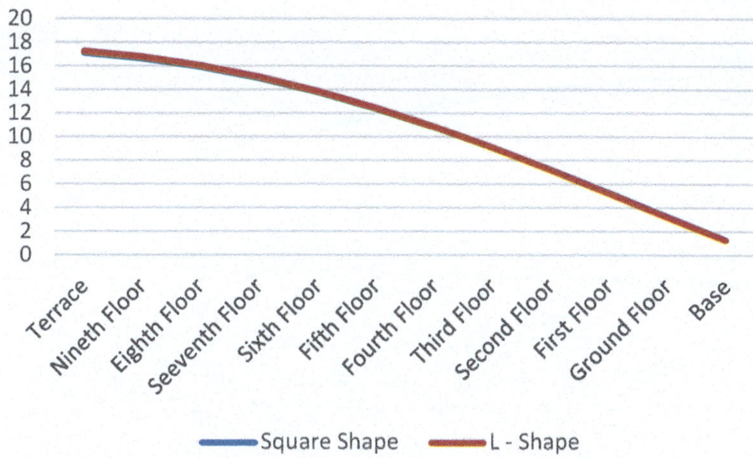

Storey Displacement for Models with LRB (Tables 29.5 and 29.6):

Table 29.5 Storey displacement in X-direction in mm

Storey number	Storey height	Square shape	L-shape
Terrace	33	32.613	38.036
Nineth F	30	32.154	37.299
Eighth F	27	31.463	36.312
Seventh F	24	30.532	35.059
Sixth F	21	29.37	33.553
Fifth F	18	27.987	31.809
Fourth F	15	26.398	29.849
Third F	12	24.616	27.694
Second F	9	22.66	25.371
First F	6	20.543	22.901
Ground F	3	18.254	20.277
Base	0	15.509	17.173

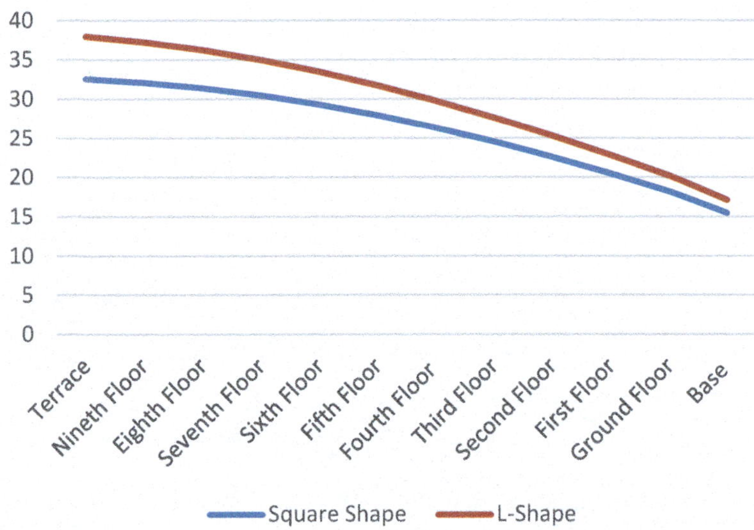

Table 29.6 Storey displacement in Y-direction in mm

Storey number	Storey height	Square shape	L-shape
Terrace	33	32.613	33.221
Nineth F	30	32.154	32.735
Eighth F	27	31.463	32.029
Seventh F	24	30.532	31.087
Sixth F	21	29.37	29.917
Fifth F	18	27.987	28.53
Fourth F	15	26.398	26.942
Third F	12	24.616	25.169
Second F	9	22.66	23.229
First F	6	20.543	21.139
Ground F	3	18.254	18.892
Base	0	15.509	16.202

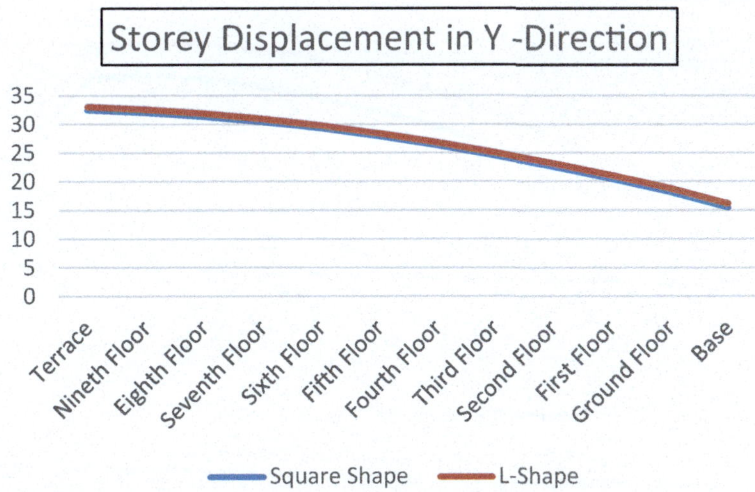

Storey Drift for Models with Fixed base (Tables 29.7 and 29.8):

Table 29.7 Story drift in X-direction in mm

Storey number	Story height	Square shape	L-shape
Terrace	33	0.000156	0.000243
Nineth F	30	0.000241	0.000343
Eighth F	27	0.000325	0.000444
Seventh F	24	0.000402	0.000536
Sixth F	21	0.000471	0.000617
Fifth F	18	0.000531	0.000685
Fourth F	15	0.000583	0.00074
Third F	12	0.000625	0.000781
Second F	9	0.000657	0.000808
First F	6	0.000675	0.000817
Ground F	3	0.00066	0.000784
Base	0	0.000445	0.000513

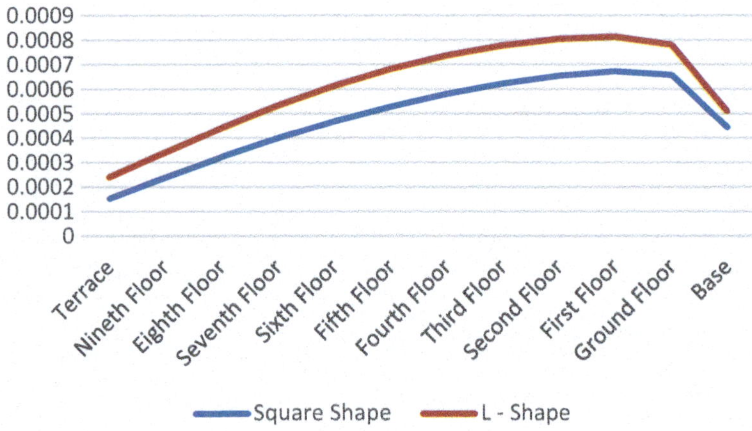

Table 29.8 Storey drift in Y-direction in mm

Story number	Story height	Square shape	L-shape
Terrace	33	0.000156	0.000161
Nineth F	30	0.000241	0.000246
Eighth F	27	0.000325	0.000332
Seventh F	24	0.000402	0.00041
Sixth F	21	0.000471	0.00048
Fifth F	18	0.000531	0.00054
Fourth F	15	0.000583	0.000591
Third F	12	0.000625	0.000631
Second F	9	0.000657	0.00066
First F	6	0.000675	0.000675
Ground F	3	0.00066	0.000657
Base	0	0.000445	0.000442

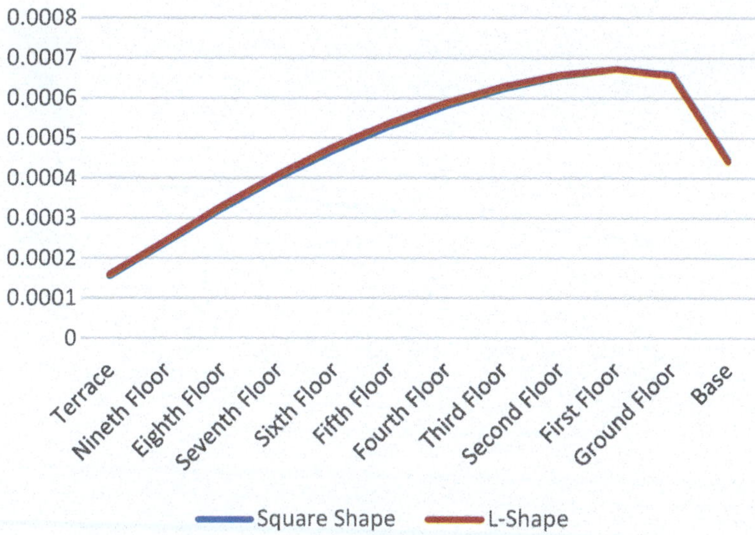

Storey Drift for Models with LRB (Tables 29.9 and 29.10):

Table 29.9 Storey drift in X-direction in mm

Story number	Story height	Square shape	L-shape
Terrace	33	0.000155	0.000248
Nineth F	30	0.000233	0.000333
Eighth F	27	0.000314	0.000423
Seventh F	24	0.000392	0.000508
Sixth F	21	0.000465	0.000587
Fifth F	18	0.000534	0.000659
Fourth F	15	0.000597	0.000723
Third F	12	0.000654	0.000778
Second F	9	0.000707	0.000826
First F	6	0.000764	0.000876
Ground F	3	0.000915	0.001035
Base	0	0.001988	0.002141

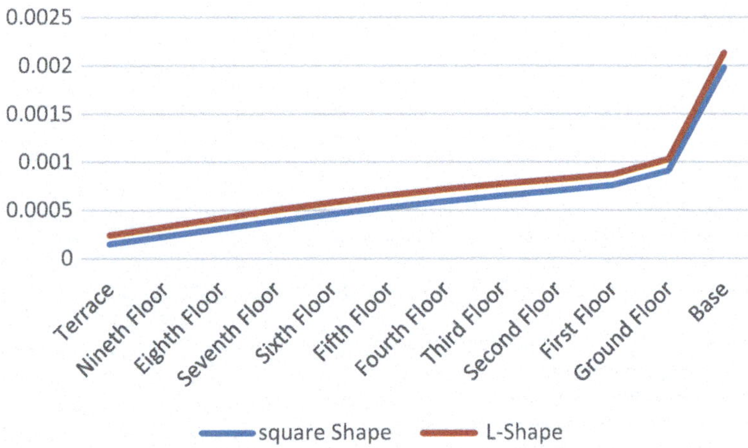

Table 29.10 Story drift in Y-direction in mm

Story number	Story height	square shape	L-shape
Terrace	33	0.000155	0.000163
Nineth F	30	0.000233	0.000238
Eighth F	27	0.000314	0.000318
Seventh F	24	0.000392	0.000394
Sixth F	21	0.000465	0.000466
Fifth F	18	0.000534	0.000533
Fourth F	15	0.000597	0.000594
Third F	12	0.000654	0.000649
Second F	9	0.000707	0.000698
First F	6	0.000764	0.00075
Ground F	3	0.000915	0.000897
Base	0	0.001988	0.001987

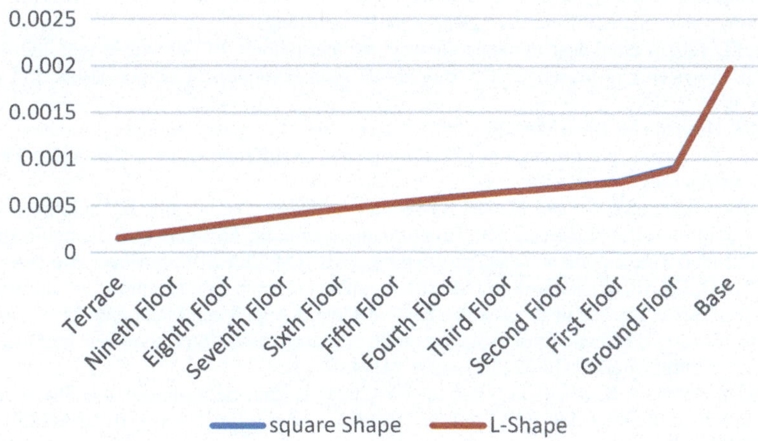

29.6 Conclusion

From the above results shown, it is concluded that.

1. Above results are shown from different building models arranged in Excel Sheet, similarly we can see the different building with fixed base and building with lead rubber bearing as a base isolator. From results we find less variation in storey displacement in building models with LRB isolator in Y Direction.
2. From the above results we seen that, the very less variation in story displacement in building models with fixed base in Y Direction

3. From the above results we seen that the Storey Displacement in Regular Buildings in both the directions is same and different in Irregular Buildings.
4. From the above results we seen that the Storey Drift in Regular Buildings in both the directions is same and different in Irregular Buildings.
5. Above results we seen that very less very less variation in Storey Drift in between regular and irregular buildings with fixed base in Y direction.
6. Above results we seen that very less variation in Storey Drift in between regular and irregular buildings with Lead Rubber Bearing as a Base Isolator in Y direction.

References

Akhare AR, Wankhade TR (2014) Seismic performance of RC structure using different base isolator. IJESRT. 3(5):724–729. ISSN: 2277–9655

Ali S, Sanghai SS (2021) Seismic vulnerability assessment of reinforced con crete buildings using pushover analysis. AIP Conf Proc 2417(1) AIP Publishing

Angelis FD (2016) Nonlinear dynamic analysis for multi-storey RC structures with hybrid base isolation systems in presence of bi-directional ground motions. Compos Struct 154 (2016) 464–492

Anwar SF, Hasmath M, Khan MA (2016) Share, study on soil-structure interaction and base isolated system for seismic performance of structures resting on different types of soils. Glob J Eng Sci Res 3(10):61–67

Bhurse PS, Sanghai SS, Kumari NL (2021) Seismic response of rectangular RC building with and without infill walls considering soil–structure interaction. In: Advances in civil engineering and infrastructural development: select proceedings of ICRACEID 2019. Springer, Singapore

Cancellara D, De Angelis F, Pasquino M (2013) A novel seismic base isolation system consisting of a lead rubber bearing in series with a friction slider. J Appl Mech Mater 256–259:2174–2184

Chandak NR (2013) Effect of base isolation on the response of reinforced concrete building. J Civ Eng. Res. https://doi.org/10.5923/j.jce.20130304.02

Chhatani Y, Dabhekar K, Khedikar I, Pawade PY (2021) Seismic performance of L-Shape building through plan irregularities. IOP Conf Ser Mater Sci Eng. https://doi.org/10.1088/1757-899X/1197/1/012050

Dhawade SM (2014) Comparative study for seismic performance of base isolated & fixed based rc frame structure. Int J Civ Eng Res vol 5

Hosseini M, Ramana NV (2015) Earthquake analysis of high rise buildings with shear walls at the center and center of each side of the external perimeter with openings. Int J Sci Res 5(12):59–71

Jain MR, Sanghai SS (2020) Seismic response control of unsymmetrical RCC framed using base isolation considering soil structure interaction, IJSART

Kalantari SM, Naderpour H, Hoseini Vaez SR (2008) Investigation of base-isolator type selection on seismic behavior of structures including story drifts and plastic hinge for mation. In: The 14th world conference on earthquake engineering, China

Khedikar I, Dabhekar K, Patil AK, Solanke AP, Agrawal R (2021) Comparative analysis of structure with and without seismic load. IOP Conf Ser Mater Sci Eng. https://doi.org/10.1088/1757-899X/1197/1/012028

Khedikar I, Dabhekar K, Patil AK, Solanke AP, Agrawal R (2021) Comparative analysis of structure with and without seismic load. IOP Conf Ser Mater. https://doi.org/10.1088/1757-899X/1197/1/012028

Kukwas A, Pandey V, Dabhekar K, Solanki T (2023) A review on seismic analysis of tall building. Advances in sustainable construction materials

Lingeshwaran N, Koushik S, Reddy TMK, Preethi P (2021) Comparative analysis on asymmetrical and symmetrical structures subjected to seismic load ELSEVIER 2021. https://doi.org/10.1016/j.matpr.2020.11.340

Mahakalkar P, Pawade P, Dabhekar K (2023) Analysis of seismic effect of dampers as energy dissipation system in multi-storied RCC building. AIP Conf Proc 2747(1), AIP Publishing

Mahakalker P, Dabhekar K, Pawade PY (2023) Analysis of seismic effect of dampers as energy dissipation system in multi-storied RCC building. In: International conference on materials engineering & manufacturing systems: ICMEMS2022. https://doi.org/10.1063/5.0132946

Mahakalker P, Dabhekar K, Pawade PY (2023) Seismic performance evaluation of high rise building in zone V with fuid viscous dampers and friction dampers. IOP Conf Ser Mater Sci Eng

Mahmoud S, Mahmoud A (2012) Building with base isolation techniques. Al-Azhar Univ Eng J Jauesvol.7

Mahore A, Patil P, Mase D (2023) Condition & life evaluation of structure by various NDT techniques for reuse. IOP Conf Ser Earth Environ Sci 1193(1) IOP Publishing

Nagpure A, Sanghai SS (2019) Effect of floor diaphragms on seismic response of RCC framed building—a review. In: Smart technologies for energy, environment and sustainable development: select proceedings of ICSTEESD 2018:323–333

Nimbekar S, Dabhekar K, Solanke SS, Sonawane TH (2021) Seismic response control of elevated water tank using base isolation: a review. IOP Conf Ser Mater Sci Eng. https://doi.org/10.1088/1757899X/1197/1/012016

Providakis CP (2007) Effect of LRB isolators and supplemental viscous dampers on seismic isolated buildings under near-fault excitations. Eng Struct 30(1187):1198

Rajput P, Dabhekar K, Pawade P (2023) Seismic performance analysis of symmetric and asymmetric structure with and without base isolation under bidirectional ground motion and subjected to torsion. In: International conference on materials engineering & manufacturing systems

Rajput P, Dabhekar K, Pawade P, Pawar N (2023) Review on seismic performance analysis of symmetric and asymmetric structure with and without base isolation under bidirectional ground motion and subjected to torsion. Advances in sustainable construction materials. https://doi.org/10.1063/5.0145502

Rodke S.J, Ghodmare S.D, Deshmukh A (2022) Critical analysis of saturation flow rate and affecting parameters. In: International conference on emerging trends in engineering and technology, ICETET. https://doi.org/10.1109/ICETET-SIP2254415.2022.9791785

Sahoo DN, Parhi PK (2018) Base isolation of residential building using lead rubber bearing technique. Int J Eng Res Technol (IJERT) 07(05) Bhubneswar, Odisha, India

Sanghai S, Gurmule C (2024) Seismic response control of RCC building with plan and vertical irregularities using dampers. J Build Pathol Rehabil 9(1):9

Sanghai S, Pawade P (2022) Performance evaluation of friction dampers for building with soil-structure interaction. Mater Today Proc 60:194–210

Sekar T, Kadappan PL (2015) Seismic analysis of multistory buildings resting on normal and sloping grounds in different seismic zones with and without base isolator. Imanager's J Struct Eng 4l(1)

Shenton HW, Lin AN (1993) Relative performance of fixed based and base isolated concrete frame. ASCE J Struct Eng 119(10):2952–2968

Somwanshi MA, Pantawane RN (2015) Seismic analysis of fixed based and base isolated building structures. Int J Multidiscip Curr Res 3:2321–3124

Chapter 30
GIS Based Approach of Surface Transport Network Analysis

Keshav Gupta and Bhalchandra V. Khode

Abstract Smart cities are a planned future for all Indian cities, big and small. This requires the proper design of surface transportation systems. Modern technologies are required to develop and create national and express highways, as well as to connect internal roads. Geographic Information Systems (GIS) might improve any country's or city's transportation systems. Today, people often use Geographic Information Systems (GIS) to study and manipulate spatially-oriented data for effective traffic management. This review demonstrates how to study road transport networks using GIS technology. The Katol Region is a heavily populated region in Egypt, making it difficult to find certain places, especially during times of crisis. The capacity to react swiftly and efficiently to life-threatening circumstances is the most critical component of any emergency response system (ERS), whether for fire, police, or healthcare. The primary goal of this article is to provide an enhanced network analysis that uses GIS capabilities to determine the best route for healthcare service providers in the Katol metropolitan area to travel from an incident location. According to the conclusions of this article, the optimal route improved travel time by 22% when compared to the shortest route. In times of crisis, arriving quickly at the site of an accident is critical to saving lives. As a result, this paper demonstrated that in emergency situations, particularly in a heavily populated city like, the GIS best route algorithm outperforms the shortest route technique.

Keywords Geographic information systems · Network analysis · Urban transportation

---

K. Gupta (✉) · B. V. Khode
G H Raisoni College of Engineering [GHRCE], Nagpur 440016, Maharashtra, India
e-mail: keshav.gupta.mtechts@ghrce.raisoni.net

B. V. Khode et al. (eds.), *Innovations in Technologies: Pioneering Sustainable Infrastructure for a Resilient Future*, Sustainable Civil Infrastructures,
https://doi.org/10.1007/978-3-031-82133-2_30

30.1 Introduction

Geographic Information Systems (GIS) have become essential tools for modern transportation planners and managers. GIS-based tools for evaluating surface transport networks may provide insights for decision-makers, engineers, urban planners, and researchers alike. This multidisciplinary technique for analyzing, developing, and improving surface transport networks integrates transportation models with geographical data and spatial analytic tools (Longley et al. 2015). GIS-based surface transport network analysis entails collecting, organizing, analyzing, and displaying geographical data related to transportation networks and trajectories. GIS can readily manage and assess a wide range of data types, including road networks, traffic volumes, land use patterns, demographic information, and environmental factors. This comprehensive data integration allows stakeholders to understand the complex interaction of transportation networks and make informed decisions (Miller and Shaw 2001).The ability to model geographical links and assess network connectedness is a significant advantage of GIS-based analysis. Geographic information systems (GIS) enable us to identify critical nodes, assess accessibility, and assess the efficiency of transportation networks by superimposing various data layers such as road networks, public transit lines, and points of interest. The use of this spatial analysis in route optimization, congestion management, and infrastructure design improves surface transportation system performance (Karimi 2003). Furthermore, stakeholders may assess the consequences of different actions and policies using GIS-enabled dynamic modelling and simulation of transportation scenarios (Li and Shibasaki 2018). Geographic information system (GIS)-based simulations help to find the best way to use resources and look at possible outcomes, like making improvements to public transportation, making freight distribution networks work better, and building new highways (Wang and Fwa 2005).Furthermore, GIS-based technologies are essential for improving the resilience and safety of transportation networks. Using GIS, we can identify issue locations, prioritise safety enhancements, and better prepare for crises by merging real-time traffic data with accident records and infrastructure risk assessments. Implementing this preventive strategy strengthens the entire transportation system, improves problem handling, and reduces risks (Jha and Shortus 2003). Geographic information systems (GIS) are important not only for surface transportation network study in conventional urban regions, but also in more rural and remote locations. Geographic information systems (GIS) combine satellite imagery, GPS tracking, and mobile data collection to map and monitor transportation networks over vast geographical regions. A geographical perspective is crucial for achieving sustainable development, optimising transportation services, and reaching out to rural regions (Moeinaddini et al. 2019).Furthermore, GIS-based transportation planning techniques facilitate collaboration among stakeholders. Geographic information systems (GIS) facilitate communication among various stakeholders, resulting in consensus-building and participatory decision-making via the use of intelligible maps, dynamic visualisations, and user-friendly data portals (Goulias 2016). This inclusive strategy promotes social equity and environmental sustainability, ensuring that transportation solutions meet the needs and wishes of local communities. To summarise, the application of geographic information systems (GIS) has revolutionised surface transportation network analysis, enabling more informed planning and decision-making (Kwan and Hong 1998). Stakeholders may utilise GIS to enhance transportation networks, increase safety, and encourage sustainable development by combining geographical data, advanced analytics, and visualisation tools. With the ever-changing transportation scene, GIS will continue to play an important role in transportation planning, encouraging efficiency and innovation in surface transportation systems all over the world (Harvey 2008).

GIS-based road analysis provides the capacity to thoroughly investigate and improve road networks. This method leverages GIS capabilities. Roads, which form the cornerstone of transportation infrastructure, play an important role in the growth of urban regions, as well as the expansion of economic activity and social relations. By combining geographical data with analytical tools, transportation experts, urban planners, and legislators may be able to get a better knowledge of the effectiveness, efficiency, and security of road networks.

This technology allows stakeholders to analyze network connections, estimate traffic patterns, optimize routes, and show data related to roadways in a geographical manner. Geographic information systems (GIS) provide for a comprehensive knowledge of the intricate interactions that occur within road networks by merging various datasets such as demographic data, land use patterns, traffic volumes, and road geometry. Geographic information systems (GIS)-based road analysis may be useful in identifying bottlenecks, evaluating alternative routes, and assisting with evidence-based decision-making in order to prioritize infrastructure investments that would increase mobility, reduce congestion, and improve road safety.

Geographic information systems (GIS)-based road research is essentially a diverse set of technologies that may enhance transportation management, operations, and planning. Stakeholders may encourage more effective, robust, and environmentally friendly transportation systems by addressing the plethora of difficulties that road networks are now facing using geographical data and analytical approaches.

30.2 Related Works

In (Miller and Shaw 2001), the authors looked for a specialty hospital in Aurangabad, Maharashtra State, India, as well as the best means to get there. To determine which hospitals were closest to the user's present location, we integrated ArcGIS with Dijkstra's method, which computes the shortest route between two locations. We determined the shortest route solely based on road length, without taking into account traffic or road conditions. To improve healthcare management in the ALMOKATAM

Zone, located in southernkatol, Egypt, the authors of Karimi (2003) created a health-care emergency response system service application based on geographic information systems (GIS). We were able to determine the best course of action by taking into consideration the distance to the nearest medical facilities. The system used a web-based interface to collect data from databases and display satellite photos' location-based aspects. As a result, a variety of end users may participate in system administration tasks such as policy formulation and decision-making. They focused only on the distance. The authors of Li and Shibasaki (2018) explored Dijkstra's approach to shortest route analysis and showed an example of a GIS-based emergency response system. They also developed a web application using GIS, web services, and asynchronous JavaScript and XML (Ajax) to determine the optimum paths between incident sites and expert response team stations. Their proposed algorithm disregards traffic congestion and road conditions in favor of choosing the quickest route based solely on distance. The authors of Wang and Fwa (2005) used geographic information systems (GIS) to create a desktop-based emergency management and planning system for Delhi, India. The primary goal of this effort was to provide for rapid response to any emergency or catastrophe situation. NAVTEQ was responsible for maintaining a complex transportation network and providing real-time traffic data in India. We discovered viable routes to the event location using near-real-time traffic data, enabling us to avoid crowded roads and reduce reaction times. GIS skills used in various research projects included network analysis, proximity analysis, buffer analysis, and determining the quickest route using an origin–destination (OD) cost matrix. To help Ghana's National Fire Service (GNFS) make choices during a fire, the authors created a GIS-based emergency response service in the Kumasi Metropolis (Jha and Shortus 2003). The best route model accounted for travel time delays, road slope, distance, and time. Moeinaddini et al. (2019) published a paper outlining the early findings of a decision-support tool that simulates traffic congestion routing and gives a backup route in the event of an emergency during rush hour. In case of an emergency, the technology can forecast traffic patterns during rush hour and provide alternative routes to hospitals. The authors utilised the network analyzer tool in ArcGIS 9.3 to perform shortest route analysis using Dijkstra's technique. The goal of this research was to determine the optimal time and distance for an ambulance to go from the scene of an incident to the nearest hospital, as well as from the event location to the closest hospital, utilising multiple routes. The authors of Goulias (2016) presented an upgraded shortest route based on Dijkstra's approach. To avoid revisiting nodes, a more current version uses the stack structure to keep the initial node's variation consistent throughout the search. This considerably enhanced the ability to find the fastest way. This technique neglects other types of real-time traffic data by focusing only on travel time.

30.3 Methodology

Figure 30.1 depicts the research flow diagram for the proposed GIS-based approach for enhanced road network analysis. We conducted a six-stage approach, starting with data collection and preparation for analysis, which included the research area base map, road network data, healthcare service provider data, and historical traffic data. The next step involves georeferencing the base map of the research area. We will then construct a geodatabase to store the processed data. In the next phases, we will create the network dataset and topology. It is now time to apply the network analysis method to the Katol Region's road network.

30.3.1 Data Preparation

We will gather the following information during this phase: a basic map of the research region, data from the healthcare provider, road network preparations, and historical traffic data. This analysis focuses mostly on the Katol metropolitan region. The coordinates 30° 11′ 10″ North to 31° 27′ 50″ East may match the place in question. The Katol region, Egypt's largest metropolitan area, is the sixteenth largest in the world and Africa's most widespread urban zone. The region covers about 1,709 km^2 and has a population of 10,400 (Kwan and Hong 1998). Thekatol Governorate, which includes parts of the Giza and Qaliobia governorates, is home to over 20,500,000 people.katol, Egypt's capital, is a thriving city. People often associate katol with

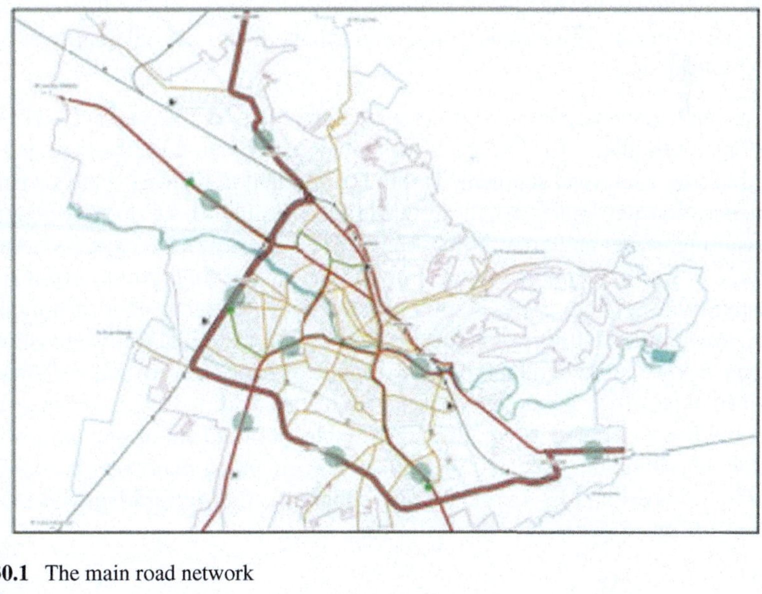

Fig. 30.1 The main road network

ancient Egypt due to its proximity to the renowned Giza pyramid complex and the long-lost city of Memphis. The Nile Delta area would be an ideal place for exploration. We created Katol using an OpenStreetMap (OSM) dataset. Subscribers to ArcGIS Online Service have free read-only access to OpenStreetMap, which serves as the basic map for GIS applications incorporated into ESRI software, such as ArcGIS Desktop. We acquired the road network of Katol using the Online ArcGIS Service. A road segment's name, length (in metres), direction (in minutes), and travel time (in TF_Minutes and FT_Minutes, respectively) make it easier to identify across networks. We chose OpenStreetMap as the data source for healthcare provider information. We stored the identities of healthcare providers and the services they provide in the database using the Name and Kind attributes. In Katol, Nagpur, the preparation of GIS data hinges on meticulous attention to detail, encompassing the collection and integration of spatial coordinates alongside other pertinent details. From administrative boundaries to infrastructure networks, cadastral information to environmental features, the acquisition of precise geographic coordinates serves as the cornerstone for accurate mapping and analysis. Through the amalgamation of GPS surveys, satellite imagery, and ground-truthing efforts, stakeholders diligently curate a comprehensive dataset reflective of Katol's dynamic landscape, ensuring that each spatial entity is accurately georeferenced and aligned within the GIS framework.

This meticulous approach to data preparation not only facilitates spatial analysis but also underpins informed decision-making processes across various sectors. By harmonizing diverse datasets and standardizing coordinate systems, the resulting GIS infrastructure empowers urban planners, policymakers, and researchers to leverage spatial insights effectively. With a robust foundation of georeferenced data at their disposal, stakeholders in Katol, Nagpur, can navigate the complexities of urban development, environmental management, and infrastructure planning with precision and confidence, fostering sustainable growth and enhancing the overall quality of life for the city's residents.

The last step in data preparation is to prepare traffic data for the road network. When compiling traffic data, it is beneficial to include information on average and peak travel times for each road segment. Traffic is important in network analysis because it influences journey times, which in turn affects results. If we do not account for traffic routing, travel and arrival times may be less accurate than expected. Another advantage of traffic planning is that it saves time by enabling you to avoid slower, congested routes. There are two sorts of ways to store traffic data: historical and current. We stored the traffic data for this study using historical traffic statistics. We used the weekly pattern of travel speeds to replicate historical traffic data. To that end, it is fair to expect that, given two consecutive instances of the same day in different weeks, the travel speeds of two different stretches of road should be quite similar. We often determine the expected velocities by aggregating many observations over a specific time period, like a year. We also adhered to the ArcGIS Network Analyst requirements when generating the historical traffic data.

Table 30.1 Daily profiles table structure

Field	Data type	Notes
Object ID	Long	A unique number assigned to each entry in the database
SpeedFactor_0000 to SpeedFactor_2300	Double	Various periods of the day indicate various free-flow scale factors

30.3.2 Geo-Processing of Toposheet

At the moment, we are georeferencing the received road network data. The georeferencing approach (Miller and Shaw 2001) facilitates the registration of the digital top sheet with the planet's surface. This is an important stage since it defines how trustworthy the road network data is.

30.3.3 Creation of Geo-Database

ArcGIS uses the Geo-database as its primary data format for editing and managing data, serving as the software's native data structure. You may have a personal, file, or commercial geodatabase. ARCGIS constructs a bespoke geo-database as the foundation for this recommended technique. A personal geo-database allows you to easily store, retrieve, and manage any kind of data, whether it is geographical or not. It will include information on healthcare providers, road networks, and traffic tables. Earlier in the approach, we discussed data from the road network and healthcare providers. The DailyProfiles and Streets_DailyProfiles tables housed the traffic information. You may find the daily speed profiles in the "DailyProfiles" section (Sect. 30.1). We break the day into one-hour intervals (Table 30.1).

30.3.4 Building Network Topology

To achieve significant discoveries and analysis, we must construct a road network topology to identify and correct any errors in the data. We applied the following topological criteria to achieve this: the road network must be free of dangles, and the roads must not overlap or link to one another.

30.3.5 Creating a Database for Networks

You can use the rectified road network defects as a dataset for network analysis. The two previously created tables and a geo-database containing a line feature class are required to generate a traffic statistics-based network dataset. One potential method for delineating the road network is by preserving the line feature class within a feature dataset. The traffic tables, in addition to providing traffic statistics, also include a connection to the road network. The network dataset facilitates an effective presentation of the transportation network. We generated the network dataset in Fig. 30.4 using the Network Analyzer plugin of ArcGIS for Desktop. Junctions denote the routes that connect two edges, whereas borders denote the routes that agents will traverse.

30.3.6 Conducting a Review of the Network

Using the ArcGIS Network Analyst Extension, we were able to examine the road networks. We assess the institution's proximity, directions, and service area (Miller and Shaw 2001). We dynamically simulate road network attributes such as daily traffic patterns, speed limits, and turn restrictions. By employing the widely recognised Dijkstra's method, the ArcGIS Network Analyst Extension determines the shortest path between any two network nodes. We also conducted network analyses to identify the most efficient routes and the closest facilities.

30.3.7 Best the Route Analysis

The optimum route analysis computes the most efficient path between two specified locations using current network traffic statistics and time of day. The network analyst plugin simplifies the process of configuring parameters for recurring analysis. To get different results depending on the day, you can change the following parameters: add travel time as an impedance factor; change the time at which the travel starts; limit the analysis to a single direction (for example, one-way or two-way road directions); and ignore invalid network locations that could cause the analysis to fail. We adjusted the optimal route analysis parameters and utilised the optimal route solver software to determine the most efficient path between two locations once the beginning and ending points had been determined. Figure 30.5 illustrates the optimal route from the beginning point (Location 1) to the terminal (Location 2).

30.3.8 *Closet Facilities Analysis*

The closet facilities analysis, taking into account route time and traffic information, pinpoints the nearest facilities accessible within a specific timeframe from the incident location. Knowing which facilities are closest to the accident site can help save lives in an emergency by reducing the amount of time, energy, and resources required to react (Karimi 2003). The network analysis addon makes the closet facility analysis simple to configure. You can specify the impedance factor, start time, reach time, number of facilities to identify, and journey direction.

30.4 Results and Discussion

This investigation juxtaposes and contrasts the outcomes of two distinct network analysis methodologies. By solving for the selected impedance factor, one can determine the optimal route (i.e., the least transit time) or the shortest path (i.e., the shortest route) between two points. The conclusion displays the quickest route from El-Maadi on the Autostrad Road inkatol, Egypt, to New El-Marg on the Ring Road. To ensure that our 3:00 PM departure time coincides with the evening traffic hour on the roads in the Katol region, the road's length will be considered throughout this investigation. The cumulative travel time of an agent increases in proportion to the length of the road segments that comprise their shortest route. The aggregate of the route durations identified through the shortest route analysis represents the time, measured in minutes, that agents will allocate to complete each individual segment of the route. Figure 30.2 illustrates the visual presentation of the quickest path. Analogous to the investigation into the quickest route, the week-long analysis considers the impedance component of road travel time. We measured the two highways at the same two locations for seven consecutive days, from Saturday to Friday. Table 30.2 presents the average travel time (in minutes) and total distance (in kilometres) (Fig. 30.3).

In order to establish the superiority of the optimal route over the shortest one, we performed identical research for both routes over a span of 120 min, beginning at 8:00 AM and ending at 12:00 AM on each day of the week. We omitted them from our calculations because there was little vehicular activity on the road network from 1:00 AM to 7:00 AM. Consequently, during specific time periods, the distinction between the shortest and optimal routes is negligible. Table 30.3 presents the average trip timings, expressed in minutes, for each analysis.

Figure 30.5 shows a graphic depiction of the results. When evaluating two alternatives, the superiority percentage (SP) determines which is preferable. It suggests that we should prioritize the ideal route over the shortest one for our purposes. We use the following formula to compute the superiority percentage for the ideal route travel time:

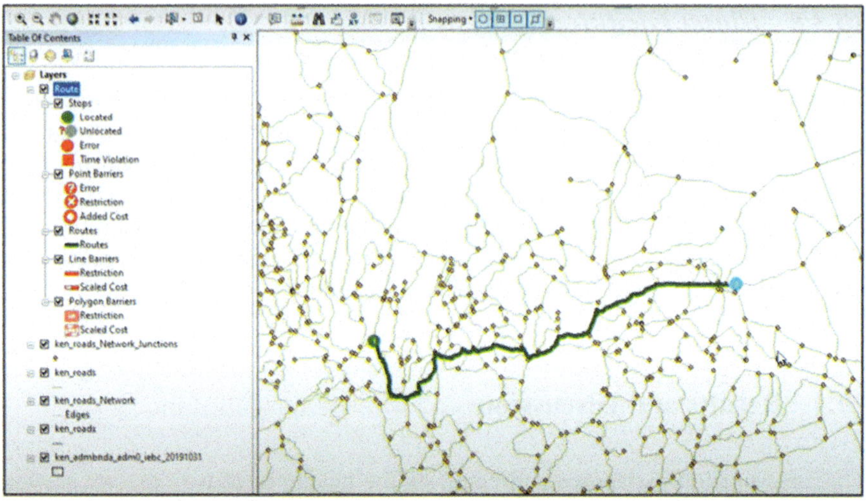

Fig. 30.2 The best route analysis results

Table 30.2 Analyze the times and lengths of the best and shortest routes

Day	Shortest route		Best route	
	Length	Time	Length	Time
Saturday	50.85	91	55.15	89
Sunday	50.85	249	55.15	213
Monday	50.85	339	55.15	215
Tuesday	50.85	173	55.15	153
Wednesday	50.85	206	55.15	165
Thursday	50.85	324	55.15	309
Friday	50.85	118	54.75	91
Avg Time		214.29		176.43

$$SP = [((\sum_{i=1}^{9} SRTi \backslash BRTi) - \frac{1}{9} * 100$$

Then, SP represents the percentage of superiority. Shortest Route Time (SRT). Best Route Time (BRT). By entering the SRT and BRT numbers from Table 30.3 into the previous computation, we obtain SP = 24%. This is a 24% reduction in travel time between the best and shortest routes. In order to save lives in emergencies, it is critical to get to the scene of an accident as soon as possi. Therefore, in emergency situations, it is crucial to choose the optimal route over the shortest one, as demonstrated by the superiority percentagage.

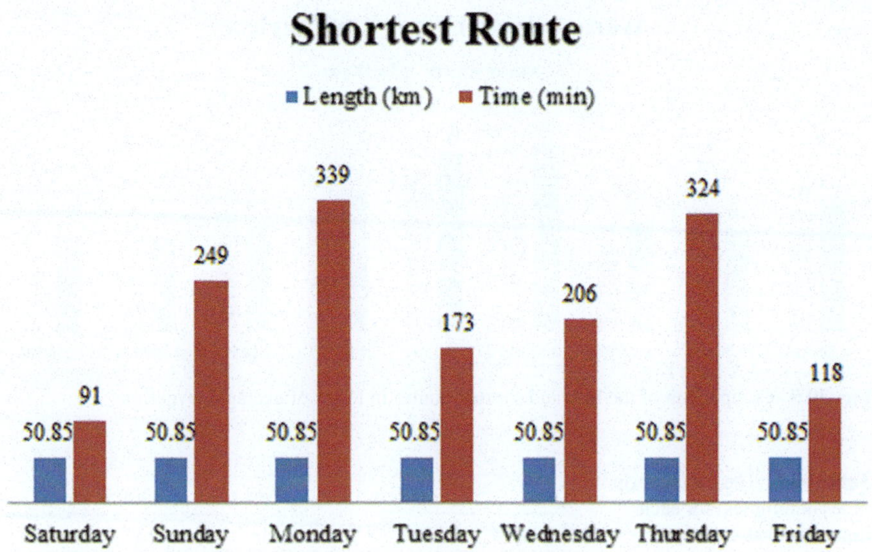

Fig. 30.3 The shortest route analysis is performed around 3:00 PM on weekdays

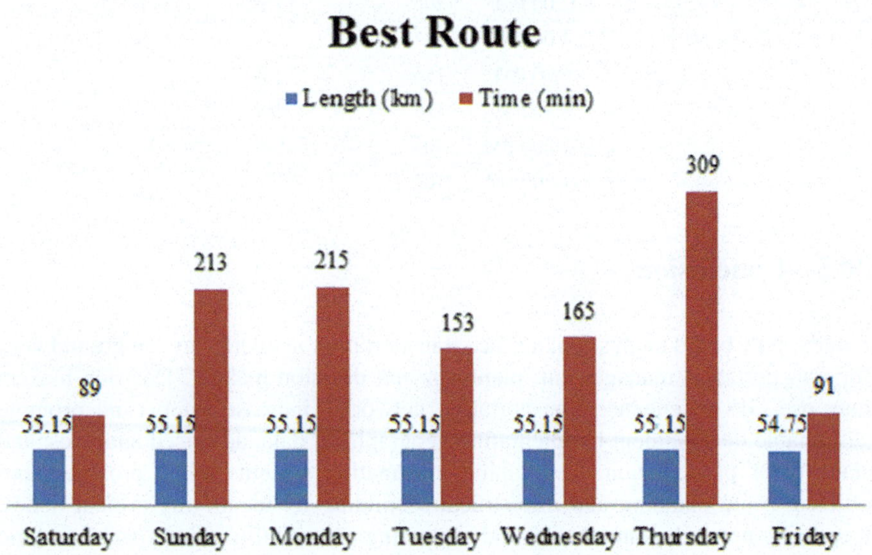

Fig. 30.4 The best route analysis is around 3:00 PM on weekdays

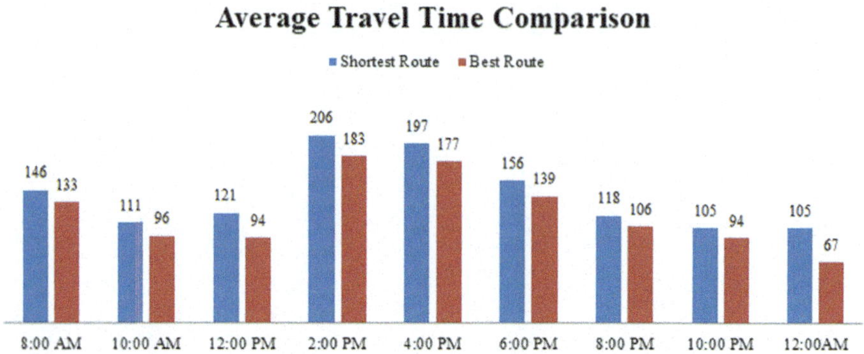

Fig. 30.5 Comparison of the best and shortest routes in terms of average travel time

Table 30.3 The average trip time (in minutes) for each time slice was calculated using the best and shortest route analysis techniques

Time	Shortest route	Best route	Difference (%)
8:00 AM	146	133	13
10:00AM	111	96	20
12:00 PM	121	94	35
2:00 PM	206	183	16
4:00 PM	197	177	14
6:00 PM	156	139	16
8:00 PM	118	106	15
10:00 PM	105	94	16
12:00AM	105	67	72

30.5 Conclusion

Finally, GIS-based analyses of surface transportation networks lay the groundwork for transportation management, planning, and decision-making. Decision-makers may use GIS to enhance transportation networks, increase safety, and promote sustainable development by integrating geographical data, advanced analytics, and stakeholder participation. Geographic information systems (GIS) provide more efficient route planning and infrastructure building by identifying critical nodes, measuring network connectivity, and evaluating accessibility using spatial analytic methods. GIS makes transportation networks more efficient and robust because of its predictive modelling capabilities, which allow stakeholders to plan ahead, analyse the consequences of actions, and effectively allocate resources. In addition to assisting with urban planning and management, GIS is effective in rural and remote areas, allowing for the administration of transportation networks in a range of contexts. For all communities, GIS provides critical insights into traffic patterns and accessibility, allowing them to improve services and boost connections. GIS promotes involvement and collaboration by facilitating easy access to data portals, interactive visualisations, and maps. By paying attention to what the locals want, this participatory strategy ensures that transportation solutions are equitable and sustainable. Global information systems (GIS) will continue to raise the bar for innovation in the transportation sector, resulting in more dependable and effective surface transportation networks around the world. To improve transportation for both present and future generations, stakeholders may make informed decisions by using GIS tools and methodologies to better understand complex transportation dynamics and build safe, efficient, and environmentally friendly systems. This study employs an upgraded GIS-based network analysis to investigate the road network in Katol. Afterwards, it determines which healthcare providers are physically closest to an event location by taking the shortest route between two sites on the road network. The suggested technique improves the findings' applicability to actual road networks by including historical traffic data in its study. In network research, the Dijkstra optimal routing algorithm of the ARCGIS program performed very well in densely populated areas such as katol. This method might save travel time by 20 to 22%, depending on the distance. To increase the accuracy of subsequent investigations, we advise, wherever possible, to use real-time traffic data rather than previous data. It is important to consider not just the total duration of travel but also the kind, quality, and width of the route. Additionally, we want to enhance the Dijkstra routing algorithm of the ArcGIS network analysis extension in order to maximise its potential.

References

Goulias KG (2016) Transportation network analysis. In handbook of regional science, pp 1–26, Springer

Harvey F (2008) GIS-based analysis of spatial patterns of road accidents in South Africa. J Transp Geogr 16(2):147–158

Jha MK, Shortus MW (2003) GIS and transportation: the applications. J Transp Stat 6(1):27–48

Karimi HA (2003) Geographic information systems: concepts, methodologies, tools, and applications. Information science reference

Kwan MP, Hong S (1998) Network-based geographic information systems for transportation planning: an overview of basic principles. Comput Environ Urban Syst 22(5):441–449

Li Z, Shibasaki R (2018) GIS-based transportation demand analysis: computational methods and applications. Springer

Longley PA, Goodchild MF, Maguire DJ, Rhind DW (2015) Geographic information systems and science. Wiley

Miller HJ, Shaw SL (2001) Geographic information systems for transportation: principles and applications. Oxford University Press

Moeinaddini M et al (2019) GIS-based surface transport network analysis for sustainable development: a review of methods and applications. Sustainability 11(4):1028

Wang JJ, Fwa TF (2005) Transportation network analysis. Wiley

Chapter 31
Design of Overlay Bituminous Pavement Layer Using Benkelman Deflection Technique

Keshav Gupta and Bhalchandra V. Khode

Abstract Road networks require meticulous construction of overlay bituminous pavement layers to ensure their optimal performance and long-term persistence. The Benkelman deflection method is an important instrument for determining the structural integrity of pavements. Engineers can accurately measure the stiffness of the present pavement and locate areas that are prone to distress by measuring the pavement's deflections under a standard wheel load. This technique allows for the adjustment of overlay design to prevent distresses such as rutting and cracking. It also allows for more informed decisions about the thickness and composition of overlay materials. The Benkelman deflection technique in pavement design processes may increase infrastructure resilience and service life, allowing road authorities to construct safer and more sustainable transportation networks. Effective pavement design is one of the most crucial aspects of project design. For vehicles, the pavement is the most visible part of the road. People often use the pavement's smoothness or roughness to determine the road's condition and usefulness. Poor pavement conditions may result in greater user costs, longer travel times, increased fuel and brake consumption, more frequent vehicle maintenance repairs, and a higher chance of accidents. The pavement will perform better in the long term if the materials have stronger strength and stiffness properties. To improve the performance of existing materials, pavement design should focus on their efficiency, cost-effectiveness, and effective use. Ecologically friendly approaches must restore unmanaged broken pavements, ensuring the preservation of roads for future generations. Understanding the various types of distress, material situations, and environmental variables may aid in the development of rehabilitation techniques.

Keywords Bituminous · Distresses · Strength · Sustainable · Transportation

K. Gupta (✉) · B. V. Khode
G H Raisoni College of Engineering [GHRCE], Nagpur, Maharashtra 440016, India
e-mail: keshav.gupta.mtechts@ghrce.raisoni.net

B. V. Khode et al. (eds.), *Innovations in Technologies: Pioneering Sustainable Infrastructure for a Resilient Future*, Sustainable Civil Infrastructures,
https://doi.org/10.1007/978-3-031-82133-2_31

31.1 Introduction

Road infrastructure is the bedrock of modern society because it facilitates mobility and commerce. We must use effective pavement design and maintenance practices to ensure the lifetime and safety of road surfaces. There are several methods Pavement engineering employs several methods, but the Benkelman deflection methodology stands out as a promising approach for evaluating pavement condition and enhancing overlay design (Huang et al. 2018). Chnique is a well-known method for measuring pavement deflections under load. Arthur Newell Talbot Benkelman, the creator of the approach, inspired its name. Engineers perform it by using portable testing equipment known as the Benkelman beam, which provides critical information on the stiffness of the pavement and the behavior of the structure (Benkelman 1954). As road networks age and traffic volumes increase, it becomes more difficult to maintain decent pavement conditions. Rutting, cracking, and fatigue are examples of stresses that may damage the structural integrity of pavements, resulting in higher maintenance costs and increased public safety issues (Darter and Wimsatt 1962). When it comes to detecting distress and choosing suitable overlay designs, typical pavement evaluation methods often lack the precision and efficiency required. The Benkelman deflection technique, which offers a systematic approach to pavement assessment (Kennedy and Theade 2003), enables engineers to make data-driven decisions for pavement rehabilitation and overlay design. Benkelman deflection is a technique that provides important information on the structural stability of pavements. This approach measures the deflections of the pavement under a standard wheel load. We can use this information to pinpoint vulnerable areas of the pavement and establish the ideal thickness and composition of the overlay bituminous layers (Masad et al. 2010). Overlay bituminous pavement layers are crucial for extending the life and usefulness of existing pavements. Proper development and execution of overlay layers can reduce road surface distress, extend the service life of road surfaces, reduce maintenance costs, and enhance road safety (Monismith and Brown 2006). There are many benefits to using the Benkelman deflection technique in pavement design procedures. These benefits include improved pavement assessment accuracy, a better understanding of pavement dynamics, and optimal overlay design solutions (Raab and Baladi 1999). This article will examine how to construct overlay bituminous pavement layers using the Benkelman deflection technique, as well as its potential applications. The goal of this study is to explore the effectiveness of the Benkelman deflection technique in terms of improving pavement performance and supporting sustainable transportation infrastructure (Asphalt Institute 2001). We will achieve this by studying the existing literature and case studies. This study aims to provide significant insights to pavement engineers, researchers, and practitioners engaged in the design and management of road infrastructure. We achieve this by thoroughly examining the concepts, techniques, and practical applications of the Benkelman deflection technique. In summary, the construction of overlay bituminous pavement layers employs the Benkelman deflection method, a promising approach to pavement engineering. It offers a methodical and data-driven approach to enhancing pavement performance and ensuring the long-term sustainability of transportation networks (AASHTO 2019).

Overlay bituminous pavement layers play an important role in road network rehabilitation and maintenance. They offer a cost-effective and efficient solution that extends the pavement's service life while also improving driver comfort (American Society of Civil Engineers (ASCE) 2020). Roads will ultimately succumb to a range of distresses, such as cracking, rutting, and surface deterioration, when faced with the persistent assault of traffic loads, weather variations, and age-related processes. Engineers use overlay bituminous layers, which consist of asphalt-based materials applied over existing pavements, to address these issues and restore pavement performance (Asphalt Pavement Alliance n.d.). When it comes to the arsenal of approaches for creating and analysing overlay bituminous pavement layers, the Benkelman Deflection Technique stands out as a time-tested method. People highly respect it for its ease of use and effectiveness in forecasting overlay thickness requirements and evaluating the structural integrity of the pavement (Federal Highway Administration (FHWA) 2001). The Benkelman Deflection Technique, named for its author, Arthur Newell Talbot, and popularised by Dwight D. Eisenhower's interstate highway system, provides a simple and instructive method for pavement inspection. The approach's core component is measuring the vertical deflection of the pavement surface under a static load. Engineers often perform this examination using a specific instrument known as the Benkelman beam (Huang 2004). By quantifying the response to loads, engineers may gain valuable insights into the pavement's structural integrity, carrying capacity, and sensitivity to distress. Engineers may utilise this data to create unique solutions that meet performance objectives while also improving cost efficiency, which serves as a foundation for the design and optimisation of overlay bituminous pavement layers (Karamihas and Barenberg 1997). Furthermore, the Benkelman Deflection Technique contains a comprehensive approach to pavement assessment and repair planning that extends beyond the act of measurement. By may locate isolated areas of weakness, uncover underlying causes of pain, and choose treatment procedures based on these results by carefully assessing deflection data along specified test sections. Engineers use the data from the Benkelman Deflection Technique to develop complex overlay designs that target specific pavement flaws and optimise resource allocation. Among other things, we can use these designs to reduce reflective cracking, subgrade instability, and load-induced fatigue (National Asphalt Pavement Association (NAPA) n.d.). Furthermore, the Benkelman Deflection Technique includes a broad set of instruments for pavement management and decision-making. This technique allows engineers to assess the efficacy of maintenance actions, monitor pavement performance over time, and verify the validity of design assumptions (Pavement Interactive n.d.). By undertaking frequent deflection measurements and ongoing data analysis, transportation authorities may improve their pavement management plans, fine-tune their repair schedules, and allocate resources more effectively. This allows them to extend the pavement's service life while lowering the costs associated with its life cycle (Transportation Research Board (TRB) 2001). In essence, overlay bituminous pavement layers are more than just a surface solution for ageing infrastructure; they represent a proactive approach to pavement preservation and asset management. Engineers can confidently administer specific treatments, accurately identify pavement ailments, and preserve our transport networks from the ravages of time and use by using procedures such as the Benkelman Deflection Technique (United States Department of Transportation (USDOT) n.d.). The utilisation of overlay bituminous pavement layers while negotiating the complex terrain of modern transportation demonstrates our ingenuity, resilience, and commitment to building a safer, smoother, and more ecologically conscious future (Witczak et al. 2013).

31.2 Techniques for Design of Overlay Bituminous Pavement Layer

1. **Falling Weight Deflectometer (FWD)**: Similar to the Benkelman Beam, the FWD measures the deflection of pavement under a dynamic load. It's a more advanced device that applies a dynamic load to the pavement surface and measures the resulting deflections. FWD data is used to assess pavement structural capacity and determine overlay thickness requirements.
2. **Ground Penetrating Radar (GPR)**: GPR is a non-destructive testing method that uses radar pulses to image the subsurface layers of the pavement. It provides valuable information about layer thicknesses, pavement structure, and the condition of underlying layers. GPR data can inform overlay design by identifying areas of distress, delamination, or debonding.
3. **Dynamic Cone Penetrometer (DCP)**: The DCP is a portable device used to assess the strength and stiffness of subgrade and base materials. It measures the penetration resistance of the soil or aggregate layers under a standard dynamic load. DCP testing helps in evaluating the support provided by the underlying layers and determining the need for subgrade stabilization or improvement.
4. **Pavement Condition Index (PCI)**: PCI is a numerical rating system used to evaluate the overall condition of a pavement based on the type, extent, and severity of distresses observed. PCI surveys involve visual inspection of pavement distresses, such as cracking, rutting, and surface roughness. The data obtained from PCI surveys can guide overlay design by identifying areas in need of rehabilitation and determining the appropriate overlay thickness.
5. **Pavement Management Systems (PMS)**: PMS software integrates various data sources, including pavement condition surveys, traffic data, and maintenance records, to assist in pavement management and decision-making. PMS tools facilitate long-term planning for pavement maintenance, rehabilitation, and overlay projects by prioritizing treatments based on the condition, performance, and cost-effectiveness of different pavement sections.
6. **Mechanistic-Empirical (M-E) Design**: M-E design methods use mathematical models to predict pavement performance based on a combination of mechanistic principles and empirical data. These models consider factors such as traffic

loading, material properties, climate conditions, and pavement structure to estimate pavement distresses, fatigue cracking, rutting, and roughness over time. M-E design software helps engineers optimize overlay thickness design and pavement rehabilitation strategies to meet performance and budget objectives.

7. **Climate and Environmental Considerations**: Consider the local climate and environmental conditions when designing overlay bituminous pavement layers. Factors such as temperature variations, freeze–thaw cycles, moisture infiltration, and environmental regulations can affect pavement performance and durability. Incorporate appropriate material selection, drainage design, and construction practices to mitigate climate and environmental impacts and enhance pavement longevity.

31.3 Field and Laboratory Investigation

1. **Dynamic Cone Penetration Test**

The Dynamic Cone Penetration Test (DCPT) is a critical component in the design of overlay bituminous pavement layers because it provides significant information on the properties of both the soil and the pavement layer. Using the DCPT, engineers may make intelligent decisions concerning the thickness and composition of overlay materials. Engineers perform this by determining the strength and compactness of the underlying subgrade and pavement layers. With this information, we can adjust overlay designs to withstand expected traffic loads and climatic conditions, leading to a more durable and efficient pavement structure. The introduction of the DCPT into pavement design procedures allows for more accurate and reliable assessments, resulting in the creation of overlay bituminous pavement layers that are suited to the specific demands of each project. This, in turn, encourages the creation of safer, more ecologically friendly road infrastructure.

2. **In-Situ Moisture Content**

The in-situ measurement of moisture content is critical in the design of overlay bituminous pavement layers. This is because moisture content has a direct impact on the behaviour and lifespan of the pavement structure. Engineers may assess the stability of the existing pavement and the layers underneath it by evaluating the moisture content of the pavement. This allows them to assess the potential risks associated with moisture-induced distresses such as rutting, stripping, and premature deterioration. If may make informed decisions about the selection of appropriate materials and building processes to prevent moisture-related issues during overlay construction if they properly measure the moisture content of the material in the field. Finally, by designing and constructing the overlay bituminous pavement layers to withstand variations in moisture and weather conditions, this proactive technique enhances the long-term performance and service life of the pavement network (Tables 31.1 and 31.2).

Table 31.1 Test for field moisture

	1 + 000	2 + 000	3 + 500
Moisture	20%	10%	8%

Table 31.2 Test for in-situ density using the core cutter technique

	1 + 000	2 + 000	3 + 500
Water content	20%	10%	8%
Dry density	1.62	1.79	1.78

31.4 Bituminous Mix Design

1. **Specific Gravity Test**

The specific gravity test is an important step in developing overlay bituminous pavement layers. This test allows you to assess the density of the asphalt binder, aggregates, and other components in the pavement mix. Engineers may use specific gravity to assess the quality and uniformity of materials, ensuring that the pavement operates optimally and lasts a long time. This information is critical for determining mix proportions and obtaining the required properties, such as stability and deformation resistance. The introduction of specific gravity testing into the design process allows engineers to improve overlay designs, resulting in more resilient and long-lasting pavement structures (Table 31.3).

2. **Penetration Test**

The penetration test is an important component in the process of generating overlay bituminous pavement layers because it provides crucial information about the consistency and hardness of the asphalt binder. Under certain conditions, engineers can determine the viscosity and consistency of an asphalt binder by measuring the depth to which a standard needle penetrates it. Under these circumstances, these qualities are critical to pavement performance. The goal of this test is to help engineers choose suitable binders and determine whether or not they are appropriate for a given environment and traffic conditions. This ensures that the overlay bituminous pavement layer is both resilient and long-lasting (Table 31.4).

3. **Softening Point Test**

Table 31.3 Specific gravity of various binders

Binder type	VG-30
Sp. gravity	1.05

Table 31.4 Penetration values of various binders used in the study

Binder type	VG-30
Penetration	72

Table 31.5 Softening point values of different binders

Binder type	VG-30	CRMB-60
Softening point	62	75

The softening point test is an important step in the process of developing overlay bituminous pavement layers because it gives vital information about the temperature susceptibility of asphalt binders. Engineers may assess the binder's durability to high temperatures and deformation under traffic loads by measuring the temperature at which the binder softens and deforms under controlled conditions. This information is required in order to choose binders that can endure the environmental conditions and traffic levels in a certain region. This will guarantee that the overlay bituminous pavement layer lasts for a long time and functions properly (Table 31.5).

4. **Aggregate**

Aggregate Impact Test: Aggregate impact value = 15%.

5. **Specific Gravity and Water Absorption**

See Table 31.6.

Bitumen Adhesion Test: The aggregate Coating Value = 97%.

6. **Shape Tests**

See Table 31.7.

7. **RAP Usage in the Mix**

Reclaimed asphalt pavement, often known as RAP, refers to materials that have been removed and/or treated and include asphalt mixed with aggregates. We form these

Table 31.6 Specific gravity and water absorption of aggregates

Location	Size (mm)	Specific gravity	Water absorption (%)
Local material	25	2.72	0.48
	20	2.76	0.56
	10	2.73	0.78
	Dust	2.59	1.85

Table 31.7 Test results of elongation and flakiness of aggregates

Location	Size (mm)	Flakiness (%)	Elongation (%)	Combined flakiness and elongation indices (%)
Local materials	25	15.245	22.945	41.8
	20	12.545	39.445	55.7
	10	74.845	65.145	143.7

materials during the process of removing asphalt pavements to construct, resurface, or gain access to underground services. High-quality, well-graded aggregates undergo proper crushing and screening before coating with asphalt cement to form RAP. RAP does its job at the DBM and WMM levels, also known as the granular layers. It is possible that high RAP content combinations will provide distinct hurdles in terms of workability and compactness. The DBM mix restricts the use of RAP to 15%, while the WMM mix allows it to surpass 35%. The Marshall Mix design.

31.5 Marshall Mix Design Procedure

When it comes to developing asphalt concrete mixes, the Marshall mix design methodology is a popular strategy, especially for building flexible pavements. We must determine the optimal asphalt binder concentration and aggregate gradation to achieve the desired performance characteristics, such as stability, durability, and deformation resistance. The method consists of many steps, which might include:

- Aggregate Selection: The first part of the process entails selecting aggregates that meet specific, predefined gradation, shape, and quality standards. We often divide aggregates into three fractions: coarse, fine, and mineral filler. Each of these fractions has a specific role in the combo's overall performance.
- Selection of an Asphalt Binder The next step is to choose an appropriate asphalt binder based on a range of characteristics, such as climate, traffic volume, and pavement performance requirements. The binder must have the necessary properties to give adequate adhesion, flexibility, and resistance to the effects of ageing and temperature fluctuations.
- A Marshall compactor compacts a mixture of aggregate and asphalt binder in a cylindrical mould to prepare Marshall specimens. This is known as the Marshall specimen preparation technique. After this, we expose the compacted specimens to a specific number of blows to achieve the required density.
- A Marshall stability testing machine applies compressive loads to compacted specimens to assess their stability and flow. The objective is to evaluate the stability and effectiveness of the specimens. The mixture's stability is its resistance to deformation while under stress, whereas the flow of the mixture displays its ability to deform and compress when exposed to traffic.
- We manufacture and test many specimens with different asphalt binder concentrations to determine the ideal binder composition. We determine the appropriate binder content based on the specimen's stability and flow properties. The idea is to find a balance between the specimen's strength and workability.
- Analysis and Verification: After identifying the optimal binder content, we scrutinize the mix design to confirm its alignment with the project parameters and performance criteria. Field conditions and the engineer's discretion may lead to adjustments to the mix proportions or binder quantity.

Table 31.8 Marshall test properties

Bitumen %	4	4.5	5	5.5	6.0
Density	2.45	2.47	2.491	2.503	2.492
Max specific gravity	2.718	2.691	2.682	2.664	2.646
Va	7.75	6.95	5.35	4.15	3.87
VMA	14.97	14.7	14.39	14.4	15.27
VFB	47.785	59.535	67.705	77.955	81.755
Stability (correction)	890	972	1085	1016	943
Flow in mm	4.2	4.5	4.9	5.2	5.4

- Control of Quality: We use quality control procedures to ensure that asphalt mixtures are uniform and meet design specifications. These precautions include evaluating the asphalt mixes at regular intervals throughout production and installation (Table 31.8).

31.6 Pavement Design

1. **Design of Flexible Pavements**

To ensure the longevity and usefulness of flexible pavements, it is vital to have a thorough understanding of the many factors involved in their design. The quantity of traffic, soil properties, temperature, and material selection are all essential considerations. Flexible pavements generally consist of numerous layers. These strata consist of a subgrade, subbase, base, and surface course. Each of these levels has a specific function in terms of load distribution and structural support. Engineers employ sophisticated methodologies throughout the design process. One example is the mechanistic-empirical (M-E) technique, which considers pavement structural reactions under various loads and environmental conditions. This approach enables the optimization of pavement thickness and material properties to withstand expected traffic loads and environmental conditions. Furthermore, advancements in pavement design tools and predictive models have enabled engineers to simulate a wide range of situations and assess the performance of flexible pavements throughout their design life. This helps to build infrastructure solutions that are both cost-efficient and ecologically beneficial. Flexible pavements can withstand the rigours of traffic and environmental conditions while still providing road users with a safe and pleasant travel experience. Meticulous planning and the application of innovative design concepts make this feasible. It was originally stated that flexible pavements are designed using the concepts mentioned in IRC 37:2012.

(1) *Irc 37:2012*

The Indian Roads Congress has established techniques for creating flexible pavements with CBR values. Only pavement designs from the 38-1984 edition of the IRC

Standards could feasibly accommodate traffic projections with up to 30 million standard axles (msa). Furthermore, the preceding code's empirical nature limits its generalisation and application. According to these characteristics, a new set of designs as well as analytical designs up to 150 msa were generated. These concepts may be used to guide the design of flexible pavements for a variety of road types, including motorways, national highways, state highways, essential district routes, and others. We are all aware that flexible pavements are bituminous surfaces, granular bases, and sub-base courses that meet the IRC and MOST requirements. Our aim is to guarantee that the new pavements meet these standards. The M-E pavement design approach simplifies analysis by representing the pavement as an elastically assumed layered structure. His construction consists of three or four horizontal layers: bituminous surface, base, subbase, and subgrade. The thickness, elastic modulus, and Poisson's ratio are the three qualities that define a layer. FBFatigue, rutting, and low-temperature cracking are often recognised as the key failure mechanisms in ituminous pavement structures. The Indian Roads Congress method uses an empirical approach to compute the pavement thickness value based on the subgrade's CBR value. By factoring in the CBR value and the cumulative standard axle load, the design table can determine the total pavement thickness. Our pavement is made up of four layers: the subgrade, the bituminous course, the WMM layer, and the DBM layer. Each of these stages needs examination.

2. **Benkelman Beam Method**

Using this method, we analyse the greatest deflection that occurs on the pavement at regular intervals and then build the pavement overlay. This approach employs the Benkelman beam method for experimental investigation. The goal of this test procedure is to estimate a pavement's rebound deflection when exposed to a typical wheel load and tire pressure, with or without temperature measurements.

3. **Equipments**

The Benkelman beam, a portable testing instrument specifically for this purpose, is one piece of equipment required to detect deflection in overlay bituminous pavement layers using the Benkelman Deflection Technique. Additionally, we affix a deflection gauge to the beam to measure the vertical displacement the pavement surface experiences under load. We apply the load using a standard weight that typically ranges between 8,000 and 18,000 pounds. The structural properties of the pavement and the design specifications govern this range. Marking tools serve to identify measurement points throughout the pavement section, while safety gear ensures the safety of workers during testing procedures. Regular calibration of testing equipment ensures the precision and reliability of readings. Engineers may utilise these sophisticated tools and equipment to effectively analyse the deflection of the pavement and make well-informed decisions on the thickness and composition of the overlay in order to maximise the pavement's performance and longevity.

4. **Procedure: Measurement of Deflection**

The Benkelman Beam Deflection Test is a critical procedure for assessing the structural behavior of flexible pavements. Let's dive into the step-by-step process for measuring deflections using the Benkelman Beam.

5. **Equipment Setup**

The Benkelman Beam consists of a slender beam that pivots at a distance of 2.44 m from the tip.

The following equipment is needed:

- The Benkelman Beam has a locking mechanism and a vibrator.
- We distribute a vehicle or trailer with two wheels, twin tyres, and an axle weight of 8.20 to 0.15 tonnes equally.
- The tread pressure gauge has graduations of 20 kPa or lower.
- The readout range is 0–6 °C, with a precision of 1 °C.
- The equipment includes a 13-mm-diameter mandrel for drilling a 100-mm-deep hole in asphalt.
- This thermometer may be filled with glycerin or oil.

1. ***Test Point Selection***
    - Preselect and mark the test point on the pavement.
    - For highway pavements, use the distances from the lane edge.
2. ***Test Procedure***
    - Position the vehicle 100–150 mm behind the test area (position A), with the test wheel pointing rearward.
    - Insert the Benkelman Beam probe, toe first, between the two tires of the test wheel.
    - After removing the locking mechanism, adjust the beam's rear so that the dial gauge makes contact with the plunger.
    - You do not need to record the precise reading; just set the dial gauge to read between 9 and 11 mm.
    - Turn on the vibrator and creep forward with the truck.

The test wheel passes over the test spot.

i. Continue advancing to position B, which is 2.7 ± 0.1 m beyond the test spot.

- While traveling, keep an eye on the dial gauge's highest reading (START READING, S).
- When the vehicle comes to a halt with the test wheel in position B, the dial gauge displays the INTERMEDIATE READING (I).

For highway pavements, use the distances from the lane edge of 28 or less, 3.0, 3.2, 3.4, 3.6, or more lane widths (meters), to 0.5, 0.6, 0.7, 0.8, and 0.9 m from the edge. Mean deflection is 0.010255 mm and Standard deviation is 0.1432 mm and Characteristic deflection is 0.24191 (Table 31.9).

Table 31.9 Dial gauge readings initial (D_o), Intermediate (D_i), Final (D_F)

Sr. No	Change and identification of lane	Dial gauge readings initial (D_o)	Dial gauge readings intermediate (D_i),	Dial gauge readings final (D_F)
1	0	0.010	0.006	0.003
2	2.7	0.007	0.007	0.007
3	9	0.005	0.005	0.006
4	2.7	0.005	0.005	0.006
5	9	0.015	0.016	0.017
6	2.7	0.016	0.016	0.017
7	9	0.024	0.024	0.025
8	2.7	0.026	0.027	0.028
9	9	0.028	0.028	0.029
10	2.7	0.015	−0.066	−0.148
11	9	0.026	−0.070	−0.167

31.7 Conclusion

When designing overlay bituminous pavement layers, the Benkelman deflection technique may provide significant benefits to pavement engineers and transportation authorities. This technique provides helpful insights into the stiffness and structural behaviour of the pavement by correctly monitoring its deflections when under stress, allowing for more informed decision-making in asphalt overlay design. Engineers may use the Benkelman deflection technique to locate areas of pavement damage and determine the optimal overlay thickness and composition. The installation of overlay bituminous pavement layers may increase the longevity and performance of existing pavements by effectively addressing these distresses. This may extend the service life of the pavements while also lowering maintenance costs.

Furthermore, the Benkelman deflection technique employs a rigorous and data-driven approach, resulting in increased efficacy and accuracy in pavement assessment and repair operations. The final effect is better pavement conditions, safer streets, and increased transportation network sustainability. Future research will improve and expand on the application of the Benkelman deflection technique. We anticipate an increasing use of this approach in pavement design procedures. Pavement engineers may continue to improve overlay designs and ensure the long-term longevity of road infrastructure by combining this technique with advances in materials and technologies. In general, designing overlay bituminous pavement layers using the Benkelman deflection method is a promising approach to pavement engineering. This technology has the potential to significantly enhance pavement performance while also contributing to the creation of safer and more environmentally friendly transportation networks. We estimate the observations using the Benkelman beam deflection technique, apply it to the necessary stretch, and determine the design thickness of the overlay. To support the increasing axle load over the next 10 years, a twelve-centimetre-thick overlay of the existing pavement will be required. We will overlay the existing flexible pavement with bituminous concrete.

References

American Association of State Highway and Transportation Officials (AASHTO) (2019) Standard specifications for transportation materials and methods of sampling and testing. AASHTO

American Society of Civil Engineers (ASCE) (2020) Transportation & development institute. https://www.asce.org/transportation-institute/

Asphalt Institute (2001) MS-17 Asphalt overlays for highway and street rehabilitation. Asphalt Institute Manual Series

Asphalt Pavement Alliance (n.d.) Overlay design. https://asphaltroads.org/design-construction/overlay-design/

Benkelman ANT (1954) Measurement of deflection in flexible pavements. Highw Res Board Proc 33:344–366

Darter MI, Wimsatt AA (1962) Deflections of concrete pavements as influenced by temperature and load. Highway Res Rec 3(65):64–78

Federal Highway Administration (FHWA) (2001) Pavement preservation checklist series 7: microsurfacing and thin hot mix asphalt overlays. https://www.fhwa.dot.gov/preservation/pub/09097/

Huang YH (2004) Pavement analysis and design, 2nd edn. Pearson Prentice Hall

Huang Y, Zeng L, Zhang Z (2018) Advances in pavement engineering: trends and strategies. Springer

Karamihas SM, Barenberg EJ (1997) Implementation of the Benkelman beam deflection test on state and local roads in South Carolina. J Transp Eng 123(6):464–468

Kennedy TW, Theade SD (2003) Benkelman beam calibration: AASHTO versus FHWA. Transp Res Rec J Transp Res Board 1846(1):139–144

Masad E et al (2010) Characterization of asphalt concrete using a portable seismic pavement analyzer. Transp Res Rec J Transp Res Board 2140(1):98–107

Monismith CL, Brown SF (2006) Design, construction, and rehabilitation of asphalt concrete overlays. Transp Res Rec J Transp Res Board 1964(1):131–141

National Asphalt Pavement Association (NAPA) (n.d.) Design & analysis. https://www.asphaltpavement.org/index.php?option=com_content&view=article&id=43&Itemid=231

Pavement Interactive (n.d.) Benkelman beam. https://www.pavementinteractive.org/reference-desk/testing/equipment/deflection-testing/benkelman-beam/

Raab C, Baladi G (1999) Highway design and traffic safety engineering handbook. McGraw-Hill Professional

Transportation Research Board (TRB) (2001) Recommended practice for Backcalculation of flexible pavement layer moduli (FHWA-RD-98-166). https://trid.trb.org/view/673749

United states department of transportation (USDOT) (n.d.) Pavement management systems. https://www.fhwa.dot.gov/pavement/preservation/101207/pavement.cfm

Witczak MW, Ryden N, Kaloush KE, Scullion T (2013) Mechanistic-empirical pavement design guide: a manual of practice, 2nd edn. National Cooperative Highway Research Program (NCHRP) Report 1-37A

Chapter 32
IOT in Construction: A Bibliographic Literature Review of Technological Trends and Research Developments

N. Hemanth Kumar and S. P. Sreenivas Padala

Abstract The construction industry is experiencing a transformative shift with the integration of Internet of Things (IoT) technologies; however, a thorough understanding of the current state and trends in IoT applications within construction remains elusive. To address this gap, this paper conducts a systematic literature review employing bibliographic analysis, co-citation network analysis of keywords, and content analysis with visualization of similarities software. A meticulous analysis of 158 articles from the top construction journals reveals four prominent scenarios of IoT applications in construction: activity-based IoT, equipment-based IoT, material-based IoT, and labor-based IoT. These scenarios emerge as pivotal, garnering the highest references in the reviewed literature. This research not only contributes to the existing knowledge base but also provides a structured framework for comprehending the diverse applications of IoT in construction. By delving into the intricacies of each scenario, this study offers insights into the current trends and gaps in IoT research within the construction domain. The presented framework evaluates the existing landscape, shedding light on the status of research published in construction journals. The systematic literature review methodology ensures a comprehensive and unbiased assessment of the field, facilitating a nuanced understanding of the current state and future directions in IoT research for construction. This work aims to inform and guide future research endeavors, fostering innovation and advancements in construction practices through the effective integration of IoT technologies.

Keywords IoT · Construction industry · Bibliographic analysis

N. Hemanth Kumar (✉) · S. P. Sreenivas Padala
Department of Civil Engineering, Faculty of Engineering and Technology, M.S. Ramaiah University of Applied Sciences, Bengaluru, India
e-mail: hemanthkumar2798@gmail.com

B. V. Khode et al. (eds.), *Innovations in Technologies: Pioneering Sustainable Infrastructure for a Resilient Future*, Sustainable Civil Infrastructures,
https://doi.org/10.1007/978-3-031-82133-2_32

32.1 Introduction

The construction industry is undergoing a significant transformation fueled by the integration of IoT technologies. However, despite this wave of innovation, there exists a substantial gap in understanding the current state and trends of IoT applications within the construction sector. Researchers have highlighted the need for a comprehensive analysis to unravel the complexities of this integration (Kanan et al. 2018). As construction projects become more intricate, the demand for advanced technologies to enhance efficiency and productivity has never been more crucial (Kim et al. 2021a). Real-time data collection is identified as a key requirement to optimize performance indicators such as cost, schedule, and safety. The use of IoT in construction enables the automation of labor tracking, monitoring equipment productivity, and ensuring worker safety. These advancements aim to improve accuracy, efficiency, and cost savings in construction projects (Navon 2005). The integration of IoT in the construction industry is a pivotal development, with researchers asserting that it holds the potential to revolutionize various facets of construction processes (Chen et al. 2017). The connectivity and data-sharing capabilities of IoT devices have the potential to significantly improve project management, monitoring, and decision-making (Zhang et al. 2023). Understanding the synergies between IoT and construction is crucial, as it can significantly transform project timelines and optimize resource utilization (Shvets and Hanák 2023).

The integration of IoT applications within the construction industry has led to a transformative impact on various fronts. One notable scenario is the implementation of activity-based IoT, which significantly enhances the real-time tracking of construction activities (Rao et al. 2022). Activity based approach has greatly improved real-time tracking of construction activities. A notable advancement is the adoption of a new automated method for monitoring construction progress, utilizing daily photos and building models to streamline project management (Golparvar-Fard et al. 2015). Safety-oriented IoT applications have also garnered significant attention within the construction sector. These applications focus on enhancing risk management and ensuring the well-being of workers. By employing IoT devices, construction companies can implement advanced safety measures and protocols, ultimately creating a safer working environment (Chung et al. 2023). Moreover, safety at construction sites has been notably boosted by the adoption of an automated method that enhances the accuracy of detecting workers wearing safety helmets. Researchers achieved this by modifying the YOLO v5 model, optimizing it for small object identification and scale differences. Through targeted data augmentation and a pre-trained model, they successfully addressed data limitations and tested their model on a challenging dataset, comparing its performance with other advanced models. The primary goal was to ensure the proper usage of safety helmets, crucial for preventing accidents resulting from worker negligence due to discomfort or lack of safety awareness (Yan et al. 2022). Equipment-based IoT applications are instrumental in providing valuable insights into the health and performance of construction equipment. This data allows for the optimization of maintenance schedules, preventing unexpected breakdowns and reducing downtime. Consequently, this contributes to overall operational efficiency and cost-effectiveness (Louis and Dunston 2018). Material-based IoT is another crucial facet in the construction industry, offering efficient tracking and management of construction materials. By employing IoT technology, companies can minimize waste, improve logistics, and ensure a more sustainable approach to resource utilization (Ren et al. 2011). Labor-based IoT applications focus on optimizing workforce management and safety. These applications address critical aspects of project execution by utilizing IoT devices to monitor and enhance the performance, safety, and well-being of construction workers (Ding et al. 2013). By synthesizing information from a diverse set of papers, this paper aims to offer a comprehensive understanding of the multifaceted applications of IoT in construction.

To address the existing gap in understanding IoT applications in construction, this paper employs a systematic literature review methodology. A meticulous analysis of 158 articles from the top construction journals is conducted, utilizing bibliographic analysis, co-citation network analysis, and content analysis with visualization of similarities software. The research identifies and categorizes four prominent scenarios of IoT applications in construction, shedding light on their significance and interconnections. This structured framework not only contributes to the existing knowledge base but also provides a foundation for comprehending the diverse applications of IoT in construction. By scrutinizing each scenario, this study aims to offer insights into current trends and gaps in IoT research, guiding future endeavors in the integration of IoT technologies for advancing construction practices.

32.2 Research Methodology

This study adopts a systematic literature review (SLR) using bibliography methodology to examine IoT technologies and their applications in construction industry for the purpose the purpose of analysis and elicitation of trends and gaps. The SLR was conducted in three phases: source identification and selection, source evaluation and source analysis (Pournader et al. 2020). The research design diagram illustrating the phases of the SLR methodology is shown in the Fig. 32.1.

32.2.1 Source Identification: Systematic Literature Review

In the first phase of this study, the SLR was initiated with the objective of identifying relevant sources that would contribute to the comprehensive analysis of IoT technologies in the construction industry. The search was strategically formulated using specific keywords, namely "Industry Internet of Things (IIoT)," in combination with the term "Construction." This deliberate choice aimed to refine the search and enhance specificity. The inclusion of the term "IoT" was further tailored to focus on applications involving sensor data, machine communications, and automation

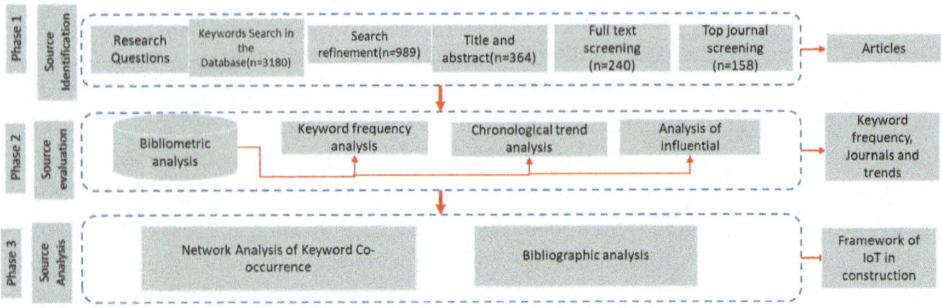

Fig. 32.1 Research design in three phases

systems. This distinction was made to ensure a more precise exploration of relevant literature, as opposed to using the broader term "IoT," which encompasses a wider spectrum of digital changes in the industry. The careful curation of search terms was a crucial step in narrowing down the scope of the review to facilitate a more in-depth and targeted analysis of IoT technologies within the context of the construction sector.

32.2.2 Source Evaluation: Bibliometric Analysis

In the second phase of the study, a comprehensive source evaluation was carried out through bibliometric analysis. This method employed quantitative analysis and statistical tools to examine publication patterns within the identified literature on IoT technologies in the construction industry. The study utilized Visualization of Similarities (VOS) viewer software to generate graphs, figures, and tables that facilitated a visual representation of key concepts and relationships within the field. The results of the bibliometric analysis were subjected to descriptive statistics, enabling the identification of trends and correlations associated with IoT concepts, thereby providing valuable insights into the evolving landscape of IoT applications in construction.

32.2.3 Source Analysis: Network Analysis and Bibliography Coupling Analysis

In the third phase of the study, a comprehensive source analysis was undertaken, employing network analysis and bibliographic coupling analysis, to delve deeper into the emerging trends and themes within the realm of IoT technologies in the construction industry. Recognizing the unique characteristics of this field, which may not exhibit the same bibliometric significance as more established areas, the research employed network analysis and bibliographic coupling to discern research themes and trends. Network analysis, specifically keyword co-occurrence mapping, was utilized to visualize clusters of keywords in IoT research publications, illustrating the dynamic developments in the field.

The methodology adopted in this study emphasizes the significance of keywords as key indicators of a study's core content and research areas. It aims to offer an informative representation of scientific knowledge production, patterns, relationships, and intellectual organization within the investigated topics. The linkages between keywords were established by analyzing the frequency of their co-occurrence in publications, providing valuable insights into the interconnected nature of research in IoT applications for construction (Eck and Waltman 2014).

32.3 Data Analysis

32.3.1 Distribution of Publications by Year

The numbers of annually published articles addressing the application of IoT in the construction industry are illustrated in Fig. 32.2.

Figure 32.2 reveals the bibliometric examination of 158 articles, spanning from 2010 to 2023. A noticeable surge in publications is observed after 2020, indicating a rising interest in IoT in construction research. This growth rate underscores the increasing attention in this field.

Fig. 32.2 Numbers of annually published articles addressing the IoT in construction

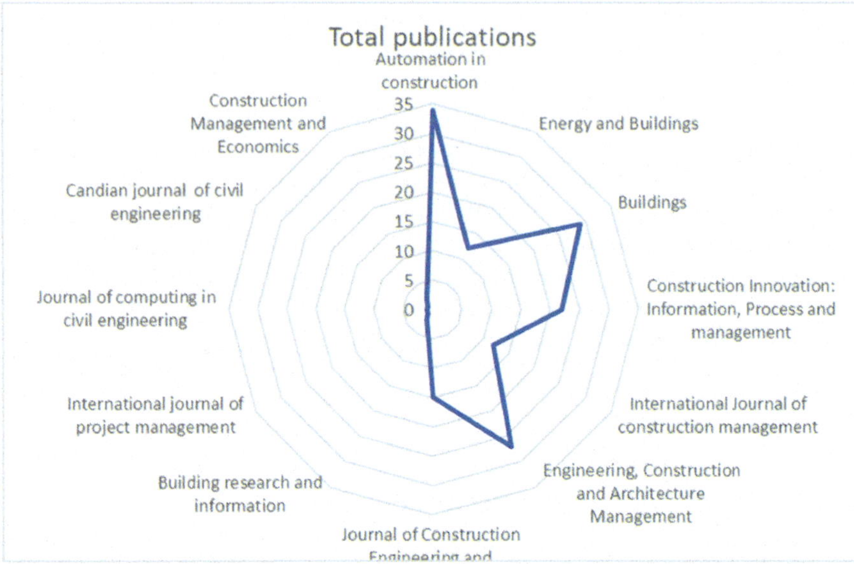

Fig. 32.3 Top peer reviewed construction journals contributing to the area of IoT in construction

32.3.2 Distribution of Journal

The field of construction has witnessed a surge in research publications, particularly in the realm of IoT as shown in the Fig. 32.3. Leading journals such as "Automation in Construction" with 34 publications, "Buildings" with 29, and "Engineering, Construction, and Architecture Management" with 27, have played pivotal roles in disseminating valuable insights. Additionally, journals like "Construction Innovation: Information, Process, and Management" with 22 publications and "Journal of Construction Engineering and Management" with 15 publications have contributed significantly to advancing knowledge in IoT applications within the construction industry. This robust publication landscape underscores the growing importance of integrating IoT technologies to enhance efficiency and innovation in construction practices.

32.3.3 Network Analysis of Keyword Co-occurrence

Figure 32.4 illustrates a color-coded keyword network that showcases the changing prominence of IoT concepts in construction from 2010 to 2023. In the SLR, the search terms "IoT" and "construction" were employed, and their outputs were excluded from the keyword network analysis. The analysis reveals that, pre-2018, IoT research predominantly concentrated on concepts such as Productivity, RFID,

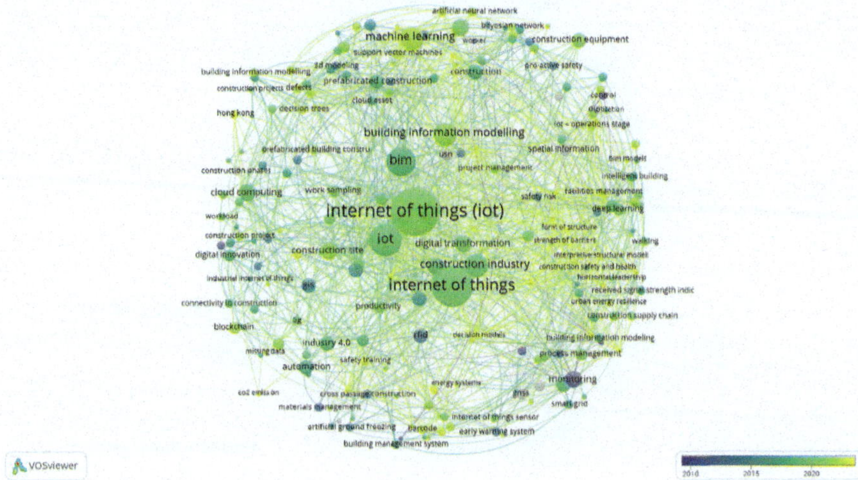

Fig. 32.4 Keyword network IoT in construction concepts popularity between 2010 and 2023

Sensors, GIS, and others. However, from 2018 to 2023, there has been a noticeable shift in research focus towards discussions on IoT in construction technologies. Noteworthy keywords during this period include building information modelling (BIM), Virtual Reality, Digital Twin, Cloud Computing, Digital Storage, and Automation. Post-2020, research has delved into the optimization of process performance and practical applications of IoT technologies. Popular keywords and themes Safety, Real-time Monitoring, Smart Cities, Digital Twin, and BIM integration with IoT.

32.3.4 Bibliographic Coupling Analysis

The VOS viewer software was employed to conduct Bibliographic Coupling Analysis (BCA) through the clustering of citations using Louvain network modularity (Graham et al. 2019). The cluster analysis aimed to capture trending keywords and their semantic network groupings from the titles, abstracts, and keywords of 158 papers that underwent the SLR process. These papers were then subjected to BCA. Following the BCA analysis, four distinct clusters related to the IoT were identified. In each cluster, a total of 22 key articles, exhibiting the highest number of co-citations, were selected and illustrated in the Fig. 32.5.

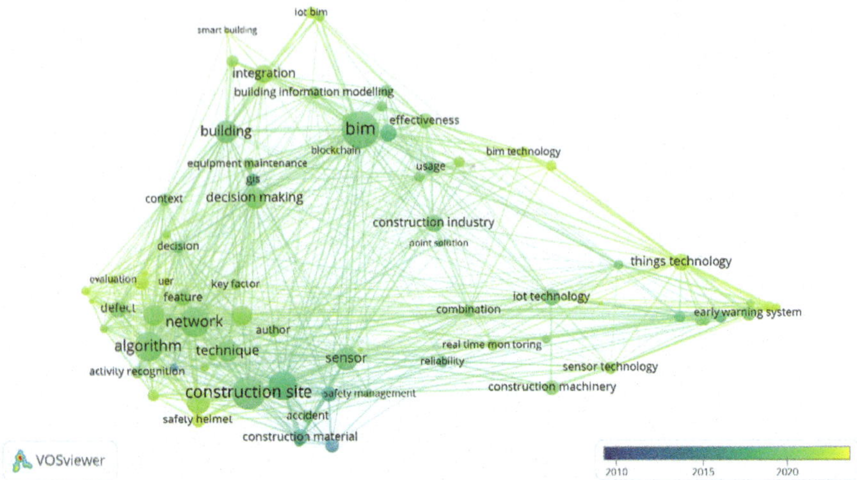

Fig. 32.5 Cluster analysis of keyword network

32.4 Findings and Discussion

32.4.1 Framework of IoT in Construction Scenarios

The key purpose of our proposed framework is to bring together the main topics and themes in IoT within construction scenarios, showcasing both the technologies and their practical applications. To make the concepts clear in terms of technology and application, we define IoT technology as data-driven technologies, decision-making support, and advanced automation as shown in the Fig. 32.6.

The four clusters identified through bibliographic coupling analysis represent distinct IoT scenarios. In our framework, these scenarios are presented with overarching themes, each serving as a cluster title. Within each scenario, IoT technologies are grouped based on their domains in the inner circle, while their applications are depicted in the outer area of the framework.

For instance, IoT technologies related to domains such as IoT, big data and analytics, cloud computing, simulation, and digital twin are applied to improve information and data management, facilitating effective construction site management. The detailed description for each scenario is illustrated in Fig. 32.5, where the four clusters are activity monitoring, labor monitoring, equipment monitoring, and material monitoring.

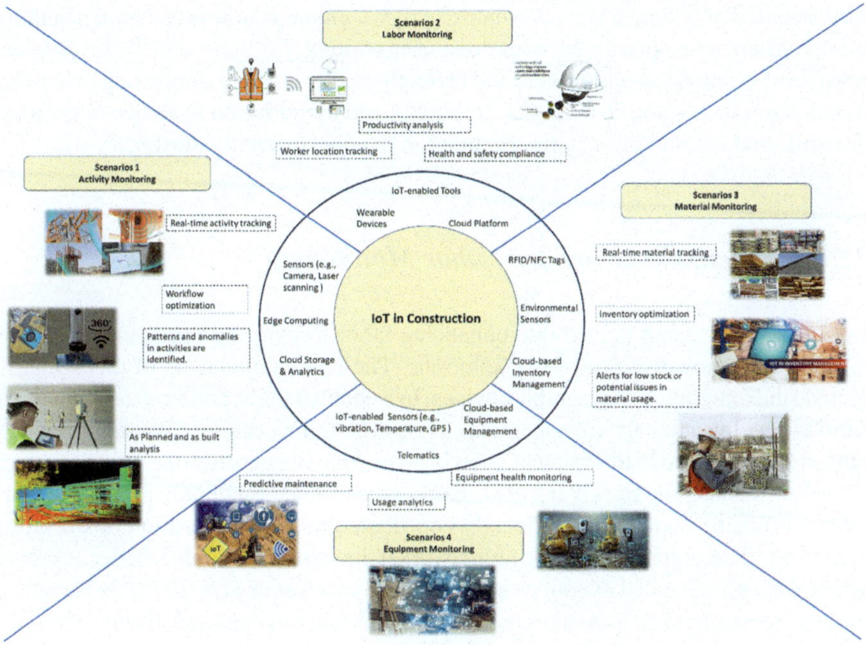

Fig. 32.6 A framework of IoT in construction scenarios: IoT technologies and their applications

32.4.2 Scenario 1: IoT for Activity Monitoring

In Scenario 1, our research in Cluster 1 delved into the profound applications of IoT solutions for activity-based monitoring on construction sites, offering an intricate understanding of the technologies involved. To begin with, a sophisticated array of sensors, including cameras and laser scanning devices, were strategically deployed. These sensors played a pivotal role in capturing real-time data, allowing for the precise tracking of construction activities as they unfolded (Golparvar-Fard et al. 2015; Lee et al. 2023). The integration of BIM facilitated a comprehensive analysis, comparing planned activities with the actual on-site progress, providing valuable insights into the efficiency and alignment of construction processes (Padala et al. 2022; Pan and Zhang 2021). Edge computing, a decentralized approach to data processing, was a cornerstone in our methodology. By processing data closer to its source, edge computing ensured quicker decision-making, contributing to the seamless operation of the entire system (Halder et al. 2022). Simultaneously, cloud storage and analytics played a pivotal role in managing the vast amounts of data generated from the construction site. This robust combination not only facilitated storage but also enabled the analysis of patterns and anomalies in real-time, adding a layer of intelligence to the construction monitoring process (Lilis et al. 2017). Furthermore, the incorporation of laser scanning and photogrammetry technology elevated our capabilities to accurately capture and assess site conditions. This integration allowed for a detailed as-planned versus as-built analysis, enabling a nuanced understanding of deviations and optimizing workflows accordingly (Li et al. 2019). In essence, Scenario 1 demonstrated a holistic and technologically advanced approach to activity monitoring, leveraging IoT, Edge Computing, and cloud-based analytics to enhance decision-making and project management in the construction industry.

32.4.3 Scenario 2: IoT for Labor Monitoring

In Cluster 2, research delved into enhancing labor monitoring in construction sites through the innovative use of IoT solutions. The core components of this scenario include the integration of wearable devices, IoT-enabled tools, and a robust cloud platform to revolutionize the management of construction workforce. Wearable devices played a pivotal role in this context, offering real-time insights into the precise whereabouts of workers on-site (Wang et al. 2021). These devices not only promoted operational efficiency but also significantly contributed to the optimization of workflow processes. The continuous tracking of worker locations facilitated better resource allocation and enhanced overall project coordination (Kanan et al. 2018). IoT-enabled tools were another key aspect of our approach. These tools were designed to provide comprehensive productivity analysis, offering a detailed understanding of labor performance (Li and Lv 2023). By collecting and analyzing data in real-time, project managers gained valuable insights into task completion rates, efficiency benchmarks, and potential areas for improvement. This data-driven approach allowed for informed decision-making, ultimately leading to enhanced project management outcomes (Yan et al. 2022). To ensure the health and safety compliance of the construction workforce, wearable devices were ingeniously utilized. These devices were equipped to monitor adherence to safety protocols, ensuring that workers operated within predefined safety parameters. This not only mitigated potential risks but also fostered a safer working environment (Chung et al. 2023; Seo et al. 2015). The deployment of a cloud platform served as the technological backbone for this scenario. It facilitated seamless communication, data storage, and real-time access to critical information for all stakeholders involved. The cloud platform not only centralized the data but also enabled stakeholders to monitor and manage labor-related aspects remotely, contributing to efficient project oversight (Yang et al. 2010).

32.4.4 Scenario 3: IoT for Material Monitoring

Cluster 3's research focuses on leveraging IoT solutions for material monitoring at construction sites, emphasizing the use of RFID/NFC tags as technological beacons. These tags enable seamless material identification, optimizing inventory management and providing real-time tracking (Kasim and Latiffi 2013). The incorporation of RFID/NFC technology enhances transparency, offering immediate insights into material location and status (Ren et al. 2011). Environmental sensors add another layer, monitoring factors like temperature and humidity to ensure optimal material storage conditions and contribute to sustainable construction practices (Wang 2018). Cloud-based inventory management system serves as the technological backbone, providing a centralized platform for comprehensive oversight. This facilitates real-time updates on material availability, usage, and conditions, promoting collaboration among construction teams. Real-time material tracking, enabled by this technology, goes beyond traditional inventory management, offering constant and accurate overviews for data-driven decision-making, efficient resource allocation, and reduced wastage (Shen 2022). The holistic approach to construction management, utilizing RFID/NFC tags, environmental sensors, and cloud-based systems, results in inventory optimization, reducing costs and promoting sustainability by minimizing material waste. The system's proactive alert mechanism for low stock or potential issues ensures timely intervention, contributing to smooth project operation. Whether restocking materials or addressing concerns, these alerts enhance proactive management, ultimately optimizing construction processes (Majrouhi Sardroud 2012).

32.4.5 Scenario 4: IoT for Equipment Monitoring

In the realm of construction site management, Scenario 4 delves profoundly into the critical aspect of equipment monitoring, harnessing the power of cutting-edge IoT solutions. At its core, this scenario is dedicated to ensuring the optimal functioning and longevity of machinery through a comprehensive suite of IoT-enabled features.

First and foremost, the integration of IoT-enabled sensors takes centre stage, offering a sophisticated array of capabilities, including vibration monitoring, temperature tracking, and GPS functionality. These sensors act as vigilant guardians, providing real-time insights into the health and performance of the tools in use. By continuously gauging vibrations and temperatures, these sensors enable construction teams to proactively identify potential issues, thereby preventing unexpected equipment failures (Louis et al. 2016; Meng and Zhu 2020). Telematics, another integral component of this scenario, facilitates seamless communication and tracking of equipment performance. Through the utilization of telematics, construction managers can access real-time data on the location, status, and usage patterns of each piece of equipment. This not only enhances operational visibility but also enables more strategic decision-making regarding equipment deployment and allocation (Kim et al. 2021b). Cloud-based equipment management serves as the technological backbone of Scenario 4, offering a centralized platform for streamlined control and comprehensive oversight. This cloud-based system not only stores vast amounts of data generated by IoT sensors and telematics but also provides accessibility to stakeholders from any location (Sherafat and Rashidi 2019). The centralized nature of this platform ensures that decision-makers can monitor and manage equipment efficiently, fostering a more responsive and agile construction site. Predictive maintenance emerges as a pivotal strategy within this scenario, utilizing the wealth of data collected by IoT sensors and telematics (Kim et al. 2020). By analyzing patterns and trends in equipment behaviour, predictive maintenance anticipates potential issues before they escalate into critical failures (Kristombu et al. 2022). This forward-looking approach minimizes downtime, enhances overall efficiency, and contributes to cost savings by preventing costly repairs and replacements (Yeon and Yeon 2019). Usage analytics, another key facet, offers a detailed understanding of how equipment is utilized on the construction site. By tracking usage patterns, construction teams gain valuable insights into the efficiency of their equipment deployment, enabling data-driven decisions for optimizing resource allocation and improving overall productivity (Louis et al. 2016). In essence, Scenario 4 stands as a beacon of innovation in construction site management, leveraging the full depth and breadth of IoT applications for equipment monitoring. This holistic approach ensures a seamless and efficient workflow, safeguarding the well-being of machinery, and ultimately contributing to the success of construction projects.

32.4.6 Drawback of Monitoring Through IoT in Construction

While the integration of IoT technologies in construction monitoring has shown great promise, several drawbacks warrant attention for a comprehensive understanding of its limitations. Firstly, concerns regarding data privacy and security pose significant challenges, as the vast amount of data collected from IoT devices, including real-time project updates and personnel information, may be susceptible to unauthorized access or cyber threats (Li et al. 2020). Additionally, the dependency on a network infrastructure for IoT communication introduces the risk of system failures or disruptions, potentially hindering real-time monitoring capabilities crucial for construction site management. The complexity of implementing and maintaining IoT systems also raises issues related to the skill gap among construction personnel, as not all workers may possess the necessary expertise to operate and troubleshoot these advanced technologies (Ali et al. 2020). Moreover, the high initial costs associated with deploying IoT solutions may act as a deterrent for smaller construction projects with limited budgets, creating a potential digital divide in the industry. Lastly, the interoperability challenges among different IoT devices and platforms may hinder seamless integration, leading to data inconsistencies and inefficiencies. Addressing these drawbacks is vital for the successful and widespread adoption of IoT in the construction sector.

32.4.7 Research Gaps and Future Potential Research Areas for Development

Research in the construction industry has made significant strides with the implementation of IoT technologies. However, there are notable gaps that warrant exploration for future development. One area of focus is enhancing the integration of various monitoring systems, such as activity and labor monitoring, to gain a more comprehensive understanding of construction site dynamics. Investigating synergies between material and equipment monitoring is also crucial for optimizing resource allocation and equipment utilization. Moreover, exploring the scalability and affordability of IoT solutions for small to medium-sized construction projects is imperative to ensure widespread adoption.

In addition to these considerations, the integration of BIM with IoT and digital twin technologies presents a promising avenue for research. Understanding how these technologies can work together could further streamline construction processes and improve overall efficiency (Navon 2005). This integration could offer a holistic view of construction projects, combining real-time data from IoT devices with the detailed modeling capabilities of BIM.

Sustainable practices in construction are another critical area for future research. Investigating the environmental impact of IoT technologies and developing eco-friendly construction methods can contribute to a more sustainable industry (Teizer et al. 2017). Furthermore, addressing cybersecurity concerns associated with IoT in construction is essential for ensuring the integrity and security of data in interconnected systems. Research efforts should aim to find innovative solutions to safeguard sensitive information and maintain the trustworthiness of IoT applications.

32.5 Conclusion

This study conducted a systematic literature review of 158 articles from top construction journals to comprehensively analyze the current state and trends in IoT applications within the construction industry. The proposed framework categorizes IoT scenarios into activity-based IoT, equipment-based IoT, material-based IoT, and labor-based IoT. The research revealed a growing interest in IoT in construction, with a notable surge in publications after 2020.

The identified scenarios shed light on practical applications, such as real-time activity monitoring, labor optimization, material tracking, and equipment health monitoring. The framework serves as a structured guide, showcasing both IoT technologies and their practical applications in construction.

Despite the contributions, there are research gaps to address. Integrating various monitoring systems for a comprehensive understanding of construction dynamics, exploring synergies between material and equipment monitoring, and investigating the scalability of IoT solutions for smaller projects are crucial areas for future development. Additionally, the integration of BIM with IoT and digital twin technologies and addressing cybersecurity concerns in IoT applications are promising avenues for further research.

It is important to note that this study has limitations. The analysis is based on a systematic literature review of articles up to the current date, and the rapidly evolving nature of IoT technology may result in emerging trends not captured in this study. Nevertheless, the proposed framework offers valuable insights and a foundation for future research endeavors in the effective integration of IoT technologies in the construction industry.

References

Ali Y, Saad TB, ur Rehman O (2020) Integration of IoT technologies in construction supply chain networks; CPEC a case in point. Sustain Oper Comput 1:28–34. https://doi.org/10.1016/j.susoc.2020.12.003

Chen B, Wan J, Shu L, Li P, Mukherjee M, Yin B (2017) Smart factory of industry 4.0: key technologies, application case, and challenges. IEEE Access 6:6505–6519. https://doi.org/10.1109/ACCESS.2017.2783682

Chung WWS, Tariq S, Mohandes SR, Zayed T (2023) IoT-based application for construction site safety monitoring. Int J Constr Manag 23(1):58–74. https://doi.org/10.1080/15623599.2020.1847405

Ding LY et al (2013) Real-time safety early warning system for cross passage construction in Yangtze riverbed metro tunnel based on the internet of things. Autom Constr 36:25–37. https://doi.org/10.1016/j.autcon.2013.08.017

van Eck NJ, Waltman L (2014) Visualizing bibliometric networks. Measuring Sch Impact 285–320. https://doi.org/10.1007/978-3-319-10377-8_13

Golparvar-Fard M, Peña-Mora F, Savarese S (2015) Automated progress monitoring using unordered daily construction photographs and IFC-based building information models. J Comput Civ Eng 29(1). https://doi.org/10.1061/(asce)cp.1943-5487.0000205

Graham P, Nikolova N, Sankaran S (2019) Tension between leadership archetypes: systematic review to inform construction research and practice. J Manag Eng 36(1):03119002. https://doi.org/10.1061/(ASCE)ME.1943-5479.0000722

Halder S, Afsari K, Serdakowski J, DeVito S, Ensafi M, Thabet W (2022) Real-time and remote construction progress monitoring with a quadruped robot using augmented reality. Buildings 12(11). https://doi.org/10.3390/buildings12112027

Kanan R, Elhassan O, Bensalem R (2018) An IoT-based autonomous system for workers' safety in construction sites with real-time alarming, monitoring, and positioning strategies. Autom Constr 88:73–86. https://doi.org/10.1016/J.AUTCON.2017.12.033

Kanan R, Elhassan O, Bensalem R (2018) An IoT-based autonomous system for workers' safety in construction sites with real-time alarming, monitoring, and positioning strategies, Autom Constr 88:73–86. https://doi.org/10.1016/j.autcon.2017.12.033

Kasim N, Latiffi AA (2013) RFID technology for materials management in construction projects—a review. Int J Constr Eng Manag 2:1–6

Kim J, Hwang J, Chi S (2020) Towards database-free vision-based monitoring on construction sites: a deep active learning approach. Autom Constr 20:1–10

Kim H, Tae S, Zheng P, Kang G, Lee H (2021a) Development of IoT-based particulate matter monitoring system for construction sites. Int J Environ Res Public Health 18(21):11510. https://doi.org/10.3390/IJERPH182111510

Kim H, Tae S, Zheng P, Kang G (2021b) Development of IoT-based particulate matter monitoring system for construction sites. Int J Environ Res Public Health 18(1):1–15
Kristombu S, Thilakarthna S, Perera JS, Arashpour M (2022) Artificial intelligence and smart vision for building and construction 4.0: machine and deep learning methods and applications—1-s2.0-S0926580522003132-main. Autom Constr 141:1–26
Lee G, Moon S, Hwang J, Chi S (2023) Development of a real-time noise estimation model for construction sites. Adv Eng Inform 58:102133. https://doi.org/10.1016/J.AEI.2023.102133
Li S, Lv Y (2023) Detection of safety helmet and mask wearing using improved YOLOv5s—s41598-023-48943-3(2)
Li J, Kassem M, Ciribini ALC, Bolpagni M (2019) A proposed approach integrating DLT, BIM, IOT and smart contracts: demonstration using a simulated installation task. In: international conference on smart infrastructure and construction 2019, ICSIC 2019: driving data-informed decision-making. ICE Publishing, pp 275–282. https://doi.org/10.1680/icsic.64669.275
Li J, Li M, Yan F, Miao D (2020) Key technologies of data monitoring for coal machinery equipment in the intelligent IOT environment. In: 10th IEEE international conference on cyber technology in automation, control and intelligent systems, CYBER 2020. Institute of Electrical and Electronics Engineers Inc, pp 364–369. https://doi.org/10.1109/CYBER50695.2020.9278951
Lilis G, Conus G, Asadi N, Kayal M (2017) Towards the next generation of intelligent building: an assessment study of current automation and future IoT based systems with a proposal for transitional design. Sustain Cities Soc 28:473–481. https://doi.org/10.1016/j.scs.2016.08.019
Louis J, Dunston PS (2018) Integrating IoT into operational workflows for real-time and automated decision-making in repetitive construction operations. Autom Constr 94:317–327. https://doi.org/10.1016/j.autcon.2018.07.005
Louis J, Dunston PS, Asce AM (2016) Methodology for real-time monitoring of construction operations using finite state machines and discrete-event operation models. https://doi.org/10.1061/(ASCE)CO.1943-7862
Meng Q, Zhu S (2020) Developing IoT Sensing system for construction-induced vibration monitoring and impact assessment—sensors-20-06120-v2. Sensors 20:1–23
Navon R (2005) Automated project performance control of construction projects. Autom Constr 467–476. https://doi.org/10.1016/j.autcon.2004.09.006
Padala SPS, Maheswari JU, Hirani H (2022) Identification and classification of change causes and effects in construction projects. Int J Constr Manag 22(14):2788–2807. https://doi.org/10.1080/15623599.2020.1827186
Pan Y, Zhang L (2021) A BIM-data mining integrated digital twin framework for advanced project management. Autom Constr vol 124. https://doi.org/10.1016/j.autcon.2021.103564
Pournader M, Shi Y, Seuring S, Koh SCL (2020) Blockchain applications in supply chains, transport and logistics: a systematic review of the literature. Int J Prod Res 58(7):2063–2081. https://doi.org/10.1080/00207543.2019.1650976
Rao AS et al (2022) Real-time monitoring of construction sites: sensors, methods, and applications. Autom Constr 136:104099. https://doi.org/10.1016/J.AUTCON.2021.104099
Ren Z, Anumba CJ, Tah J (2011) RFID-facilitated construction materials management (RFID-CMM)—a case study of water-supply project. Adv Eng Inform 25(2):198–207. https://doi.org/10.1016/j.aei.2010.02.002
Sardroud JM (2012) Influence of RFID technology on automated management of construction materials and components. Sci Iran 19(3):381–392. https://doi.org/10.1016/j.scient.2012.02.023
Seo J, Han S, Lee S, Kim H (2015) Computer vision techniques for construction safety and health monitoring. Adv Eng Inform 29(2):239–251. https://doi.org/10.1016/J.AEI.2015.02.001
Shen W (2022) Application of BIM and internet of things technology in material management of construction projects. Adv Mater Sci Eng 2022. https://doi.org/10.1155/2022/5381252
Sherafat B, Rashidi A (2019) A hybrid kinematic acoustic system for automated activity detection of construction equipment. Sensors 19:1–21

Shvets Y, Hanák T (2023) Use of the internet of things in the construction industry and facility management: usage examples overview. Procedia Comput Sci 219:1670–1677. https://doi.org/10.1016/J.PROCS.2023.01.460

Teizer J et al (2017) Internet of Things (IoT) for integrating environmental and localization data in building information modeling (BIM). In: ISARC 2017—proceedings of the 34th international symposium on automation and robotics in construction. International Association for Automation and Robotics in Construction I.A.A.R.C), pp 603–609. https://doi.org/10.22260/isarc2017/0084

Wang M (2018) Framework of IoT-based shop floor material management system for a panelized home builder

Wang Z, Wu Y, Yang L (2021) Fast personal protective equipment detection for real construction sites using deep learning approaches. Sensors 21:1–21

Yan K et al (2022) Deep learning-based substation remote construction management and AI automatic violation detection system. IET Gener Transm Distrib 16(9):1714–1726. https://doi.org/10.1049/gtd2.12387

Yang J, Arif O, Vela PA, Teizer J, Shi Z (2010) Tracking multiple workers on construction sites using video cameras. Adv Eng Inform 24(4):428–434. https://doi.org/10.1016/J.AEI.2010.06.008

Yeon S, Yeon C (2019) Smart construction monitoring for disaster prevention based on spatial information and GNSS/USN/IoT

Zhang Y, Zhang J, Wang C, Ren X (2023) An integrated framework for improving the efficiency and safety of hydraulic tunnel construction. Tunn Undergr Space Technol 131:104836. https://doi.org/10.1016/J.TUST.2022.104836

Chapter 33
A Comprehensive Review of Sewer Network Design by SewerGEMS

R. M. Bhagat, A. S. Borse, Onkarnath Mukherjee, J. M. Raut, A. S. Kurzekar, A. R. Gajbhiye, M. S. Bhagat, S. W. Dhengare, and Y. P. Kherde

Abstract The study discusses the SewerGEMS V8i, an innovative software application designed for modeling and designing sewage networks. This advanced tool streamlines project completion by enhancing efficiency and reducing costs. SewerGEMS V8i focuses on the fundamental element of sewer systems, the sewer network, with an emphasis on cost-saving design. It is the first multi-platform, completely dynamic system that supports CAD, GIS, and standalone environments for modelling combined and sanitary sewers. The application's hydraulic design capabilities enable the calculation of flow and hydraulic modeling for pipe diameters and slopes, with the added benefit of generating comprehensive reports and graphical representations. SewerGEMS V8i employs AutoCAD technology for its graphic system, offering tools and features that empower designers to create, label, dimension, and plot sewage network drawings efficiently. These software studies simplifies sewerage network design and Safe Time. This software is a valuable asset for professionals seeking a comprehensive and cost-effective solution for sewage network design.

Keywords AutoCAD · Sewage flow · Sewer network · SewerGEMS V8i · CPHEEO

R. M. Bhagat · A. S. Borse (✉) · J. M. Raut · A. S. Kurzekar · A. R. Gajbhiye · M. S. Bhagat · S. W. Dhengare · Y. P. Kherde
Civil Engineering Department, YCCE, Nagpur, India
e-mail: ashwiniborse2407@gmail.com

O. Mukherjee
ONM Consultant, Nagpur, India

B. V. Khode et al. (eds.), *Innovations in Technologies: Pioneering Sustainable Infrastructure for a Resilient Future*, Sustainable Civil Infrastructures,
https://doi.org/10.1007/978-3-031-82133-2_33

33.1 Introduction

Water is vital for life, but managing wastewater is equally important. Sewage management is crucial for maintaining public health and environmental well-being. Modern technology, like SewerGEMS, is now commonly used for designing sewage systems. Sewerage systems are essential for carrying away sanitary waste efficiently, and computer software like SewerGEMS helps design cost-effective sewer networks (Bhavsar 2017; Shivapur and Bhavimani 2017). Sewerage networks are integral to public health and environmental preservation, but their cost and complexity pose challenges. Sewerage networks, designed with tools like SewerGEMS, ensure the safe disposal of wastewater (Tonde et al. 2020). SewerGEMS V8i software provides solutions for cost-effective sewerage network design (Kothai et al. 2018). SewerGEMS V8i is a versatile software solution that offers efficiency and cost-effectiveness in designing sewer networks. It enables analysis, design, and maintenance of sewer systems (Punam Harising 2016; Prajapati et al. 2016).

33.1.1 SewerGEMS

SewerGEMS with SewerGEMS V8i, projects can be completed quickly, efficiently, and affordably (Swamee 2001; Gulati and Paul 2014; Katti and Kumar 2015). The GIS, CAD, and Stand-Alone platforms are the only fully dynamic, combined sewer and sanitary modeling solutions available, and Bentley SewerGEMS V8i is the first among them (Tiwari and Bhorkar 2020; Virupakshi and Ravikiran 2016; Kannur et al. 2015; Nisha Patel 2019). Bentley SewerGEMS V8i, which allows us to analyze all sanitary and combined sewer system elements in a single pack-age (Murugesh Katti and Manoj Kumar 2015; Roy et al. 2015; Rai 2020). Hydraulic design includes hydraulic modelling for network pipe diameters and slopes, transit and total flow estimates, and hydraulic modelling. SewerGEMS V8i's dedicated tools and features give the designer a wide range of options for sketching, labelling, dimensioning, and plotting the drawings (Katti 2015; CPHEEO Mannual 2013; Rai et al. 2014).

33.1.2 Applications of SewerGEMS V8i

SewerGEMS manage your model effectively so that you can focus on making the best engineering decisions. The programme offers reports, layouts, and transversal or longitudinal cross sections of the pipe network, all of which are shown in a sophisticated AutoCAD-based graphic system. With the help of this programme, tasks may be completed quickly, cheaply, and with great efficiency. SewerGEMS V8i software is main advantages is it is freely version software, so that it is easily downloaded. Final drawings produced into a dwg format which is the most generally accepted file format for engineers (Rao et al. 2013; Scarlat et al. 2009).

33.2 Methodology

33.2.1 Sanitary Sewer Network for Vasana Village

In the context of designing a wastewater management system, the determination of design flows is of paramount importance to ensure the efficiency and effectiveness of the system. These design flows are calculated based on several critical factors, with the primary factor being peak flows. Peak flows are determined by multiplying the average weather flows by a peak factor, reflecting the highest anticipated wastewater flow rates. Guidelines established by the Central Public Health and Environmental Engineering Organization (CPHEEO) suggest that approximately 80% of the water supply eventually becomes wastewater that flows into sewers, with an additional 10% accounting for non-domestic flows, including infiltration. In the design of the sewer system, it is assumed that during peak flows, any potential silting that might occur at lower flow rates will be effectively flushed out. This assumption simplifies the hydraulic analysis of the sewers, which operates under steady flow conditions, even though actual flow conditions may vary, such as during the morning hours (Bhavsar et al. 2017).

For the sewer design, specific criteria are considered, including the choice of sewer diameter and gradient. These choices are made with the requirement that the depth of the cut should not exceed 6 m. Reinforced Cement Concrete (RCC) with NP3 specifications is selected as the material for the sewer line due to its cost-effectiveness and availability in various diameters, with the minimum diameter set at 150 mm. An essential parameter in sewer design is the self-cleansing velocity, which ensures that no solid material deposits in the invert of the sewer. To prevent solid accumulations, the sewer's velocity must be maintained at a level that ensures self-cleansing action.

For separate sewers, a minimum velocity of 0.6 m per second (m/s) is recommended. To optimize the network design, the village is divided into eight (Fig. 33.1), enabling a precise calculation of the population and area covered by each zone. These data are then used to determine the number of manholes needed in each zone, facilitating the development of a sewer network that effectively meets the wastewater management needs of the community. This meticulous approach ensures a sustainable and well-designed wastewater management system for the village.

The AutoCAD functionality integrated within SewerGEMS V8i offers a seamless and user-friendly experience by mirroring the structure and features of the base product. This system is designed for creating sewer networks that connect residential collection points to a central Sewage Treatment Plant, serving multiple Zones-1 (Fig. 33.1). To enhance maintenance and cleaning, manholes are strategically placed

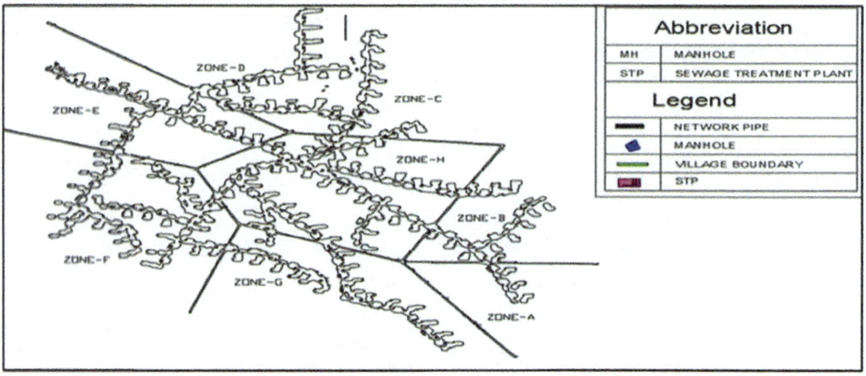

Fig. 33.1 Sewerage network design in AutoCAD (Bhavsar et al. 2017)

at around 30 m intervals, especially at points where the sewer line changes direction or the ground's elevation abruptly shifts. These manholes are systematically numbered to match the flow direction, simplifying the identification and monitoring of each manhole's cumulative discharge. This comprehensive design ensures efficient management of the sewer system and maintains accurate records of its performance and maintenance requirements (Bhavsar et al. 2017).

33.2.2 Sanitary Sewer Network for Zone-10, Belagavi City

To develop a sewer network for Belagavi city, in the Karnataka Urban Water Supply and Drainage Board (KUWDB) embarked on a comprehensive data collection effort. This involved gathering vital information such as population trends spanning the past five decades (Fig. 33.2).

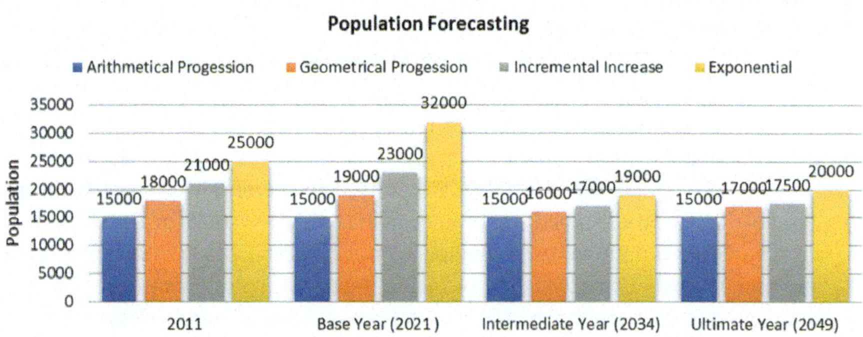

Fig. 33.2 Population projection of Ramatheerth Nagar (Zone-10) (Shivapur and Bhavimani 2017)

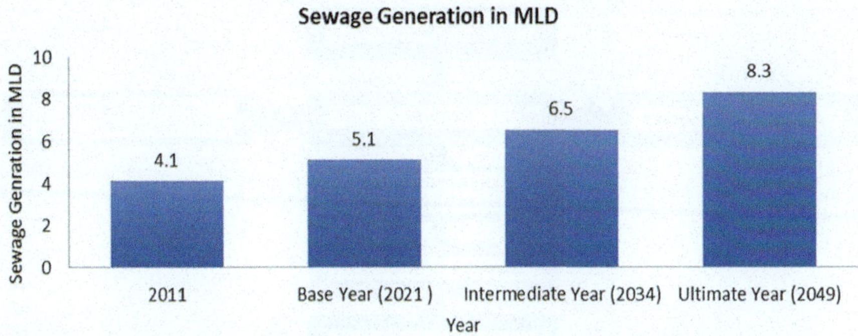

Fig. 33.3 Sewage generation of Ramatheerth Nagar (Zone-10) (Shivapur and Bhavimani 2017)

Existing water supply, details topographical and road maps, as well as data associated with the city's previous sewer network, including the invert levels of existing manholes. This extensive data set forms the foundation for designing a highly functional and efficient sewer system that can adapt to the evolving needs of the city, ensuring effective wastewater management and sanitation for its residents (Fig. 33.3).

The primary design considerations for this project include per capita sewage generation at 135 L per capita per day (80% of CPHEEO recommendations), a minimum sewer diameter of 150 mm using concrete pipes, and a minimum peak flow velocity of 0.6 m/s. The peaking factor varies based on cumulative population data. The sewer network adheres to CPHEEO manual requirements and was efficiently designed using Sewergemv8i software, which automates link and node connectivity, simplifying the design process and eliminating the need for manual labeling (Katti 2015; CPHEEO Mannual 2013; Rai et al. 2014). Bentley SewerGEMS V8i offers users three distinct operational modes: Modeler mode, AutoCAD coordinated mode, and ArcGIS mode, each tailored to specific functionalities. Your choice of mode determines the available features, as some functions are mode-specific. ArcGIS mode stands out for its seamless integration into the GIS environment, enabling users to work without data import or export, edit multiple scenarios in the same geodatabase, custom querying, flexibility in model building, and powerful reporting capabilities. The integration of AutoCAD into Bentley SewerGEMS V8i mirrors the standalone version's user-friendliness and comes with several advantages. It provides users with a fully scaled mode for layout creation and design within the familiar drafting environment. Additionally, it leverages native AutoCAD features like precise positioning, making it easier to align components with other elements. Users can also use AutoCAD commands such as move, delete, and rotate on Bentley's SewerGEMS elements, with automatic updates and synchronization, enhancing overall efficiency and compatibility in an AutoCAD environment (Shivapur and Bhavimani 2017).

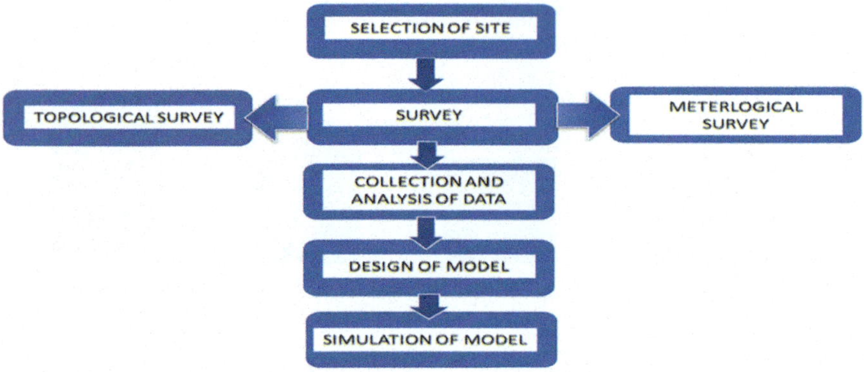

Fig. 33.4 Methodology for Bhilarewadi village (Tonde et al. 2020)

33.2.3 Design of Drainage Line of Bhilarewadi Village

SewerGEMS is a specialized software by Bentley Systems designed for sewerage system planning and analysis, particularly in areas poised for substantial population growth. This tool facilitates the efficient creation of plans and profile sheets for multiple sewer networks simultaneously (Fig. 33.4).

The software follows a multi-step approach to calculate flow within a study area: it forecasts the area's future population, determines daily sewer flow based on per capita sewage flow rates, factors in peak flow, industrial and commercial sewage allowances, and infiltration. SewerGEMS simplifies this intricate process, serving as a valuable asset for engineers and urban planners (Tonde et al. 2020).

33.2.4 Design of Underground Sewerage System for Kalapatti

The methodology for designing a sewer network is a systematic and strategic approach that considers both the geographical and demographic characteristics of a specific area. It begins with surveys and studies to evaluate ground elevation and population density. This information is crucial for determining the ideal placement of trunk mains and branch sewers. The process also involves selecting appropriate pipe sizes and capacities, designing the network's alignment, and accounting for future growth and expansion. The ultimate aim of this methodology is to create a well-planned sewer network that efficiently handles wastewater while minimizing adverse environmental effects, ensuring the sustainable management of this critical. In the design and implementation of the sewer network system, meticulous planning was employed to strategically position the outfall in Ward Number 34, taking advantage of the lowest ground elevation point to facilitate the natural flow of wastewater through gravity (Fig. 33.5) (Kothai et al. 2018).

Fig. 33.5 Location of outfall (Kothai et al. 2018)

To further enhance the wastewater management system, a sewage treatment plant was proposed to treat the discharged water, making it suitable for agricultural and gardening purposes. The network included trunk mains, large pipes responsible for the conveyance of wastewater from branch sewers and laterals, ultimately leading to either lift stations or the outfall. Branch sewers, situated at the center of streets, played a critical role in collecting wastewater from house connections and ensuring equitable service to both sides of the road. Manholes were strategically placed every 30 m and at junctions, referred to as nodes, to facilitate maintenance and access. These trunk mains spanned from the highest elevation node at 401.841 m to the lowest at 387.805 m in Zone-1 (Fig. 33.6).

Fig. 33.6 Location of pumping station for zone-1 (Kothai et al. 2018)

Lift stations, also known as pump stations, served as pivotal components in the wastewater management system, efficiently pumping sewage from lower to higher elevations when the natural gradient was insufficient for gravity flow or when utilizing gravity conveyance would result in extensive excavation and increased construction costs. In a region featuring seven distinct drainage zones leading in different directions, a smart and innovative approach has been put into action to consolidate these zones within a low-lying area. This comprehensive solution involves the utilization of a sewage storage structure, known as a wet well, alongside a strategically placed network of pumping stations at four key locations. Notably, Ward Number 34 has been selected as the ideal site for the lift station servicing Zone-1. The design of this sewer system adheres to the guidelines outlined in the CPHEEO manual, taking into account several critical parameters. These parameters include the design period, typically set at 30 years, to determine the sewer's long-term capacity, the calculation of peak flows, essential for hydraulic design, and based on a peak factor, often set at 3, which relies on the contributory population. Assumptions regarding water supply and sewage generation are made to establish flow rates.

To ensure optimal functionality, the sewers are designed to run at 80% capacity, with a minimum velocity of 0.6 m/sec to prevent erosion. The depth of starting manholes and minimum cover depth are influenced by topography and road planning details, while pipe diameter and recommended slopes play a vital role in determining velocity Table 33.1. These considerations are paramount in crafting an efficient and sustainable sewer system tailored to the specific needs of the region (Kothai et al. 2018).

In the study area, which boasts a current population of fewer than 100,000 residents, a meticulous approach has been taken to sewage management. Circular pipes have been deliberately selected for the construction of both trunk mains and branch sewers, demonstrating a well-considered choice in infrastructure design. The minimum diameter of 150 mm for these pipes has been carefully determined to match the specific requirements and capacity needs of this relatively small community. This tailored selection not only ensures efficient and effective sewage management but also contributes to cost-effectiveness in the maintenance and operation of the local

Table 33.1 Adopted slope for minimum velocity (Kothai et al. 2018)

Sewer size (mm)	Minimum slope	
	(%)	1 in
150	0.6	170
200	0.40	250
250	0.28	360
300	0.22	450
375	0.15	670
450	0.12	830
>525	0.10	1000

sewage system. This method underscores the importance of aligning infrastructure choices with the unique characteristics and demands of a given region.

33.3 Results and Discussions

33.3.1 Sanitary Sewer Network for Vasana Village

The results of this study provide crucial insights into the manhole population determination and the calculation of manhole discharge. The study calculated the population of each manhole by dividing the population of each zone by the number of manholes in that zone. This population information was essential for the discharge calculation using the given equation. Furthermore, the study involved a comparison of various parameters, including the Ratio of Discharge, Ratio of Velocity, and Depth Ratio. The elevation of each manhole was determined using "Google Earth" software, which is vital for future work related to sewer line gradient determination. Elevation data was utilized to derive the Depth of Cut and Invert Levels for each manhole. Calculations and analyses collectively contributed to the final design, as detailed in the full report (Bhavsar et al. 2017).

33.3.2 Sanitary Sewer Network for Zone-10, Belagavi City

In summary, the network's final outcomes, generated through the 'SewerGEMSV8i' software and adhering to CPHEEO parameters, successfully met the prescribed d/D ratio of 80% set for this study. These results not only met but also exceeded the defined constraints and specifications. The accompanying table provides a clear and concise overview of the study's findings, demonstrating its full compliance with CPHEEO guidelines (Table 33.2).

The schematic representation of the sewerage network in Ramateerth Nagar Zone-10 provides a clear and comprehensive overview of the system's intricate pipe network, with color coding for different pipe diameters. This visual aid simplifies identification and ensures efficient network functioning, crucial for sewage management. Additionally, the distribution of manhole depths, as presented in Table 33.3

Table 33.2 Summary of sewer network of the Ramateerth Nagar (Zone-10) (Shivapur and Bhavimani 2017)

Sr. no	Parameters	Quantity
1	Area (m^2)	103,843.75
2	Manhole in number	201.00
3	Sewer line length (m)	5,645.99

through analysis with SewerGemV8i, is valuable for assessing infrastructure conditions, facilitating maintenance and repair planning, and ensuring the safety and functionality of the sewer network (Shivapur and Bhavimani 2017).

The data in the table highlights key findings regarding manhole depths and pipe diameters in the specific design. Notably, over half of the manholes (52.24%) are shallower than 1 m, indicating a prevalence of relatively shallow manholes. Furthermore, 38.3% of manholes fall within the 1–2 m depth range. In contrast, only 7.46% of manholes have a depth between 2 to 3 m, and a mere 1.99% are in the 3 to 4 m depth range Table 33.4 (Shivapur and Bhavimani 2017). The table also provides insights into the percentage and total length of pipes with various diameters, which is crucial for project planning and management.

The sewer system, with a total length of 5646 m, features a thoughtfully distributed range of pipe diameters. Roughly 93.13% of the pipes have a diameter of 150 mm, 3.76% have a diameter of 200 mm, and 1.99% has a diameter of 300 mm. Table 33.4 (Shivapur and Bhavimani 2017). This allocation optimizes the system' efficiency, effectively managing sewer loads at various nodes. The preponderance of 150 mm pipes accommodates typical or lower loads, while strategically placed 200 and 300 mm pipes handle higher loads in specific areas. This well-balanced distribution ensures efficient waste management within the sewer system.

The study provides significant insights into manhole flow velocities in the sewerage system of Ramateerth Nagar, Zone-10 of Belagavi city. The analysis indicates that 50.7% of manholes have velocities less than 0.3 m/s, 29.85% fall within the range of 0.3–0.6 m/s, and 19.4% exceed 0.6 m/s Table 33.5 (Shivapur and Bhavimani 2017). The lower velocities, particularly those below 0.3 m/s, are attributed to low load generation at branch connection points. In compliance with CPHEEO guidelines, it is recommended that manholes with velocities exceeding 0.3 m/s should be 49.25%, and this study-work successfully meets that requirement. Additionally, the utilization of specialized software for network design has proven highly beneficial, offering an accurate and efficient approach to determine friction factors and hydraulic discharge. The study recommends a total of 201 manholes in the network,

Table 33.3 Depth wise manhole details in the study area network (Shivapur and Bhavimani 2017)

Diameter (mm)	150	200	300
Length (m)	5258	211	177
Percentage (%)	93.13	3.736	3.13

Table 33.4 Details of sewer lines in the study area network (Shivapur and Bhavimani 2017)

Sr. no	Depth of manhole (m)	Number of manhole	Percentage (%)
1	<1	105	52.24
2	1–2	77	38.30
3	2–3	15	07.46
4	3–4	04	01.99

Table 33.5 Velocity in sewer line and number of manhole experiencing the velocity

Sr. no	Diameter (mm)	Length (m)	Percentage (%)
1	450	2630	75.07
2	600	873	24.93
3	Total	3503	

with 52.24% having a depth of less than 1 m. Furthermore, 30% of manholes exhibit velocities between 0.3 and 0.6 m/s, while 19.4% exceed 0.6 m/s. The use of software streamlines the sewerage network design process, emphasizing the importance of software tools in modern infrastructure planning and design for effective urban sewerage system management.

33.3.3 *Design of Drainage Line of Bhilarewadi Village*

The comprehensive survey of the Mangadewadi area employed theodolite technology to align roads with precision. The survey data was meticulously converted into an AutoCAD format and subsequently transformed into a shapefile for integration with SewerGems, a specialized software. This facilitated efficient planning and design for sewer systems and infrastructure in the surveyed region. Population forecasting using various methods yielded an estimated population of 20,461 by 2049. The sewer network analysis, guided by CPHEEO design parameters, met all specifications, ensuring its suitability for implementation in the field. Visual representations, including tables, graphs, and profiles, were employed to provide a comprehensive understanding of the sewer network's characteristics. The d/D ratio of 80% was adhered to, and the results confirm the successful application of recommended standards in the analysis (Table 33.6).

Table 33.6 Summary of sewer network the village (Tonde et al. 2020)

Sr. no	Parameters	Total	Units
1	Manholes	178	Numbers
2	Sewer line length	3503	Meter
3	Area	377	Hectare

33.3.4 Design of Underground Sewerage System for Kalapatti

In this study, a sewer network design process was conducted, taking into account specific design parameters. This process utilized a macro-enabled Excel spreadsheet for a thorough analysis. The resulting findings have been meticulously documented, offering valuable insights and data that can be used to inform the design and implementation of the sewer network (Table 33.7).

SewerCAD is a highly efficient software designed for the planning and design of storm and sanitary sewer networks. This user-friendly tool simplifies the process of designing sewer systems, offering advanced features. In a specific project, SewerCAD successfully managed 138 manholes connecting the trunk mains, with a total of 522 manholes in Zone-1. The sewer system incorporated a combination of RCC and stoneware pipes with varying diameters ranging from 150 to 300 mm. The depth of excavation for these pipes corresponded to their respective diameters as outlined in Table 33.8. SewerCAD proves invaluable in streamlining the design and analysis of sewer systems, making it a crucial resource for engineering projects (Table 33.8).

In the sewer network design process, two methods are used to depict the pipe layout: schematic and scaled models. Schematic drawings require manual input of pipe lengths, while scaled drawings automatically calculate pipe lengths based on pipe and node positions. Figure 33.7 displays the pipe network layout for Zone-1. Using SewerCAD and adhering to CPHEEO design standards, the sewer network was analyzed, meeting all design parameters for practical field implementation. There are graphical representations in Fig. 33.7, such as elevation invert level versus length, illustrating the typical analysis of the sewer network.

Efficient analysis and design of a sewer network for a proposed area were achieved through the combined use of a macro-enabled Excel sheet and the SewerCAD software. This approach enabled the determination of the optimal pipe size for the network, resulting in highly satisfactory results with minimal variations in sewer network slope, thereby minimizing excavation depth. In the case of larger areas, where manual calculations can be labor-intensive, such tools prove invaluable when

Table 33.7 Details of the diameter with the length and velocity in sewer lines (Tonde et al. 2020)

Sr. no	Diameter (mm)	Length (m)	Percentage (%)
1	450	2630	75.007
2	600	873	24.93
3	Total	3503	

Table 33.8 Population forecasting (Kothai 2018)

Dia (mm)	150			200			250			300		
Depth range (m)	0–2	2–3	3–4	0–2	2–3	3–4	0–2	2–3	3–4	0–2	2–3	3–4
Length (m)	7869	1424	300	290	20	30	226	710	70	1221	665	210

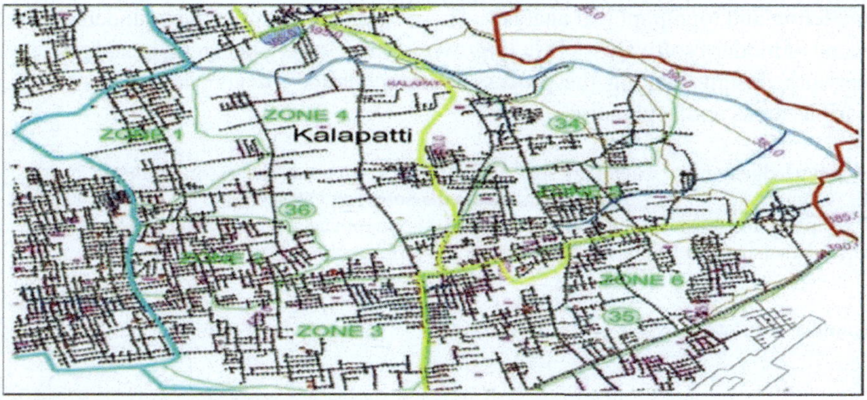

Fig. 33.7 Profile of trunk main from MH1 to MH138 (Kothai 2018)

coupled with engineering expertise. To promote environmental sustainability, a sewage treatment plant was recommended for Ward Number 34 to treat wastewater and repurpose the treated water for agricultural use, fostering resource conservation.

33.4 Conclusion

Waste water management for various villages has been thoroughly analyzed, and sewerage networks have been designed using software tools, such as MS Excel, Google Earth, SewerGEMS, and AutoCAD. This approach is particularly beneficial with inadequate waste water management systems, offering an economical and adoptable solution. The study on the sewerage system design demonstrates the usefulness of software tools for determining hydraulic parameters, making it an accurate and convenient alternative. The design process is straightforward, and the results are practical for the proposed area. The software used for sewerage network design proves to be simple and helpful, especially for larger networks. This study suggest that software simplifies sewerage network design, ensuring accuracy, 52.24% manholes < 1 m depth, 29.85% velocity 0.3–0.6 m/s, 19.4% > 0.6 m/s. The minimum velocity requirements are met, and the software's accuracy and time-saving capabilities make it a valuable tool for network design. The results are easily accessible and comprehensible. The analysis and design of the sewer network, along with the optimization of pipe sizes, were effectively carried out using software tools like macro-enabled Excel and SewerCAD. These tools simplify the design process, especially for larger areas, and can be applied with engineering judgment. Additionally, a sewage treatment plant suggestion for water treatment and agricultural use is made. These tools offer an economical and adaptable solution for areas with inadequate wastewater management systems, ensuring accurate and practical results. SewerGEMS V8i and AutoCAD Civil 3D emerge as particularly convenient software options for hydraulic modeling and topographical analysis. These conclusions collectively underscore the value of modern software tools in improving waste water management and sewerage network design, making the process more accurate, efficient, and accessible for various areas and applications.

Acknowledgements I extend my sincere thanks to Er. Bhavana Thubri, Er. Shrikant Choudhary, Er. Varun Kapate for providing guidance about the Software. I also thank to Dr. Harshal Warade Sir for support and technical writing guidance and a special thanks to Nagar Palika, Nagar Parishad, Gram Panchayat of mention area in the paper for their support.

Conflict of Interest None.

References

Bhavsar Y, Mehta R, Patel K, Bhavsar V (2017) Design of sewer network for Dolarana Vasana village using SewerGEMS software. Int J Sci Technol Eng 4(2):1–4

Chaudhary U, Kumari T, Bamiri S, Robert BRG (2020) Sanitary gravity sewer design using sewer GEMS software connect edition for Utsav Vihar, Karala. Int J Recent Technol Eng 9(2):598–602

CPHEEO Mannual (2013) Manual on sewer and sewerage system. Ministry of urban development government of India New Delhi

Fryd O, Dam T, Jensen MB (2012) A planning framework for sustainable urban drainage systems. Water Policy 14:865–886

Gulati M, Paul A (2014) Design of sewerage system for Jaffarpur area in southwest New Delhi. Int J Civ Eng Res 29–34

Kannur SS, Patil S, Kannur SS (2015) Hydraulic design and analysis of underground drainage system: for a zone in Tumkur city. Int Res J Eng Technol 2(5)

Katti M (2015) Comparative study of sewer version 3.0 and sewergGEMS V8i software for sanitary sewer network design. Int J Sci Eng Technol 3(4)

Katti M, Kumar M (2015) Design of sanitary sewer network using sewer GEMS V8i software. Int J Sci Technol Eng 2(01):254–258

Kothai A, Priya RM, Rajkumar A (2018) Design of underground sewerage system for Kalapatti, east zone Coimbatore. Int J Eng Res 7(7):127–132

Murugesh Katti KBM, Manoj Kumar B (2015) Design of sanitary sewer network for district 2B, Vijayapur city using sewerGEMSV8i software. Int J Sci Res Dev 4(08):2278–7798

Nisha Patel P (2019) Design of sewer network: a review. J Emerg Technol Innov Res 6(3):286–290

Prajapati RR, Joshipura NM, Patel B (2016) Design of underground drainage for Anklav town. Int J Sci Res Dev 4(4):735–739

Punam Harising R (2016) Design of sanitary sewer network for Gandhinagar city using sewer gems V8i software. Sci J Impact Factor 3(2):254258

Rai K (2020) Sewerage system assessment using sewer gems V8i and AutoCAD Civil 3D. Int J Eng Sci Inven 9(5):24–29

Rai RK, Nagoshe SR, Sadigale SG (2014) Trends in engineering technology and architecture

Rao D RM, Ahmed Z, Reddy KRM, Raj E (2013) Selection of drainage network using raster GIS—a case study. Int J Eng Sci Inven 2:35–40

Roy PK, Mallick S, Pal S, Banerjee G, Mazumdar A, Chakraborti R (2015) Comparative study of gravity waste water network development with different load pattern helps of software modeling. Eco Env Cons 303–308

Scarlat N, Cazanescu S, Malureanu S (2009) Modeling of sewage networks with computer aided design software bulletin UASVM Agriculture 66(2)

Shivapur AV, Bhavimani HS (2017) Design of sanitary sewer network for zone 10, Belagavi city using SewerGEMSV8i software. J Res Trends Innov 2(8):83–88

Swamee P (2001) Design of sewer line. J Environ Eng 776781

Tiwari S, Bhorkar M (2020) Design of underground drainage system in rural area using SewerGEMS software. Int Res J Eng Technol 07(06):1211–1219

Tonde M, Gavale L, Jotawar P, Admane S (2020) Design of drainage line of Bhilarewadi village in Pune area by using Bentley's SewerGEMS. Int Res J Eng Technol 7(4):5736–39

Virupakshi AS, Ravikiran K (2016) Design of sewer network system for the Janwad village using SewerCAD V8i. Int Res J Eng Technol 3(4):1375–1378

Chapter 34
Seismic Response Determination of a Seismically Isolated Reactor Building Using Frequency Domain Approach

Vijay A Salunkhe, R. S. Jangid, and Rajiv Ranjan

Abstract This paper studies the influence of Soil-Structure Interaction (SSI) on the seismic response of the base isolated Reactor Building (RB) of the Pressurized Heavy Water Reactor (PHWR) resting on the hard rock founding media. 1.6 m diameter elastomeric rubber bearing with a lead core (ERBL) are selected for seismic isolation of RB. SSI analysis of RB is performed using 0.36 g peak ground acceleration severity earthquake level which corresponds to the Beyond Design Basis Earthquake Event (BDBE) for a representative Indian Nuclear Power Plant (NPP) founded on rigid rock. The nonlinear behavior of the base isolator is modeled using the constant effective stiffness based on the design displacement of the ERBL isolator. The dynamic response of seismically isolated RB using standard and sub-structuring approaches of SSI analysis is evaluated. Seismic responses in terms of accelerations at key locations in RB structure are studied thoroughly. The results of this research show that the application of seismic isolation works effectively in reducing the structural frequencies and floor accelerations to a greater extent than the non-isolated condition. The standard approach of assuming fixed base condition and SSI analysis using sub-structuring approach for base isolated RB founded on rigid founding media and by using constant effective stiffness of the isolator shows a minimal variation in the floor accelerations. Hence, it is concluded that the SSI effect for base isolated structures founded on rigid founding media is very minimal, however seismic motion incoherency, varying founding media and incorporation of nonlinear behavior of isolator during analysis needs to be studied to evaluate the SSI effect on the seismic response of base isolated RB.

Keywords Soil-structure interaction · Seismic isolation · Bilinear curve · Reactor building · Coherent motion · Elastomeric rubber bearing with lead core

---

V. A Salunkhe (✉) · R. S. Jangid
Department of Civil Engineering, Indian Institute of Technology Bombay, Mumbai, Maharashtra 400076, India
e-mail: vijaysalunkheiitb@gmail.com

V. A Salunkhe · R. Ranjan
Nuclear Power Corporation of India Ltd., Mumbai, Maharashtra 400094, India

B. V. Khode et al. (eds.), *Innovations in Technologies: Pioneering Sustainable Infrastructure for a Resilient Future*, Sustainable Civil Infrastructures,
https://doi.org/10.1007/978-3-031-82133-2_34

34.1 Introduction

Following the Fukushima nuclear accident, new codal guidelines are proposed that demands the seismic qualification of Nuclear Power plant Structures (NPPs) for high intensity seismic events. Considering the latest codal requirements regarding qualification of NPPs against BDBE loadings and experience gained from the performance of Fukushima Daiichi NPPs during strong earthquakes, it is prudent to establish the design margins of NPPs against such high levels of seismic events or implement advanced measures which will ensure the safety of the NPP's during these unpredictable events in an economically viable way. Seismic isolation is one of the most effective seismic control techniques employed for enhancing the safety of critical NPPs. Seismically isolated building housing emergency response center for Fukushima Dai-Chi nuclear plant demonstrated and ensured the safety of the Emergency headquarters without any damage (Hijikata et al. 2011) during a strong earthquake event. It was also observed that the seismic excitation expected for the reactor vessel of the IRIS steam supply system building is reduced to a great extent by seismic isolation (Forni et al. 2009). As a result, there is a great inclination towards employing seismic isolation systems for the protection of the important safety-related NPP structures during disastrous seismic events (Perotti et al. 2013; Forni et al. 2009; Grandis et al. 2009; Zhou et al. 2016). Presently, nuclear power stations Koeberg in South Africa, Cruas, the Jules Horowitz Reactor, and ITER in France are the only NPPs supported by seismic isolation devices all over the world. Various strategies for the implementation of seismic protection systems in the UK's NPPs was studied by Medel-Vera and Ji (2016). Different isolation system responses was studied for assessing the impact of isolation damping under the action of near fault motion, and an increase in responses was observed for increased damping of the isolation system (Jangid and Kelly 2001). Presently, research efforts are in progress in Japan to develop the seismic isolation system for standardization of NPPs for high seismicity events (Kubo et al. 2014; Imaoka et al. 2015). The limited practical implementation of seismic isolation in nuclear structures an be attributed to the fact that studies are in progress at the theoretical stage and the availability of very few test results about the behavior of large isolators subjected to extreme earthquake loadings (Medel-Vera and Ji 2014). In short, the above studies indicate that research in NPPs base isolation is limited and needs additional studies in all the fields affecting the seismic isolation characteristics of the structures. SSI effect is one of the crucial components that must be taken into consideration while analyzing the seismic performance of base isolated structures.

The design and analysis of seismically isolated structures are generally performed as fixed base structures, and the SSI effect is discounted. Given the well-documented SSI impacts on NPPs seismic response, the disregard for SSI effects on base isolated structures raises concerns about the effectiveness of the isolation scheme (Zhou et al. 2016). Generally, the SSI effect leads to a reduction in the predominant frequencies of the heavy structures and modified input ground motions felt by these structures

(Drosos and Sitar 2015). On the other hand, seismic isolation also leads to reductions in structural frequencies and base shear (Constantinou and Kneifati 1988). Few studies suggest that, in comparison to fixed base structures, the impact of SSI on the seismic response of base-isolated structures is quite small (Zhou et al. 2004). One study showed that SSI effects are significant for stiff, squat structures with low mass ratios but have relatively little effect on system damping (Spyrakos et al. 2009). All of the aforementioned findings showed that the SSI effect is a crucial parameter for the assessment of the seismic response of base isolated NPPs.

This paper evaluates the response of base isolated RB of PHWR in relation to SSI impact using a 3D Finite Element model in ACS SASSI software (Ghiocel Predictive Technologies Inc. 2022). This study uses frequency domain approach for SSI analysis of base isolated RB of PHWR under the action of BDBE ground motion.

34.2 Approach of Seismic SSI Analysis

SSI analysis is executed using the frequency domain approach of the ACS SASSI software to account for the kinematic interaction. This method accounts for wave propagation through infinite soil media. Flexible Volume Sub-Structing Method (FV) is used in this study. Coupled excavated soil-structure system is described by the differential complex dynamic stiffness produced by deducting the excavated soil from the basement in the FV sub-structuring approach. FV method (or Direct method), all the excavated soil dynamic degrees of freedom are considered to be interaction nodes for computing the SSI solution. Two connected subsystems dynamics, the structure and the excavated soil, varies in terms of their dynamic complex stiffness, which defines the SSI interaction effects for an embedded foundation problem. The graphical presentation of sub-structuring methods of the SSI analysis problem into simpler domains in ACS SASSI is shown in Fig. 34.1.

34.3 Problem Statement

34.3.1 Finite Element Model of RB of PHWR

RB of PHWR with a base isolation system is used for finite element simulation, analysis, and seismic response evaluation. The RB structure is made up of four different structures. All these structures are horizontally independent and constructed over a common base raft of 100 m by 100 m in plan. The bottom elevation of RB raft is 84.5 m and ground elevation is 100 m. The RB structure is 56 m above ground and embedded below ground by 15.5 m. However, RB is assumed to be resting on the ground for FE simulation and SSI analysis, thus the embedment effect is discounted. Reinforced concrete slab and walls are modelled using four nodded shell elements.

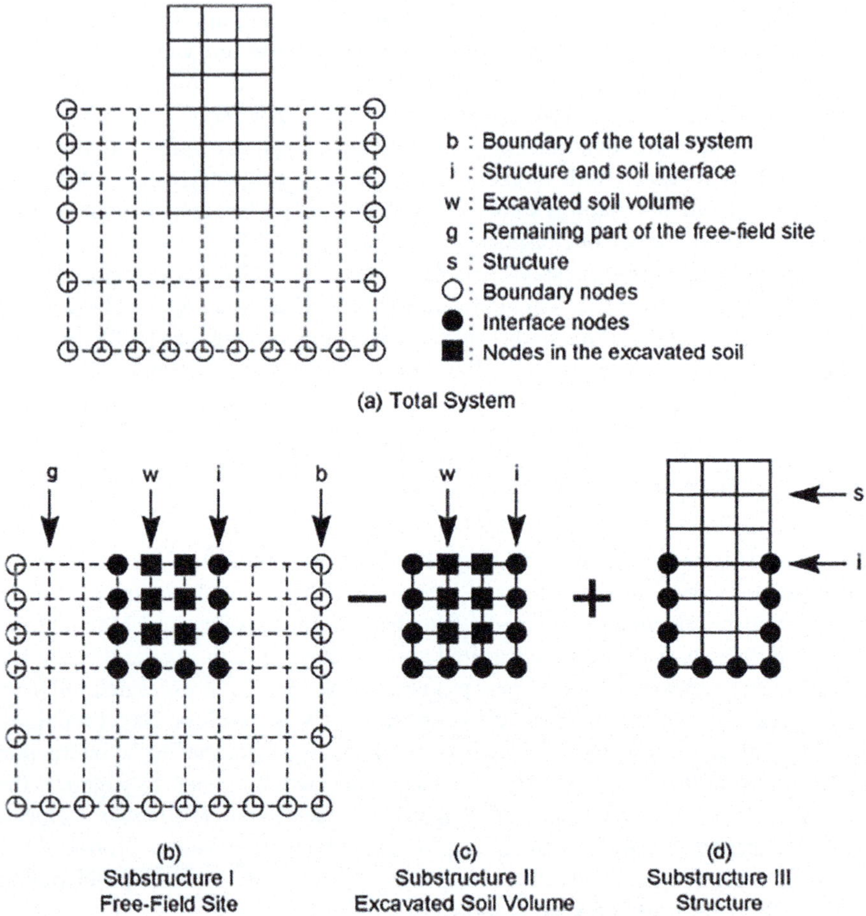

Fig. 34.1 Sub-structuring approach (ACS SASSI user manual, 2022)

The raft of RB is 7 m thick hence modeled using eight nodded solid elements. Beams and column elements are modelled using beam elements. The 3D view of the RB is given in Fig. 34.2.

34.3.2 ERBL Isolator Properties

RB is seismically isolated using 829 ERBL of 1.6 m-diameter located below the RB raft, as shown in Fig. 34.3a. The design of the ERBL is performed and design values are verified with available experimental results (Imaoka et al. 2015). The ACS SASSI tool works on the principle of equivalent linear approach, and it is a major task to simulate the ERBL by elastic springs. ERBL's effective stiffness and damping ratio

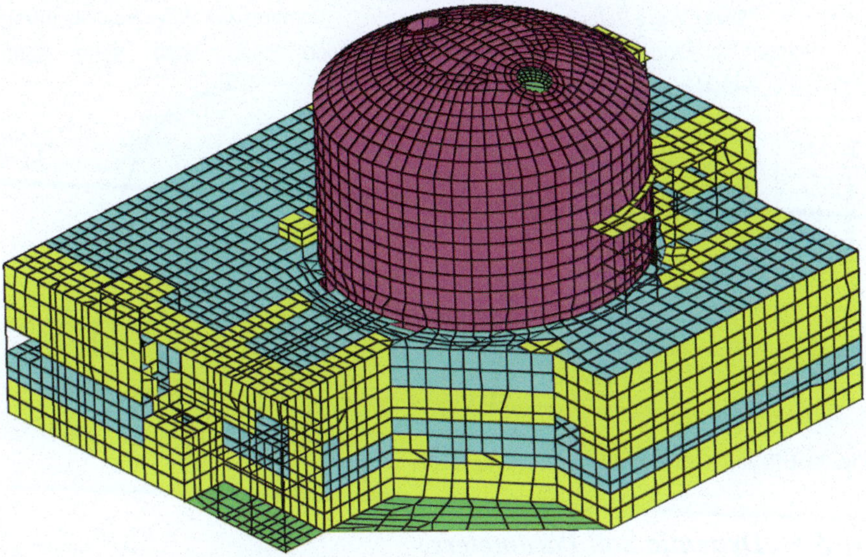

Fig. 34.2 FE model view of RB and its substructures

depend on the load vs. displacement pattern of the bearings. However, for present study the ERBL isolators are simulated as linear springs by using constant effective stiffness of isolator. ERBL characteristic parameters are axial stiffness (K_1), initial stiffness (K_0), post yield stiffness (K_2), and characteristic load (Q_d), which are 9343 kN/mm, 41.526 kN/mm, 3.069 kN/mm, and 962 kN, respectively. The load versus displacement for ERBL is shown in Fig. 34.3b.

(a) ERBL layout (b) Hysteretic curve for ERBL

Fig. 34.3 Placement of ERBL isolator below RB raft

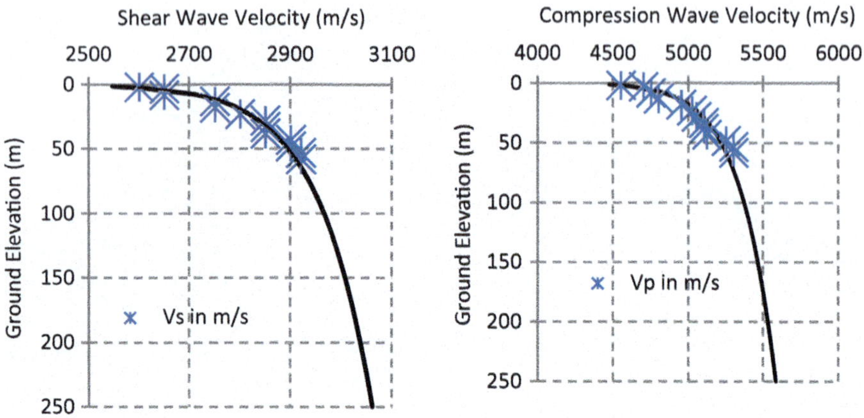

Fig. 34.4 S and P wave profile for founding media

34.3.3 Dynamic Soil Parameters

This study examines SSI's impact on isolated RB's seismic response using rock and very rigid base (to simulate fixed base) soil profiles characterized by shear (Vs) and primary (Vp) wave profiles shown in Fig. 34.4. Rock density of 27,000 N/m^3, Poisson's ratio of 0.26 and 5% of damping is assigned to the founding media. The given soil profile corresponds to the representative hard rock founding media of Indian NPP. The depth of soil profile considered in the analysis is 250 m.

34.3.4 Input Ground Motion

For each SSI analysis, unidirectional design-based ground motion serves as an input. Seismically isolated RB is subjected to a beyond-design earthquake with a PGA of 0.36 g. The synthetic time history used in the analysis is presented in Fig. 34.5. Synthetic time history illustrated in Fig. 34.5 corresponding to horizontal excitation is synthetically generated which is consistent with US NRC RG 1.60. The interval of 0.005 s is used to develop time history for 24 s.

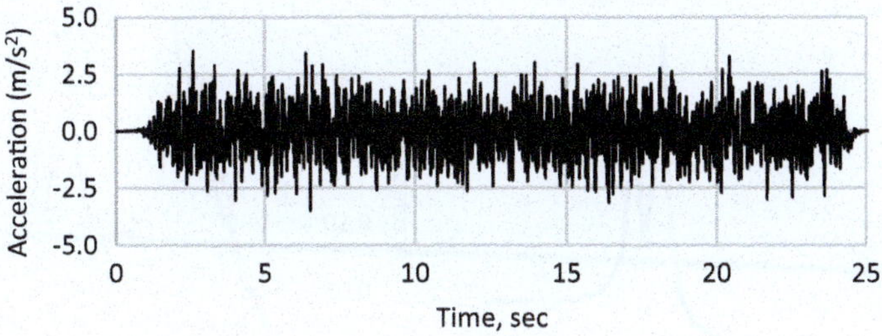

Fig. 34.5 Input ground motion corresponding to BDBE event

34.4 Seismic SSI Analysis Results and Discussion

34.4.1 Seismic Response with and Without SSI Effect on a Seismically Isolated RB under Horizontal Excitation

Dynamic analysis of RB is performed for three different cases: (i) without seismic isolation; (ii) with seismic isolation (fixed base, ignoring SSI effect); and (iii) with seismic isolation (with SSI effect). Figure 34.6 shows the comparison between 5% damped Response Spectra (FRS) in the north–south direction (horizontal) for non-isolated and isolated condition at elevation 153 m of the RB. Isolation frequencies for RB structure is 0.52 Hz. The FRS at EL. 153 m is presented in Fig. 34.6a and its enlarged view is presented in Fig. 34.6b. Figure 34.6a indicates substantial shift in structural frequencies and reduction in spectral accelerations on account of seismic isolation. SSI effect consideration shows no variation at isolation frequencies and slightly amplified spectral accelerations at the natural frequencies of the structure; however, there is no shift in the isolation or structural frequencies due to SSI consideration as seen in Fig. 34.6b.

The structural acceleration variation along the RB elevation is shown in Fig. 34.7. The FRS is derived at different elevations of the RB structure. Figure 34.7a represents the case of base isolated RB for fixed base condition which ignores the SSI effect and Fig. 34.7b corresponds to the case of base isolated RB which accounts the SSI effect. It is seen that the SSI effect enhances the floor accelerations at the natural frequencies of the structure by very minimal amount of 3.6% and no change in floor accelerations is observed at structures isolation frequencies.

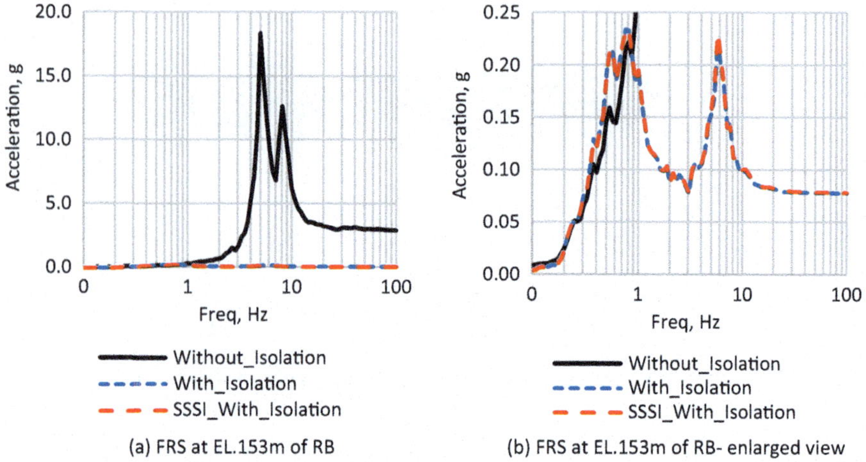

Fig. 34.6 FRS comparison for horizontal excitation for IC

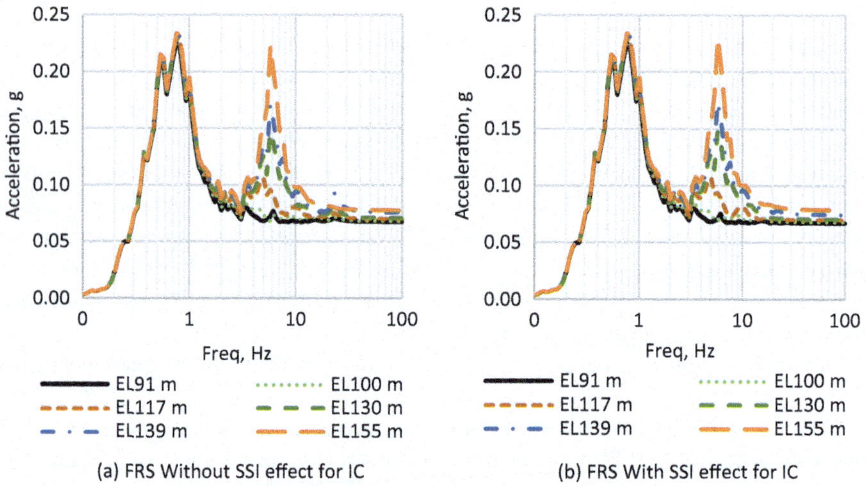

Fig. 34.7 FRS along the elevation of IC structures in horizontal direction

34.5 Conclusion

End results of studies conducted to determine SSI impact on the seismic response of base-isolated RB of PHWR are presented. In this direction, a 3D FE model of seismically isolated RB is developed in ACS SASSI followed by frequency domain SSI analysis of RB. ERBL are simulated as equivalent linear springs of constant stiffness distributed below the RB raft. RB is subjected to horizontal excitation. Following can be concluded based on the above study:

- Seismic isolation effectively de-amplified seismic accelerations at all locations of the RB under lateral excitation at all frequency ranges.
- The comparison between with and without SSI analysis results shows that the SSI effect amplifies the FRS accelerations at the natural frequencies of the structures by very minimal amount, and there is a no change in FRS accelerations at isolation frequencies.
- In the current study, the constant effective stiffness of the isolator is considered while performing the SSI analysis. The RB is assumed to be resting on the ground surface, and the incoherency effect of seismic input motion is also neglected. Hence, it is very important to simulate non-linear hysteresis behavior of isolator, embedment and the incoherency effect during SSI analysis to arrive at the realistic and more accurate seismic responses of seismically isolated RB.

References

Constantinou MC, Kneifati MC (1988) Dynamics of soil-base isolated-structure systems. J Struct Eng 114(1):211–221

De Grandis S, Domaneschi M, Perotti F (2009) A numerical procedure for computing the fragility of NPP components under random seismic excitation. Nucl Eng des 239(11):2491–2499

Drosos VA, Sitar N (2015) Soil-structure interaction effects on seismically isolated nuclear power plants. IFCEE. ASCE

Hijikata K, Takahashi M, Aoyagi T, Mashimo M (2012) Behavior of a base-isolated building at Fukushima Dai-Ichi nuclear power plant during the great east Japan earthquake. In: Proceedings of the international symposium on engineering lessons learned from the 2011 great east Japan earthquake, Tokyo Japan, pp 1542–1551

Imaoka T, Kosugi S, Kanazawa K, Nakayama T, Yamamoto T, Jimbo M, Umeki Y (2015) Development of evaluation method for seismic isolation systems of nuclear power facilities–break test of full-scale lead rubber bearings for nuclear facilities, part 1 outline of break test of LRB of 1.6 m in diameter. SMiRT-23, Division V, Paper ID 344, Manchester United Kingdom

Ghiocel Predictive Technologies Inc (2022) ACS SASSI–an advanced computational software for 3D dynamic analyses including SSI effects, revision 8, user manual. Ghiocel Predictive Technologies, Inc., NY USA

Jangid RS, Kelly JM (2001) Base isolation for near-fault motions. Earthq Eng Struct Dyn 30:691–707

Kubo T, Yamamoto T, Sato K, Jimbo M, Imaoka T, Umeki Y (2014) A seismic design of nuclear reactor building structures applying seismic isolation system in a high seismicity region—a feasibility case study in Japan. Nucl Eng Technol 46(5):581–594

Forni M, Poggianti A, Bianchi F, Forasassi G, Lo Frano R, Pugliese G, Perotti P, dell'Acqua LC, Domaneschi M, Carelli MD, Ahmed MA, Maioli A (2009) Seismic isolation of the IRIS nuclear plant. In: Proceedings of the ASME pressure vessels and piping conference (PVP '09). Prague, Czech Republic, pp 289–296

Medel-Vera C, Ji T (2014) Seismic protection technology for nuclear power plants: a systematic review. J Nucl Sci Technol 52(5):607–632

Medel-Vera C, Ji T (2016) Seismic risk control of nuclear power plants using seismic protection systems in stable continental regions: the UK case. Nucl Eng des 307:377–391

Perotti F, Domaneschi M, De Grandis S (2013) The numerical computation of seismic fragility of base-isolated nuclear power plants buildings. Nucl Eng des 262:189–200

Spyrakos CC, Koutromanos I, Maniatakis C (2009) Seismic response of a base isolated buildings including soil-structure interaction. Soil Dyn Earthq Eng 29(4):658–668

Zhou ZG, Wong J, Mahin S (2016) Potentiality of using vertical and three-dimensional isolation systems in nuclear structures. Nucl Eng Technol 48(5):1237–1251

Zhou L, Zhao R, Zhao J (2004) Analysis of the response to earthquake of the pile-soil-isolated structure interaction. Chin J Geotech Eng 26(6):782–786

Chapter 35
Critical Review on Causes of Accidents on Highway

Prayag Deshmukh and Sujesh D. Ghodmare

Abstract Highway accidents are a global problem, incurring significant human and economic costs. This detailed examination dives into the various causes of such incidents, giving significant insights into the multiple elements that lead to their occurrence. By throwing light on the complexity underlying highway accidents, the study intends to improve our knowledge of these incidences, opening the way for informed policies and actions to lessen their impact on public safety and economic well-being globally. The review encompasses various elements, including driver behaviour, road infrastructure, vehicle design, and environmental conditions, while emphasizing the importance of a holistic approach to understanding and addressing these issues. Key findings emphasize the significant involvement of human variables in accidents, such as distracted driving, speeding, and intoxicated driving. Furthermore, poor road design, insufficient signage, and badly maintained road surfaces have been recognized as key infrastructure-related concerns. Vehicle-related issues include technical flaws and design flaws. This critical review underscores the need for comprehensive policies and initiatives aimed at mitigating these causes, emphasizing the importance of education, engineering, and enforcement strategies. An integrated approach, that considers all these aspects, is crucial for reducing accidents on highways and enhancing road safety.

Keywords Driver behaviour · Road infrastructure · Vehicle issues · Environmental conditions

P. Deshmukh (✉) · S. D. Ghodmare
G H Raisoni College of Engineering, Nagpur, India
e-mail: prayag8055@gmail.com

B. V. Khode et al. (eds.), *Innovations in Technologies: Pioneering Sustainable Infrastructure for a Resilient Future*, Sustainable Civil Infrastructures,
https://doi.org/10.1007/978-3-031-82133-2_35

35.1 Introduction

35.1.1 General

Highway accidents are a global concern with disastrous effects on public health and economics. This comprehensive analysis tackles the critical need for a thorough understanding of the complex causes of these accidents, laying the groundwork for the creation of effective preventive measures. Aside from the obvious toll on life, injuries, and property, these incidents place a strain on healthcare systems, impose financial obligations, and cause incalculable pain. Highway accidents, contrary to popular belief, are rarely isolated events; rather, they result from a complex interplay of human behavior, road infrastructure, vehicle features, and environmental conditions.

Recognizing the multifaceted nature of these incidents, the review emphasizes the interdependence of various elements and aims to unravel the intricate web of circumstances surrounding highway accidents. It acknowledges that a single factor rarely causes these accidents; rather, it's the dynamic interplay of human actions, road design, maintenance practices, and vehicle-related issues that contribute to their occurrence.

The study goes beyond traditional research methods to examine the critical impact of driver behavior, looking into aspects such as distracted driving, speeding, and impaired driving. It acknowledges the symbiotic relationship between these behavioral aspects and road conditions, as well as vehicle-related issues. The review aims to provide a detailed knowledge of the linked variables that contribute to highway accidents by evaluating this aggregation holistically.

This thorough study's ultimate purpose is to support evidence-based measures targeted at reducing the occurrence of highway accidents and raising road safety standards. By diving into the complicated network of causes, the research hopes to pave the path for tailored treatments and policies that address the underlying concerns, thereby contributing to worldwide efforts to reduce the human and economic toll of highway accidents.

Figure 35.1 displays a comparative examination of accidents that occurred on the Mumbai-Pune Expressway and the Yamuna Expressway between 2020 and 2023. The Mumbai-Pune Expressway had 198 accidents, 92 fatalities, and 269 injuries in 2022, whereas the Yamuna Expressway had 262 accidents, 91 fatalities, and 559 injuries. Over the indicated years, the overall number of accidents on the Yamuna Expressway appears to be substantially greater than those on the Mumbai-Pune Expressway.

35.1.2 Literature Review

The methodology used by the multiple authors in the research demonstrates a lot of possibilities for investigating traffic accidents. Accident data from the Hunan

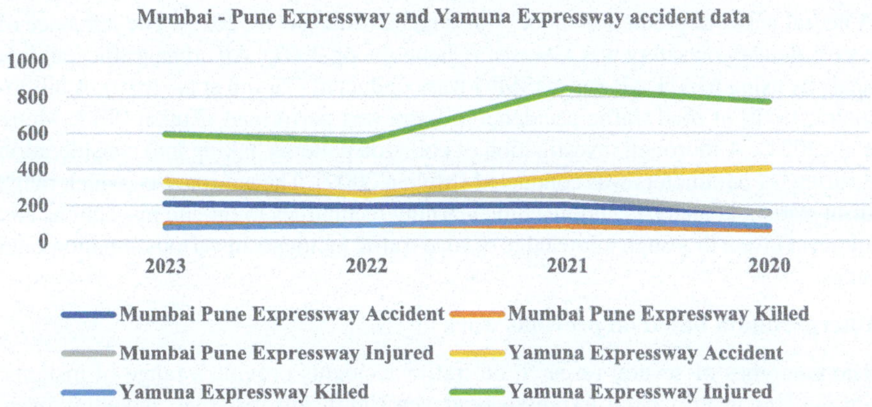

Fig. 35.1 Comparison of accidents that occurred on Mumbai—Pune expressway and Yamuna expressway from 2019 to 2023

Provincial Expressway from 2012 to 2018, estimating economic losses and finding significant variables was examined (Liu et al. 2012; Mane et al. 2020). On-site inspections, traffic device analysis, and speed surveys to uncover shocking facts concerning commercial vehicle accidents were carried out (Takey et al. 2023). Surveys, spot analysis, and rating approaches to identify unsafe places on the Mumbai-Pune Expressway and the Pune-Solapur National Highway were conducted. On the Mumbai-Pune Expressway, a mix of surveys and spot investigations to identify unsafe areas was performed (Khetade et al. 2023). A ranking approach based on the Accident Severity Index and Accident Density Method to examine the Pune-Solapur National Highway was used (Patil et al. 2013). Investigation of data from the Pune-New Mumbai Expressway was done to find accident features (Kumar et al. 2022; Sarkar et al. 2020). A binary logistic regression analysis to identify risk factors on Baghdad Expressways was performed (Joni and Jasim 2022). A quantitative survey on age and gender at RML Hospital's Emergency Department was Executed (Kaur et al. 2017). Identification of blackspots on highways using a rating technique was accomplished (Yadav et al. 2022a). A frailty model for highway accidents was developed using binary integer programming on censored data (Chung and Recker 2015). Synthetic Minority Over-sampling Technique (SMOTE) techniques were used for Advanced Traveler Information Systems (Parsa et al. 2019; Ghodmare et al. 2021). For traffic accidents, kernel density estimation and cluster significance evaluation were used (Bíl et al. 2013). A quantitative technique to evaluate spatial analysis methods was used (Yu et al. 2014). Optimal and medium placement approaches in speed experiments were evaluated (Kianfar and Edara 2010). The Maximum Likelihood and Maximum Posteriori approaches for detecting highway accidents were examined (Liyanage et al. 2022). Validation of accident analysis methods for sensitivity, consistency, and dependability were performed (Cao et al. 2023). Accident trends using data mining statistics were examined (Chen et al. 2020). Concentration on truck rear-end collision speed and headway disparity was done (Zou et al. 2022).

A neural network prediction model was created (Shang et al. 2022). The influence of access density on safety was assessed (Cheng-cheng 2007). An urban traffic incident analysis using MATLAB and VISSIM was conducted (Ya-qin et al. 2019). A bibliometric study of road traffic accidents in India was performed (Ruikar 2013; Metre et al. 2021). A thorough investigation of collision patterns, taking into consideration a variety of parameters was conducted (Mohan 2009). The approaches, which range from statistical analysis and machine learning techniques to on-site inspections and surveys, give a thorough knowledge of road traffic incidents in various situations and areas.

Interpretation based on previous work

The outcomes of several research on traffic accidents provide a range of insights. On the Hunan Provincial Expressway eleven significant variables, including overspeeding and inappropriate driving practices were found (Liu et al. 2012). Frightening numbers, such as 57% of commercial vehicles breaking speed limits and 88% of incidents due to mechanical defects were reported (Takey et al. 2023; Yadav et al. 2022b). A horrifying 90° abrupt turn on the Mumbai-Pune Expressway was discovered (Khetade et al. 2023). An accident density of 2.9 per 1000 m on the Pune-Solapur National Highway, indicating possible black spots was recorded (Patil et al. 2013). Investigation of underage driving, speeding, traffic law violations, cell phone usage, and illegal parking on the Pune-New Mumbai Expressway (Kumar et al. 2022). Violation of speed limits was the primary cause of accidents on Baghdad Expressways (Joni and Jasim 2022). 40% of incidences occurred between 12 and 6 p.m., predominantly impacting men (Kaur et al. 2017). The need to employ the Safe System Approach for successful mitigation was stressed (Yadav et al. 2022a). Six statistically significant factors that contribute to the maximum congested distance caused by highway accidents were discovered (Chung and Recker 2015). The Probabilistic Neural Network (PNN) outperforms Support Vector Machines (SVM) in road safety applications (Parsa et al. 2019; Rodke et al. 2022) 840 potentially risky locations account for 46.5% of all traffic accidents (TAs) on the network (Bíl et al. 2013). The advantage of spatial analysis methodologies was demonstrated (Yu et al. 2014). The optimal placement technique beat the halfway placement method (Kianfar and Edara 2010). Using the Maximum a posteriori (MAP) technique, considerable gains in accident detection accuracy were achieved (Liyanage et al. 2022). The impact of road geometry, weather circumstances, and accident length on assessment findings was emphasized (Cao et al. 2023). Ramp accidents to rainy weather between 12:00 and 18:00. For truck rear-end collisions were linked (Chen et al. 2020). A significant association between predicted accident rate data and probabilistic risk data was found (Zou et al. 2022). A neural network prediction model for several types of accidents was created (Shang et al. 2022). A growing access density is associated with a significant rise in overall accident rates (Cheng-cheng JtJ 2007; Mendhe et al. 2022). The present assessment technique for analyzing the traffic effect of accidents was viable (Ya-qin et al. 2019). The need for improved data collection processes to improve the accuracy of research on road traffic events in India was emphasized (Ruikar 2013). A linear drop in the yearly growth rate of road deaths to 0% by 2030 was forecasted

(Mohan 2009). The wide variety of findings emphasizes the complexities of road traffic accidents and the necessity for various methods to understand and tackling this vital issue.

Issues required to be considered for analysis

While research on traffic accidents gives useful information, they have certain drawbacks. A serious issue owing to the long 1000-km stretch for data collecting on the Hunan Provincial Expressway, which resulted in lengthy data collection and processing times (Liu et al. 2012). Concentration on commercial vehicles was a drawback, highlighting the necessity for more thorough research that includes other vehicle types (Takey et al. 2023). The single focus was on the driver's point of view, implying the need for a broader study to gain a more comprehensive knowledge of the causes leading to events on the Mumbai-Pune Expressway (Khetade et al. 2023; Nimburkar et al. 2022). A significant flaw in the study's concentration was only on four causes of accidents on the Pune-Solapur National Highway was pointed out, underlining the necessity for future research to broaden the scope and depth of knowledge about road safety problems (Patil et al. 2013). A shortcoming in the narrow emphasis, highlights the need for a more comprehensive study to properly understand the black spots on the Pune-New Mumbai Expressway (Kumar et al. 2022). Weaknesses in the assumptions concerning wet road events, underlining the significance of future research that includes real data were demonstrated (Joni and Jasim 2022). A limitation in the study's concentration on only two criteria, age, and gender, and called for future research to examine a more comprehensive collection of factors influencing road traffic injuries (Kaur et al. 2017; Khanorkar et al. 2014). Limitation in providing post-implementation assessments of the identified blind spots, indicating a possible path for additional research and progress in road safety measures (Yadav et al. 2022a). Constraints in owing a significant number of missing data, underlining the necessity for enhanced data collection approaches in future freeway studies (Chung and Recker 2015). The study's scope is limited by focusing solely on machine learning models for Advanced Traveller Information Systems (ATMS) and Variable Message Signs (VMS) (Parsa et al. 2019). Conclusion of limitations, such as the possibility of unavailability or significant interdunal variation in Average Annual Daily Traffic (AADT) data (Bíl et al. 2013; Bari and Khode 2021). Flaws in the single focus on locating hotspots, perhaps ignoring a more comprehensive knowledge of accident spatial patterns were highlighted (Yu et al. 2014). Admittedly that the limitations in the study only focus on speed studies for clustering, implying that further research is needed to evaluate the applicability and usefulness of these placement techniques over a larger variety of parameters (Kianfar and Edara 2010). Limitations in loosening the assumption that speed measurements are independent and identically distributed (iid) and presenting highway accident detection as a multi-objective optimization task were endorsed (Liyanage et al. 2022). Recognized flaws, notably the dependence on a spatial effect model for accidents (Cao et al. 2023). Limitations in the sole emphasis on truck and bus accidents, indicating that future studies should include a broader spectrum of vehicles (Chen et al. 2020). Due to their exclusive emphasis on truck rear-end incident constraints, highlighting the necessity for future research to include a broader spectrum of accident circumstances (Zou et al. 2022). Limitations in the study's concentration on multi-bridge and multi-tunnel segments, recommending that future research broaden the evaluation to diverse road designs were noted (Shang et al. 2022). The study's modest sample size, underlines the need for future research with larger sample sizes (Cheng-cheng JtJ 2007). Limitations of the study were narrow concentration on urban traffic accidents and a small sample size, urging future research to use bigger and more diverse datasets (Ya-qin et al. 2019). Shortcomings such as a lack of precise national-level traffic collision data and underestimating injury figures, underlining the need for improved data collection techniques were noted (Ruikar 2013). The limits of their analysis's decade-long period (Mohan 2009). The many constraints throughout the studies highlight the problems and complexities inherent in investigating road traffic accidents, underlining the need to resolve these limits in future research attempts.

35.2 Methodology

A concise review of major studies and research on the topic of accidents. Examine the methods used in previous studies, taking into account their advantages and disadvantages. Identifying National Highways and Expressways in India that are Relevant to Research Work. Primary and secondary data collection—Primary data collection entails direct data collection, such as surveys, interviews, or observations. Secondary data is information obtained from governmental publications, databases, and scientific studies. Driver behavior, road infrastructure, and vehicle problems are all being studied. Over speeding, Fatigue, Aggressive Driving, Collision, and Alcohol are some of the driver behavior characteristics examined. Road infrastructure features such as road quality, parking, visibility, alignment, skid, and occupancy are all investigated. Mechanical failure, low fuel, tire issue, brake failure, loading, vehicle type, and overturn are some of the variables evaluated for vehicle troubles. Use charts and graphs to glean important insights and conclusions from graphical representations of collected data. Summarize the important findings from the graphical analysis (Fig. 35.2).

35.3 Graphical Analysis

In the context of this study, graphical representation entails the use of visual aids to display data and evaluate trends connected to accident causes. To depict the proportionate distribution of accidents caused by variables such as vehicle-related difficulties, driver conduct, and infrastructure deficiencies, bar charts and pie graphs will be used. These visualizations will provide a clear and comprehensive summary, making it easier to compare and identify prominent elements.

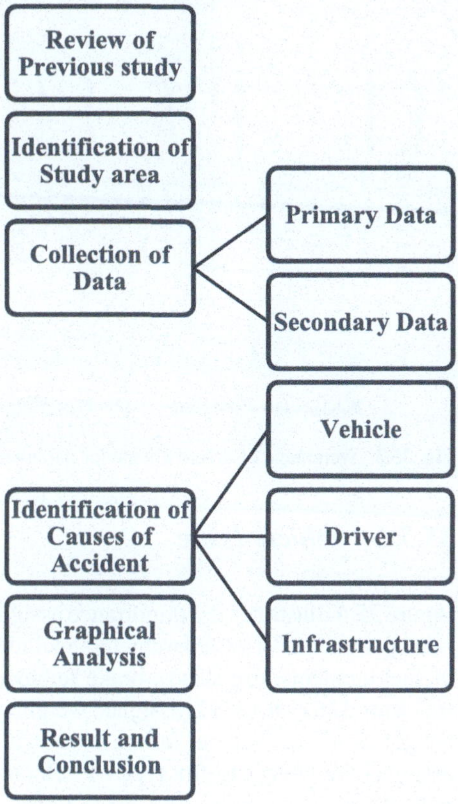

Fig. 35.2 Methodology adopted

35.3.1 *Driver's Behavior*

Figure 35.3 depicts the priority attributed by various writers to several characteristics linked to driver behavior. Several research, including those done by Liu et al. (2012), Takey et al. (2023), Joni and Jasim (2022), Zou et al. (2022), and Ruikar (2013), have investigated Aggressive Driving, assigning varying degrees of value. Liu et al. (2012), Prabhakar Khetade et al. (2023), Kumar et al. (2022), Chang et al. (2015), Parsa et al. (2019), and Shang et al. (2022) have all investigated the relevance of over speeding to varied degrees. Furthermore, research by Liu et al. (2012) and Joni and Jasim (2022) have assigned weightage to Alcohol as a driving behavior measure.

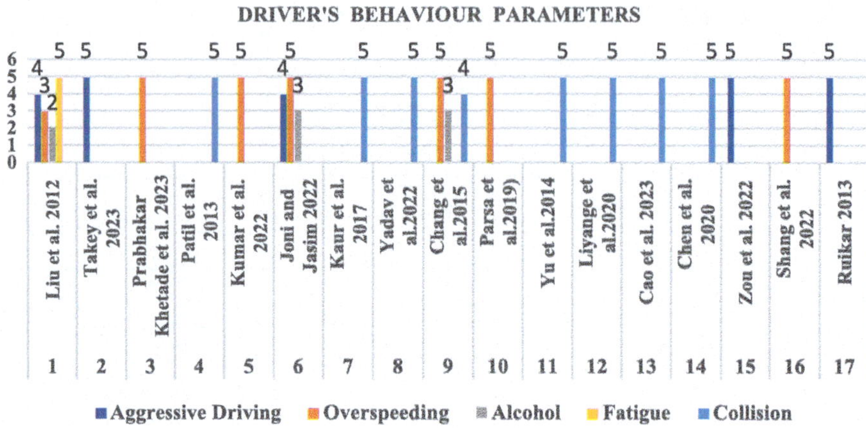

Fig. 35.3 Weightage of driver's behavior parameters considered by authors

35.3.2 *Vehicle Issue*

Figure 35.4 illustrates the importance assigned by various authors to different vehicle issue parameters. These parameters include factors such as mechanical failure, out-of-fuel incidents, tire issues, brake failure, overloading, vehicle type, and overturn incidents. Liu et al. (2012) assigned weightage to loading. Takey et al. (2023) assigned weightage to mechanical failure, out of fuel, and tyre issue. Khetade et al. (2023) assigned weightage to the type of vehicle. Patil et al. (2013) assigned weightage to overturn. Joni and Jasim (2022) assigned weightage to the tyre issue and brake failure. Chen et al. (2020) assigned weightage to the tyre issue. Zou et al. (2022) assigned weightage to type of vehicle. Ruikar (2013) assigned weightage to the type of vehicle.

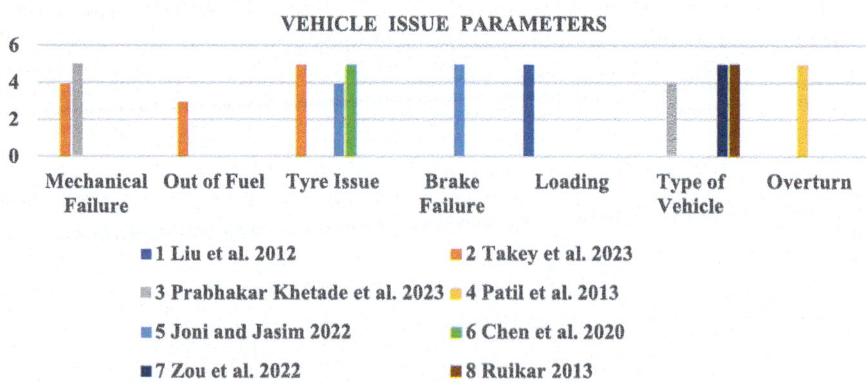

Fig. 35.4 Weightage of vehicle issue parameters considered by authors

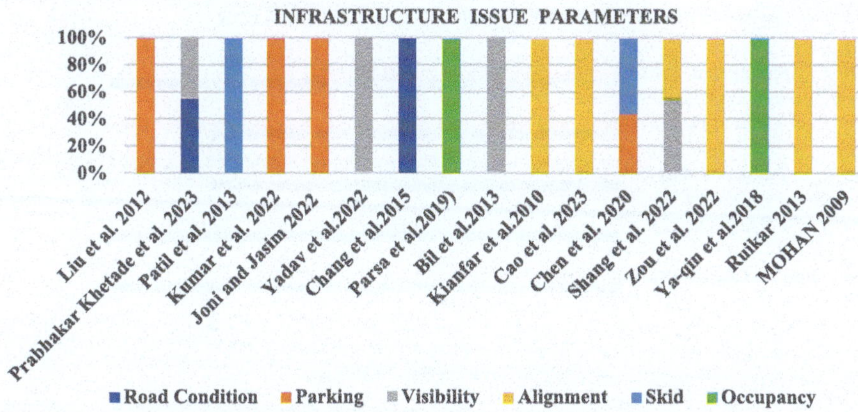

Fig. 35.5 Weightage of infrastructure issue parameters considered by authors

35.3.3 Infrastructure Issue

Figure 35.5 demonstrates the emphasis attributed to key infrastructure issue parameters as examined by numerous authors. Aspects like road conditions, parking facilities, visibility, alignment, skid resistance, and occupancy. Various authors, including Khetade et al. (2023) and Chang et al. (2015) have emphasized road conditions, attributing different degrees of weightage. Parking is considered an infrastructure issue parameter by Liu et al. (2012), Kumar et al. (2022), Joni and Jasim (2022), and Chen et al. (2020). Visibility is given weightage by authors such as Khetade et al. (2023) and Yadav et al. (2022), while alignment is deemed important by authors including Kianfar et al. (2010), Cao et al. (2023), Shang et al. (2022), Zou et al. (2022), Ruikar (2013) and Mohan (2009). Skid resistance is acknowledged by authors like Patil et al. (2013) and Chen et al. (2020). Occupancy is assigned weightage by Parsa et al. (2019) and Ya-qin et al. (2019).

35.3.4 Weather Condition

Figure 35.6 shows the relative priority accorded by various writers to various meteorological condition parameters. Rain, fog, and snow are some of the weather conditions explored by many authors in their papers. Joni and Jasim et al. (2022) emphasized the significance of rainy and foggy weather conditions. Chang et al. (2015) assigned importance specifically to rainy conditions. Parsa et al. (2019) considered both rainy and snowy conditions as important factors. Additionally, Chen et al. (2020) prioritized the analysis of in rainy and foggy conditions, while Shang et al. (2022) extended their focus to include snowy weather conditions.

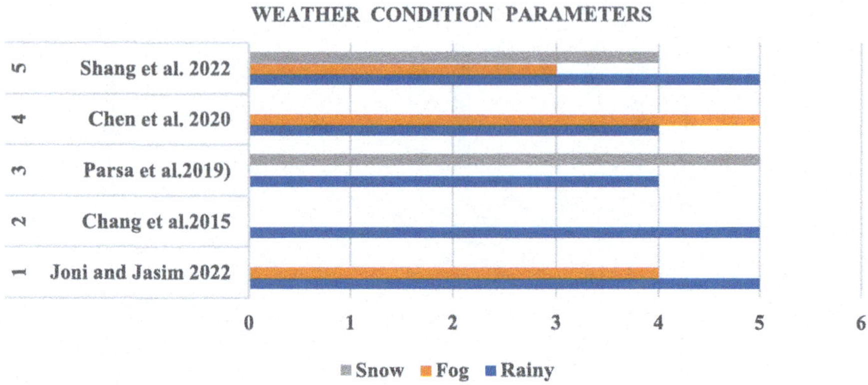

Fig. 35.6 Weightage of weather condition parameters considered by authors

35.3.5 Other

Figure 35.7 presents the relative priority ascribed to several criteria in a study by various authors. Fire, animal participation, age, gender, time, and pedestrians are among the criteria. The fire was given a weightage by Takey et al. (2023). Pedestrians were given a weightage by Liu et al. (2012), Khetade et al. (2023), and Kumar et al. (2022). Time was given a weightage by Chang et al. (2015) and Zou et al. (2022). Time and age were given weightage by Kaur et al. (2017). Mohan (2009) attributed rankings to age, time, pedestrians, and gender. Ruikar (2013) weighted age, gender, and pedestrians.

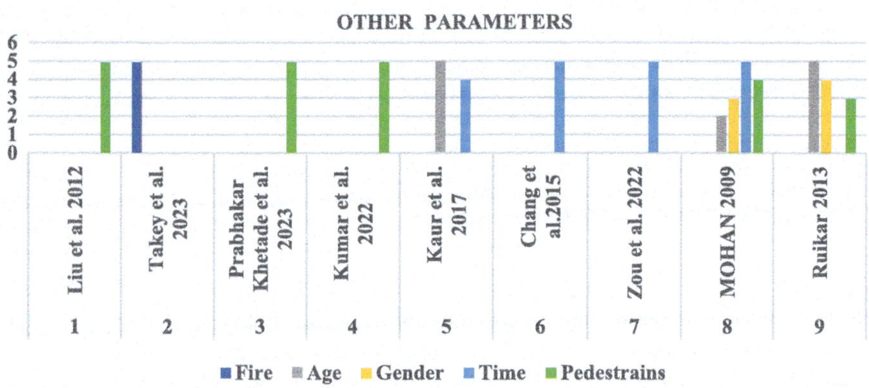

Fig. 35.7 Weightage of other parameters considered by authors

35.4 Holistic Approach to Road Safety

35.4.1 *Driver's Behavior*

Education and Awareness: Launch extensive education programs to increase public knowledge of safe driving procedures, the risks associated with careless driving, and the significance of obeying traffic laws.
Enforcement: Boost traffic law enforcement by implementing strategies like automated speed cameras, more police presence, and severe punishments for infractions.
Driver Training: Improved driver education programs should have more of an emphasis on danger awareness, defensive driving strategies, and the repercussions of risky driving habits like speeding, intoxicated driving, and distracted driving.
Incentives: Offer incentives for safe driving behavior, such as reduced insurance premiums or rewards for completing defensive driving courses.

35.4.2 *Road Infrastructure*

Road Design: Improve road design to allow for safer driving behaviors, such as wider lanes, clear signs, appropriate lighting, and correctly designated pedestrian crossings.
Safety Barriers: Install crash barriers and guardrails along hazardous stretches of road to minimize the risk of collisions and protect vulnerable road users.
Traffic Calming Measures: Implement traffic calming features such as speed bumps, roundabouts, and traffic islands to limit vehicle speeds in residential and high-risk locations.
Pedestrian and Cyclist Facilities: To improve safety for vulnerable road users, provide dedicated lanes and walkways for walkers and bicycles that are kept distinct from vehicular traffic wherever feasible.

35.4.3 *Vehicle Design*

Safety Features: Encourage the use of modern safety technologies in automobiles such as automated emergency braking, lane departure warning systems, blind-spot detection, and adaptive cruise control.
Crashworthiness: Set high safety criteria for vehicle design to ensure crashworthiness in the case of a collision, such as reinforced passenger compartments, airbags, and seat belt reminders.
Vehicle Maintenance: Encourage frequent vehicle maintenance and inspections to ensure that safety–critical components such as brakes, tyres, and lights are in optimal working order.

Innovation: Support R&D projects targeted at integrating new technologies such as vehicle-to-vehicle communication and self-driving systems to improve road safety.

35.5 Conclusion

It is possible to deduce from the information supplied that various authors researched and gave varying weights to various characteristics linked to driver behaviors, vehicle faults, infrastructure concerns, weather conditions, and other aspects in the context of road safety. The weights represent the estimated importance of various characteristics in contributing to traffic accidents. Several writers' opinions underline that a few investigations focused on weather conditions and other external elements. The most of research concentrated on driver behavior, followed by infrastructural problems and vehicle problems. According to prior research, the most common causes of accidents include aggressive driving, overspeeding, collision, tire issues, vehicle type, alignment of the road, visibility on the road, and road quality.

References

Bari CS, Khode BV (2021) PM10 dispersion modelling at unsignalised intersection using caline4—a case study. In: Lecture notes in civil engineering, pp 531–539

Bíl M, Andrášik R, Janoška Z (2013) Identification of hazardous road locations of traffic accidents using kernel density estimation and cluster significance evaluation. Accid Anal Prev 55:265–273. https://doi.org/10.1016/j.aap.2013.03.003

Cao Q, Bai Q, Li Z, et al (2023) Optimal deployment of sensors along freeway corridors for traffic accident detection. J Transp Eng A Syst 149. https://doi.org/10.1061/jtepbs.teeng-7588

Chen L, Huang S, Yang C, Chen Q (2020) Analyzing factors that influence expressway traffic crashes based on association rules: using the Shaoyang–Xinhuang section of the Shanghai–Kunming expressway as an example. J Transp Eng A Syst 146. https://doi.org/10.1061/jtepbs.0000425

Cheng-cheng JJ (2007) Safety impact on accesses in two-lane highways

Chung Y, Recker WW (2015) Frailty models for the estimation of spatiotemporally maximum congested impact information on freeway accidents. IEEE Trans Intell Transp Syst 16:2104–2112. https://doi.org/10.1109/TITS.2015.2394798

Ghodmare SD, Khode BV, Ladekar SM (2021) The role of artificial intelligence in industry 4.0 and smart city development. In: Lecture notes in civil engineering, pp 591–604

Joni H, Jasim M (2022) Identifying risk factors influencing traffic accidents for Baghdad Expressways. Eng Technol J 40:1–8. https://doi.org/10.30684/etj.2021.131184.1011

Kaur H, Ahwal S, Kurian RN (2017) A study to determine the nature and risk factors for road traffic injuries. Int J Contemp Surg 5:38. https://doi.org/10.5958/2321-1024.2017.00035.6

Khanorkar AR, Ghodmare SD, Khode BV (2014) Impact of lane width of road on passenger car unit capacity under mix traffic condition in cities on congested highways

Kianfar J, Edara P (2010) Optimizing freeway traffic sensor locations by clustering global-positioning-system-derived speed patterns. IEEE Trans Intell Transp Syst 11:738–747. https://doi.org/10.1109/TITS.2010.2051329

Kumar R, Ubhe S, Mane S, et al (2022) Black spot analysis from Pune to New Mumbai expressway optimum and safe design of rubble mound breakwater for fisheries Harbor view project sediment transport view project black spot analysis from Pune to New Mumbai expressway

Liu X, Huang S, Chen Q (2012) Analysis of the causes of expressway traffic accidents based on statistics of accidents on the Hunan provincial expressway in. https://doi.org/10.1093/tse/tdad010/7049137

Liyanage YW, Zois D-S, Chelmis C (2022) Near real-time freeway accident detection. IEEE Trans Intell Transp Syst 23:1467–1478. https://doi.org/10.1109/TITS.2020.3027494

Mane H, Ghorpade P, Bahel V (2020) Computational intelligence based model detection of disease using chest radiographs. In: 2020 international conference on emerging trends in information technology and engineering (ic-ETITE). IEEE, pp 1–5

Mendhe PD, Ghodmare SD, Titarmare AS, Wadhai PJ (2022) Sustainable transportation using automated highway system for intercity connectivity. In: 2022 10th international conference on emerging trends in engineering and technology–signal and information processing (ICETET-SIP-22). IEEE, pp 1–6

Metre K, Baghel H, Suman G, et al (2021) Compact city and related impact on sustainable development in urban areas. In: Lecture notes in civil engineering, pp 523–530

Mohan D (2009) Road accidents in India. Int Assoc Traffic Saf Sci Res 33:75–79. https://doi.org/10.1016/s0386-1112(14)60239-9

Yadav DD, Khode BV, Ghodmare SD (2022b) Metro follows an optimized speed trajectory to save energy. In: 2022 10th international conference on emerging trends in engineering and technology–signal and information processing (ICETET-SIP-22). IEEE, pp 1–6

Nimburkar AS, Ghodmare SD, Tenpe A (2022) Traffic signal synchronization system for highway corridor in urban area. In: 2022 10th international conference on emerging trends in engineering and technology–signal and information processing (ICETET-SIP-22). IEEE, pp 1–6

Parsa AB, Taghipour H, Derrible S, Mohammadian AK (2019) Real-time accident detection: coping with imbalanced data. Accid Anal Prev 129:202–210. https://doi.org/10.1016/j.aap.2019.05.014

Patil JR, Pune A, Bobade SU, Patil JR (2013) Accidental study on Pune–Solapur national highway quality management system in construction projects view project analysis and design of bored piles in rock using neural networks View project Jalindar Patil accidental study on Pune–Solapur national highway

Khetade PA, Naik RT, Nangare RS, et al (2023) Causes of accident on Mumbai–Pune expressway. Int Res J Eng Technol

Rodke SJ, Ghodmare SD, Deshmukh A (2022) Critical analysis of saturation flow rate and affecting parameters. In: 2022 10th international conference on emerging trends in engineering and technology–signal and information processing (ICETET-SIP-22). IEEE, pp 1–6

Ruikar M (2013) National statistics of road traffic accidents in India. J Orthop Traumatol Rehabil 6:1. https://doi.org/10.4103/0975-7341.118718

Sarkar M, Subudhi B, Ghosh S (2020) Unified Smith predictor-based wide-area damping controller to improve the control resiliency to communication failure. IEEE/CAA J Autom Sin 7:584–596. https://doi.org/10.1109/JAS.2020.1003066

Shang T, Tang J, Huang Z-D, et al (2022) Study on roadside accident prediction of multi-bridge and multi-tunnel section on expressway in mountainous area

Takey G, Parwate A, Nagarnaik P, Parbat D (2023) Road safety audit: a case study on Samruddhi Mahamarg, Nagpur–Mumbai expressway (ME-02). Int J Nov Res Dev 8:433–440

Yadav DK, Ghodmare SD, Naveen Kumar N (2022a) Mitigation of blackspots on highways by the application of safe system approach. Mater Today Proc 52:1228–1235. https://doi.org/10.1016/j.matpr.2021.11.042

Ya-qin H, Yu-lun R, ZU-peng L, Sheng-pin D (2019) Traffic influence degree of urban traffic accident based on speed ratio. J Highw Transp Res Dev 13:96–102

Yu H, Liu P, Chen J, Wang H (2014) Comparative analysis of the spatial analysis methods for hotspot identification. Accid Anal Prev 66:80–88. https://doi.org/10.1016/j.aap.2014.01.017

Zou H-Y, Liu K, Zhang C, et al (2022) Study on the risk of truck rear-end collision in continuous downhill section of expressway

Chapter 36
Performance Evaluation of Reinforced Flyash and Silica Fume Concrete in Aggressive Environment

Sheetal Sahare

Abstract Corrosion of steel reinforcement results in early deterioration of reinforced concrete (RC) structures reducing its strength, durability and service life. Porous nature of concrete is responsible for chloride and moisture ingress initiating corrosion in RC elements. Materials like flyash and silica fume; by-product of industrial processes otherwise threat to the environment, are now-a-days commonly used in the construction industry for the preparation of concrete. Previous researches proved that it enhances mechanical properties of concrete. In the present study performance of concrete prepared by using flyash and silica fume is experimentally investigated for corrosion resistance of RC elements when exposed to aggressive environment for different exposure durations. The experimental results witness that use of silica fume and flyash as cement replacement material and increased curing duration not only enhances mechanical properties of concrete but also improves performance of RC element there by resisting corrosion in aggressive environment.

Keywords Reinforced concrete · Corrosion resistance · Flyash · Silica fume

36.1 Introduction

Reinforced Concrete (RC) is worldwide used as a structural material in the construction industry. However, corrosion of reinforcing steel causes distress to the concrete structures. It causes cracking and spalling of concrete cover reducing strength and service life as well as affects durability of RC structures (Zhang et al. 2021; Kearsley and Joyc 2014; Apostolopoulos and Koulouris 2021; Ranjith et al. 2016). Different methods like surface treatment to rebar and concrete surfaces, use of mineral admixtures and corrosion inhibitors in concrete, cathodic protection, etc. are used to control

S. Sahare (✉)
Vishwakarma Institute of Technology, Pune 411048, Maharashtra, India
e-mail: sheetal.sahare@vit.edu

B. V. Khode et al. (eds.), *Innovations in Technologies: Pioneering Sustainable Infrastructure for a Resilient Future*, Sustainable Civil Infrastructures,
https://doi.org/10.1007/978-3-031-82133-2_36

corrosion in RC structures (Kamde and Pillai 2020; Medeiros et al. 2016; Cao et al. 2022; Akshatha et al. 2015; Cheytani and Chan 2022).

Flyash (FA) is generated as a by-product from the thermal power plants, have particle size ranging from 10 to 100 micron. Whereas Silica fume (SF) is a waste product of the ferrosilicon industry having very fine particle of size approximately 100 times smaller than that of the average cement particle. Both these waste products discharged randomly poses threat to the environment and raises health issues. On the other side FA and SF are commonly used by many sectors in the construction industry; cement production, and as supplementary cementitious materials (SCMs) for economy and thus can protect the environment by lowering the amount of CO_2 emission. When cement was partially replaced with FA and SF found to enhance mechanical properties like compressive, tensile and flexural strength of concrete (Sundararaman and Azhagarsamy 2016; Jamali and Gupta 2558; Pan et al. 2019). It also reduces permeability of concrete (Nagataki and Ujike 1986; Nath and Sarker 2011). Previous researches showed that SF with 10% and FA with 20% replacement with cement provide optimum performance of concrete (Sundararaman and Azhagarsamy 2016; Joshi 2017; Kosior-Kazberuk and Lelusz 2007). Some researchers found that increased curing duration helps to enhance strength of Flyash concrete (FAC) and Silica Fume Concrete (SFC) (Dong et al. 2020).

The main focus of this study was to experimentally investigate behaviour of RC elements under aggressive environment prepared with partial replacement of cement using 20% FA and 10% SF. The RC test specimens were cast using FAC and SFC and water cured for 28 days and 45 days followed by accelerated corrosion for 7, 13, 19 and 25 days. The aggressive environment was simulated in laboratory by accelerating corrosion in test specimens. Effect of corrosion duration on performance of FAC and SFC test specimens in aggressive environment was also investigated.

36.2 Experimental Program

36.2.1 Materials and Mix Proportion

Silica fume of size 25 μm, class F-type flyash of 45 μm and 53 grade ordinary Portland cement (OPC) of fineness 320 m^2/kg was used in the present experimental work. Crushed stone of maximum size 20 mm and river sand having size less than 4.75 mm as course and fine aggregates respectively, and potable water was used in the preparation of mix of concrete. Physical and chemical properties of cement, FA and SF are reported in Tables 36.1 and 36.2 respectively. Physical properties of fine and coarse aggregates are shown in Table 36.3.

Three types of concrete mixes, Conventional Concrete (CC), FAC and SFC were designed as per IS: 10,262-2009 corresponding to target compressive strength of M25 grade concrete. In FAC and SFC cement was replaced by FA and SF in the

Table 36.1 Physical properties of cement, FA and SF

Physical property	Cement (OPC)	FA	SF
Specific gravity	3.14	2.07	2.10
Bulk density (kg/m^3)	1440	1100	416

Table 36.2 Chemical composition of cement, FA and SF

Oxides	Concentration (wt.%)		
	Cement (OPC)	FA (Class F)	SF
CaO	62.85	12.34	0.34
SiO_2	19.07	72.08	92
Al_2O_3	5.32	5.15	1.28
Fe_2O_3	4.62	0.57	1.65
MgO	0.83	4.04	0.66
SO_3	2.50	5.06	1.97
Chloride	0.0028	–	0.03
Loss on ignition	1.05	0.76	2.10

Table 36.3 Properties of coarse and fine aggregates

Properties	Fine aggregate	Coarse aggregate
Specific gravity	2.69	2.93
Fineness modulus	3.17	6.70
Water absorption (%)	0.15%	0.91%

Table 36.4 Concrete mix proportions

Concrete mix	Cement (kg/m^3)	FA (kg/m^3)	SF (kg/m^3)	FA (kg/m^3)	CA (kg/m^3)
CC	350	–	–	704	1243
FAC (20%)	280.0	70.0	–	704	1243
SFC (10%)	315.0	–	35.0	704	1243

proportion of 20% and 10% respectively. Water-cement ratio was kept as 0.45. The mix proportions obtained are tabulated in Table 36.4.

36.2.2 Preparation of Concrete Cube Specimens

Cement, FA or SF, fine and coarse aggregates, and water were properly mixed in a tilting rotating mixer to prepare concrete. To determine compressive strength of concrete, nine replica concrete cubes of size 150 mm × 150 mm and depth 150 mm were cast using each of the three mixes. Three cubes from each set were water cured for 7, 28 and 45 days at a room temperature (28 ± 2 °C) and 100% relative humidity.

36.2.3 Preparation of RC Test Specimens

Six replica of concentrically reinforced concrete specimens of size 100 mm diameter and height of 200 mm were cast using all the three concrete mixes. Steel rebars of diameter 20 mm were placed into the concrete cylinder in such a way that it provides a concrete cover of 40 mm around the rebar. Total length of the rebar was 750 mm. The bonded length of the rebars was kept 160 mm (eight times the rebar diameter). Apart from bonded length and top 200 mm, remaining length of the bar was sealed with double layer of epoxy coating. The top 200 mm length of the bar was kept uncoated to facilitate electric connection. Figure 36.1 shows schematic view of the RC test specimen.

Out of six, three specimens cast using each concrete mix were water cured for 28 days and remaining three for next 45 days at room temperature (28 ± 2 °C) and 100% relative humidity.

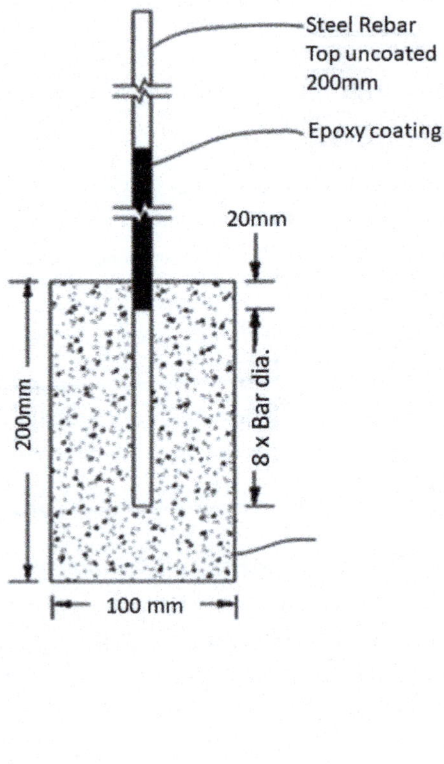

Fig. 36.1 Schematic view of test specimen for accelerated corrosion test

36.2.4 Compression Test

Cube specimens after curing for 7, 28 and 45 days were tested for compressive strength using compression testing machine of capacity 2000 kN. Three cube specimens from each concrete mix were tested and average compressive strength values were calculated.

36.2.5 Accelerated Corrosion Test Program

After 28 days and 45 days of curing, the RC test specimens were immersed in 5% NaCl solution for 24 h for complete saturation of concrete, then subjected for accelerated corrosion by impressing constant voltage of 4 V for 7, 13, 19 and 25 days. The positive and negative terminals of the external DC source were connected to the steel rebar and Stainless Steel (SS) mesh respectively. Th SS mesh was also immersed in the NaCl solution. A schematic view of the accelerated corrosion test setup is presented in Fig. 36.2. Variation in the current was continuously monitored and recorded in especially developed data monitored and acquisition system.

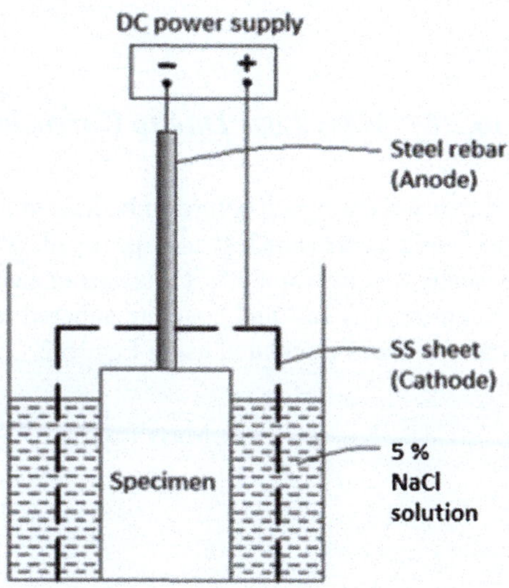

Fig. 36.2 Schematic view of accelerated corrosion test setup

36.2.6 Visual Inspection and Gravimetric Mass Loss Calculation

During the accelerated corrosion, external surface of the concrete specimen was continuously monitored for the effect of corrosion activity. Time required for development of signs of corrosion i.e. development of corrosion generated brown stain and first distinct crack on the concrete surface was noted.

36.2.7 Corrosion Current Density (I_{corr})

During accelerated corrosion, current generated in the test specimens was continuously monitored. Considering uniform corrosion on bonded length of the bars, corrosion current density in mA/cm^2 was calculated using Eq. 36.1.

$$I_{corr} = \frac{i_c}{\pi . D . L} \tag{36.1}$$

where i_c is corrosion current in mA, D is diameter and L is bonded length of the bar in cm.

36.2.8 Mass Loss Due to Corrosion

Rebars were weighed before embedded in the concrete (m_1). After completion of the accelerated corrosion test, test specimens were opened and the rebars were inspected visually for corrosion effect. Surface of each rebar was cleaned off using wire brush to remove any rust and concrete adhered to it and weighed (m_2). Then percentage mass loss was calculated using Eq. 36.2.

$$\% \text{ Mass loss} = \frac{m_1 - m_2}{m_1} \times 100 \tag{36.2}$$

36.2.9 Rate of Corrosion

The rate of corrosion in steel bar was calculated in mm per year using Eq. 36.3 as reported in previous study (Harilal et al. 2021). Where 87.6 is a conversion factor, density of steel was considered in g/cm^3, surface area of bar exposed to corrosion in cm^2, time of exposure to aggressive environment in hours and weight loss in mg.

$$\text{Corrosion rate} = \frac{87.6 \times \text{weight loss}}{\text{Density} \times \text{area} \times \text{time of exposure}} \quad (36.3)$$

36.3 Results and Discussion

36.3.1 Compressive Strength of Concrete

Average compressive strength of all the concrete mixes obtained after 7 days, 28 days and 45 days of curing is presented in the Fig. 36.3. The compressive strength in CC, FAC and SFC was found to be 20.46, 21.22 and 21.37 N/mm^2; 29.27, 30.74 and 31.82 N/mm^2; and 29.29, 31.79 and 33.25 N/mm^2 respectively after 7, 28 and 45 days of curing. Negligible increase in strength was observed in CC after 28 and 45 days of curing. After 7 days of curing, 3.71 and 4.45%; after 28 days 5.02 and 8.71%; and 8.74 and 13.52% increase in strength was observed in FAC and SFC respectively as compared to CC. It indicated that increased curing duration, helps in densification of the microstructure in concrete, thereby enhancing its compressive strength. Highest gain of strength was found in SFC both after 28 and 45 days of curing.

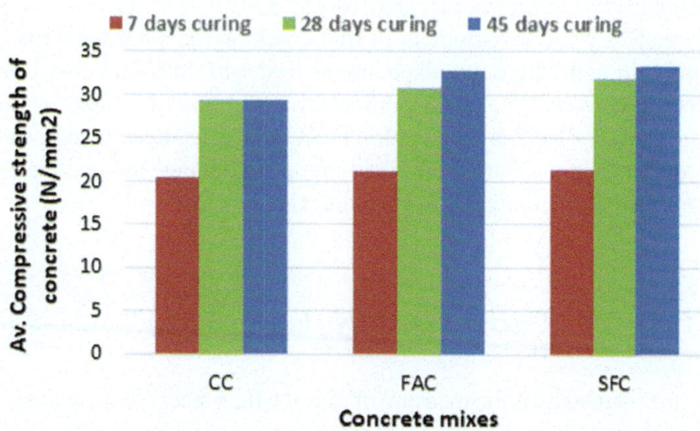

Fig. 36.3 Compressive strength versus curing period for different concrete mixes

Table 36.5 Time required for appearance of corrosion stain and crack on concrete surface

Concrete mix	Duration for stain development (days)		Duration for crack development (days)	
	Curing period		Curing period	
	28 days	45 days	28 days	45 days
CC	8	8	9	9
FAC	11	13	13	15
SFC	13	15	16	18

36.3.2 Time for Appearance of Rust Stain and Crack on Concrete Surface

From the day accelerated corrosion test started, corrosion effects on the concrete surface were visually observed rigorously. The time required for development of corrosion stain and appearance for first distinct crack on concrete surface is shown in Table 36.5. Corrosion generated (rust) brown stain was observed on 8th day and first distinct crack was developed on 9th day on concrete surface in specimens cast using CC and cured for 28 and 45 days. Rust stains were observed on 11th day and 13th day on FAC and SFA specimens respectively cured for 28 days. Whereas FAC and SFC specimens cured for 45 days showed the rust stain on 13th day and 15th day respectively. FAC specimens cured for 28 days and 45 days required 13 and 15 days respectively for development of first crack on the other hand the crack was appeared on 16th and 18th day on specimens cast with SFC and cured for 28 days and 45 days respectively. The results indicate that addition of FA and SF delayed the corrosion initiation as well as propagation in RC elements. Results indicated that use of SFC cured for longer duration efficiently reduced damage to concrete caused due to corrosion of steel as compared to FAC and CC.

36.3.3 Corrosion Current Density (I_{corr})

Corrosion current density is the measure of rate at which steel rebar corrodes in a given environment. Variation in the values of corrosion current density (***I**corr*) of different concrete mixes to be a function of time exposure to corrosion was calculated based on impressed voltage technique and gravimetric mass loss as presented in Fig. 36.4. During initial period of corrosion, all the specimens CC, FAC and SFC cured for 28 days and 45 days showed similar values of I_{corr}. But CC specimens exhibit highest value of 0.20 mA/cm^2. A reduction in values of I_{corr} was observed 0.18 mA/cm^2 in FAC and 0.15 mA/cm^{2}in SFC specimens. The values of I_{corr} found to be continuously increased with the progress of corrosion. After 25 days of corrosion I_{corr} value were 0.64, 0.59 and 0.57 mA/cm^2 in CC, FAC and SFC specimens respectively cured for 28 days. Lesser I_{corr} values in FAC and SFC specimens cured for 45 days than in

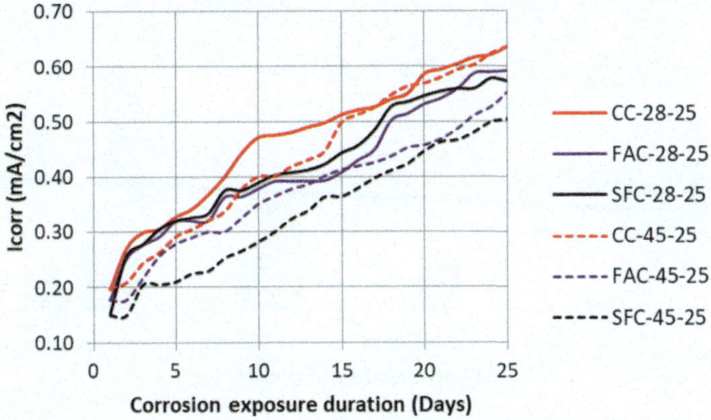

Fig. 36.4 Variation of I_{corr} with corrosion exposure duration

specimens cured for 28 days indicates contribution of pozzzolanic activity of FA and SF in resistance of corrosion in RC structures.

36.3.4 *Gravimetric Mass Loss*

Due to corrosion of steel, loss of original bar material occurs. It reduces diameter of the rebar leading to reduced strength and load carrying capacity and eventually affects structural performance of the RC element. The loss of mass of the bar occurred due to progress of corrosion is calculated in terms of percentage and expressed in the Fig. 36.5. It is found that loss of mass increases with the progress of corrosion. Maximum loss of mass (12.75%) was found in specimens cast using CC whereas minimum (1.88%) in specimens cast using SFC and cured for 45 days. Percentage mass loss in FAC specimens was less as compared to SFC specimens but more than CC specimens. Addition of FA and SF and increased curing duration improves performance of RC element under aggressive environment.

36.3.5 *Corrosion Rate*

Corrosion rate indicates the speed at which rebar deteriorates in the given environment. The corrosion rates calculated using Eq. 36.1 is presented in Table 36.6. The highest rate of corrosion 0.15 mm/year was observed in specimens cast using CC. As CC contains pores large in number which allows easy access to the chloride ions progressing corrosion at faster rate. In FAC specimens cured for 28 days and 45 days, the corrosion rate found to be 0.07 and 0.06 mm/year respectively. Corrosion rate of 0.03 and 0.02 mm/year was found in SFC specimens cured for 28 days and 45

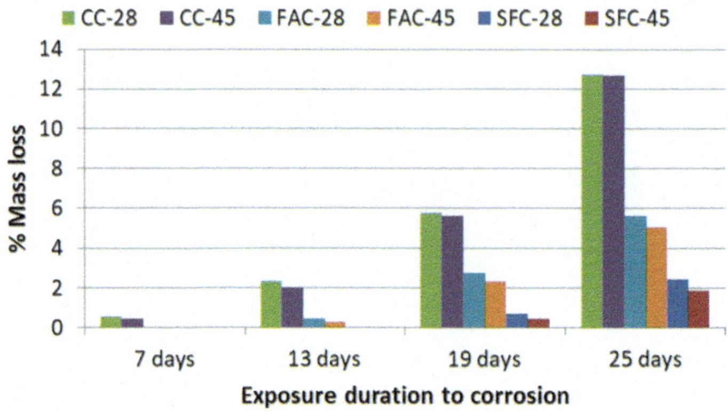

Fig. 36.5 % mass loss versus corrosion exposure duration

Table 36.6 Corrosion rate (mm/year)

Concrete mix	Corrosion rate (mm/year)	
	Curing period 28 days	Curing period 45 days
CC	0.15	0.15
FAC	0.07	0.06
SFC	0.03	0.02

days respectively. It is because addition of FA and SF resulted in pore filling effect producing impermeable concrete and blocked the access to the chloride ions. Also longer curing period helped in generation of C-S-H gel through secondary hydration reaction produces dense concrete. As SF particle size is least as compared to FA and cement particles, it occupies maximum pore space in concrete, thereby providing denser concrete as compared to CC and FAC, hence resulted in least corrosion rate.

36.4 Conclusions

Performance of FAC and SFC with 20% and 10% replacement of cement by FA and SF respectively under corrosion conditions was experimentally investigated for M25 grade of concrete. The results were compared with CC. Based on the results obtained from experimental work, following conclusions are drawn:

(1) On 25th day, corrosion current density in SFC cured for 45 days was least as compared to FAC and CC.
(2) Corrosion rate was found maximum in CC and minimum in SFC with 45 days curing.
(3) After 25 days of corrosion, minimum mass loss was observed in specimens cast with SFC and cured for 45 days.

(4) Time required for corrosion initiation and crack development found less in SFC than FAC and CC.
(5) FAC and SFC shown better corrosion resistance when cured for longer duration, indicating contribution of pozzolanic activity of FA and SF.
(6) SFC cured for longer duration showed better performance in resistance of corrosion as compared to FAC and CC.
(7) SFC cured for longer duration significantly enhances service life of RC structures in aggressive marine environment.

References

Akshatha GM, Jagadeesha Kumar BG, Pushpa (2015) Effect of corrosion inhibitors in reinforced concrete. Int J Innov Res Sci Eng Technol 4(8):6794–6801

Apostolopoulos C, Koulouris K (2021) Special issue of corrosion effects on durability of RC structures. Metals 11(11):1812

Cao Z-L, Su Z-C, Hibino M, Goda H (2022) Effects of mineral admixtures on macrocell corrosion behaviors of steel bars in chloride-contaminated concrete. Int J Corros. Article ID 3332123, 2022:1–12

Cheytani S, Chan SLI (2022) Data analysis of the long-term residual effect of cathodic protection on reinforced concrete structures. MATEC Web Conf 361:02022

Dong PS, Tuan NV, Thanh LT, Thang NC, Cu VH, Mun J-H (2020) Compressive strength development of high-volume fly ash ultra-high-performance concrete under heat curing condition with time. Appl Sci 10(20):7107

Harilal M, Kamde DK, Uthaman S, George RP, Pillai RG, Philip J, Albert SK (2021) The chloride-induced corrosion of a fly ash concrete with nanoparticles and corrosion inhibitor. Constr Build Mater 274:122097

Jamali M, Gupta S (2023) Utilization of silica fume with fly ash and properties of Portland cement-silica fume-fly ash-concrete. In: AIP conference proceedings, advancements in civil engineering COSMEC, vol 2558, no 1

Joshi R (2017) Effect on compressive strength of concrete by partial replacement of cement with fly ash. Int Res J Eng Technol 4(2):315–318

Kamde DK, Pillai RG (2020) Effect of surface preparation on corrosion of steel rebars coated with cement-polymer-composites (CPC) and embedded in concrete. Constr Build Mater 237:117616

Kearsley EP, Joyc A (2014) Effect of corrosion products on bond strength and flexural behaviour of reinforced concrete slabs. J S Afr 56(2):21–29

Kosior-Kazberuk M, Lelusz M (2007) Strength development of concrete with fly ash addition. J Civ Eng Manag 8(2):115–122

Medeiros MHF, Pereira E, Quarcioni VA, Helene PRL (2016) Surface treatment systems for concrete in marine environment: effect of concrete cover thickness. REM–Int Eng J 69(3):287–292

Nagataki S, Ujike I (1986) Air Permeability of concretes mixed with fly ash and condensed silica fume. Int Concr Abs Portal 91:1049–1068

Nath P, Sarker P (2011) Effect of fly ash on the durability properties of high strength concrete. In: Proceedings of twelfth east Asia-Pacific conference on structural engineering and construction, vol 14, pp 1149–1156

Pan Z, Zeng X, Hu Z (2019) Effect of fly ash and silica fume on mechanical properties of high-performance FRCC. IOP Conf Ser: Mater Sci Eng 562:012048

Ranjith A, Balaji Rao K, Manjunath K (2016) Evaluating the effect of corrosion on service life prediction of RC structures—a parametric study. Int J Sustain Built Environ 5:587–603

Sundararaman S, Azhagarsamy S (2016) Partial replacement of cement with fly ash and silica fume for sustainable construction. Int Res J Eng Technol 3(5):752–755
Sundararaman S, Azhagarsamy S (2016) Partial replacement of cement with fly ash and silica fume for sustainable construction. Int Res J Eng Technol 3(5). ISSN: 2395-0056
Zhang X, Zhang Y, Liu B, Liu B, Wu W, Yang C (2021) Corrosion-induced spalling of concrete cover and its effects on shear strength of RC beams. Eng Fail Anal 127:1–14

Chapter 37
A Novel Efficiency Enhancement Techniques from the Deployment of New Perovskites Solar Material

Pankaj Ramtekkar, Harikumar Naidu, and Prema Daigavane

Abstract Perovskites are around 100 times lighter than traditional solar cells per kilogramme. It produces 18 times more power. Therefore, the latest material usage perovskites is reviewed for deployment in future. The term "soiling" refers to the buildup of dust on solar panels, which lowers their efficiency and the amount of solar radiation that reaches the photovoltaic cells underneath. In certain areas, there is a shortage of water, which makes cleaning more difficult and expensive. Numerous parameters including sun intensity, module area, semiconductor, and tracking mechanism influence the efficiency of solar photovoltaic technology, which is extensively utilized for efficient energy conversion. Because of how much dust and dust accumulation affect the productivity, these topics have emerged as crucial study issues which will be of immense benefit in the future. Hence a method is proposed to detect the accumulation of dust using Node MCU Method to measure the amount of dust accumulated and to trigger the cleaning mechanism for effective and efficient operation of solar panel.

Keywords Perovskites · Solar cell · Technology

37.1 Introduction

A team of researcher from MIT have engineered ultra-light flexible solar thin-film cell which weigh around a hundred times less than a typical solar cell. It generates about 18 times more power per kilogram than the solar cell. It is ink-based material which uses scalable fabrication techniques. Further to create a solar cell, researcher used a layer of electronic ink-like nanoparticles on a ready-to-peel backing. When the panel surface is clean and there is enough solar radiation, photovoltaic panels produce their nominal power. Due to dirt and dust, solar panels' power generating input might

P. Ramtekkar (✉) · H. Naidu · P. Daigavane
G H Raisoni College of Engineering, Nagpur, India
e-mail: pankaj.ramtekkar@raisoni.net

B. V. Khode et al. (eds.), *Innovations in Technologies: Pioneering Sustainable Infrastructure for a Resilient Future*, Sustainable Civil Infrastructures,
https://doi.org/10.1007/978-3-031-82133-2_37

be lowered by up to 30% (Al-Dhaheri et al. 2010). The need for solar power systems can be met globally since there is an abundance of sun radiation (Al-Qubaisi et al. 2009). The world's energy demands might be met 10,000 times over by the solar radiation that penetrates the earth's surface. The globe is greatly impacted by solar panels, which may improve. Environmental conditions without requiring the utilization of more damaging power plants for the environment (Rankovska and Goranov 2020). However, everyday maintenance is necessary for solar power facilities, which increases their cost significantly (Al-Qubaisi et al. 2009). In addition to providing benefits that are used during the day, solar plants can be operated off-grid, on-grid, with or without energy storage, and they can even just offset large consumers' use, which lowers the cost of electricity (Aly and Gandhidasan 2015).

37.1.1 How Dust Affects the Efficiency of Solar Panels

The most common cause preventing a solar panel from performing as intended is dust. Due to dust buildup, the panel's power output might decrease by up to 45–50%.

It is likely that the panel accumulates sand and fine particles in arid environments (Radoi and Bejan 2014). The amount of sunlight that reaches the solar panels is reduced as an outcome of the dust layer's gradual and final buildup (Electrodynamic cleaning technology for solar panels 2020). Comprehension criteria for dust are as follows:

1. A wind's speed.
2. Properties of dust.
3. Surroundings.
4. Heat and humidity levels.
5. Composition of solar panels.

37.1.2 Solar Panels May Be Cleaned Using a Variety of Techniques

To remove dust and grime from solar panels, many methods are often employed, including manual cleaning, sprinkler systems, Robot cleaning, Gekko solar, and PV-spin module cleaning solar systems (Olmez and Ergezer 2021). With this cleaning technique, users utilize pure water to clean the solar panel at least once a month. The device's flip-flop action is utilized to clear dust particles from the display exterior (Jagtap and Chandrakar 2021). One drawback of the manual cleaning method is that it requires personnel to clean for longer lengths of time, which reduces power production due to erroneous time determination (Al-Dhaheri and Lamont 2010). A robot that cleans photovoltaic cells using brushes, demineralized water, and steady pressure was created by GEKKO Solar (Motwani and Adware 2023; Naidu and Ramtekkar 2023). With the exception of replacing and refilling the water filters,

Heliotex requires no maintenance (Edke and Adware 2017). Electrostatic surface technology in automatic cleaning systems makes them efficient due to their low power consumption and continuous operation.

37.2 Methods

Easy screen-printing methods involve adding an electrode to the peak of the solar cell that gathers the power created. The photovoltaic cell can then be separated from the substrate and bonded onto a range of materials, including flexible textiles. This type of lightweight solar cell could open up a world of applications.

- By hand cleaning:– Soap or industrial cleaning detergents are the most popular methods for cleaning solar panels, but the discharge can damage plants and animals as well as leave stains on homes. Because solar panel frames are especially vulnerable to some cleaning agents, it's critical to select safe chemicals that won't harm the environment over time (Electrodynamic cleaning technology for solar panels 2020). While renewable energy sources, such as solar panels, work to lessen greenhouse gas emissions, it's equally critical to think about how cleaning goods and techniques affect the environment (Al-Qubaisi et al. 2009).
- Sprinkler systems:– In arid regions, they can be an affordable way to maintain clean solar panels. These products, which are available on websites like Heliotex, function by simulating the cleansing impact of rainfall (Aly and Gandhidasan 2015). When cleaning solar panels, sprinkler systems can be a more environmentally friendly choice than cleaning solutions that can damage nearby plants, trees, and wildlife or contaminate the soil with hazardous substances. In arid and sandy locations, Heliotex's sprinkler system for cleaning solar panels works well. The system can be improved with water treatment for hard water in addition to having a soap dispenser and water filter. But it's possible that the system won't clean as well as a brush system.
- Robotic Tidying:– A solar panel cleaning robot called "Wash Panel" was created by an Italian business. It employs a vertical brush to clean PV panel arrays. The robot is powered by a 12 V battery and can be deployed automatically. It has two preset functions that work with water jets and a rain sensor to provide consistent and reliable cleaning. It is completely self-sufficient. Because it is modular, the system can be deployed on a variety of surfaces, including as roofs, buildings, and ground systems (Radoi and Bejan 2014). The robot may be remotely operated and delivers text messages for ongoing monitoring, as depicted in Fig. 37.1.
- Gekko Solar: -

  Robotic cleaning devices known as "Gekko Solar" and "Gekko Solar Farm" have been created by Gekko Solar-Sabot Swiss Innovations for movable installation on solar photovoltaic panels. (5) Gekko Solar uses demineralized water and a spinning brush to clean up to 400 m^2/h. It has a move-able lifting platform for support, and an operator can operate it using a joystick. The robot can move

Fig. 37.1 Wash panel cleaning over SPV

flexibly in whatever direction it chooses thanks to hoover technology, which is based on its feet. With a joystick, it can also be radiocontrolled.

Specialising in cleaning huge solar parks, the GEKKO Solar Farm is a robot that works well. Large solar panel rows present the biggest obstacle, necessitating effective cleaning equipment. With its ability to clean 2000 m^3/h, the GEKKO Solar Farm is the perfect choice for this application. It has the capacity to transverse cleaning up to 60 cm and climb as much as 30°. Along with an incorporated crane arm for extra adaptability, the robot is controlled by radio control with the use of a joystick. Maintaining massive solar installations is made easier with the help of the GEKKO Solar Farm (Olmez and Ergezer 2021).

- The Ecoppia E4 A completely autonomous robot called the Ecoppia E4 was created to clean long rows of solar panels in arid and sandy conditions. Equipped with spinning microfiber brushes that travel downward, generating an airflow that removes dust as shown in Fig. 37.2 the robot's built-in solar panel and energy storing batteries allow it to function at night. The robot's bottom-mounted solar panel allows the brushes to clean the rail as well. Even in arid environments, The E4 disinfectant preserves solar panels operating at maximum efficiency by eliminating 99% of dust every day.

  Eliminating physical work and wastewater from water, the fleet of E4 robots for cleaning may be absently controlled via a dashboard or smartphone software. Robots from E4 employ an environmentally friendly, water-less system that clean itself both throughout and after cleaning procedures. The robot's battery life is greatly increased by the Eco-hybrid system. All things considered, the Ecoppia E4 is a practical and environmentally friendly way to clean solar panels under challenging conditions (Rankovska and Goranov 2020).

Fig. 37.2 Ecoppia E4 robot

- The NOMADD:– Solar panel cleaning in arid settings can be done entirely automatically with the NOMADD (NO Water Mechanical Automated Dusting Device) technology. Using just three moving parts and a special mechanical motion, it performs a cost-effective daily clean while eliminating 99.6% of dust. The system is made up of one specialized device per row of panels, an affordable guiding rail system that ensures no pressure is applied to the panel surfaces, and onboard power and control systems. Up to 200 m long and 5 m wide rows can be cleaned by the device; plans call for a 300-m range. With one tool and less than 20 min per machine, inexperienced staff can complete maintenance by replacing inexpensive, basic parts on a 2–4-year cycle. With eight patents filed for its distinct mechanical operations, the NOMADD is a creative answer to the problem of cleaning solar panels in arid environments (Naidu and Ramtekkar 2023).

Maintenance requirements, and limited place, were taken into account during the system's brief construction and testing. The electrical foundation of the design was based on microcontroller technology because of its many benefits, one of them being low power consumption. Because microcontroller technology was incorporated into the design, power losses were calculated and discovered to be insignificant. In general, the cleaning system with an automated microcontroller PIC is a good way to lessen the sheading's effects on PV panels outcome power (Wath et al. 2022). The automatic cleaning device with PLC microcontroller is intended to raise the power outcome of photovoltaic modules, especially for offshore applications where dependability and efficiency are crucial. A programmable logic controller, which enables low power usage and program-me upgrades, powers and regulates the system as shown in Fig. 37.3. The primary component of the gadget, which has been built and evaluated, is the PLC, which powers likewise regulates the electrical and mechanical elements of the blueprint as shown.

The system is fully automated and doesn't require human input, save from initial activation and maintenance. The device is effective in clearing the PV panels of dust and bird droppings, which improves output and boosts efficiency and dependability for offshore applications. All things considered, the automatic cleaning device with a PLC micro-controller is a useful way to boost PV module performance in offshore settings (Qureshi et al. 2023).

Mechanism of electrodynamics (EDS) Unwanted material scan be removed from surfaces using electromagnetic forces by an Electrodynamics System (EDS). Electrodynamic cleaning and electrodynamic separation are other names for this procedure. EDS technology uses low-power, three-phase pulsed voltages to clean digitalized electrodes composed of Zinc Oxide within an observable dielectric sheet as shown in Fig. 37.4. By supplying a high voltage to a metal plate or wire, an electric field is produced in an EDS. The material that needs to be removed interacts with the flow of charged particles created by this electric field, including ions and electrons.

A force known as electromagnetic force is produced by the charged particles, pushing the undesirable material away from the surface and towards a point of collection. They work especially well at eliminating small particles like dust and grime that are challenging to clean with conventional techniques. Since EDS systems don't

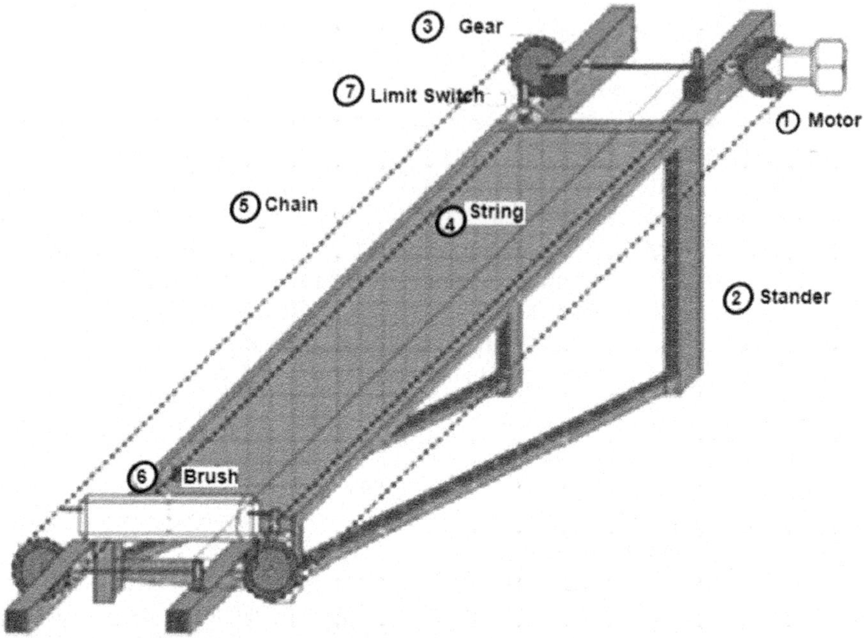

Fig. 37.3 Automated cleaning systems with PLC control

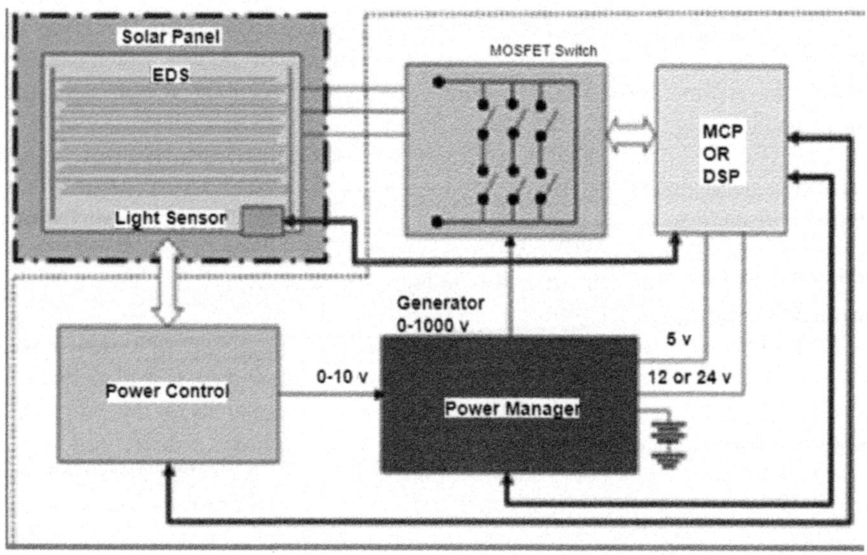

Fig. 37.4 Electro-dynamic system

require chemicals or generate hazardous byproducts, they are also environmentally benign. Even though the EDS needs an external high-voltage power source to function, the PV cell's power output can be used to make the device self-sustaining. In just a few minutes, the system can restore reflectivity up to 90% and be sustainable by combining a transparent EDS with a PV array. All things considered, EDS systems provide a practical and eco-friendly means of cleaning PV solar panels and getting rid of undesirable materials. They are a crucial piece of technology for many different sectors and uses.

Proposed Scheme to enhance the performance

Node MCU is an open source IoT platform. It includes firmware which runs on the ESP8266 Wi-Fi SoC from Espress if Systems, and hardware which is based on the ESP-12 module. The term "NodeMCU" by default refers to the firmware rather than the DevKit. The firmware uses the Lua scripting language. It is based on the eLua project, and built on the Espressif Non-OS SDK for ESP8266. It uses many open source projects, such as lua-cjson.

The Fig. 37.5 depicts the working principle, where when the dust accumulation is above the certain level set value or below certain value or some sensor or other sensors then the information is passed through ESP 8266 Wi-fi to the DDAS Server and tells the cleaning to operate and reduce the dust on the solar panel. Code 532 if 532 means decrease so this data comes back to the Wi-fi and stop the process if the dust is below a certain value. However, the loop is continuously operation the proposed scheme by monitoring and removing the dust from the panel.

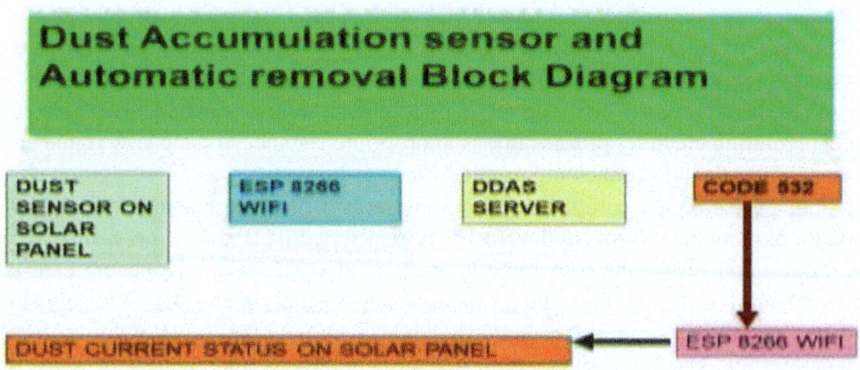

Fig. 37.5 Block diagram for automatic dust removal scheme

37.3 Discussions

This could be attached to a boat's sails to give propulsion while at sea or used to attach to traps frequently used in disaster recovery efforts. Most crucially, the researchers state that there are few installation requirements for lightweight solar technology, making it simple to incorporate into constructed surroundings. Upon testing the apparatus, researchers discovered that the solar panel made of high-strength fabric could produce approximately 370 W/kg. This is around eighteen times the power per kilogramme. Next are the standard silicon solar cells. An average silicon solar array would raise the roof's weight by about a thousand pounds. However, a little more than fifty pounds of this solar cloth would be sufficient. Additionally, they examined the device's mechanical durability and discovered that, even after more than 500 rolls and unrolls of the cloth solar panel, the cells maintained over 90% of their original power-generating capacity. The annual yield may not be significantly affected by manual cleaning if it is not done on a regular basis. These cleaning robots have certain drawbacks despite their many advantages over conventional manual cleaning techniques. For instance, some of them need to be fed or have their water and cleaning supplies refilled on a frequent basis due to their poor performance. Furthermore, certain robots can find it challenging to clean panels with intricate surfaces or steep slopes. These restrictions should be overcome as technology advances, though, and cleaning robots should become even more effective and efficient in the future. It minimizes water waste by cleaning the panels with a low pressure water spray and collecting and reusing the water. The robot may be operated remotely or be set to operate autonomously. It has sensors that sense when the panels need to be cleaned. Robots from Heliotex can clean both level and slanted panels, and they can be tailored to meet the unique requirements of any solar array. To maintain optimal performance, continuous maintenance is required and a sizable initial investment is required. Micro-controller-based methods provide a fully automated, adaptable, and self-regulating cleaning process that is appropriate for special panels. A framework that explains the different elements that influence dust settling and assimilation. Dust settlement is a highly complicated process that is difficult to manage in real-world situations. The cleaning system effectively removes dust from a variety of surfaces without endangering the protective layer. It is also effective whether an external power source is used or not. The problem of dust formation on solar panels must be addressed in order to preserve their efficiency, even with the limits of conventional cleaning methods. Because to cutting-edge technologies and creative designs, the market is still able to provide a variety of efficient cleaning solutions for various uses. Increasing solar PV panel output and encouraging the usage of renewable energy sources is the ultimate objective.

37.4 Conclusion

Even though perovskites are significantly lighter and more flexible than conventional solar cells, they still need to be made since air interaction could harm the cells, and covering them with thick glass would be counterproductive. Thus, the team is currently working on creating incredibly thin plastic wrapping that would just slightly increase the device's weight. According to the researcher, their goal is to eliminate as much of the plastic material as they can without compromising the solar material's qualities. After reviewing the many solar panel cleaning methods discussed in this paper, a brush-type cleaning system seems to be a workable substitute due to its inexpensive cost and room for domestic growth. In addition to being a useful addition to existing solar PV systems, it requires less water to clean. A thorough testing programme must be implemented to evaluate the brush-type cleaning system's power consumption, dust removal effectiveness, frequency of operation, cleaning mechanism speed, dependability, durability, and overall economic feasibility. Before conducting extensive testing, it won't be able to determine whether this method of cleaning solar panels is a practical strategy to boost the output and efficiency of solar PV systems. However, the proposed scheme of deployment of Node MCU will be a productive solution to enhance the efficiency of solar panels in future. This will help the R&D and Industrial experts in utilizing the proposed methods in their future endeavours.

References

Al-Dhaheri SM, Lamont LA (2010) Automated design for boosting offshore photovoltaic performance. P O Box: 2533, Abu Dhabi, UAE

Al-Dhaheri S, Lamont L, El Chaar L, Al-Ameri O (2010) Automated design for boosting photovoltaic (PV) performance offshore. In: Proceedings of 2010 transmission and distribution conference and exposition, Abu Dhabi

Al-Qubaisi E, Al-Ameri M, Al-Obaidi A, Rabia M, ElChaar L, Lamont LA (2009) Microcontroller based dust cleaning system for a standalone photovoltaic system, Abu Dhabi. In: Proceedings of the 2009 electric power and energy conversion systems

Aly SP, Gandhidasan P (2015) Novel dry cleaning machine for photovoltaic and solar panels. Qatar environment and energy research institute (QEERI), Hamad BinKhalifa University (HBKU), Qatar foundation PO Box 5825, Doha, Qatar. IEEE

Bhawoorjar I, Jagtap P (2022) Grid-connected hybrid PV power system performance analysis using unified power flow controller. In: 2022 6th International conference on intelligent computing and control systems (ICICCS). IEEE, pp 136–140

Edke RK, Adware RH (2017) Improvement in dynamic performance of an three phase induction motor by using ultra capacitor. In: 2017 Third international conference on advances in electrical, electronics, information, communication and bio-informatics (AEEICB). IEEE, pp 515–519

Electrodynamic cleaning technology for solar panels (2020) Int J Energy Environ Eng

Jagtap P, Chandrakar V (2021) Comparative study of UPFC controllers to improve transient and dynamic stability of power system. In: 2021 IEEE 4th International conference on computing, power and communication technologies (GUCON). IEEE, pp 1–7

Khandelwal V, Ramtekkar P, Chauhan M, Bhute Y, Kouthekar R (2022) Sensor based vibration analysis of motor using MATLAB software. In: 2022 10th International conference on emerging trends in engineering and technology-signal and information processing (ICETET-SIP-22). IEEE, pp 1–4

Motwani D, Adware R (2023) Fast charging EV battery technology: a step towards sustainable future. In: 2023 14th International conference on computing communication and networking technologies (ICCCNT). IEEE, pp 1–6

Naidu H, Ramtekkar P (2023) An innovative affordable robot to defuse landmines using IoT and wireless communication technique to save precious life. In: 2023 11th International conference on emerging trends in engineering & technology-signal and information processing (ICETET-SIP). IEEE, pp 1–5

Olmez BB, Ergezer O (2021) Autonomous solar panel cleaning robot wheeled and air absorbing motor. Int J Energy Appl Technol 8(4)

Qureshi N, Balpande C, Tembhurne H, Khubalkar S, Daigavane P, Naidu H (2023) Enhancing battery life of electric vehicle with super-capacitor. In: 2023 IEEE renewable energy and sustainable e-mobility conference (RESEM). IEEE, pp 1–6

Radoi R, Bejan M (2014) Mechatronic drive system for cleaning machine of photovoltaic panels. Hidraulica (No. 4/2014). ISSN 1453-7303

Rankovska VV, Goranov GD (2020) PIC microcontroller module in a flexible microprocessor development system. In: 2020 55th International scientific conference on information, communication and energy systems and technologies (ICEST), Niš, Serbia, pp 3–6. https://doi.org/10.1109/ICEST49890.2020.9232775

Vidhate AA, Nangrani S (2019) Multi PV array analysis and real time monitoring using two WIRE interface. In: 2019 3rd International conference on trends in electronics and informatics (ICOEI). IEEE, pp 1095–1098

Walde K, Ramtekkar P (2022) Designing parameters and MATLAB model of 11 KV/440-volt mobile distribution station. In: 2022 2nd Asian conference on innovation in technology (ASIANCON). IEEE, pp 1–5

Wath AK, Daigavane PM, Naidu HK (2022) A technique of image and fuzzy logic to solve traffic congestion at signal. In: 2022 International conference on smart generation computing, communication and networking (SMART GENCON). IEEE, pp 1–6

Chapter 38
A Review on Soil-Structure Interaction on Composite Structure

P. S. Ganorkar and Sanket Sanghai

Abstract Soil-structure interaction (SSI) plays a crucial role in the performance and safety of composite structures subjected to various environmental loads. This review provides an overview of recent research advancements and methodologies concerning SSI in composite structures. The study encompasses both theoretical developments and practical applications, focusing on the interaction mechanisms between soil and composite elements. Beginning with an introduction to the significance of SSI in composite structures, the review discusses the key factors influencing this interaction, including soil properties, structural characteristics, and loading conditions. Various analytical and numerical approaches for modeling SSI are examined, highlighting their advantages, limitations, and applicability to different types of composite structures. Furthermore, the review addresses the implications of SSI on the dynamic behavior, stability, and seismic performance of composite structures. It explores the effects of soil-structure interaction on natural frequencies, damping ratios, and response spectra, emphasizing the importance of accurate modeling for reliable predictions and design optimization. Additionally, case studies and experimental investigations are presented to illustrate real-world applications and validate analytical predictions. The review identifies gaps in current research and proposes future directions for enhancing the understanding and prediction of SSI in composite structures, such as incorporating advanced material models, considering nonlinear soil behavior, and accounting for site-specific conditions.

Keywords Structure-soil-structure interaction · Seismic performance · Soft soil

P. S. Ganorkar (✉) · S. Sanghai
G H Raisoni College of Engineering, Nagpur, India
e-mail: prajwal.ganorkar6@gmail.com

S. Sanghai
e-mail: sanket.sanghai@raisoni.net

B. V. Khode et al. (eds.), *Innovations in Technologies: Pioneering Sustainable Infrastructure for a Resilient Future*, Sustainable Civil Infrastructures,
https://doi.org/10.1007/978-3-031-82133-2_38

38.1 Introduction

Considerable research has been conducted in the area of soil and structure intersections. An examination of the interactions between various design conditions—such as soft ground beneath foundations or adjacent buildings is a mandatory component of the design process. Numerous variables, including the building's length, the material, and the material itself, affect the damage pattern during an earthquake (such as shear wave propagation speed). The building's height in relation to the foundation's length, the frequency structure's height above ground, the foundation's excitation rate, and, lastly, the superstructure's relative stiffness in relation to the ground are all significant additional factors. The force of the ground on the behavior of the structure (and vice versa) is divided into two types: inertial effects and kinematic effects. Numerous studies have led to the concept of inertial interaction, which describes the interaction between the superstructure and the ground as a result of building movement. The variance between the baseline ground motion and the anticipated response in the absence of the model is a common definition of the interaction of motions. Three main reasons make it necessary to examine soil interactions. First, when considering CAE (Computational Approach Element), the duration of this model is generally longer than the duration of the fixed model. This will cause the critical moment of the model to decrease in the longitudinal region of the design, which generally has lower pseudo-acceleration response values. However, sometimes it is pushed further into the so-called acceleration zone, as in fragile buildings (Malhotra 2015). Second, due to its ductile strength and nonlinear response, the subsoil itself will function as a damping mechanism. Third, the reaction method is used to lessen the structure's seismic input load, with the assumption that the structure will enter the elastic region in an excited state (Adeel 2009; Sanghai and Gurmule 2024). This article's goal is to provide an overview of scientific and computational studies on the interfaces between soil, rock, and structure, enabling readers to closely examine these studies.

Composite Structures

Composite structures in buildings involve combining different materials to create structures that jointly carry loads and resist energy. Generally, these models combine materials such as stone, metal, wood, or composite materials (such as fiber and polymer) to get the best results from each product.

Following are few types of combinations in buildings:

Composite Beams: These join steel and concrete as beams, taking advantage of the high strength of steel and the durability of concrete. The power line can be embedded in the stone or connected to the stone to make it more efficient.

Composite Columns: Similar to composite lines, composite lines use steel and concrete to create lines that can be constructed. These lines will have a metal core surrounded by concrete or other materials to provide greater strength and stability.

Fiber-Reinforced Polymer (FRP) Composites: FRP materials such as carbon fiber or fiberglass are used in materials such as beams, columns and steel bars.

They provide strength-to-weight ratio, corrosion resistance and durability and often replace traditional steel bars in concrete.

Composite Slabs: They offer different materials such as concrete, steel panels, and rebar to create the floor system. In addition to providing tensile strength when molded to form the concrete panel, the steel deck serves as formwork during construction.

Hybrid Structures: They provide a variety of information and optimization methods. For example, a combination of steel frames and concrete can be used to make a building strong and durable.

Advantages of composite structures in buildings include increased strength, reduced material consumption, improved durability and sometimes cost savings through proper construction and construction. To guarantee the integrity and safety of the samples, however, the design of composite structures necessitates careful consideration of materials, relationships, structure, and durability.

Soil Structure Interaction on Building

Soil-architecture the term "interaction" describes how a building (or other structure) interacts and plays with the ground or foundation that it is built upon. Understanding this effect is mandatory for the design of safe and robust structures, especially in areas with seismic activity or unstable soil.

Some important points about ground interaction are:

Foundation Design: The foundation of the building is the connection between the structure and the soil. Different soils (such as clay, sand, rock) have different properties that affect how they support the weight of the structure. Engineers examine the soil to create suitable foundations (e.g., shallow foundations, deep foundations) to distribute building materials evenly.

Load Transmission: Buildings apply vertical loads (gravity loads) and lateral loads (wind, earthquake) to the ground. These loads are transferred through the building structure and finally to the foundation and ground. Understanding how soils respond to these loads is important to ensure stability and minimize settlement or displacement.

Soil Properties: Soil properties such as bearing capacity, hardness, consistency and shear strength affect its interaction with the structure. Professionals examine these features to determine suitability and design for a location.

Seismic Effects: In earthquake zones, building-soil interaction is important during an earthquake. Ground motion and the behaviour of the soil (liquefaction, expansion) influence how the structure reacts. Architects use special seismic design techniques, such as foundation isolation or damping systems, to reduce these effects.

Soil Improvement: Sometimes the soil may not be suitable to support the structure. Engineers can use a variety of soil improvement methods, such as compaction, tamping, or the use of piles, to increase the load-bearing capacity of the earth soil.

Monitoring and Analysis: It is important to regularly monitor and review building performance, especially in areas where soil conditions change. This helps evaluate any changes in soil-structure interactions and allows adjustments or reinforcements to be made to maintain structural integrity.

More significantly, in order to guarantee that the building is robust, secure, and able to endure a range of conditions, engineers and architects must take into account how soil interacts with the structure during the design, construction, and maintenance phases of the project.

38.2 Analytical Models of Soil in Literature

The theory of soil interaction (SSI) has garnered interest from a diverse group of scholars owing to its pertinence to soil and mechanical engineering concerning structural engineering and earthquake engineering. With the advent of nuclear-powered planets in the late 20th century, SSI research was first carried out. Efforts have been made recently to simulate soft soils using dynamic SSI (Rahgozar 1993; Sanghai and Pawade 2022; Horvath and Rachlew 2015).

In general, lands with shear wave speeds above 600 m/s will cause significant changes in the seismic zone of buildings due to the impact of foundations and structures. This is especially true for models that will stand the test of time. These changes are stated below:

A. Make natural design and damping of model.
B. Cause external change.
C. Vary the base shear force (based on the frequency content as well as the type and force of the input shock).

SSI effects during seismic analysis usually this is done in one of two ways:

A. Full analysis of the deep structure and the field of change and interaction of the same.

38.2.1 Structure-Soil-Structure Interaction Under Static and Dynamic Loading Conditions

Many studies have focused on the effects of SSI in static products. These studies (Martel 1940; Tanabashi and Ishizaki 1953) show that the magnitude of the force changes as the static properties change. Numerous research works have examined SSI in three dimensions. It is known that studies of soil interaction models in two dimensions could overestimate or underestimate the response. ("Martel RR, Tanabashi R, Ishizaki H") Dynamic load research has demonstrated that a building structure's basic support can reduce the structure's overall stiffness heavy. Lessen the structure's overall stiffness. The structure's seismic response will alter as a result of its lengthening period (Menglin et al. 2011).

In SSI, two methods are introduced to create a simple model describing soil data: the Winkler model and the continuous model. However, since every model has shortcomings, these theories have undergone numerous revisions, some of which are listed below.

38.2.1.1 Winkler Model

In this case, many resources determine the character of the soil. Springs are similar but independent, rigid and linearly elastic. It is assumed that base deformations are caused by arbitrary loads and are limited to the load-bearing area. Figure 38.1 shows a good example.

The development of the Winkler model is as follows (Menglin et al. 2011):

A. Filonenko-Borodich principle: The connection between individual springs in the Winkler model is altered as a result; this connection is made by a thin, elastic membrane T, which is constantly under tension. The top of the spring membrane is where the spring membrane is attached. The spring-spring interaction is created by tension in the membrane.
B. Hetenyi basis: This formula can be seen as a concept that connects the Winkler basis and the isotropic continuum process. Each bow is connected to a string or spring. The only calculation made is for the beam's (or plate's) bending deformation. There are several instances provided in Hetenyi (1950); Passsternakpl 1954).
C. Pasternak Foundation: Method includes, the end of the spring is connected only to the beam or plate that is exposed to shear deformation. Details of this method and analysis can be found in Kerr (1964), Wang et al. (2001), Kerr (1965).
D. Generalized basis: The main idea of this method is every single point of contact's moment is related to the rotational angle, which is Winkler's hypothesis. Readers can refer to Wang et al. (2001) for brief details about this method.

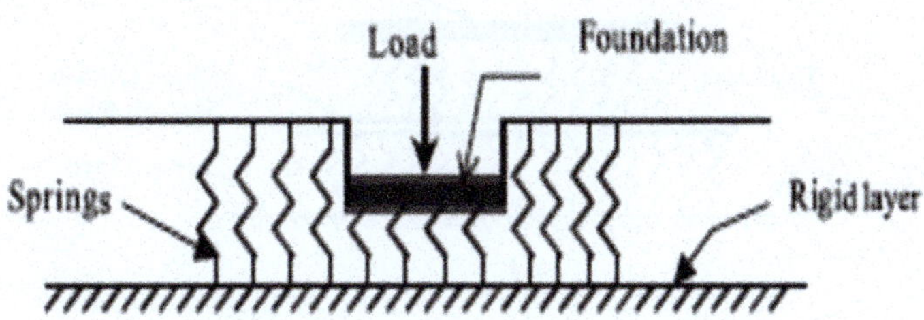

Fig. 38.1 Winkler idealization of soil-based media (Menglin et al. 2011)

E. Kerr Foundation: A variation of this method is to presume that the shear layer and the spring constant in the Winkler model are different at the top & bottom of the layer (Kukwas et al. 2023; Horvath 1993).
F. Beam-column analogy model: This method resolves the traditional issue of beams on elastic foundations by the use of a fresh subgrade model (Jakovlevjuv 1954).
G. New extended Winkler model: This method makes the connection through intersection (there is no arc in the Winkler model) (Kurian and Manojkumar 2001; Ali and Sanghai 2021; Vlasov 1949). The diagrammatic of (A–D) methods is displayed in Fig. 38.2.

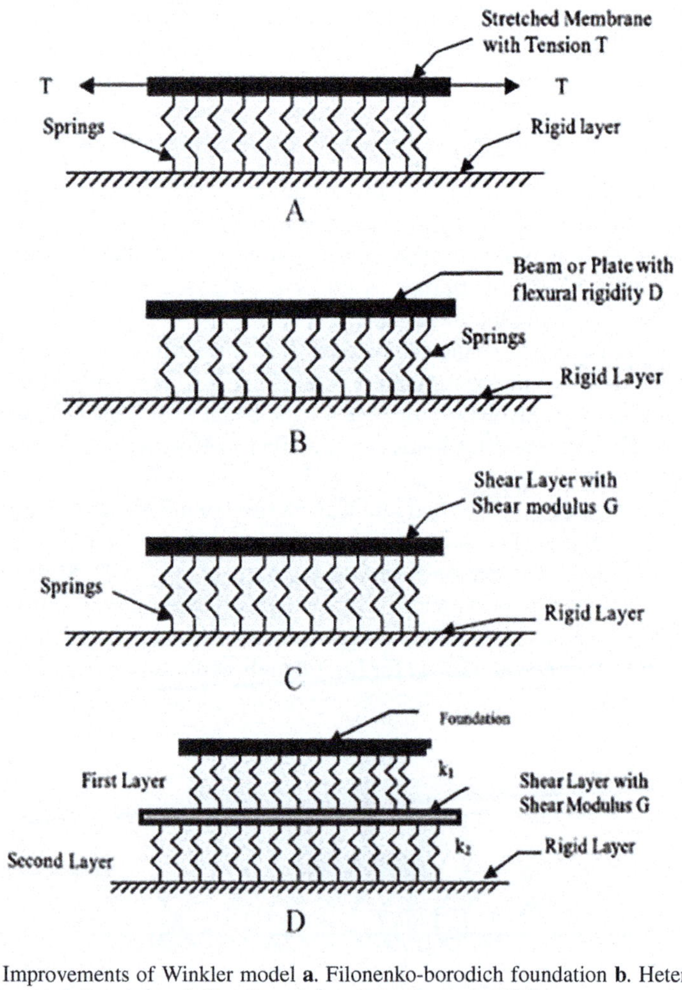

Fig. 38.2 Improvements of Winkler model **a**. Filonenko-borodich foundation **b**. Hetenyi's foundation **c**. Pasternak foundation **d**. Kerr foundation (Menglin et al. 2011)

38.2.1.2 Elastic Continuum Model

The idea will be launched from the Businsk project. Soil contains a particle compressed by interparticle impacts. This method represents soil using the principles of continuum mechanics. The theory represent, Soil is regarded as an isotropic substance. In this way, the input parameters' simplicity and the important instructions they provide for stresses and deformations make it more attractive than Winkler's aim. However, the inaccuracy of this method must be taken into account when calculating the root boundary response.

Evolution of the continuum model:

A. Vlasov Foundation:
B. Reissner Foundation.

In the first method, different elements are used to create continuous models (Vlasov et al. 1960; Wen et al. 2016). Some limitations with this method are restricted to the elastic layer's potential deformations. In the second method, the correlation between the pressure deflection where interface between the slab and the ground is calculated according to the following assumptions: The initial presumption is that the plane stress of the entire base is small and negligible, the second assumption is. Horizontal displacement of the lower and upper part of the foundation level Stress is about zero.

More information on the development of models of Winkler and continuum models is located in the reference (Menglin et al. 2011).

38.2.2 Dynamic Soil-Structure Interaction

In 2000, C. B. Crouse conducted a study on SSI's energy distribution. The importance of this study is in the use of dampers or separators like friction dampers, liquid dampers or separators. As mentioned before, SSI is beneficial because it reduces the seismic response of the building. Experiments in the field of SSI also show that nuclear power plants and mid-rise buildings are examples of constructions, have large damping ratios, consistent with theoretical calculations.

Power distribution models in non-nuclear analysis The SSI Power plant model is a model for dynamic analysis in the software industry. The diagrammatic of the ground connection model is mention in Fig. 38.3 (Nagpure and Sanghai 2018) Root and ground. At this stage the fixed basis is estimated but the estimated basis is not included. However, the combination of Soil structure is influenced by damping technique and foundation. And the soil in foundation interface. When calculating vertical vibrations and horizontal vibrations, highest base damping in this case, lowest when moving in a rolling motion. In this instance, the basic damping ratio exceeds the commonly recommended 5% damping ratio for buildings. This study presents a simple method to provide benchmarks for SSI systems. The reader is presented with a detailed example of the seismic tank design taking SSI energy dissipation into account.

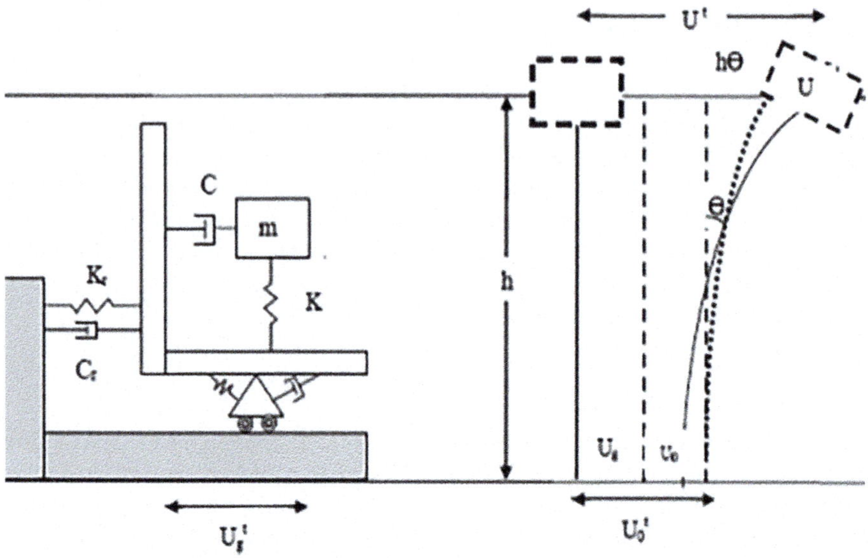

Fig. 38.3 Free body diagram of soil-structure interaction

Dynamic CAE is especially useful for low frequency features. Therefore, the system's modal damping ratio increases, which leads to the formation of complex-valued frequencies. The critical frequency and the coupling system's modal damping ratio is highly dependent on the soil's hardness. Interestingly, video links are not appreciated on difficult sites (Chandak and Pawade 1913).

Gives and Stewart proposed that feedback excitations using soil-foundation-structure systems can be modelled as free moments to provide kinematic interactions. Be a representative. Guidelines for measuring the performance of foundation impedance interaction are described in Foundation-Ground Impedance and Standards (NEHRP 2012; Mahore et al. 2023; Garcia 2008).

Garcia investigates the impact of interactions between soil and structure on analysis and design. He looked on the impact of building seismic design and the relationship between soil and structure on the structure's dynamic response (Gerolymos and Gazetas 2006b). He said that in his research, the interaction with the ground was taken into account in the analyses to increase the period of the vibration and damping system and to reduce the horizontal spectral acceleration values. 6-storey reinforced concrete frame building.

In order to investigate how caisson foundations behave under static, cyclic, and dynamic loading scenarios, Gerolymos and Gazetas created the Winkler spring nonlinear approach. They developed a mathematical method to evaluate the results obtained from experiments and test data (Chang-Liang 1974). Previous studies have investigated the dynamic response of caissons under inertial and kinematic loads in an elastic half-space. Figure 38.4 represents the design. The prediction model and 3D wave propagation were evaluated to obtain Winkler spring stiffness and damping parameters (Sung 1953). When doing a seismic study of a structure, two commonly utilized approaches that consider the ground's effect are: batch size (also known as the batch parameter) and elastic half-point theory (Bhurse et al. 2019). Numerous academic works assert that, in comparison to other procedures, the first method yields better and more comprehensive results. Chen et al. an analysis was conducted to determine how the interplay of the soil, structural, and inertia systems affected the seismic response of buildings and, considering the SSI effect, it was found that the natural moment of inertia system controls the suspended structure and its changes (Xuj et al 2004). They also found that inert systems were less efficient in terms of energy consumption due to the effect of the sample-to-soil-to-sample ratio. Figure 38.5 shows their mechanical structure.

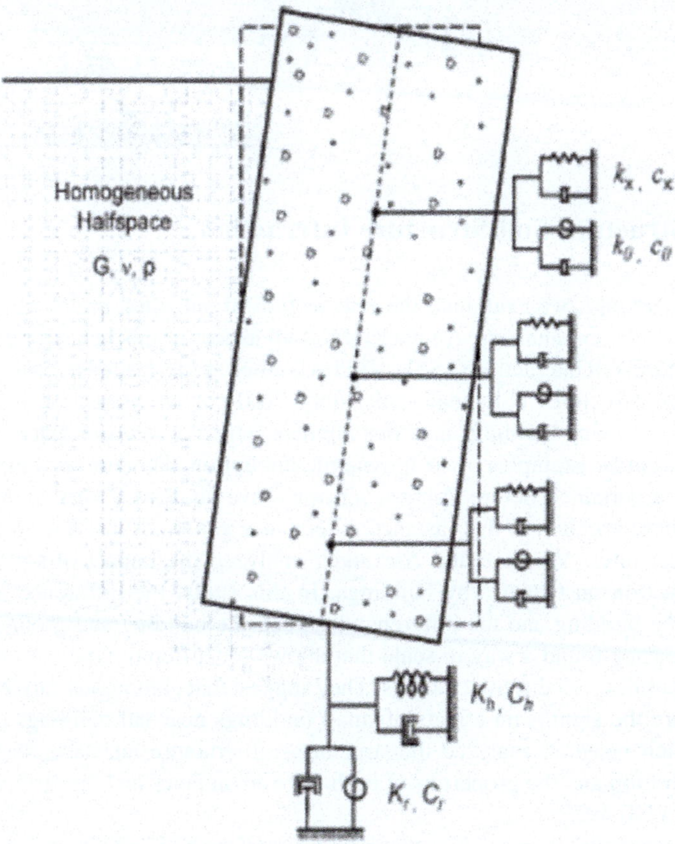

Fig. 38.4 The four varieties of dashpots and springs for the study of kinematically and inertially loaded caissons (Given et al. 2012)

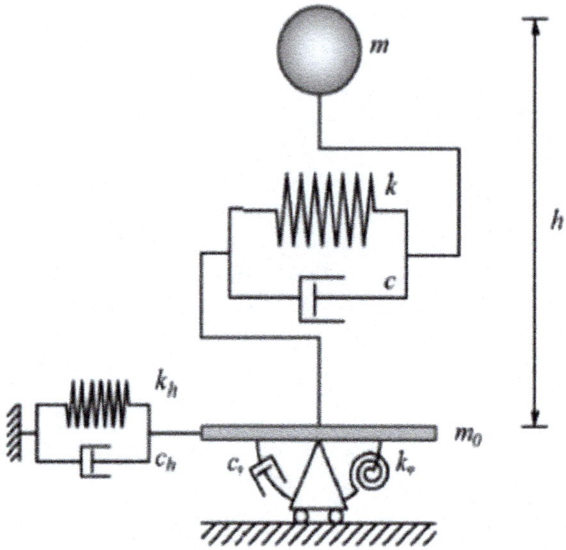

Fig. 38.5 Mathematical representation of the SDOF structure on soft soil (Gerolymos and Gazetas 2006b)

38.3 Structure-Soil-Structure Interaction

In the earlier part, we examined the intersection of soils and briefly reviewed the models used in soil analysis. Another important aspect of this behavior arises from soil-structure-soil interaction (SSSI). SSSI is defined as the ground interplay between two nearby structures (Clouteau and Aubry 2001). This interaction involves the interaction of many buildings near one another. Seismic waves or external stresses exerted might be examples of such dynamic disruption. The motion's input matrix has to be ascertained before doing a seismic wave analysis. Every component of the wavefield problem in this instance is seen as a generator that has an impact on nearby structures. Vicencio and Alexander analyzed the impact of soil-structure-soil interaction on two nearby buildings. In this study, the difference in seismic input to the building and the difference between the building and ground type are analyzed. They found a way to scale that allows for different types of interactions between buildings of different heights. They suggest that continuous variable models can capture the significant effects of small buildings near tall buildings (Xuj et al. 2004). Their research analyzed the structure-soil-structure interface, showing that soil penetration and the properties of each have advantages and disadvantages (Lee IK, Brown PT, 1972).

Some efforts has been made to better understand the SSSI effects. An experiment, mathematical experiments, calculations, tests and comparisons have been made and are summarized below.

38.3.1 Application of Numerical and Infield Methods in the Study of SSSI

Because there are several mathematical techniques that may fully represent the gamut of possible results when modeling different scenarios, these methods are of greater interest than analytical methods when testing model-prey-intercept models. The advantage of this approach is the ability to model behavioral problems such as abnormal stress behaviors, inconsistent data, and changes in the elderly. There are several ways to numerically simulate the SSSI issue, including the boundary approach, hybrid method, finite element method, and standard finite element software (Chang-Liang 1974).

38.3.1.1 Finite Element, Boundary Element and Finite Difference

The physical properties of the soil are used to represent it as a material in these programs (FEM, BEM, and FDM). (Such as modulus, stiffness, damping ratio). Regarding soil and how it interacts with structures, several techniques have been created, such as the boundary approach, the finite element technique, and the combination of the two approaches (Gerolymos and Gazetas 2006a). All of these processes are related to the management of system variables. The equal path is estimated to be different on each path.

38.3.1.2 Eulerian and Lagrangian Grid Points

Two ways to define deformed grid points are Eulerian grid points and Lagrangian grid points. Deformations are not permitted in the first technique since the mesh is specified. It can pass through the mesh cells if the material deforms, but the overall mesh geometry remains unchanged. In contrast, in the Lagrangian method, the mesh is allowed to deform according to the behavior of the object relative to the mesh points. Therefore, from a product flow perspective, the network will also flow.

38.3.2 Computer Programs for SSSI Studies

Due to the rapid development of computers, many methods have been adopted and developed to analyze SSSI. A popular approach is the finite element method (FEM). The continuum is divided into several points of finite size by this solution. FEM covers complicated geometries and materials and models soil and structural mechanics the best (Menglin et al. 2011). But it requires a lot of context and can lead to errors. The development of some computer programs has limited the use of finite element methods in SSSI research. These services are: SASSI, CLASSY, FLUSH (fast LUSH), ALUSH (axial symmetric LUSH) and SASSI. General purpose finite element programs such as ANSYS are also available as finite element solutions.

Some of the research in SSSI is done by examining responses to measurement instruments (called survey models). These studies provide a better understanding of the model's predictive performance for upcoming earthquakes.

CLASSY, FLUSH and ALUSH on Fast Fourier Transform (FFT) analysis. ALUSH and FLUSH cannot handle heavy objects, and the accuracy of 3D analysis needs to be determined for both services.

It's important to remember that SASSI is arguably the most trustworthy instrument for SSSI study (Menglin et al. 2011), which makes use of split half-space modified 3D substructures. However, the recognition program needs to be taken into account before evaluating in SASSI, Additionally, when modeling is required, even basic implementation will not be sufficient. So, before doing an SSI or SSSI evaluation, designers should have a solid grasp of the SASSI program.

Analysis of the results of SSSI effects in the nuclear industry compared with the results obtained with FEM in SASSI (Sung 1953) Boundary Element Method (Xuj et al. 2004).

38.4 SSI for Composite Structure

Soil-Structure Interaction (SSI) expands due to the ease of foundation. The extension period causes the P-Δ effect, which results in excessive force/moment in the member, and this effect increases as the soil displacement increases. In addition to reducing motion and interaction history, when it comes to improving the structure's performance under seismic loads, the SSI control model's support needs will be helpful. Providing support in high-rise buildings can effectively control and reduce damage during earthquakes (Chen et al. 2018) (Fig. 38.6).

Reliable and detailed composite material models have been developed and developed for the study of materials and their behavior. Additionally, the design parameters examined are layer thickness, actuator size and locations. The dynamic behavior of a combination of linked plates and connected piezoelectric actuators and sensors was simulated using an analytical model based on the Heaviside function and δ-Dirac distribution (Kalyanshetti and Hirave 2016). Tallah and Khemissa, on the other hand, examined the behavior of geotechnical structures in one of the soil interaction and solid monatomicities as direct experimental and numerical tests. Utilizing the Plaxis finite element algorithm, numerical simulations of the experimental system are included in the computations. Once the mesh is created and the initial stress (the Ko process) is initiated, calculations include the normal stress set and then the shear level using the transform on the top half of the box. They explained that the behavior is caused by the increase in shear strength, followed by a phase dependent on the transport to more or less sand thickness and the ultimate shear strength for loose sand. A good analysis of the experiment and According to the findings, there is a greater tendency for rough interception than smooth interception, and the dense sand

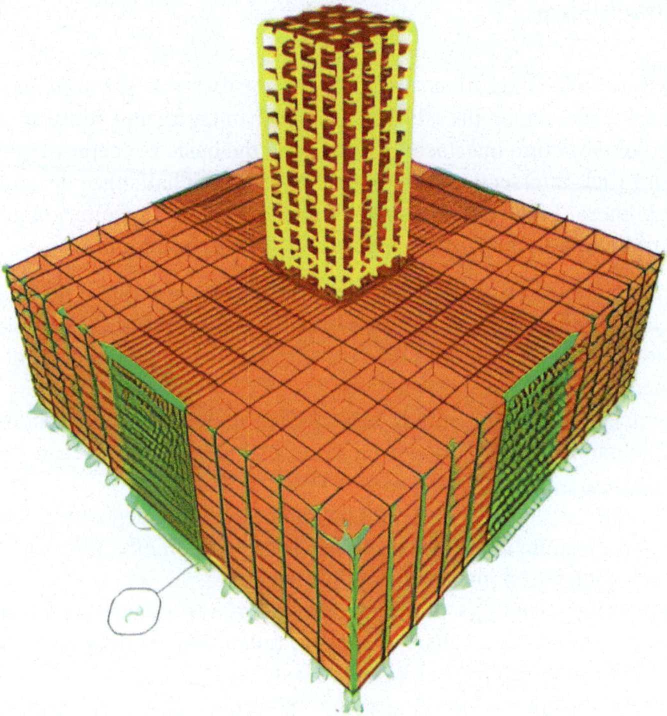

Fig. 38.6 Typical view of elastic continuum model (Kalyanshetti and Hirave 2016)

is higher than the loose sand. The ideal Rint of a rough interface is higher than that of a smooth interface, and the image quality of coarse sand is higher than that of loose sand (Correia et al. 2000). Lee et al. got a numerical analysis of the soil-structure-soil interaction (SSSI) during seismic excitation is performed on a race table. Based on the results of the competition table, this paper presents a 3D finite element numerical simulation method using ANSYS software. The simulations utilized a comparable soil structure model, and contact points were employed to accommodate the nonlinearity of the foundation-surrounding soil interface. It is possible to evaluate the test bench's dynamic reaction by contrasting the analytical findings with the information gathered from the test bench. Analyzing the dynamic behavior of surrounding structures helps to comprehend the consequences of SSSI (Li et al. 2017).

38.5 Conclusion

This article examines SSSI effects that can alter the response of buildings, especially in urban areas. Understand the advantages, disadvantages, and strategic advantages of the soil-rock-structure interface by discussing the basic concepts of soil-structure and structure-rock-intersection. Focus on the different challenges designers face in the design process. Finally, a summary of each program's specifics is given, along with an introduction to the current SSI and SSSI study programs.

1. It is recommended that SSSI effects be taken into account at the design stage, especially for soil samples in dense or soft areas, as failure to comply with soil samples may result in serious damage to the structure or loss of work. The brief summary is as follows:
2. The lumped parameter method can be used to analyze the soil-structure interface under dynamic loading. However, finite element modelling can describe the model better.
3. It is generally accepted to consider the interaction between soil structures under static and dynamic loading conditions. Therefore, an effective but simple soil-structure-foundation interface model is needed.
4. Despite its shortcomings, the Winkler model is a straightforward approach that generates acceptable results. As a result, rather than explaining fixed systems, theory should be applied to real-world issues.
5. Analyzing the impacts of anisotropy, inhomogeneity, and nonlinearity un a scenario may be done effectively by dissecting a process into its constituent parts.
6. All models are subject to dynamic soil-structure interactions. Even in hard soils, vertical patterns are more affected when soil layers are interconnected.
7. Dynamic soil-structure interaction will reduce the characteristic frequency of connected soil-structure systems. It may also lead to differences between variables and the emergence of useful designs.
8. It can be seen that there is a difference in the experimental results at the sample-soil-sample interface. Most of the existing SSSI literature uses only theoretical derivation and calculation.
9. The type of soil feedback can influence the linear model-soil-structure interaction behavior.
10. Displacement response does not follow velocity response. Therefore, when the pressure between buildings is different, modal coupling should be taken into account during the analysis.
11. In order to avoid the settlement-land-intersection problem, the difference between buildings must be determined according to the layout plan of adjacent buildings.
12. Considering the SSI influence on the inertial system structure's seismic response; the inertial system's inherent vibration period continuously regulates the structure and modifies its dynamic qualities.

13. For buildings with inert structural systems, the vibration reduction efficiency of the inert system can be reduced by considering how soil-structure interaction affects things.

References

Ali S, Sanghai SS (2021) Seismic vulnerability assessment of reinforced concrete buildings using pushover analysis. In: AIP conference proceedings, vol 2417, no 1. AIP Publishing

Adeel Z (2009) Response modification factor of reinforced concrete moment resisting frames in developing countries. Master's Thesis, University of Illinois at Urbana-Champaign, USA

Bhurse PS, Sanghai SS, Kumari NL (2021) Seismic response of rectangular RC building with and without infill walls considering soil–structure interaction. In: Advances in civil engineering and infrastructural development: select proceedings of ICRACEID 2019. Springer Singapore

Chandak M, Pawade PY (2021) Influence of Metakaolin and steel fibers on stress strain behaviour of concrete. In: Journal of physics: conference series, vol 1913, no 1. IOP Publishing

Chang-Liang V (1974) Dynamic response of structures in layered soils. Research Report R74-10. Department of civil engineering. M.I.T

Chen Q, Zhao Z, Zhang R, Pan C (2018) Impact of soil structure interaction on structures with inerter system. Int J Sound Vib 433:1–15

Clouteau D, Aubry D (2001) Modifications of the ground motion in dense urban areas. J Comput Acoust 9:1659–1675

Correia VMF et al (2000) Modelling and design of adaptive composite structures. Comput Methods Appl Mech Eng 185(2–4):325–346

Garcia J (2008) Soil structure interaction in the analysis and seismic design of reinforced concrete frame buildings. In: 14th World conference on earthquake engineering, Beijing, China

Gerolymos N, Gazetas G (2006a) Development of Winkler model for static and dynamic response of caisson foundations with soil and interface nonlinearities. Int J Soil Dyn Earthq Eng 26:363–376

Gerolymos N, Gazetas G (2006b) Winkler model for lateral response of rigid caisson foundations in linear soil. Int J Soil Dyn Earthq Eng 26:347–361

Given MJ, Stewart JP, Haselton CB, Mazzoni S (2012) Assessment of soil- structure interaction modeling strategies for response history analysis of buildings. In: 15th World conference on earthquake engineering, Lisbon, Portugal

Hetenyi M (1950) A general solution for the bending of beams on an elastic foundations of arbitrary continuity. Int J Appl Phys 21:55–58

Horvath JS (1993) Beam-column-analogy model for soil-structure interaction analysis. J Geotech Eng 119(2):358–364

Horvath A, Rachlew E (2015) Nuclear power in the 21st century: challenges and possibilities

Jakovlevjuv (1954) On the analysis of heat exchange apparatus. Trudi Khakovskoaviatsionnogo Institute

Kalyanshetti MG, Hirave VV (2016) Seismic response of composite building frame with alternate bracing system considering soil structure interaction (SSI) 05(06)

Kerr AD (1964) Elastic and viscoelastic foundation models. J Appl Mech. Trans ASME 31(4):491–498

Kerr AD (1965) A study of a new foundation model. Acta Mech I/2:135–147

Kukwas A et al (2023) A review on seismic analysis of tall building (G+10). In: AIP conference proceedings, vol 2759, no 1. AIP Publishing

Kurian NP, Manojkumar NG (2001) A new continuous model for soil-structure interaction. J Struct Eng 27(4):269–276

Li P et al (2017) Numerical analysis of a shaking table test on dynamic structure-soil-structure interaction under earthquake excitations. Struct Des Tall Spec Build 26(15):E1382

Lou M, Wang H, Chen X, Zhai Y (2011) Structure-soil-structure interaction: literature review. Soil Dyn Earthq Eng 31:1724–1731

Mahore A, Patil P, Mase D (2023) Condition & life evaluation of structure by various NDT techniques for reuse. In: IOP conference series: earth environmental science, vol 1193, no 1. IOP Publishing

Malhotra PK (2015) Normalized response spectrum of ground motion

Martel RR (1940) Effect of foundation on earthquake motion. Civ Eng 10(1):7–10

NEHRP consultants joint venture (2012) Soil-structure interaction for building structures, report NIST/GCR 11-917-14, prepared by the applied technology council and consortium of universities for research in earthquake engineering for the national institute of standards and technology, Washington, D.C

Nagpure A, Sanghai SS (2019) Effect of floor diaphragms on seismic response of RCC framed building—a review. In: Smart technologies for energy, environment and sustainable development: select proceedings of ICSTEESD 2018, pp 323–333

Passternakpl (1954) On a new method of analysis of an elastic foundation by means of two foundations constants. Moscow, USSR

Rahgozar M (1993) A seismic soil-structure interaction analysis of structural base shear amplification. M. Eng. Thesis, department of civil and environmental engineering, Carleton University, Ottawa, Ontario, Canada

Sanghai S, Gurmule C (2024) Seismic response control Of RCC building with plan and vertical irregularities using dampers. J Build Pathol Rehabil 9(1):9

Sanghai S, Pawade P (2022) Performance evaluation of friction dampers for building with soil-structure interaction. Mater Today: Proc 60:194–210

Sung TY (1953) Vibrations in semi-infinite solids due to periodic surface loading. ASTM STP No. 156. Harvard University, pp 35–68

Tanabashi R, Ishizaki H (1953) Earthquake damages and elastic properties of the ground. Bulletin No. 4, Disaster Prevention Research Institute, Koyoto University

Vlasov VZ (1949) Structural mechanics of thin walled three dimensional systems. Moscow, Stroizdat

Vlasov VZ, Leontiev NN (1960) Beams, plates and shells on an elastic foundation. Moscow, USSR: Fizmatgiz

Wang CM, Iang YX, Wang Q (2001) Axisymmetric buckling of Reddy circular plates on Pasternak foundation. J Eng Mech. ASCE 127(3):254–259

Wen X, Zhou F, Zhu H (2016) Influence of an adjacent structure on the horizontal and torsional input motion of an embedded foundation. J Adv Struct Eng

Xuj et al (2004) Seismic response prediction of NUPEC's field model tests of NPP structures with adjacent building effect. In: Proceedings of the 2004ASME/JSME pressure vessels and piping conference. San Diego, California, USA, pp 1–11

Chapter 39
Integrated Spatial Analysis for Urban Fire Risk Assessment: A Case Study of Nagpur's Ten Zones

Kirti Vijayrao Thakare and Kiran Megharaj Tajne

Abstract This study focuses on evaluating the urban fire risk landscape in Nagpur, India, through a comprehensive analysis of fire incidents from 2011 to 2020 and spatial distribution of property types across the city's ten municipal zones. The investigation employs a field data collection method using formats tailored to various property categories and GPS techniques for accurate geospatial mapping. The analysis integrates GIS technology to map the spatial distribution of different structures like high-rise buildings, industries, and healthcare facilities. The results reveal that May is the peak month for fire incidents, driven largely by climatic factors, with residential and mercantile fires making up 80% of occurrences. Zone-wise spatial mapping demonstrates distinct fire risk profiles, with educational institutions predominantly located in Zones 2 and 10, commercial sectors densely present in Zones 6 and 10, and significant industrial activity in Zones 1 and 8. Additionally, high-rise buildings and healthcare facilities pose unique fire safety challenges due to their high-density occupancy. The study concludes that Nagpur's urban landscape requires zone-specific fire safety strategies to address the unique risks posed by different property types. The meticulous data collection and GIS mapping provide a robust foundation for targeted fire prevention protocols and resource allocation, ensuring enhanced emergency preparedness and response strategies tailored to each municipal zone's distinct risk profile.

Keywords Urban fire risk assessment · ArcGIS · Satellite imaging · Nagpur's ten zones

K. V. Thakare (✉)
Priyadarshini College of Engineering, Near CRPF, Hingna Road, Nagpur 440019, Maharashtra, India
e-mail: kirtithakare@gmail.com

K. M. Tajne
Government College of Engineering, Sector-27, Mihan Rehabilitation Colony, Khapri, Nagpur 441108, Maharashtra, India

B. V. Khode et al. (eds.), *Innovations in Technologies: Pioneering Sustainable Infrastructure for a Resilient Future*, Sustainable Civil Infrastructures,
https://doi.org/10.1007/978-3-031-82133-2_39

39.1 Introduction

Hazards, including urban fire disasters, can result in physical harm, loss of life, and damage to properties or the environment. These incidents often stem from factors like inadequate heating methods, socio-economic issues, lack of education, and poor urban infrastructure. Particularly in high-density areas like central business districts, fire risks are exacerbated, complicating fire control and rescue efforts (Mtani and Mbuya 2018).

The International Association of Fire and Rescue Services reported an average fire incidence rate in India of 1.18 per 1000 people annually, with a mortality rate of 1.04 per 100 fires from 2014 to 2018. The 2018 Indian Risk Survey ranked fire as the third highest risk, with a noted increase in physical asset damage over three years. By 2019, fire risk was the tenth highest concern. In 2019, the National Crime Records Bureau of India reported 11,037 accidental fires, predominantly in structures, causing 10,915 fatalities and 441 injuries (Singh et al. 2021).

McKinsey and Company projects that by 2030, India's urban population boom will lead to at least thirteen cities with over four million residents each and sixty-eight cities with over a million inhabitants (Sankhe et al. 2010). His surge necessitates enhanced fire service resources and informed decision-making based on comprehensive data for effective fire safety management (Zhang et al. 2020).

Understanding fire patterns and severity is essential for sustainable urban planning. This includes systematic risk assessments and mitigation measures based on historical fire incident data and geographical variations.

This study aims to precisely map regions affected by fires in Nagpur Municipal Corporation (NMC) using satellite data and ArcGIS. This approach will enable a more effective urban fire risk assessment, specific to NMC's context.

39.2 Methodology

39.2.1 Study Area

Nagpur, positioned at 21.90° N and 79.50° E in Maharashtra, India, had a population of about 2.45 million according to the 2011 census, ranking as the third-largest city in Maharashtra and thirteenth in India. It spans 225.08 km^2 with a high density of 10,873 individuals per square kilometer. The Nagpur Municipal Corporation (NMC) divides the city into 10 zones: Laxmi Nagar, Dharampeth, Hanuman Nagar, Dhantoli, Nehru Nagar, Gandhibagh, Satranjipura, Lakadganj, Aasi Nagar, and Mangalwari, each with distinct characteristics and demographics (see Fig. 39.1).

Nagpur experiences a tropical savannah climate, classified as 'Aw' in the Köppen climate classification. Summer temperatures can reach 47.8 °C, while winter temperatures range between 10–12 °C. The city receives an average annual rainfall of 1161.54 mm.

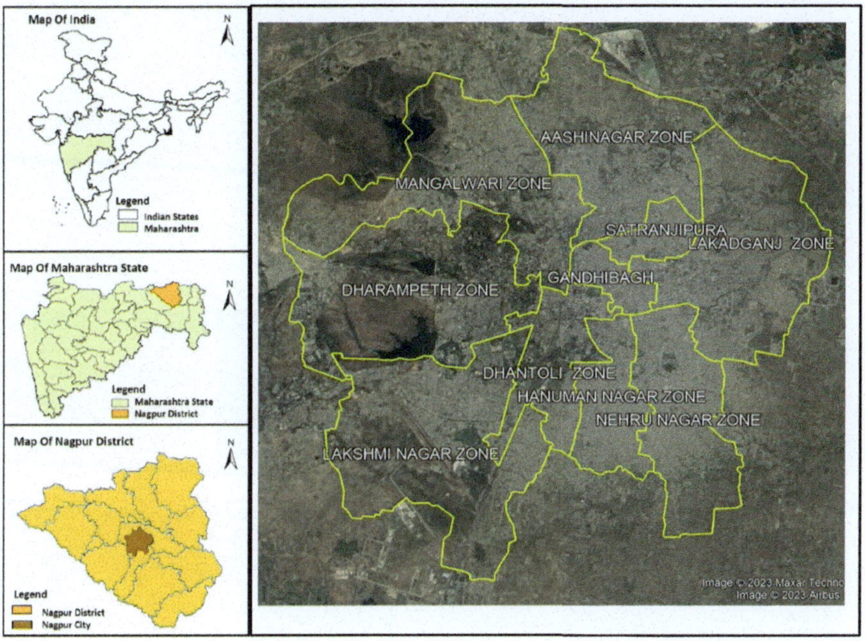

Fig. 39.1 Administrative zonal map of Nagpur: spatial distribution and division

Fire incident data from 2011 to 2020 was obtained from the NMC Fire Service Department, showing an upward trend in fire occurrences, with a peak in May. This month exhibited a z-score value exceeding 2.01, indicating a significant correlation between climatic conditions and fire incidents (see Fig. 39.2). This data forms the basis for an in-depth analysis to identify and assess fire hotspots (Singh et al. 2021; India Meteorological Department (IMD) 2022).

A total of 570 fire events during May from 2011 to 2020 were analyzed. These incidents were categorized by occupancy types according to the National Building Code of India (NBCI), 2016. Residential fires accounted for 48% of these incidents, followed by mercantile fires at 32%. Other types included industrial, business, assembly, storage, educational, and institutional fires (Bureau of Indian Standards 2005; India Meteorological Department (IMD) 2022).

39.2.2 Field Data Collection Method

The field data collection method for industries, commercial entities, and high-rise buildings within the municipal limits is a critical process due to the significant impact these structures have on the urban landscape and their associated fire risks. It begins with the collection of NMC zone boundary data to understand the municipal zoning.

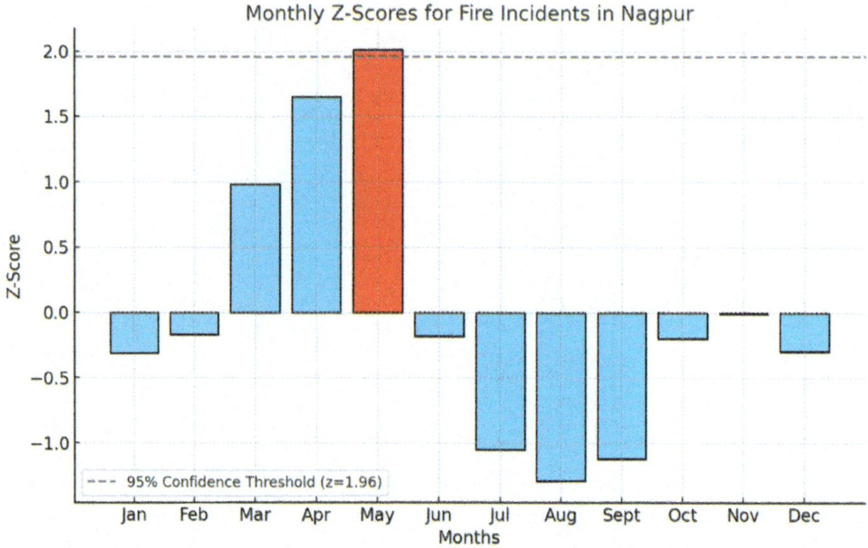

Fig. 39.2 Monthly Z-scores for fire incidents in Nagpur

Formats for data collection are then prepared, ensuring that the data captured will be consistent and accurate. The process includes learning GPS data capturing techniques for precise location tracking.

Data is then collected from sites, zone-wise, to ensure detailed and area-specific information. After collection, the data is compiled and converted into a CSV format for ease of use and analysis. The dataset is then reviewed to identify any inadequacies or ambiguities that may affect the analysis' integrity. If data inadequacies are found, a field revisit is conducted for additional data collection to fill the gaps. Once the dataset is deemed complete, a rigorous quality check is performed to validate the data. Finally, the data is processed zone-wise in GIS software, which allows for the spatial analysis necessary to identify potential fire hotspots, assess the effectiveness of current fire prevention and response measures, and inform the development of improved safety protocols and urban planning strategies. This meticulous approach ensures a robust foundation for fire risk analysis in urban settings.

Data Collection Categories. The urban fire risk assessment in the study is comprehensive, focusing on a variety of urban structures categorized by their function and fire risk profile. Industries are included due to the inherent risks in manufacturing and production processes. Educational institutions like schools and colleges are recognized for their varying occupancies and the unique challenges this presents in a fire. Restaurants, with their commercial kitchens, have distinct hazards that need to be accounted for, such as cooking equipment and flammable materials.

The study also examines commercial spaces which encompass diverse business operations and can range from offices to retail spaces. High-rise buildings are of particular interest; their complex evacuation procedures and the potential scale of an

Table 39.1 Data collection parameters for urban structure fire risk assessment

Category	Data collected
High rise	Geospatial coordinates, total number of stories, and total floor area
Industries	Location coordinates and spatial footprint
Commercial	Precise coordinates and area covered
School	Location details and building floor area
Petrol pump	Coordinates and extent of the facility
Hospitals	Geolocation and total floor area
Restaurants	Exact positioning and total floor area
Gas godown	Storage location coordinates and coverage area

incident make them critical to the study. Hospitals are included for their vital role and need for stringent safety measures due to the vulnerability of occupants. Petrol pumps and gas godowns are high-risk due to the flammable substances they contain. Finally, shops and small boutiques are considered for the fire risks associated with their often cramped and confined spaces.

Each category is selected based on the specific risks it poses, with data collection tailored to address and understand these risks comprehensively. This targeted approach ensures that each structure's unique fire risk profile is thoroughly evaluated in the urban fire risk assessment.

Data Collection Parameters. In the evaluation of urban fire risk, the meticulous gathering of data from various urban structures is vital. The data collection is customized to the unique fire safety challenges each type of structure presents. Comprehensive and precise data collection is crucial to understanding the risk factors inherent to each structure. The following Table 39.1 encapsulates the specific data collection parameters aligned with each category of urban structure, laying the groundwork for a robust urban fire risk assessment model.

39.3 Results

The results of the study provide a quantitative analysis of the distribution of property types that are critical to urban fire risk across the ten administrative zones of Nagpur. The distribution is depicted in a series of bar graphs, each representing a different category of urban structure and their respective quantities within each zone. The data visualized is instrumental for identifying potential hotspots for fire risks and the strategic allocation of fire safety resources.

39.3.1 Distribution of Property Types

The study presents a quantitative analysis of the distribution of urban properties across Nagpur's ten zones, crucial for urban fire risk assessment. The following Table 39.2 presents a comprehensive tabulation of the area percentages for various property types, including high-rise buildings, industries, commercial spaces, educational institutions, petrol pumps, gas godowns, hospitals, restaurants, and shops.

Complementing

Table 39.2, the Fig. 39.3 offers a stacked bar graph visualizing the distribution of property types in each zone, based on their respective area percentages. This graphical representation provides a clear and immediate understanding of how different property types are proportionally distributed within each zone. It effectively illustrates the relative prominence of specific property types in each zone, thereby aiding in the identification of areas that may require more focused fire safety measures.

The values on the Y-axes represent the percentage of the total area of each zone occupied by different urban property types.

Zone 1 (Laxmi Nagar zone) showcases a diverse mix with educational institutions taking precedence at 0.02026%, followed by industries at 0.01990%. The presence of high-rise buildings at 0.01478% introduces complexities in fire safety due to factors such as evacuation challenges and potential for vertical fire spread. Restaurants and shops, with respective area percentages of 0.00563 and 0.00413%, emphasize the need for fire safety measures in commercial kitchens and retail spaces, which are often high-traffic environments with varied fire risks.

In Zone 2 (Dharampeth zone), the dominance of schools and colleges at 0.05603% necessitates fire safety protocols tailored to the dynamic and densely populated nature of educational settings. The commercial sector, accounting for 0.01862%, underscores the need for vigilant fire risk management due to the diversity of goods and the high volume of electrical equipment typically found in such locales.

Zone 3 (Hanuman Nagar zone) and Zone 4 (Dhantoli zone) display a modest industrial footprint, yet the latter distinguishes itself with the highest area percentage dedicated to healthcare facilities at 0.00839 and 0.03474%, respectively. This underscores the imperative for fire safety in environments where both life and critical infrastructure must be protected.

The analysis of Zone 5 (Nehru Nagar zone) reveals a concentration of educational and healthcare properties at 0.04027 and 0.00794%, respectively. These institutions are critical to community resilience, and their protection is paramount in fire safety planning.

Zone 6 (Gandhibag zone), with its significant area percentage for commercial spaces (0.15949%) and shops (0.12682%), is marked as a hub of commercial activity, necessitating a comprehensive approach to fire safety management to address the elevated risks associated with high occupant loads and dense urban development.

Zone 7 (Satranjipura zone), although not characterized by high percentages of any particular property type, still requires fire safety considerations, especially for

Table 39.2 Comprehensive area percentage distribution of urban properties across Nagpur's zones (2023)

Zones	Industries area %	High rise area %	Commercial area %	School/ College area %	Petrol pump area %	Gas godown area %	Hospital area %	Restaurants area %	Shops area %
Zone 1 (Laxmi Nagar zone)	0.01990	0.01478	0.01349	0.02026	0.00126	0.00017	0.01215	0.00563	0.00413
Zone 2 (Dharampeth zone)	0.00734	0.00779	0.01862	0.05603	0.00081	0.00004	0.00417	0.00425	0.00411
Zone 3 (Hanuman Nagar zone)	0.00000	0.00399	0.01845	0.01885	0.00207	0.00000	0.00839	0.00168	0.00512
Zone 4 (Dhantoli zone)	0.00000	0.01905	0.03554	0.08430	0.00341	0.00065	0.03474	0.00329	0.01197
Zone 5 (Nehru Nagar zone)	0.00629	0.00023	0.00600	0.04027	0.00075	0.00000	0.00794	0.00401	0.00574
Zone 6 (Gandhibag zone)	0.00000	0.09091	0.15949	0.04886	0.00420	0.00049	0.02011	0.00796	0.12682
Zone 7 (Satranjipura zone)	0.00000	0.00338	0.01282	0.00588	0.00013	0.00000	0.00839	0.00088	0.01347
Zone 8 (Lakadganj zone)	0.02771	0.00749	0.00160	0.00020	0.00110	0.00017	0.00335	0.00336	0.01781

(continued)

Table 39.2 (continued)

Zones	Industries area %	High rise area %	Commercial area %	School/ College area %	Petrol pump area %	Gas godown area %	Hospital area %	Restaurants area %	Shops area %
Zone 9 (Aashi Nagar zone)	0.00060	0.00515	0.01322	0.00373	0.00089	0.00036	0.00335	0.00131	0.00632
Zone 10 (Mangalwari zone)	0.00560	0.00835	0.07857	0.03183	0.00082	0.00009	0.00556	0.00380	0.00410

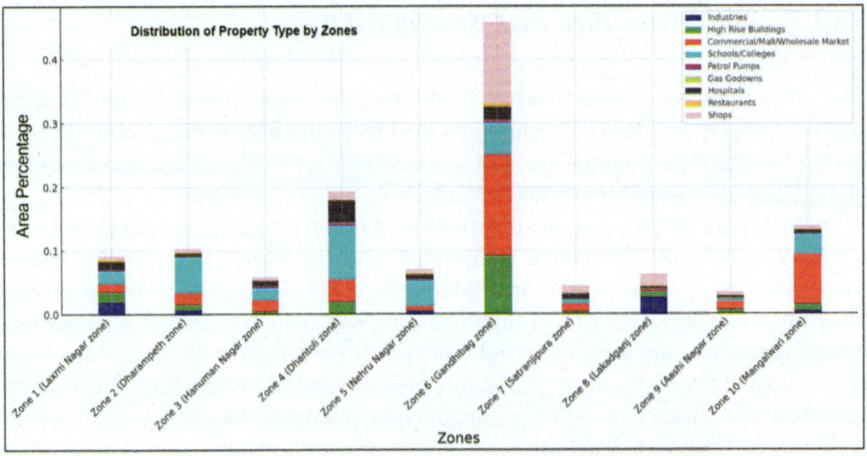

The values on the Y-axes represent the percentage of the total area of each zone occupied by different urban property types.

Fig. 39.3 Zone-wise distribution analysis of urban properties in Nagpur (2023)

its retail spaces at 0.01347% and healthcare facilities at 0.00839%, to ensure the safety of its residents and the continuity of essential services.

The industrial landscape of Zone 8 (Lakadganj zone) is prominently defined by its saw mills, which constitute a notable area percentage of 0.02771%. The predominance of such industries, which inherently process and store combustible materials, underscores an elevated fire risk.

In Zone 9 (Aashi Nagar zone), the area percentage of commercial spaces stands at 0.01322%, and the modest presence of restaurants (0.00131%) and shops (0.00632%) calls for fire safety measures that cater to the unique risks present in these establishments.

Finally, Zone 10 (Mangalwari zone) emerges as a commercial and educational hub, with 0.07857% of the area dedicated to commercial activities and 0.03183% to educational institutions. The diverse usage patterns and high density typical of such areas necessitate specialized fire risk assessments and targeted fire safety protocols.

Collectively, the spatial distribution of property types across Nagpur's zones, as depicted in Fig. 39.3, is indicative of the multifaceted nature of urban development and the corresponding fire safety challenges. This granular distribution necessitates a stratified approach to fire safety, where the unique characteristics of each zone are considered in the development of fire risk management strategies. This approach is pivotal in formulating effective fire safety regulations, ensuring preparedness and response strategies are appropriately calibrated to the distinct urban tapestry of Nagpur.

39.3.2 Zone-Wise Fire Risk Spatial Analysis

The 2023 study on Nagpur's urban landscape undertook a spatial analysis using satellite imagery and ArcGIS technology to identify fire risks across the city's zones. Each zone presents a unique array of urban structures with varying area coverages that necessitate specific fire risk assessments and safety measures.

In the Laxmi Nagar zone, the distribution of urban properties is characterized by a notable density of educational institutions covering 81,631 m^2, which is indicative of the zone's prioritization on academic facilities. Industries, while spanning a wide range, occupy 80,189 m^2, highlighting a substantial industrial presence that demands rigorous fire safety oversight due to the potential for hazardous materials. High-rise buildings, covering 59,582 m^2, present specific challenges for fire safety, particularly in evacuation and fire containment strategies. The commercial spaces, including malls and wholesale markets, stretch across 54,358 m^2, necessitating robust fire prevention and response frameworks in high-traffic commercial centers. Dharampeth zone is predominantly occupied by educational institutions, which sprawl over an extensive area of 207,877.8 m^2, reflecting the necessity for fire safety protocols that cater to large, dynamic populations. Commercial spaces cover 69,068 m^2, representing a significant portion of the zone's landscape, where a variety of goods and electrical equipment increase the complexity of fire risk management. In Hanuman Nagar zone, the absence of industries is contrasted by the presence of high-rise buildings and commercial spaces, which cover 4,384.5 and 20,294 m^2 respectively. This distribution underscores the need for specialized fire safety solutions in residential and commercial high-rise structures. Dhantoli zone is characterized by an absence of industries but a significant area covered by high-rise buildings and healthcare facilities, totaling 25,522.4 and 112,964.2 m^2 respectively. The high concentration of healthcare facilities necessitates a heightened focus on fire safety protocols to protect both life and critical medical infrastructure. In the Nehru Nagar zone, there is a marked presence of educational and healthcare facilities, covering areas of 83,361.7 and 16,429.4 m^2 respectively. This emphasizes the importance of fire safety measures in spaces that are crucial for community welfare and where the population density can vary greatly throughout the day (Fig. 39.4).

Gandhibag zone (Zone 6) exhibits a dense commercial sector, with shops alone occupying 53,265 m^2. This concentration of commercial activities requires a comprehensive fire safety strategy to mitigate the risks associated with high foot traffic and dense urban development. Satranjipura zone may not have large industrial or high-rise profiles but still requires attention to fire safety due to its retail and healthcare spaces, which occupy 6,332 and 3,942 m^2 respectively. These figures indicate the need for vigilance in protecting smaller, yet critical, urban structures. Lakadganj zone has a significant industrial presence, with industries covering 88,964 m^2. The nature of industrial activities, particularly in sawmills, increases the potential for fires, making it imperative to focus on stringent fire safety regulations and preparedness in this zone. In Aashi Nagar zone, the commercial spaces, while covering 37,684 m^2, are complemented by a moderate presence of restaurants and shops. Even with a

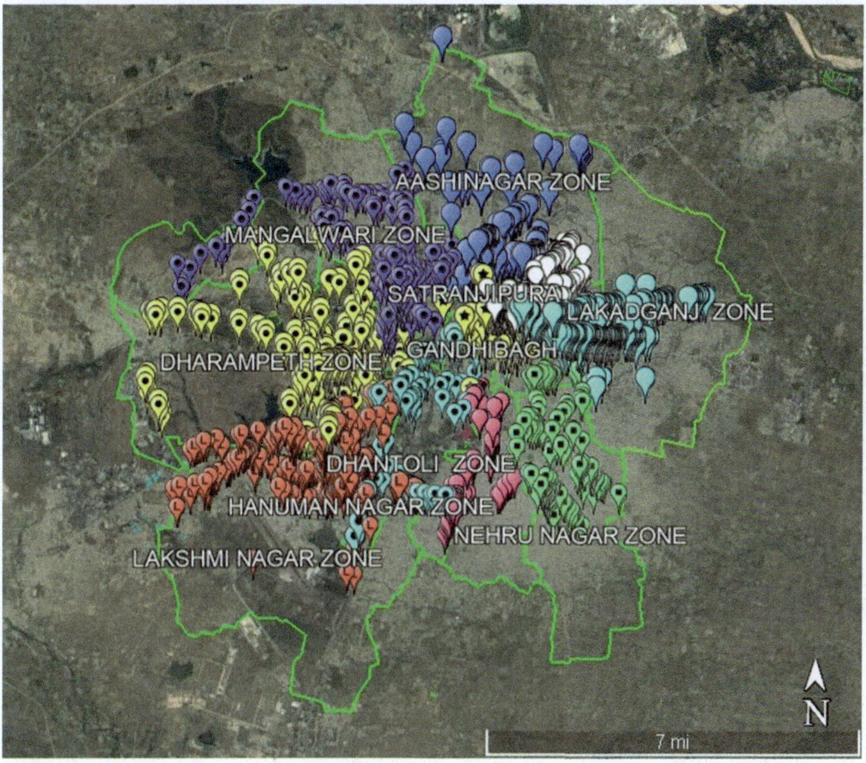

Fig. 39.4 Comprehensive GIS mapping of urban structures in Nagpur for fire risk assessment (2023)

smaller footprint, the potential fire risks in these establishments necessitate tailored fire safety measures. Mangalwari zone, highlights a substantial focus on commercial activities and educational institutions, occupying 276,565 and 112,054 m^2 respectively. This combination of high-density commercial and educational environments calls for specialized fire risk assessments and targeted fire safety protocols to ensure the safety and preparedness of these areas.

The spatial distribution of these zones, as visualized in the satellite imagery, underscores a diverse range of urban structures, each with unique fire risk profiles. The need for zone-specific fire safety strategies is evident, with particular emphasis on high-rise buildings, industrial areas, and densely populated commercial and educational spaces. The comprehensive data provided by the GIS mapping enables a more informed approach towards enhancing fire safety measures, ensuring that they are well-suited to the specific needs and risks of each zone within Nagpur's urban landscape. This targeted approach is critical for improving the efficacy of fire risk management, emergency preparedness, and response across the city.

39.3.3 Historical Trends

In Nagpur's urban fire hazard assessment, causes of fires are classified into five categories. Gas cylinder leakages represent 14%, electrical short circuits 31%, nearby garbage fires 2%, and other causes 6%, including lamps, cooking, mechanical defects, and electrical iron issues. A significant 48% of fires are unspecified. Electrical short circuits are common in May due to the increased use of electrical appliances, indicating a weather-related pattern.

Figure 39.5 displays the annual trends of fire causes from 2011 to 2020, showing a minor increase in unknown sources and electrical short circuits, with R-squared values of 0.083 and 0.0333, respectively, and a decrease in gas cylinder leakages with an R-squared value of 0.0332. The prevalent unspecified cause underlines the importance of detailed investigations for targeted fire prevention strategies (Singh et al. 2021).

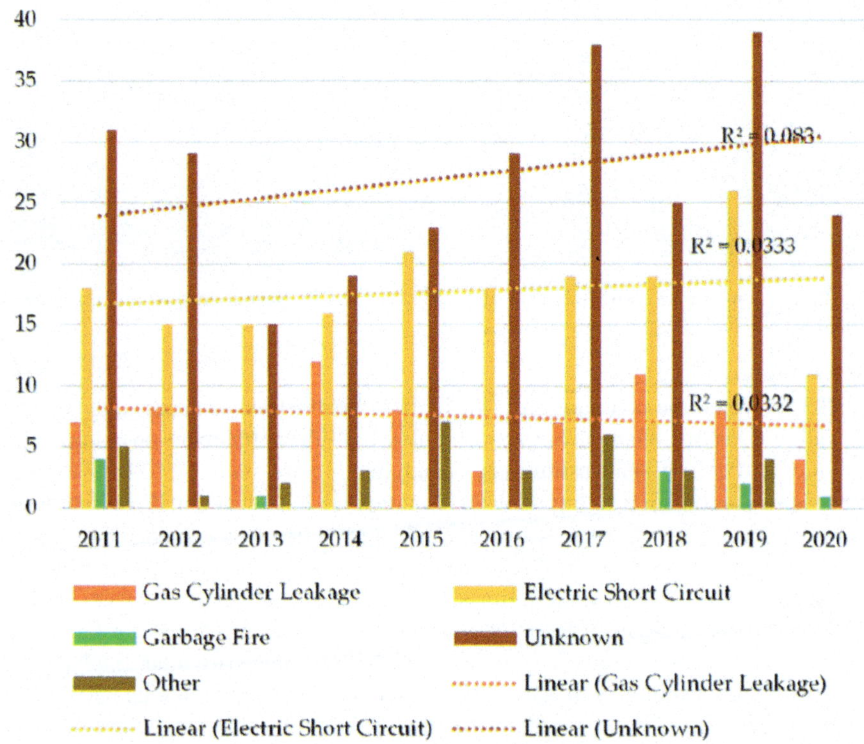

Fig. 39.5 Annual distribution and trends of urban fire causes in Nagpur (2011–2020)

39.4 Discussion

The spatial analysis conducted in the NMC zones reveals a distinct correlation between urban structure types and fire incidents. High-rise buildings, industrial areas, and zones with dense commercial activities exhibit a higher fire risk. This finding aligns with the historical trends observed, where increased urbanization and population growth in Nagpur have escalated fire risks. Particularly, the surge in fire incidents during the hotter months indicates a strong relationship between climate conditions and fire occurrences.

The study also points to socio-economic factors playing a significant role in fire risk. Zones with lower socioeconomic status often lack adequate fire safety infrastructure and awareness, contributing to higher fire risks. This disparity necessitates targeted interventions in fire safety infrastructure development in these areas.

The effective use of GIS and satellite data in this study highlights the potential of technology in identifying and managing urban fire risks. This approach could be a model for other urban areas, especially in leveraging technology for more proactive fire safety planning and response.

39.5 Limitations and Scope for Future Research

While the study provides significant insights, it is limited by its focus on structural fires and does not extensively cover wildfires or fires in informal settlements, which are also pertinent to Nagpur. Future research could expand to these areas, employing a similar spatial analysis approach.

39.6 Conclusion

The study of Nagpur's fire incidents and spatial distribution of property types has revealed critical insights into the urban fire risk landscape across its ten municipal zones. Data analysis from 2011 to 2020 highlighted May as the peak month for fire incidents, significantly influenced by climatic conditions. Residential and mercantile fires accounted for 80% of the total fire occurrences.

Spatial analysis using GIS software demonstrated the diverse nature of fire risk profiles across the different zones. Key findings include the prominence of educational institutions in Zones 2 and 10, the commercial sector's dense presence in Zones 6 and 10, and significant industrial footprints in Zones 8 and 1. High-rise buildings and healthcare facilities also exhibited unique fire safety challenges.

The results underscore the importance of zone-specific fire safety measures, tailored to the distinct property distributions and inherent risks within each zone. Comprehensive data collection and spatial mapping ensure accurate identification of fire hotspots and support the strategic allocation of fire safety resources. Ultimately, this study forms the basis for developing targeted fire prevention protocols and enhancing emergency response strategies across Nagpur.

References

Bureau of Indian standards: national building code of India. Bureau of Indian Standards, New Delhi, India

India Meteorological Department (IMD) (2022) Ministry of earth sciences: climate application and user interface; climatology of smart cities, climate smart city-Nagpur (Sonegaon). https://imdpune.gov.in/caui/smartcities.html. Last accessed 23 Mar 2022

Mtani IW, Mbuya EC (2018) Urban fire risk control: house design, upgrading, and replanning. Jàmbá: J Disaster Risk Stud 10(1):1–8

Sankhe S, Vittal I, Dobbs R, Mohan A, Gulati A, Ablett J et al (2010) India's urban awakening: building inclusive cities, sustaining economic growth. McKinsey Global Institute, Mumbai

Singh PP, Sabnani CS, Kapse VS (2021) Hotspot analysis of structure fires in urban agglomeration: a case of Nagpur city India. Fire 4(3):38

Zhang X, Yao J, Sila-Nowicka K, Jin Y (2020) Urban fire dynamics and its association with urban growth: evidence from Nanjing China. ISPRS Int J Geo-Inf 9(4):218

Chapter 40
Static and Dynamic Analysis of SPSW with and Without Opening Using Finite Element Analysis Software

Shantanu N. Shelke and Shrikant Solanke

Abstract A steel plate shear wall (SPSW) is an efficient and effective lateral load-resisting system used in buildings and structures to resist the effects of lateral loads such as wind and earthquakes. The SPSW system consists of steel plates that are bolted together to form a vertical panel, which is then anchored to the building's foundation and framing system. The objective of current research is to conduct structural investigation of SPSW using techniques of FEA. The modelling and FEA simulation is conducted in ANSYS simulation package. The analysis show that mid-section region of SPSW is prone to failure. The deformation is obtained to be maximum at the top left end of SPSW and reduces along the depth.

Keywords SPSW · FEA

40.1 Introduction

A steel plate shear wall (SPSW) is a lateral load-resisting system used in buildings and structures to resist the effects of earthquakes, wind, and other lateral loads. It is composed of steel plates that are bolted together to form a vertical panel, which is then anchored to the building's foundation and framing system. SPSWs are typically used in tall buildings and structures, where the forces generated by lateral loads can be significant (Fig. 40.1).

They provide high strength, stiffness, and ductility, which enables them to absorb and distribute the forces generated by lateral loads, thus reducing the risk of structural damage or collapse. The advantages of using SPSWs include their ability to provide excellent seismic performance, low maintenance requirements, and ease

S. N. Shelke (✉) · S. Solanke
Department of Civil Engineering, GHRCE, Nagpur, India
e-mail: shantanu.shelke.mtechstr@ghrce.raisoni.net

S. Solanke
e-mail: shrikant.solanke@raisoni.net

B. V. Khode et al. (eds.), *Innovations in Technologies: Pioneering Sustainable Infrastructure for a Resilient Future*, Sustainable Civil Infrastructures,
https://doi.org/10.1007/978-3-031-82133-2_40

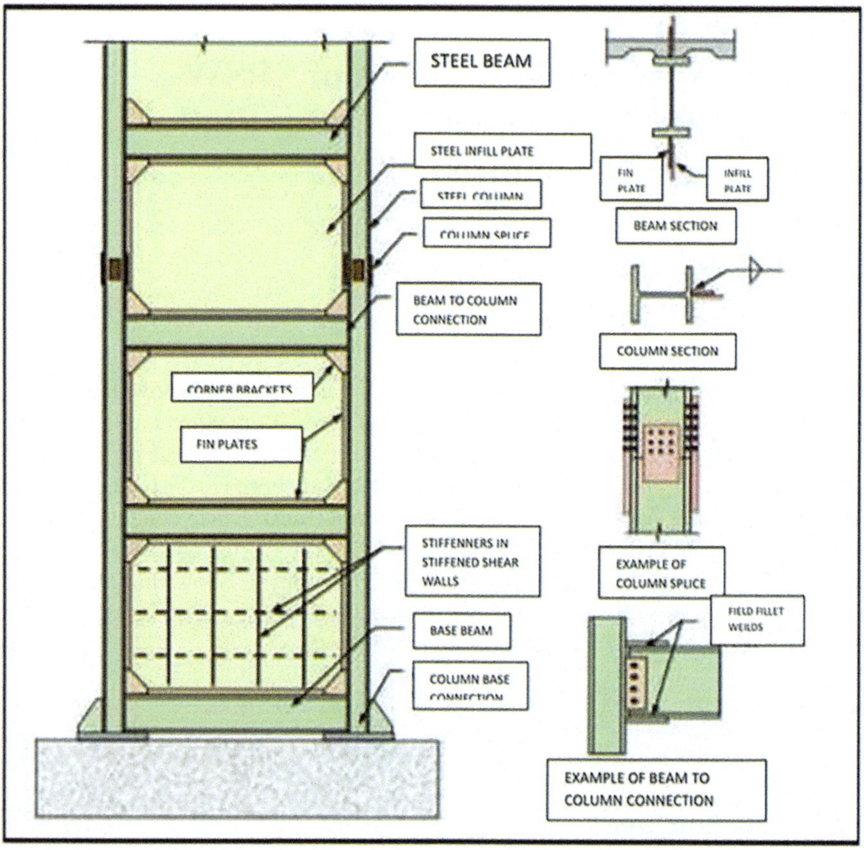

Fig. 40.1 Components of SPSW (Honarmand and Homami 2020)

of construction. They are also efficient in terms of space usage and can provide significant savings in building material cost.

40.2 Literature Review

Astaneh et al. (2001) provides a comprehensive overview of the development, design, behavior, and applications of SPSW's. It includes a discussion of the research that has been conducted on SPSWs, design guidelines, and practical considerations for the use of SPSWs in building structures.

Chandurkar et al. (2013) provides an overview of the state-of-the-art in the design and retrofit of building structures using SPSW's. It includes a discussion of the behavior and design principles of SPSWs, design guidelines, and retrofit techniques for existing structures.

Varsha et al. (2014) provides a comprehensive overview of the design, behavior, and applications of SPSWs in high-rise buildings. It includes a discussion of the benefits and limitations of SPSWs, design principles, and practical considerations for the use of SPSWs in high-rise buildings.

Anshuman et al. (2011) presents the results of an experimental study on the behavior of SPSWs under simulated seismic loading. The study includes a discussion of the influence of various design parameters on the behavior of SPSWs and provides insights into the seismic performance of these systems.

Kameswari et al. (2011) presents the results of an experimental study on the behavior of SPSWs with circular openings under lateral loading. The study includes a discussion of the influence of the size and location of the openings on the behavior of SPSWs and provides insights into the design of these systems with openings.

Zhao et al. (2024) presents a study on the design and behavior of high-strength SPSW's. The study includes a discussion of the design principles for high-strength SPSWs and provides insights into the behavior and performance of these systems under lateral loading.

40.3 Objectives

The objective of current research is to conduct structural investigation of SPSW using techniques of FEA. The modelling and FEA simulation is conducted in ANSYS simulation package.

40.4 Methodology

The model of SPSW is developed in imported in ANSYS simulation package. The model is evaluated for any geometric errors or imperfections (Fig. 40.2).

The model of SPSW is meshed. The meshing characteristics depends upon geometry type, topological consistency and other factors. The meshed model of SPSW is shown in Fig. 40.3. The model is meshed with hexahedral/brick element type (Fig. 40.3).

After meshing, the structural boundary conditions are applied on SPSW. The bottom face of SPSW is applied with fixed support whereas top left face of SPSW is applied with force along lateral direction. After applying structural boundary condition, the FEA simulation is run using standard conditions (Fig. 40.4).

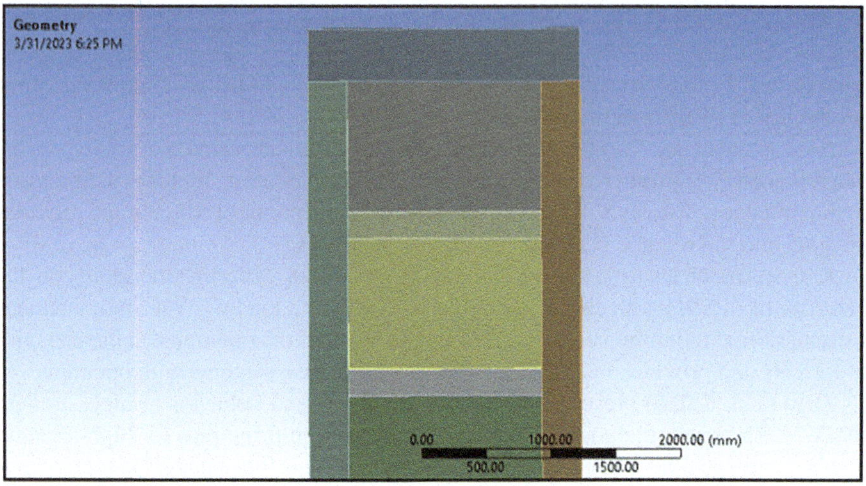

Fig. 40.2 3-story model of SPSW

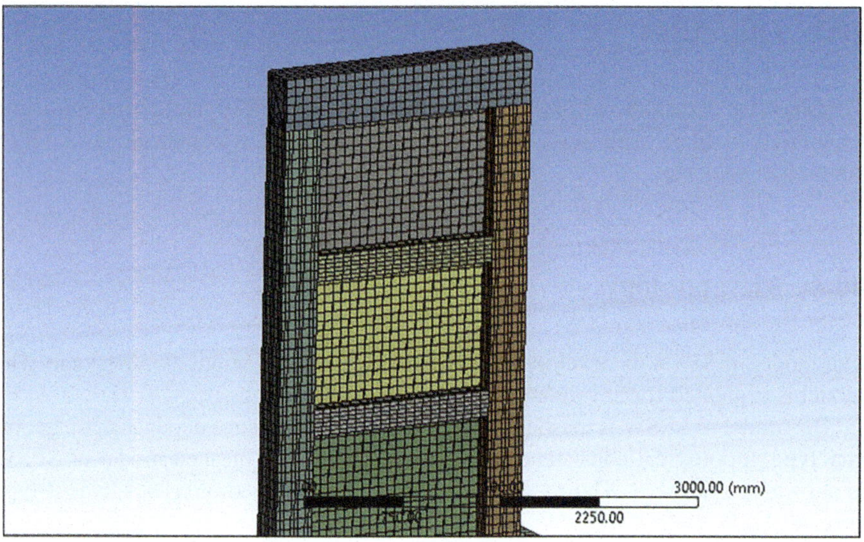

Fig. 40.3 Meshed model of SPSW

40.5 Results and Discussion

After running FEA simulation, the stresses and deformation plots are obtained. The deformation plot of SPSW is shown in Fig. 40.5. The deformation is maximum at the topmost region of SPSW and reduces along the base of SPSW. The maximum deformation of SPSW is 0.006085 m (Fig. 40.6).

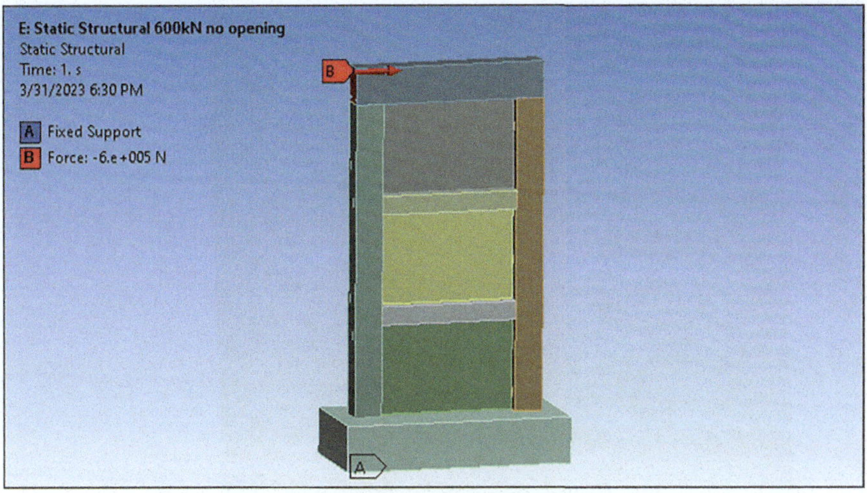

Fig. 40.4 Structural boundary condition on SPSW

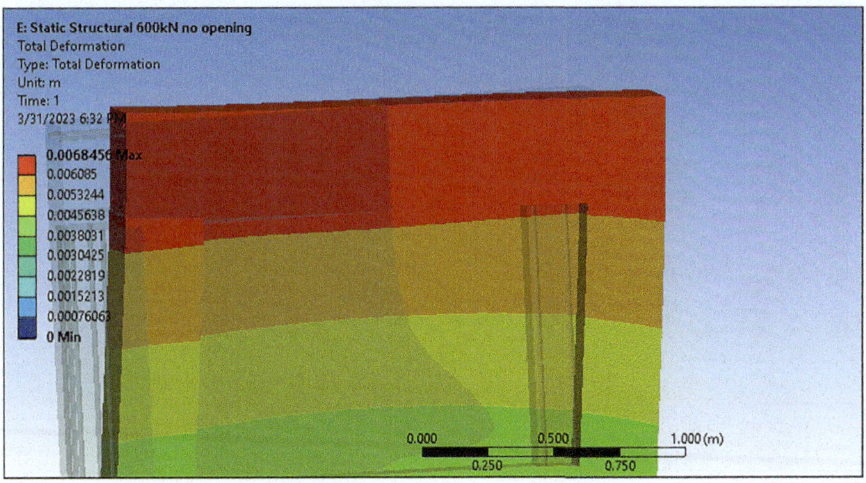

Fig. 40.5 Total deformation on SPSW

The shear stress distribution plot of SPSW is obtained from the analysis. The maximum shear stress is obtained near the edges of SPSW with magnitude of more than $3.93 * 10^6$ Pa. The shear stress is almost uniform for mid-section zone of plate (Fig. 40.7).

The equivalent stress is observed to be higher at the mid-section region of SPSW and is lower at the other corner region of SPSW. The maximum equivalent stress obtained is more than $2.63 * 10^8$ Pa.

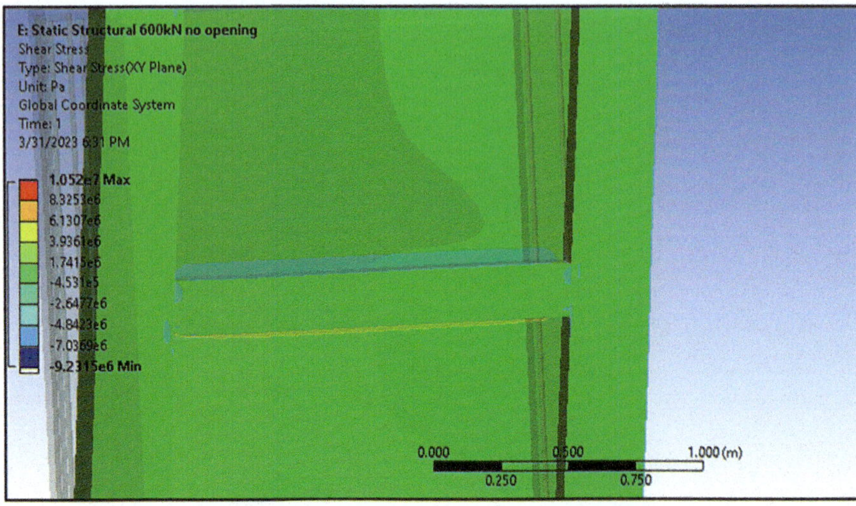

Fig. 40.6 Shear stress on SPSW

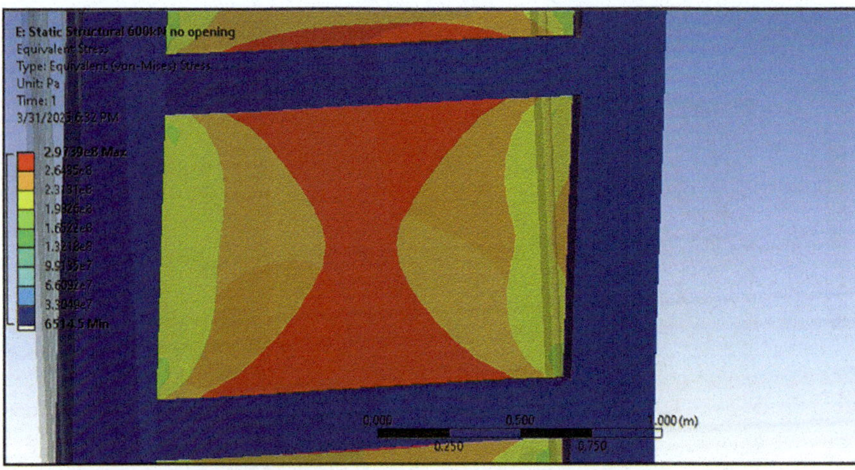

Fig. 40.7 Equivalent stress on SPSW

40.6 Conclusion

The FEA analysis is conducted on SPSW to determine critical regions of high stress. The static pushover analysis enabled to determine failure zones of steel plate shear walls. The analysis show that mid-section region of SPSW is prone to failure. The deformation is obtained to be maximum at the top left end of SPSW and reduces along the depth.

References

Anshuman S, Dipendu B, Bhavin R (2011) Solution of shear wall location in multistorey building. Int J Civ Struct Eng 2(2)

Astaneh-Asl A (2001) Seismic behavior and design of steel shear walls. Steel tip 37, structural steel educational council, Moraga, CA

Chandurkar PP, Pajgade DPS (2013) Seismic analysis of RCC building with and without shear wall. Int J Multidiscip Educ Res 3(3):1805–1810

Harne VR (2014) Comparative study of strength of RC Shear wall at different location on multistoried residential building. Int J Civ Eng Res 5(4):391–400. ISSN 2278-3652

Honarmand S, Homami P (2020) A study on the significance of the design parameters of steel plate shear walls subjected to monotonic loading. Civ Environ Eng Rep, Sciendo

Kameswari B, Elangovan G, Sivabala P, Vaisakh G (2011) Dynamic response of high rise structures under the influence of discrete staggered shear wall. Int Res J Eng Technol

Zhao Q, A Astaneh-Asl (2024) Cyclic behaviour of traditional and innovative composite shear walls. In: 13th World conference on earthquake engineering

Chapter 41
A Review of Hot Rolled Section and Cold Formed Section

Prachi Chaudhari and P. Y. Pawade

Abstract This article examines cold and hot rolled steel for use in construction, based on a review of the literature. In our work, we examine this idea with materials other than regular steel (hot rolled). Cold-formed steel structures (CFS) are commonly employed due to their many advantages, such as their lightweight and speed. The creation, design, and analysis of cold steel systems have all improved. This paper will assist with future company development research and development, since it is based on the most recent study on cold formed steel (CFS) and hot rolled steel (HRS).

Keywords Hot rolling · Cold forming · Manufacturing process · Strength · Durability

41.1 Introduction

41.1.1 General

Steel sections are essential components in various industries, serving as structural elements in buildings, bridges, vehicles, and machinery. The methods used to produce these sections significantly impact their properties and applications. This review provides a comparative analysis of two primary manufacturing processes: hot rolling and cold forming. Hot rolled sections are shaped at high temperatures, resulting in sections with distinct properties, while cold formed sections are manufactured at room temperature using different techniques. Understanding the differences between these methods is crucial for selecting the most suitable sections for specific engineering and construction needs.

P. Chaudhari (✉) · P. Y. Pawade
G. H.Raisoni College of Engineering, Nagpur, India
e-mail: prachi.chaudhari.mtechstr@ghrce.raisoni.net

B. V. Khode et al. (eds.), *Innovations in Technologies: Pioneering Sustainable Infrastructure for a Resilient Future*, Sustainable Civil Infrastructures,
https://doi.org/10.1007/978-3-031-82133-2_41

Globally, profile steel finds extensive application in both non-commercial and commercial settings. This is a unique idea in India. In an attempt to solve this issue, this research analyzes the applications of hot rolled and cold formed steel profiles in industrial design. The construction industry is growing at an extremely rapid pace. To increase living conditions, the country invests much in development. As the economy grows, demand for housing, machinery, infrastructure, and other goods increases. To sustain development and progress, resources must be made available and protected. The country makes investments to improve the living standards of its people. The ability to swiftly and cheaply erect, store, transport, and then dump temporary buildings or structures when their useful life are over makes them a common tool in the construction industry. It has been demonstrated that metal is the best material for temporary constructions.

41.1.2 Hot Rolled Sections

Hot rolling is the process of rolling metal at a temperature higher than its recrystallization temperature, usually exceeding 1000°F. When the metal is heated over its recrystallization point, it softens and becomes flexible. It produces more iron as well. Subsequently, room temperature is applied to the metal to bring it back to normal and relieve any concerns with material tensions during the hardening or quenching process. The uneven contraction of the metal during cooling results in less control over the final size and shape of the hot-rolled product. Hot-rolled steel is a solid choice for situations where exact tolerances are not a concern. Numerous choices are at one's disposal. Its value is greater than reality. (Fig. 41.1)

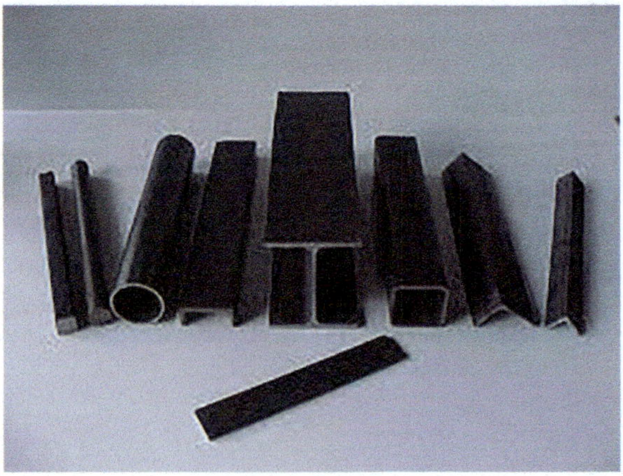

Fig. 41.1 Hot rolled section

41.1.3 Cold Formed Steel Section

As one may expect, the process for making cold rolled steel (CFS) differs from that of hot rolled steel (CFS). The steel is pressed at room temperature using pressure rollers, despite the name of the method. Because cold rolled steel may harden, it is roughly 20% stronger than hot rolled steel. Cold rolled steel is made via molding, demounting, semi-finishing, sizing, semi-roughing, roughing, and threading. Cold rolled steel may be precisely formed into forms. Unlike hot rolling, this process is done at ambient temperature, thus the metal does not shrink when it cools. The best choice for projects where visual impact and appeal are important factors is cold-rolled steel siding (Fig. 41.2).

41.2 Objectives of Study

- To investigate the concept of different steel sections.
- Analyze the general features of different steel sections in compliance with the applicable codes.
- To review state-of-the-art studies carried out by different steel sections.
- Analysis of the flexural behavior and comparison of hot rolled steel (HRS) and cold formed steel (CFS) under load.
- To look at how well the hot rolled steel (HRS) and cold formed steel (CFS) portions of the structure function seismically.

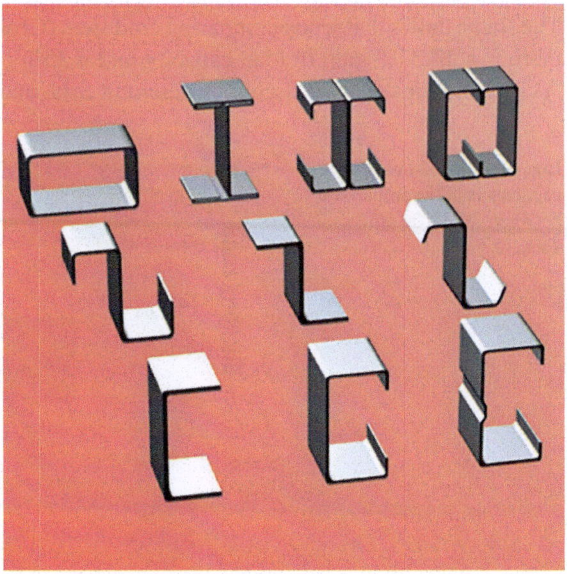

Fig. 41.2 Cold formed section

41.3 Literature Review

41.3.1 JayaramanA, N Sathyakumar, V Senthilkumar (2018)

The project demonstrates the use of extra connections in commercial roof construction, which are composed of standard and cold-formed steel. The load capacity of the ISMC 100 mm long 1 m and 0.5 m long cold section is 55.55%, 46.67%, 53.85%, and 50.00% lower, respectively, than the ISMC 100 mm section with or without connection. The load capacity of linked and disconnected cold-formed steel profiles is 42.85.55%, 77.77%, 68.08%, and 49.10% more, respectively, than that of the 1 m long ISMC 75 mm conventional steel profile. The deflections of cold-formed steel profiles were observed when they were linked and detached. The bonded and disconnected cold-formed steel had deflections of 53.70% and 52.94%, respectively, compared to 56% and 54.16%, respectively. The typical steel profiles that are ISMC 75 mm and ISMC 100 mm. The weight of the sample and the jointed and non-jointed steel as a whole were all less than the weight of conventional steel by 66.60%, 70.00%, 88.88%, and 51.22%, in that order. In general, cold steel should be 83.66% and 88.46% less expensive than steel for joints. Cold-formed steel profiles have a simple shear force that is 51.79% and 54.46% lower, respectively, than normal steel profiles, whether they are linked or not.

41.3.2 Salokhe S. A, Patil P. S (2015)

The study shows that when comparing cold steel (CFS) profiles to hot rolled profiles, there is a 17.37% gain in load capacity and a 33% decrease in external distortion. Axial deviations are also slightly different in cold- and hot-worked steels (Fig. 41.3).

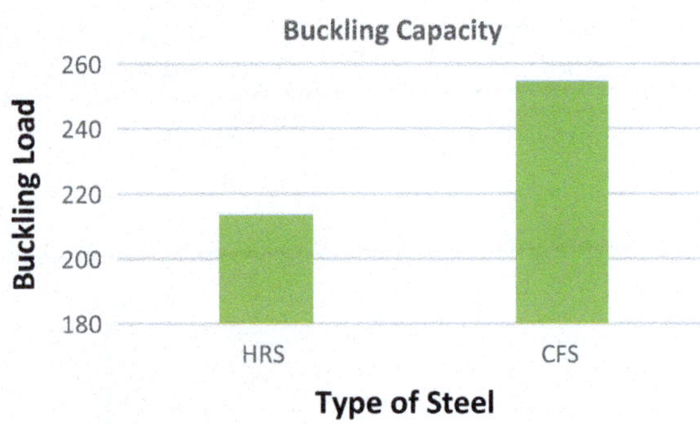

Fig. 41.3 Load buckling capacity in KN comparison between ISLB100 and 140C-2

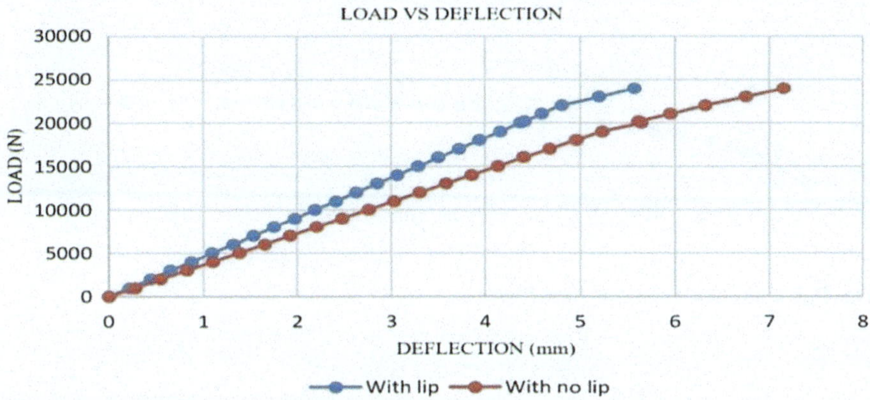

Fig. 41.4 Load vs deflection for HAT Section with and without lips

41.3.3 *Akula VenkataPhani Manoj. Dr. S. Senthil Selvan (2017)*

In this study work, the bending characteristics of Z-shaped steel and HAT-shaped steel with and without lips were theoretically compared, leading to the following conclusions. The maximum load bearing capacity of cold-formed Z profiles with lips is 11% higher than that of cold-formed Z profiles without lips. The former's ultimate load bearing capacity is 29% higher than that of cold-formed lipless HAT profiles. The ultimate deflection of the cold-formed Z profile with lips is 4.416 mm, whereas the final deflection of the cold-formed Z profile without lips is 4.96 mm. The cold lip HAT profile's final 1.4705 mm and 1.5438 mm deflections have both been measured. A fashionable lip-free hat silhouette. The lips in this portion increase load bearing capacity while reducing member deflection (Fig. 41.4).

41.3.4 *Drishya S Krishna, Binu P (2020)*

This article examines each model's behavior in the ANSYS program using buckling analysis. Compare Ordinary Cee with Supa Cee, Ordinary Zed and Supa Sigma, and Supa Zed and Ordinary Sigma. There is no web fold in the conventional paradigm to enable product switching. As a result, the valve material is more robust than usual. The article states that the load capacities of Supa Cee, Supa Zed, and Supa Sigma are 35.4%, 31.7%, and 31.5% greater, respectively, than Ordinary Cee, Ordinary Zed, and Ordinary Sigma.

Supa Zed is the finest discovery. The payload capacity surpasses that of Supa Sigma and Ordinary Zed by 31.5%. It is true that this is regarded as the finest. In this article, buckling analysis is used to analyze (Figs. 41.5 and 41.6).

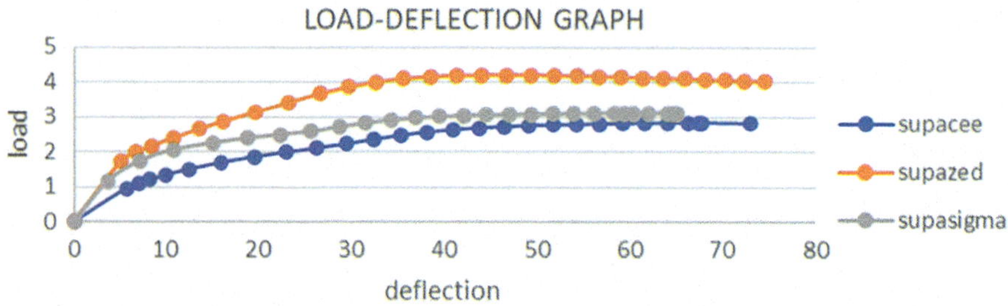

Fig. 41.5 Supa Cee, Supa Zed, Supa Sigma

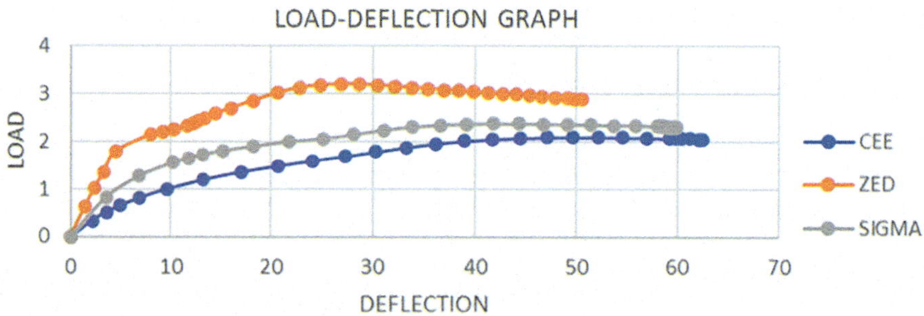

Fig. 41.6 Load Deflection Graph Between 3 Material CEE, ZED, SIGMA

The ANSYS program's depiction of each model. Compare Supa Zed and Ordinary Sigma, Ordinary Cee and Supa Cee, and Ordinary Zed with Supa Sigma. The traditional type lacks web folding, which would allow for product change. The valve material is therefore stronger than typical. The carrying capacities of Supa Zed, Supa Cee, and Supa Sigma are reported in the article to be 31.7%, 35.4%, and 31.5% greater, respectively, than those of Ordinary Zed, Ordinary Cee, and Supa Zed.

The greatest discovery is Supa Zed. The payload capacity is 31.5 percent more than Supa Sigma and Ordinary Zed. Supa Zed is said to be the best.

41.3.5 *Ajeet Sharma, et al. (2016)*

This article describes how to use data analysis counts and Indian Standard and British Standard specification standards to find the flexural strength or bending and torsion capacity of cold-formed lipless and lip-filled profiles. The chosen water channel is 230 mm deep, which is 2% less than the expected value, according to the research. Clinical study indicates that the divergence of the 180 mm canal cross-section from the predicted value is 3.5%. The deeper it goes, the stronger and more potent the beam. When a beam is 230 mm deep, its maximum bending moment is 30.57 KN·m; when it is 180 mm deep, its maximum bending moment is 9.97 KN·m. Local buckling

Fig. 41.7 Relationship Between the Displacement and Shear Strength of Different Sample

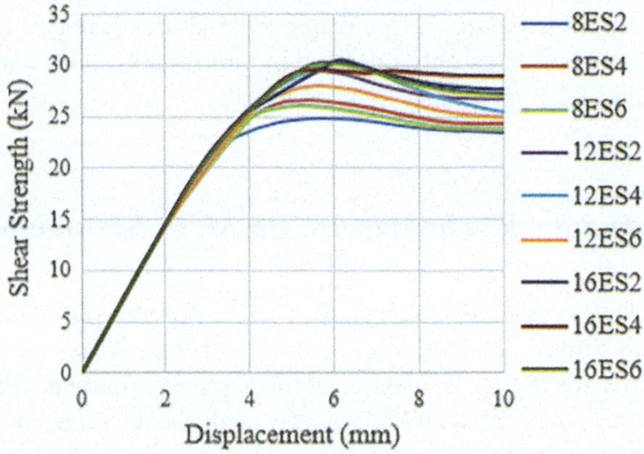

causes damage to the cold-formed steel top flange locally. Cold-formed products are prone to distortion. We can see from experiments that the deflection increases progressively until the local structure bends. Check this amount against the shipping cost. The angular component of the photometer can be used as a continuous line with these values.

41.3.6 *Sandra Raphy, Sreedevi Bhasin, K. P. Lakshmipriya (2021)*

The benefits of cold-formed steel (CFS) over hot-rolled steel are highlighted in this article, including its high strength-to-weight ratio (H/S), flexibility, and simplicity of prefabrication. several uses. It is used in construction. Every inspection of an edge-hardened mesh tunnel steel pipe concrete foundation with bending and shear effects reveals a t difference of 15%–17%, with the greatest deformation occurring around the edge of the reinforcement. Models with different side lengths and beams perform worse than lines with a side length of 12 mm and an infill of 4 mm (Fig. 41.7).

41.3.7 *Pranali V. Lambat, Prof. Sanjay Denge (2021)*

This article investigates the economic feasibility of two-story light steel structures in seismic zones—areas vulnerable to large earthquakes. The two-story building functioned as a benchmark for other service buildings. There are two methods for creating the structure, estimating its price, and doing an analysis. Research indicates that heavy metal constructions cost 22% more and are 4.45 times lighter than concrete structures. In areas prone to earthquakes, light steel constructions are 22% less expensive than RCC structures. The RCC structure weights 4.45 times more than the substructure. While concrete structures take two to three months to create, light steel buildings only take two to three weeks. As a result, strong concrete structures cost 22% more than light steel ones.

41.3.8 R. Thenmozhi, Dr. M. C. Sashikkumar (2017)

This article describes how to use ANSYS to calculate a CFS Sigma profile's maximum load bearing capacity with a thickness of 2 mm and a span of 1000 mm. ABAQUS may be used to calculate the maximum payload capacity for a given length; the result is 25.06 KN. Two-way analysis of variance was performed on the Z quotient and multiple sigma. CUFSM software was used to analyze the behavior of the CFS sigma and Z sections. The ultimate load bearing capability of the two surfaces was determined using the software packages ANSYS and ABAQUS.

41.3.9 R. B. Kulkarni, Shweta B. Khidrapure (2014)

This article states that buckling usually denotes the maximum strength of the structure. The average profit in cold running might increase from 15 to 30%. In compliance with IS: 801–1975 (the Indian Standard Specification) and BS 5950–5: 1998 (the British Standard Specification), this research conducts a parametric and comparative test of the flexural strength of cold steel. While the BS 5950–5: 1978 specification utilizes limit state, the IS: 801–1975 specification employs operating pressure. It was found that the two designs had about the same duration.

41.3.10 Benjamin W. Schafer (2011)

Cold-formed steel is extensively used in many structural applications and has also been utilized to manufacture novel materials, such as mid-rise buildings. The purpose of this part and the robust equipment is to offer many recommendations. The finite strip approach and generalized beam theory provide unique potential for the transformation and analysis of thin structures; nevertheless, a significant amount of work needs to be put into developing efficient sample analysis tools, of which there can be as many as eight. Since the section stability is utilized to decide whether to apply elasticity directly, the substance of the model has changed significantly. The new device attempts to replicate this performance by using cooling.

41.3.11 Shweta S. Kumbhar, H. S. Jadhav (2021)

This article explains how Cold Steel Structures (CFS) are becoming more and more popular due to their many advantages, including their light weight and speedy construction. The analysis, design, and development process for CFS is always evolving. This paper presents new CFS research that may be useful for future investigations on housing change. Analyze the behavior of members, such as CFS columns and beams. effects of environmental stress, mechanical strength, and temperature increase.

41.3.12 Navyashree P. M., Sowjanya G. V., Dr. T. V. Mallesh (2020)

This article describes the behavior of many cold-formed steel profiles (final analysis). Using the software packages ANSYS and ABAQUS, we can evaluate the behavior and performance of CFS partitions under various loads based on research data. The aim of this study is to provide suggestions for the use of cold steel in construction.

41.3.13 S. A. Kakade et al. (2014)

801:1975 This article is completely false in the field of stress research. On rare occasions, I.S. 800:2007 uses the entire section of members for design. Road should be used in place of M.K.S. and Labor. Indeed. Limit State Method (LSM), Systems 801: 1975, and S.I.

41.3.14 Deepti D. Katkar, Prof. N. P. Phadtare (2018)

This website compares and explains prefabricated and conventional structures, and it includes information on the trials and research done in this area. Research indicates that steel structures are more energy-efficient, stronger, and simpler to construct than other kinds of buildings.

41.3.15 Mr. Swapnil R. Kalbhor, Prof. Nagesh Shelke, Prof Vishwajeet Kadlag (2020)

This article's illustration of the business model's optimization leads to the following conclusions: ——1. They can be used quickly and don't require large tools or equipment to operate. It's ideal for woodworking because of its simple hand loading mechanism. 2. Everything can be well-made, sealed, and insulated with a variety of materials. 3. This type is easily replaceable or adjustable as needed.

41.3.16 Deepla N and Venkat R (2018)

This article shows that plate buckling, specifically localized buckling of the material, is the main reason why the load-carrying capacity decreases as the b/t ratio increases. Similarly, the slenderness effect causes the load capacity of an element to decrease as its length increases. As belt depth increases, the load capacity on the balance sheet decreases. With length, the maximum load-bearing capacity decreases. The weight bearing capacity of angle steel constructions is lower than that of channeled steel profiles. The low rod length will cause angular damage along the length of the steel extension, damaging the hinged bracket.

41.3.17 S. Prabhakaran (2020)

This article shows the maximum bearing capacity of cold-formed steel connectors at different angles. Real data shows that changing the row spacing within the same area can enhance the capacity of each row. Local buckling is the nearest type of buckling. By casting and examining this application, the column's predicted axial load capacity may be confirmed. It also demonstrates how to calculate the theoretical bearing capacity of a column.

41.3.18 Apurv Rajendra Thorat, Santosh K. Patil (2017)

This article describes how the Kirby report, which is based on ASCE-07 research methodology, may be used for prefabricated building design and inspection. Two instances were selected for this study. A shift of 34% in the X direction, 13% in the Y direction, and 23% in the Z direction is the final result.

41.3.19 *Sivani Meher, et al. (2018)*

This article addresses the applicability of the thesis and its implications for teaching mathematics in the Mumbai region through the use of freight and transportation. Once completed, the AUTO-CAD design may be uploaded. This informs the selection and construction of numerous sections, including cages, columns, and purlins. The whole process is based on Indian norms. The completed map shows that commercial warehouses may be constructed using designs that comply with state regulations.

41.3.20 *Athira V. V., Sruthy S. (2018)*

This article examines the properties of cold-formed C and Z sections, which are used as columns and purlins. For this, a special design was made using ANSYS software. In order to determine which holes buckled the most, six factors were examined. When it comes to frames and structures, use the best hole design and type for each location. Independent investigations were conducted to determine the buckling stresses on framed and frameless structures. The field equation is used to study linear buckling and stress surrounding the hole.

41.3.21 *Gizachew Markos Makebo, Dr. A. Paulmakesh (2021)*

This paper covers lipless and lipped 1.5, 1.6, 2, 3.4 mm one corner and 2 mm two corners. For numerical investigations, finite element computations were carried out using ANSYS software. Output voltage arises when the connection reaches its maximum capacity. Test load and deflection curves show that cold-formed angle steel's ultimate load-carrying capability increases with cross-sectional area and the number of connecting bolts. The joints appear to becoming more rigid as the stiffness rises.

41.3.22 *Hao Liang, Krishanu Roy, Zhiyuan Fang, James BP Lim (2022)*

The essay emphasizes how the global building sector is primarily responsible for the majority of energy consumption and CO2 emissions related to power. Modern construction techniques such as Cold Steel Fabrication (CFS) can mitigate this situation. The advantages of CFS channels include recyclability, reusability, and a high strength-to-weight ratio. Cold-formed steel sections may be rolled into a wide variety of forms, and the structure can be made more flexible and efficient by optimizing these shapes. But because of its high heat conductivity, steel can produce thermal bridges that reduce the strength and thermal efficiency of the building.

41.4 Conclusion

After studying the above research paper, we conclude that:

Hot Rolled Sections:

- Offer higher strength and durability due to the hot rolling process.
- Ensure uniformity in shape and size, suitable for heavy-duty structural components.
- Initial production costs may be higher, but long-term maintenance costs are often lower.

Cold Formed Sections:

- Manufactured at room temperature, providing precise shaping and dimensional control.
- Cost-effective option with less energy consumption, ideal for lightweight and intricate sections.
- Flexible design and customization for specific geometries and weight-sensitive applications.

Conclusion:

Hot rolled sections are preferred for heavy-duty structural needs, emphasizing strength and uniformity, while cold formed sections excel in precision, cost-efficiency, and customization. The choice between them depends on project requirements, balancing factors such as structural performance, cost, and design flexibility.

References

Athira VV, Sruthy S (2018) Investigation on the use of cold formed perforated steel sections as columns and purlins. Int J Eng & Technol 7(4.5)

Deepla N, Venkat R (2018) Comparitive study on cold formed light gauge steel angle sections and channel sections of compression members. Res & Rev: J Appl Sci Innov: RRJASI 2(2)

EN 1993–1–3 Design of steel structures - General rules - Supplementary rules for cold-formed members and sheeting (Euro code) 41.1.3

Indian Standards IS: 800: 2007 (2007) Code of practice—general construction steel. New Delhi: BIS 41.1.1

IS 2062–2011: Hot rolled medium & High Tensile Structural Steel 41.1.2

IS 801 (1975): - Code of practice for use of cold formed light gauge steel structural members in general building construction 41.1.3

IS 811(1987): - Cold formed light gauge structural steel sections. 41.1.3

Jayaraman A, Athibaranan S, Mohanraj A (2015) Flexural behaviour of light gauge cold formed steel members : comparison of is code and Euro code. IJRET: Int J Res Eng Technol 04(06)
Jayaraman A, Sathyakumar N, Murugananthan K, Gowtham T, Siva Sankar K (2020) Development of light weight and seismic performance of light gauge cold formed steel structures by different types of moment resistant steel connection. Int J Recent Technol Eng (IJRTE) 8(6)
Journal-A Beginner's Guide to Cold-Formed Steel Framing-Build steel USA.
Journal-Rapid Building System (Light Gauge Steel Buildings)- Volta Green Structures Hyderabad, Telangana
Kakade SA, Bhandarkar BA, Sonar SK, Samare AD (2014) Study of various design methods for cold formed light gauge steel sections for compresive strength. IJRET: Int J Res Eng Technol 03(09). NCETCE2014
Kalbhor SR, Shelke N, Kadlag V (2020) Optimization of industrial structure using light gauge steel section. Int Res J Eng Technol (IRJET) 07(08)
Katkar DD, Phadtare NP (2018) Comparative study of an industrial pre—engineered building with conventional steel building. Int Res J Eng Technol (IRJET) 05(10)
Konkong N (2017) An investigation on the ultimate strength of cold-formed steel bolted connections. Eng, Technol & Appl Sci Res 7(4)
Drishya S Krishna, Binu P (2020) Comparative study of ordinary and Supa sections of cold formed steel. Int J Eng Res & Technol (IJERT) 9(6)
Kulkarni RB, Khidrapure SB (2014) Parametric study and comparison of Indian standard code with British standard code for the Design of Light gauge cold formed flexural members. Int J Eng Tech Res (IJETR) ISSN: 2321–0869, 2(11)
Kumbhar SS, Jadhav HS (2021) Current research on cold-formed steel structures: a review. Int Res J Eng Technol (IRJET) 08(01)
Lambat PV, Denge S (2021) Comparative study on seismic analysis of light gauge cold formed steel and Rc building for severe seismic zone. Int Res J Mod Eng Technol Sci 03(07)
Landolfoa R, Mammana O, Portioli F, Di Lorenzo G, Guerrieri MR (2009) Experimental investigation on laser welded connections for built-up cold-formed steel beams. J Constr Steel Res 65
Markos Makebo G, Paulmakesh A (2021) Analysis of light gauge steel by using fem subjected to tension load. J Univ Shanghai Sci Technol 23(9)
Manoj AVP. Senthil Selvan S (2017) Flexural behaviour of light gauge cold formed steel 'Z' And 'Hat' sections with and without lips. Int J Civ Eng Technol (IJCIET) 8(3)
Meher S, Nar R, Jagadale S, Kalal G, Chandanshive V (2018) Design of industrial warehouse. Int J Eng Res & Technol (IJERT) 7(02)
Navyashree PM, Sowjanya GV, Mallesh TV (2020) Analysis of cold formed steel sections: review. Pap-Int J Eng Res & Technol (IJERT) 8(14)
Prabhakaran S (2020) Study on load carrying capacity of cold formed steel built-up column. J Xidian Univ 14(5)
Raphy S, Bhasin S, Lakshmipriya KP. Shear strength of CFS channel edge stiffened at web hole. Int J Creat Res Thoughts (IJCRT) 9(6)
Salokhe SA, Patil PS (2015) Comparative Study Of Behaviour Of Cold-Formed Steeland Hot Rolled Steel Section Under Compressive Loading. IJRET: Int J Res Eng Technol 04(04)
Schafer BW (2011) Cold-formed steel structures around the world a review of recent advances in applications, analysis and design. Steel Constr 4
Sharma A, Senthil Selvan S, Suresh Babu S, Elango D (2016) Experimental study on the flexural-torsional behavior of cold-formed steel channel section connected back to back. Indian J Sci Technol 9(39)
Thenmozhi R, Sashikkumar MC (2017) Study on capacity of cold formed steel sections. Int J Innov Res Sci, Eng Technol 6(5)
Thorat AR, Patil SK (2017) A study of performance of pre-engineered building of an industrial warehouse for dynamic load. Int Res J Eng Technol (IRJET) 04(06)

Chapter 42
Various Techniques for Pesticide Detection to Improve Rainwater Harvesting

Manav M. Wasekar and Harikumar Naidu

Abstract Rainwater harvesting requires the detection and removal of pesticides for plant growth. Pesticides are widely used in the agriculture sector as well as the non-agriculture sector for improving the quality of food products and their production yield. Heavy use of pesticides contaminates the biosphere directly and indirectly. Soil gets affected by the kinds of pesticide residues sprayed on plants for yield production. The soil loses its respiration property, which reduces its quality. Sprayed pesticides reach water bodies by surface runoff or other ways and contaminate water, which is then consumed by living organisms. Automatic monitoring of pollutants is essential for human health and air, water, and soil quality. Various methods are present to detect pesticide residues. This paper focuses on on-spot detection of residues with electroanalysis techniques for pesticides in soil and water, which can be communicated with the proposed setup at a remote location automatically. This paper is a summary of different methods used for electrochemical-type detection. Though conventional techniques detect the residues accurately, they have some cons like being costly and chemical-intensive, requiring skilled personnel for lab work, and being time-consuming, etc. Electrochemical sensors will save time, as the material is less costly and requires no skilled person to use it. This paper brings out the types of pesticide detection techniques used and instruments using various sensor-based approaches to help the plant growth environment.

Keywords Electrochemical detection · Pesticide sensor · Amperometry · Conductometry · Potentiometry · Voltammetry · Electrochemical impedance spectroscopy · Chronocoulometry · Arduino

M. M. Wasekar · H. Naidu (✉)
Department of Electrical Engineering, G H Raisoni College of Engineering, Nagpur, India
e-mail: harikumar.naidu@raisoni.net

M. M. Wasekar
e-mail: manav.wasekar.mtechips@ghrce.raisoni.net

B. V. Khode et al. (eds.), *Innovations in Technologies: Pioneering Sustainable Infrastructure for a Resilient Future*, Sustainable Civil Infrastructures,
https://doi.org/10.1007/978-3-031-82133-2_42

42.1 Introduction

Pesticides can be categorized into various groups according to their chemical composition, including organophosphates, carbamates, pyrethroids, neonicotinoids, organochlorines, triazines, and substituted phenols (Asif et al. 2022). Additionally, their usage can determine their classification as bactericides, fungicides, herbicides, and insecticides.

Around 2 billion kilograms of pesticides are utilized worldwide each year to control weeds, insects, and pests. India uses just 3.75% of this global annual pesticide use for crop protection (45% in Europe, 25% in the US, and 25% in other countries) (De et al. 2014). Herbicides account for 40% of all pesticides used worldwide, followed by insecticides (17%), fungicides (10%), and other modest uses for the remaining chemicals (Alavanja 2009). Studies indicate that 80% of directly sprayed pesticides negatively impact the ecosystem. Additionally, 64% of the agricultural land on the planet was very susceptible to various forms of contamination due to pesticides (Tang et al. 2021). The review highlighted that pesticide deposition is contributing to the degradation of soil and water quality (Rajmohan et al. 2020). Extensive irrigation or heavy rainfall can increase seepage of pollutants into groundwater (Dietrich et al. 2015).

42.1.1 Methods to Analysis Pesticide Residues

Several extraction techniques, including solid-phase micro-extraction, solid-phase extraction, and liquid–liquid extraction, have been employed over time. These methods are often followed by different chromatographic methods, including both liquid as well as gas chromatography (LC and GC) options for quantitative analysis. Examples include GC–MS (Gas Chromatography-Mass Spectrometry), UHPLC-MS/MS (Ultra-High-Performance Liquid Chromatography-Tandem Mass Spectrometry) (Corazza et al. 2019), LC–MS/MS (Liquid Chromatography-Tandem Mass Spectrometry) and another popular method for detection is HPLC (High-performance liquid chromatography) with UV detection (Tuzimski and Sherma 2018). Optical techniques that depend on the interactions between light and matter, including calorimetric and spectroscopic approaches, are more susceptible to interact with other substances in the system and are more influenced by noise levels in unfavorable detecting environments (Chen et al. 2021). Electrochemical techniques yield more precise and dependable findings that can be achieved with basic tools like a potentiostat, allowing for on-site dynamic detection.

42.1.2 Electrochemical Sensor

A sensor is composed of two components: a transducer and a receiver. Transducers enable the complete analysis and electronic presentation of analytical signals and precious energy, while receivers accept physical, biological, or chemical stimuli and make it possible to translate them into electrical signals or energy. Electrochemical (EC) sensors, a subtype of chemical sensors, use a chemical reaction to determine an analyte's concentration. EC sensors detect information related to chemical processes, which they then transform into quantifiable voltage or potential, conductivity, current, frequency, etc. (Sharma and Palatty 2023). In electrical measurements, additional electrodes are also utilized, such as reference electrodes and occasionally a counter electrode.

42.1.3 Communication Framework

The suggested system includes sensors that detect signals, an A/D converter, a microprocessor, and a transmitter or receiver to depict the concentration of pesticides present in the samples (Sao et al. 2022). This is depicted in the block diagram at the remote master station industry away from the testing facility, as shown in Fig. 42.1.

Figure 42.2 illustrates the basic input/output instructions, loops (for, do-while, while), and decision-making structures (if-else, switch-case) of Arduino programming. Instructions in a program are detailed steps or commands guiding a computer or device like Arduino, written in a programming language and executed sequentially to achieve desired outcomes (Naidu and Ramtekkar 2023). Each instruction serves a purpose, like calculations, data manipulation, or hardware interaction, forming program building blocks for efficient goal achievement (Padwad et al. 2023). The resulting success is displayed on the monitor.

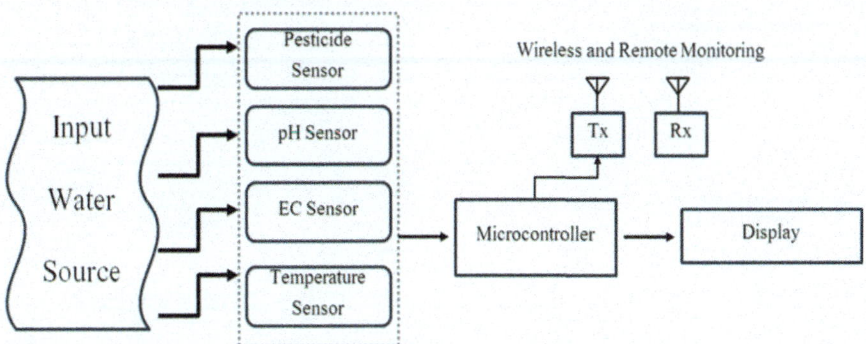

Fig. 42.1 Proposed industrial automation block diagram

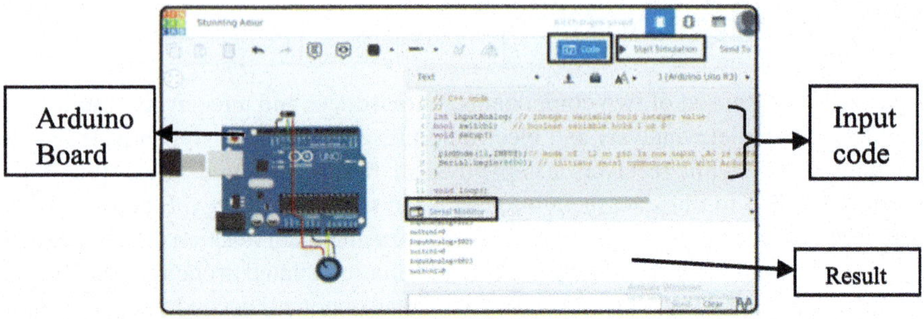

Fig. 42.2 Basic input and output instruction with sensor

42.2 Electrochemical Detection Method

The basis of analytical methods is the measurement of certain electrical characteristics that emerge when two distinct electrodes are kept in contact with an electrolyte that contains the desired chemicals. Examples of interfacial methods that are employed to examine the electrochemical processes occurring at the electrode-solution interface, as shown in Fig. 42.3.

42.2.1 Conductometry

The process of measuring conductance involves determining the resistance of a sample solution between two parallel electrodes, as shown in Fig. 42.4.

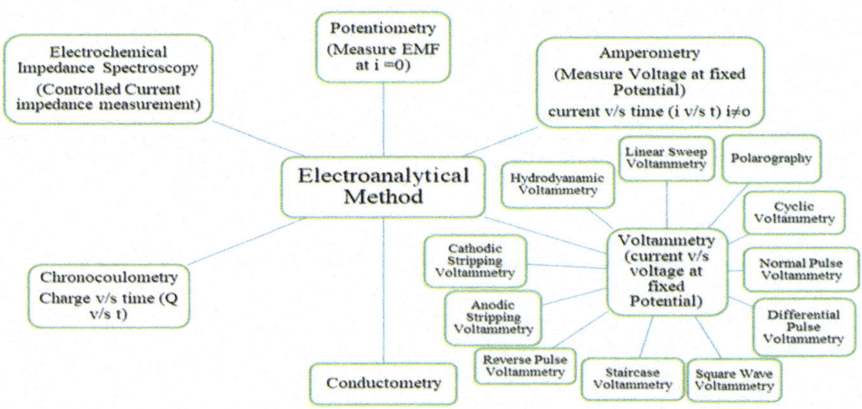

Fig. 42.3 Types of electrochemical detection method

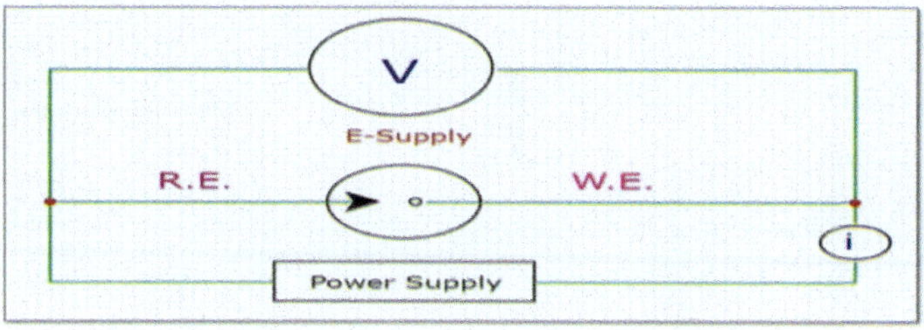

Fig. 42.4 Two electrodes conductometry simplified electric diagram

42.2.2 Amperometry

The reaction rate of redox agents at the electrode contact determines the current. Initially, the current will increase from the base current I_1, then stabilize at a steady-state value I_2, which is defined by the chemical potential of electroactive particles (Guo and Kang 2013). The working principle of amperometric detection involves applying an initial voltage V_1 to the electrode when there is no chemical reaction, and a subsequent voltage V_2 to the electrode when there is no chemical reaction, and a subsequent voltage V_1 and V_2, the corresponding current responses are I_1 and I_2 respectively, as shown in Fig. 42.5.

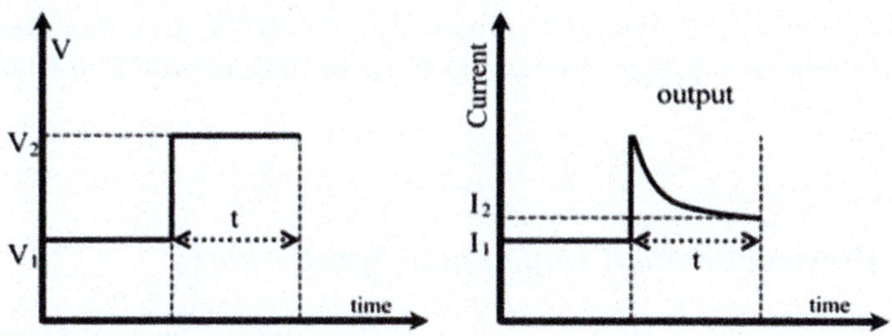

Fig. 42.5 Amperometry characteristic graph

42.2.3 Potentiometry

The electrical potential difference between the electrodes is subsequently measured, and these measurements are utilized to determine the quantity of the specific compound present in the sample. Consequently, potentiometric analysis involves the use of a reference electrode, an indicator electrode, and equipment for measuring electrical potential, as shown in Fig. 42.6.

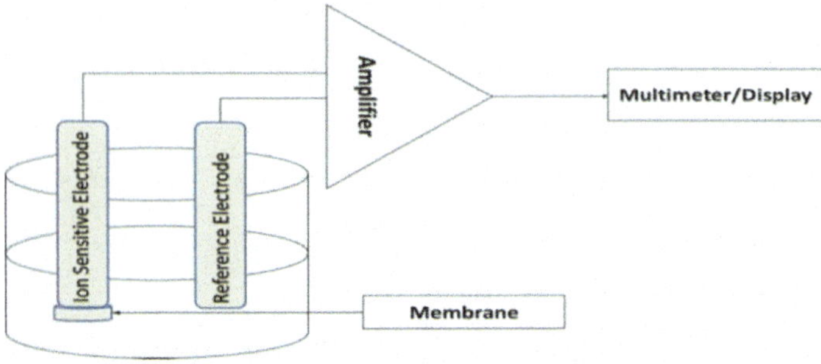

Fig. 42.6 Two electrode potentiometry method with electronic components

At zero current, potentiometric sensors detect the electrical potential between the working and reference electrodes (Sadak 2023).

42.2.4 Chronocoulometry

Chronocoulometry, a signal rise rather than a reduction in time is observed; the reduction of noise through signal integration produces a refined hyperbolic curve, making the contributions of absorbed or double-layer charged species plainly visible. The electrochemical setup involved a modified glassy carbon working electrode, SCE as R.E., and a Pt wire for the counter electrode (or a Pt mesh CE in RDE studies) (Wei et al. 2018).

42.2.5 Electrochemical Impedance Spectroscopy

Electrochemical reactions within an electrolytic cell are portrayed in EIS through an electrical equivalent circuit (EEC). The current flowing from the electrified interface as a result of an electrochemical reaction causes charge transfer along the contact, resulting in both faradaic and non-faradic components. Figure 42.7. depicts an idealized electrical equivalent circuit for an electrochemical process that includes these faradaic and non-faradic components.

The C = Cd represents a double-layer capacitor; R2 = Rs represents the resistance of the solution; R1 = Rct represents the charge transfer resistance; and Zw is the Warburg impedance. Predicting the analytical concentration in an electrolytic solution involves calculating the impedance parameter and other RC properties (resistive-capacitive) of the EEC. Impedance spectra, as a resultant signal, can be visualized through two methods. The Bode plot uses a logarithmic frequency scale on the x-axis and a logarithm of impedance (Z) on the y-axis, with an additional y-axis for phase shift (Φ). The Nyquist plot is especially sensitive to alterations in the electrode/

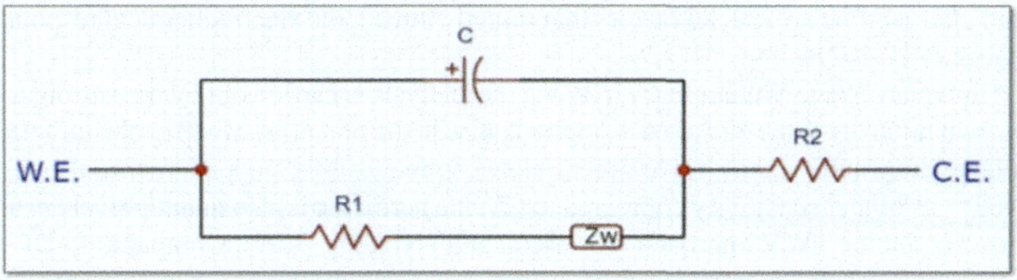

Fig. 42.7 Randles EEC representation for electrochemical reaction

solution interface, making it a preferred method in pesticide analysis (Souza et al. 2020).

42.2.6 Voltammetry

The voltammetry profile, which takes selectivity and sensitivity into account, is defined by the potential variation mode. This produces variations in the analytical data. Forms of voltammetry are shown in Fig. 42.8.

Differential pulse voltammetry techniques are employed to identify analyte substances in a variety of settings where they are present in trace amounts. In this voltametric approach, modulation amplitude, step potential, modulation duration,

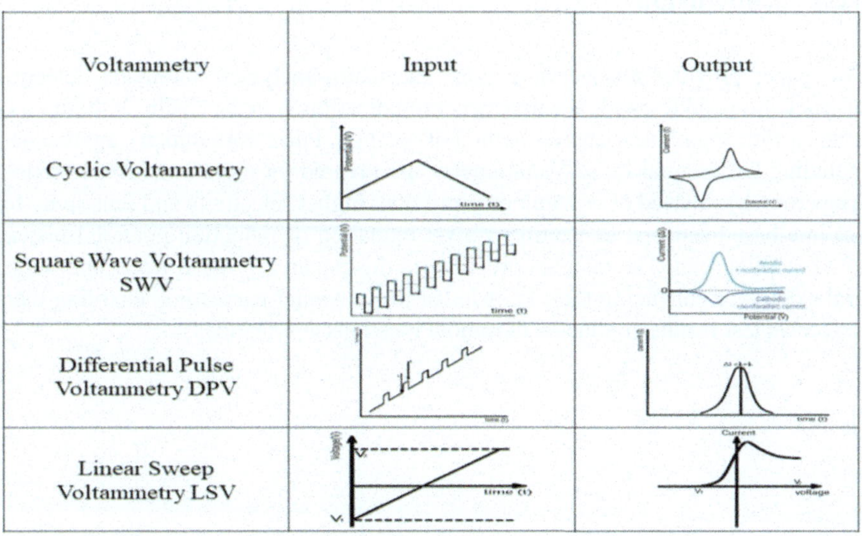

Fig. 42.8 Types of voltammetry input and output signals

and interval time are crucial factors that directly impact the electrochemical response (Gege et al. 2021).

In Square Wave Voltammetry (SWV), the analysis is carried out by measuring the current intensity at the end of each applied pulse in the potential sweep cycle, utilizing a large-amplitude square wave potential sweep featuring a stepped potential ramp. Due to its higher sensitivity compared to CV, the herbicide is also quantitatively evaluated using the SWV approach (Akinbulu and Nyokong 2009). Compared to DPV, SWV is often quicker, more sensitive, and more selective. Notably, studies by Zaouak et al. used the DPV approach to study electrochemical changes of five nitro diphenyl ethers (Zaouak et al. 2011), but the LOD for aclonifen was higher compared to the results achieved by Inam's approach (İnam and Çakmak 2013) (0.6 mg mL^{-1} vs. 0.073 mg mL^{-1}). Additionally, Garcia et al. utilized a multi-walled carbon nanotube-modified electrode in conjunction with SWV and DPV for paraquat detection in natural water samples, with SWV demonstrating superior detection performance and outcomes, highlighting its efficacy (Garcia et al. 2013).

Swept voltammetry or pulsed voltammetry measures the concentration of the analyte by observing the redox response of the sample while adjusting the potential window.

CV and LSV utilize a linear potential wave with a constant scan rate throughout the voltametric sweep to extract information such as analyte concentrations, diffusion constants, and rate constants of a redox system. The existing applied techniques for the detection of pesticides and their outcomes are shown in Table 42.1.

42.3 Conclusion

This paper provides an overview study of electro-analytical automatic detection of pesticide residues with a wireless communication system. Cyclic Voltammetry offers a simple and inexpensive method of analysis. Pulse Voltammetry approaches, including Differential Pulse Voltammetry, are favored for more sensitive detection. Amperometry, instead of voltammetry, provides higher selectivity and sensitivity for enzyme-based sensors, as it utilizes a set oxidation or reduction potential instead of a potential scan. The insights provided in this paper will assist R&D and industrial experts in enhancing their knowledge for rainwater harvesting, enabling them to receive test results conducted at remote locations automatically.

Table 42.1 Example of pesticide detection using electrochemical technique, sensor, material, and instrument with result

Pesticide/reference	Samples./instrument	Analysis Technique	Sensor or W.E	R.E	C.E	Analytical parameters	Electrical parameters used
Fenitrothion/ El-Hallag et al. (2023)	Potentiostat model 273A-PAR equipped with a C-2 stand and magnetic heating stirrer. Operation software 270/250-PAR	CV and LSV	Modified CPE was constructed by applying a safranine layer on its surface	Ag/ AgCl	pt wire	LR = 1 to 15 μM R^2 parameter = 0.99906 LOD = 0.1 μM LOQ = 0.34 μM	CV Potential window = +0.6 to −1.3 V Scan rate = 100 mV/s Peak current = −0.73 V LSV Potential range = −0.4 to −1 V Sweep rate = 100 mV/s
Arsenic/Hamid Kargari et al. (2023)	Tap water and rice flour samples./AUTOLAB PGSTAT30	CV and DPV	AAGO-PDDA-PA/ GCE	Ag/ AgCl	Pt	Sensitivity = 1.79 A/M LOD = 0.12 μM LOQ = 0.38 μM	CV Potential range = −1 to +2 V DPV Potential range = −1 to 0 V Pulse amplitude = 50 mV
Flubendiamide/ Aghris et al. (2023)	White rice./metrohm/ potentiostat coupled to 797 VA computrace software	CV	rGO/CHI/PGE	Ag/ AgCl	Pt Wire	LR = 0.7 to 150 μM Sensitivity = $47.3 \mu A \mu M^{-1} cm^{-2}$ LOD = 0.24 μM LOQ = 0.84 μM	CV Potential range = −1.5 to −0.2 V Potential scan rates range = 10–500 mVs^{-1}

(continued)

Table 42.1 (continued)

Pesticide/reference	Samples./instrument	Analysis Technique	Sensor or W.E	R.E	C.E	Analytical parameters	Electrical parameters used
Diuron/Serge et al. (2021)	Orange juice, mineral and tap water./DY 2300 electrochemical analyzer with the DY 2300 EN software	CV and SWV	Carbon Paste Electrode (CPE) with nanocrystalline Cellulose (NC)	SCE	Pt Wire	LR = 4.2–47 μmol/L LOD = 0.35 μM	SWV Frequency = 50 Hz Pulse height = 60 mV Scan rate = 20 mV
Carbaryl/Mashuni et al. (2022)	Potentiometer (SANFIX DM-888C)	Potentiometer	Au–Ag/rGO/CS	Ag/AgCl	Pt wire	LR = 1×10^{-8} to 1.0 mgL^{-1} LOD = 1.0×10^{-9} $\mu g\ mL^{-1}$	–
Diuron/Qin et al. (2020)	Pear juice, Mineral water, and dry tea./CHI-660 electrochemical workstation	DPV, EIS, and CV	Nafion/AuNPs/RGO/GCE	Ag/AgCl	Pt wire	LR = 1×10^{-12}M to 1.0×10^{-9}M LOD = 4.1×10^{-13} M	CV Potential range = –0.2 to +0.8V Scan Rate = 0.1V/s EIS Frequency range = 0.1 MHz to 0.1 Hz AC amplitude = 5mV DPV Potential range = 0.70–1.40 V Scan Rate = 0.1V/s

(continued)

Table 42.1 (continued)

Pesticide/reference	Samples./instrument	Analysis Technique	Sensor or W.E	R.E	C.E	Analytical parameters	Electrical parameters used
Phenyl Carbamates/Della Pelle et al. (2018)	Wheat and Maize./Portable Palm-Sens 4 Analyzer using commercial screen-printed electrodes (Drop-Sens C110)	CV, DPV, and EIS	CB-SPE	Ag/AgCl	–	LR = 1.0×10^{-7} and 1.0×10^{-4} $molL^{-1}$ Sensitivity $\geq$ 0.390 A M^{-1} cm^{-2} LODs $\leq 8.0 \times 10^{-8}$ $molL^{-1}$	CV Potential range = −0.50 V to + 0.80 V (vs. pseudo-Ag/AgCl) Scan rate = 0.050 Vs^{-1} DPV Potential range = 0 to +0.6 V Pulse width = 50 ms Pulse amplitude = 20 mV
Carbofuran/Ayyalusamy et al. (2018)	Real sample from farm field./electrochemical work station (Autolab PGSTAT 30, Eco Chemie, Netherland)	Amperometry and CV	PET-AC/GCE & bare GCE	Ag/AgCl	Large Pt foil	LOD = 0.03 µM LR = 1–10 µM Sensitivity = 0.11 $\mu A \mu M^{-1} cm^{-2}$	CV Scan rate = 50 mVs^{-1}
Carbofuran (CBF) and Diuron (DU)/Wong and Sotomayor (2014)	Soil and carrots./palm sense potentiostat interfaced with a microcomputer	Amperometry and CV	Modified electrode	Ag/AgCl	Pt wire	CBF LR = 5×10^{-6} to 1.4×10^{-4} mol L^{-1} LOD = 1.7×10^{-6} mol L^{-1} DU LR = 5×10^{-5} to 1×10^{-3} mol L^{-1} LOD = 2×10^{-5}mol L^{-1}	CV Potential Range = −0.2 to 1 V Amperometric Potential = 450 mV and 800 mV (vs. Ag/AgCl) for the redox reaction of CBF & DU

(continued)

Table 42.1 (continued)

Pesticide/reference	Samples./instrument	Analysis Technique	Sensor or W.E	R.E	C.E	Analytical parameters	Electrical parameters used
Carbofuran/Qi et al. (2018)	Cowpea and Pakchoi/ Autolab PGSTAT 204	DPV and CV	MIP/AuNPs/GCE	SCE	Pt wire	LR = 5×10^{-8} to 4×10^{-4} mol/L LOD = 2.4×10^{-8} mol/L	CV Potential range = 0.2–0.6 V Scan rate = 0.05 V/s DPV Scan range = 0.6–0.2 V Pulse width = 50 ms Time interval = 0.5 s Step potential of = 0.005V Amplitude = 0.05 V

rGO—reduced graphene oxide, CPE—Carbon Paste Electrode, CHI—Chitosan, PGE—Pencil Electrode, CS—Chitosan, Au—Gold, Ag—Silver, AgCl—Silver Chloride, AuNPs—Gold Nanoparticles, CB-SPE—Carbon Black Modified Screen-Printed Electrode, PET—Polyethylene Terephthalate, GCE—Glassy Carbon Electrode, AC—Activated Carbon, MIP—Molecularly Imprinted Polymers, AAGO—Acrylic Acid-functionalized Graphene Oxide, PDDA—Poly (Diallyl Dimethyl Ammonium Chloride), PA—Polyaniline, Pt—Platinum, LR—Linear Range, LOD—Limit of Detection, LOQ—Limit of Quantification, SCE—Saturated Calomel Reference Electrode

References

Aghris S et al (2023) An electrochemical sensor for flubendiamide insecticide analysis based on chitosan/reduced graphene oxide. Sens Diagn 2(2):398–408

Akinbulu IA, Nyokong T (2009) Characterization of polymeric film of a new manganese phthalocyanine complex octa-substituted with 2-diethylaminoethanethiol, and its use for the electrochemical detection of bentazon. Electrochim Acta 55(1):37–45

Alavanja MC (2009) Introduction: Pesticides use and exposure, extensive worldwide. Rev Environ Health 24(4):303–310

Asif M, Sajid H, Ayub K, Ans M, Mahmood T (2022) A first principles study on electrochemical sensing of highly toxic pesticides by using porous C4N nanoflake. J Phys Chem Solids 160:110345

Ayyalusamy S, Mishra S, Suryanarayanan V (2018) Promising post-consumer PET-derived activated carbon electrode material for non-enzymatic electrochemical determination of carbofuran hydrolysate. Sci Rep 8(1):13151

Chen Q et al (2021) Colorimetric and fluorescent sensors for detection of nerve agents and organophosphorus pesticides. Sens Actuators B Chem 344:130278

Corazza G, Merib J, do Carmo SN, Mendes LD, Carasek E (2019) Assessment of a fully optimized DPX-based procedure for the multiclass determination of pesticides in drinking water using high-performance liquid chromatography with diode array detection. J Braz Chem Soc 30:1211–1221

De A et al (2014) Worldwide pesticide use. Target Deliv Pestic Using Biodegrad Polym Nanopart 5–6

Della Pelle F, Angelini C, Sergi M, Del Carlo M, Pepe A, Compagnone D (2018) Nano carbon black-based screenprinted sensor for carbofuran, isoprocarb, carbaryl and fenobucarb detection: application to grain samples. Talanta, vol 186, pp 389–396

Dietrich AM et al (2015) Partitioning, aqueous solubility, and dipole moment data for cis-and trans-(4-methylcyclohexyl) methanol, principal contaminants of the West Virginia chemical spill. Environ Sci Technol Lett 2(4):123–127

El-Hallag IS, Moharram YI, Selim SM (2023) Poly (safranine) modified carbon paste electrode as a sensor for the determination of fenitrothion pesticide. Sci Rep 13(1):14332

Garcia LLC, Figueiredo-Filho LCS, Oliveira GG, Fatibello-Filho O, Banks CE (2013) Square-wave voltammetric determination of paraquat using a glassy carbon electrode modified with multi-walled carbon nanotubes within a dihexadecyl hydrogen phosphate (DHP) film. Sens Actuators B Chem 181:306–311

Gege Ü, Karakaya S, Dilgin Y (2021) Sensitive electrochemical determination of trifluralin at a disposable pencil graphite electrode. Electroanalysis 33(6):1547–1556

Guo J, Kang Y (2013) Amperometric detection. In: Li D (ed) Encyclopedia of microfluidics and nanofluidics. Springer US, Boston, MA, pp 1–8. https://doi.org/10.1007/978-3-642-27758-0_25-2

Hamid Kargari S, Ahour F, Mahmoudian M (2023) An electrochemical sensor for the detection of arsenic using nanocomposite-modified electrode. Sci Rep 13(1):8816

İnam R, Çakmak Z (2013) A simple square wave voltammetric method for the determination of aclonifen herbicide. Anal Methods 5(13):3314–3320

Mashuni M, Ritonga H, Jahiding M, Rubak B, Hamid FH (2022) Highly sensitive detection of carbaryl pesticides using potentiometric biosensor with nanocomposite Ag/r-graphene oxide/chitosan immobilized acetylcholinesterase enzyme. Chemosensors 10(4):138

Naidu H, Ramtekkar P (2023) An innovative affordable robot to defuse landmines using IoT and wireless communication technique to save precious life. In: 2023 11th international conference on emerging trends in engineering and technology-signal and information processing (ICETET-SIP). IEEE, pp 1–5

Padwad D, Naidu HK, Kokate P, Agrawal P (2023) Smart IoT system for indoor plants with augmented artificial lighting. In: 2023 11th international conference on emerging trends in engineering and technology-signal and information processing (ICETET-SIP). IEEE, pp 1–6

Qi P et al (2018) Sensitive and selective detection of the highly toxic pesticide carbofuran in vegetable samples by a molecularly imprinted electrochemical sensor with signal enhancement by AuNPs. RSC Adv 8(45):25334–25341

Qin J et al (2020) Electrochemical sensor based on Nafion/gold nanoparticle/electrochemically reduced graphene oxide composite-modified glassy carbon electrode for the detection of diuron. Int J Electrochem Sci 15(11):11203–11214

Rajmohan KS, Chandrasekaran R, Varjani S (2020) A review on occurrence of pesticides in environment and current technologies for their remediation and management. Ind J Microbiol 60:125–138

Sadak O (2023) Potentiometric sensors. In: Fundamentals of sensor technology. Elsevier, pp 99–121

Sao S, Dudhe S, Naidu H (2022) A Novel IoT based sample collection method for baby meconium aspiration syndrome and analyzing suction pressure of the system. In: 2022 10th international conference on emerging trends in engineering and technology-signal and information processing (ICETET-SIP-22). IEEE, pp 1–5

Serge M et al (2021) Electrochemical determination of diuron in soil using a nanocrystalline cellulose modified carbon paste electrode. Int J Electrochem Sci 16(5):210552

Sharma SS, Palatty S (2023) Advances in functionalized polyaniline nanocomposites for electrochemical sensing and energy storage applications. In: Applications of multifunctional nanomaterials, Elsevier, pp 177–196

De Souza D, Gonçalves-Filho DL, Franco DL (2020) Pesticide residues analysis by electroanalytical techniques. Sustain Agric Rev 48; In: Pesticide occurrence, analysis and remediation, vol 2. Springer, pp 1–75

Tang FH, Lenzen M, McBratney A, Maggi F (2021) Risk of pesticide pollution at the global scale. Nat Geosci 14(4):206–210

Tuzimski T, Sherma J (2018) Determination of target xenobiotics and unknown compound residues in food, environmental, and biological samples. CRC Press

Wei P, Gan T, Wu K (2018) N-methyl-2-pyrrolidone exfoliated graphene as highly sensitive analytical platform for carbendazim. Sens Actuators B Chem 274:551–559

Wong A, Sotomayor MDPT (2014) Determination of carbofuran and diuron in FIA system using electrochemical sensor modified with organometallic complexes and graphene oxide. J Electroanal Chem 731:163–171

Zaouak A, Matoussi F, Dachraoui M (2011) Electrochemical study of diphenyl ether derivatives used as herbicides. Int J Electrochem 2011

Chapter 43
Effect of Bitumen Content and Polypropylene Fiber on Performance of Bituminous Mix Properties

Ashraf Mathakiya, Hemantkumar Sonkusare, Mayursinh Jadeja, and Hardik Pujara

Abstract This research paper investigates impact of varying bitumen content and different percentages of polypropylene fibre on properties of bituminous mixes. Bituminous mixes are commonly used in road construction, and their performance is inclined by various factors such as bitumen substance and the addition of reinforcing fibres. In this particular investigation different bitumen content levels are considered to assess their impact on the overall properties of mechanical and durability of bituminous mixes. Additionally, various percentages of polypropylene fibre are incorporated into the mix to analyse their effect on cracking resistance and tensile strength. The research methodology involves laboratory testing, including Marshall Stability, flow, stripping value, ITS test (indirect tensile strength) and TSR (tensile strength ratio). The results are analysed to identify optimal bitumen content and polypropylene fibre percentages are 4.9 and 0.3% respectively for Marshall Stability and 5.4 and 0.3% respectively for tensile strength to enhanced bituminous mix performance. The following outcome from this explores the optimization of bituminous mix design, offering valuable insights for road construction and maintenance practices.

Keywords Bituminous mix · Marshall Stability · Indirect tensile strength · Polypropylene fibre

43.1 Introduction

For the development of economic and social road like a flexible pavement is more and highly important in country for transport of any things like passenger/people, goods, etc. (Slebi-Acevedo et al. 2022). Ensuring the longevity of road surfaces is a crucial concern with significant implications for the economy and various other

A. Mathakiya (✉) · H. Sonkusare · M. Jadeja · H. Pujara
Department of Civil Engineering, Atmiya University, Rajkot, India
e-mail: ashrafmathakiya@gmail.com

B. V. Khode et al. (eds.), *Innovations in Technologies: Pioneering Sustainable Infrastructure for a Resilient Future*, Sustainable Civil Infrastructures,
https://doi.org/10.1007/978-3-031-82133-2_43

factors (Bala et al. 2018). Road which is made by bitumen is the well known type of pavement of flexible and entire world including India are using it (Bala et al. 2018). Categories of bituminous mixtures are various (Tapkin et al. 2010); to make stronger and durable flexible pavements, high density mixture like dense mixture of bitumen and aggregate is developed where accurately maintain the proportion of bitumen as a binder and filler, fine and coarse aggregate are heated and mixed it properly as per standard requirement (Tapkin et al. 2010). Flexible pavement road failure day by day due to combination impact of vehicular movement and atmospheric condition, different types of distresses could be identified in flexible pavement road and lifetime of pavement reduce by it, and these because of various parameter like sensitivity of moisture is one of the parameter that caused to premature pavement deterioration (Klinsky et al. 2018). Modified versions of asphalt mix are commonly employed to mitigate the impact of rising traffic loads and challenging weather conditions (Klinsky et al. 2018; Koçkal and Köfteci 2016). These additives enhance the characteristics of bitumen, thereby improving the overall properties of hot mix asphalt (Klinsky et al. 2018) various authors worldwide have worked and found that bitumen performance and The incorporation of modifiers can enhance bituminous mixes performance (Vishnuvardhan et al. 2020). Polymers and some other wastage materials which are environmentally degradable have been incorporated as a modifiers Enhancing the experiment result of bitumen and mixture of bituminous is imperative for optimizing their overall effectiveness (Slebi-Acevedo et al. 2022; Vishnuvardhan et al. 2020). Traditional bitumen is not much performed more than the modified bitumen in present, where the stone aggregates are prone to coating aggregate (IRC 2010). Because of it good rheological and mechanical properties, these could also be applicable at heavy traffic intersections, bridge and its decks and roundabouts for increased life span of the pavement surface (IRC 2010). In sort many authors have worked on bituminous mix with different fibre at optimum content of bitumen but the aim of current researcher is to analyse and compared the impact of various bitumen content and different percentage of the polypropylene fibre on bituminous mix's properties shown in below Fig. 43.1

43.2 Literature

In this paper, the neural networks (NN) utilization is explored for forecasting various parameters of Marshall test outcomes in asphalt mixtures modified with polypropylene (PP). M-03 type (fibre length 3 mm), M-09 type (fibre length 9 mm) and waste fibers were utilized. Based on workability criteria, fibre length 3 mm were identified most suitable modifiers. The neural network prediction models on parameter like stability, flow and marshal quotient have been identified strong agreement with experimental results (Tapkin et al. 2010). Author has focussed on technology of compaction method like gyratory compaction for preparation of marshal mould using addition of fibre of multi filament polypropylene doses from 0 to 7% of interval of 0.5%, and identified 5.5% best from marshal stability experiment (Tapkin 2013). In

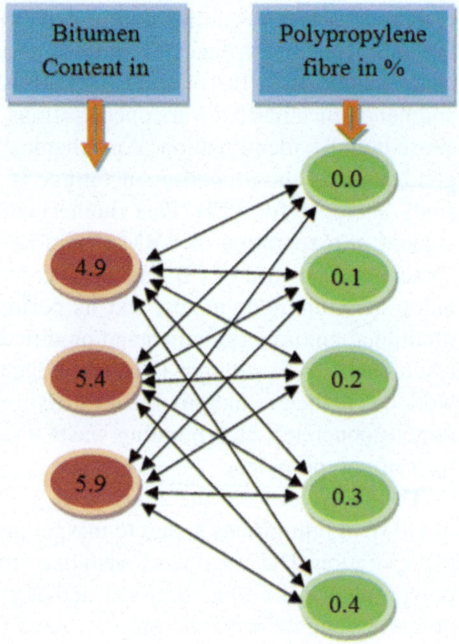

Fig. 43.1 Combination of bitumen content and PP fibre 1

this investigation, 12 mm polypropylene fibers length was incorporated into hot mix asphalt (HMA) at different proportion of 0, 5, 7.5, and 10% by weight of binder. Subsequently, these mixtures underwent exposure to impact of moisture, as well as $CaCl_2$, NaCl, and Na_2SO_4 solutions for a specified duration. The findings indicated a continuous increase in stability for mixtures containing up to 7.5% polypropylene fiber, followed by a decline in stability observed in mixtures reinforced with 10% fiber. Furthermore, the study identified result of moisture impact for durability and Na_2SO_4 solution as the least and most detrimental conditions respectively (Koçkal and Köfteci 2016). In this research, laboratory work was completed for identify the experiment result of pp fibre and brick as a aggregate in bituminous concrete mix, result shows that adding PP fibre improved the mechanical properties of mix of bituminous with containing brick as a aggregate, also result shows that without modify of bitumen not found better tensile strength ratio with only brick added in mix but modified bitumen result fulfil the all requirement of tensile strength ratio (Zachariah et al. 2018). In this scholarly article, the author has been examined the potential of mixing of fibres of polypropylene and aramid on the modified asphalt mixture to see the any changes in performance or characteristics of it. The study utilizes a typical dense HMA (hot mix asphalt) employed in Sao Paulo, Brazil. Results from laboratory testing program indicate that incorporating a blend of polypropylene fibre and aramid fiber into hot mix asphalt improves its mechanical properties. The findings suggest that the application of fiber-modification hot mix asphalt (HMA) in for pavement construction or pavement strengthening has the potential to delay the onset of bitumen pavement failure, thereby extending it life span (Klinsky et al. 2018). In this research endeavour, the author delves into the impact of nanosilica material on the improvement attributes of bitumen as a modified by polymer. To achieve this, a control grade of 80/100 bitumen binder is subjected to modification with pp polymer and nano-silica particles at concentrations ranging from 0 to 4%. Laboratory work carried out for identification of mechanical properties of bituminous mix and found good result in the properties of fatigue in modified specimen (Bala 2018). In this study, shredded form FB (Flex Banner) and PF were utilized to determine optimum content in preparation of RBM (Reinforced Bituminous Mixes), bitumen mixture containing both 6% of FB and PF was found as the optimal blend for the formulation of bituminous mixes, and its performance was evaluated in comparison to traditional mixing of bitumen and modified bitumen mix and modified bitumen mix result show good strength (Vishnuvardhan et al. 2020). The focus of the author's work lies in the evaluation of resistance to rutting, moisture impact, etc. in BC (bituminous concrete), incorporating waste material of CBW (crushed brick waste) and fiber of polypropylene.

The study involved the analysis of 7 different mixes to assess the effectiveness of CBW in bituminous concrete mix, employing both wet and dry mixing for fiber incorporation. The outcomes reveal that bituminous concrete samples modified with polypropylene and containing crushed over-burnt brick waste demonstrate durability properties and deformation properties with control samples (Zachariah et al. 2021). Authors has worked on coir fiber for preparation of bituminous mix in which they was mainly performed the experiment like marshall stability test to determine the optimum binder and its fond 5.6% and they have continue the experiment to find optimum fibre content is 0.4% (Pratul et al. 2021).

43.3 Material and Methods

43.3.1 Materials

VG-30 viscosity grade bitumen is used in this research work, further on that conducted various properties of the bitumen softening point, ductility, viscosity, penetration and specific gravity. The tests have been determined laboratory according to Indian Standards (IS:1201-1220 1978; Ministry for Road Transport and Highways 2013). Table 43.1 showing the various bitumen results conducted in laboratory. The filler, fine and coarse aggregates used for study and various experiments were performed to examine aggregate's properties as per MoRTH standard for Marshall samples preparation (Ministry for Road Transport and Highways 2013; IS:2386 2016). Below Table 43.2 indicated aggregate's properties. Polypropylene fibre added to made modified bitumen for preparation of mould of Marshall. Table 43.3 showing the polypropylene fibre properties incorporated in the work.

Table 43.1 Physical properties of bitumen

Property	Result
Penetration at 25 °C/100 gm/5 s, mm	64 mm
Ductility, cm	81.02 cm
Softening point, °C	50 °C
Viscosity at 60 °C, poise	1057
Specific gravity	1.01

Table 43.2 Physical properties of aggregate

Property	Result
Aggregate impact value %	10.29%
Abrasion value %	27%
Water absorption %	0.45%
Specific gravity	2.74

Table 43.3 Physical properties of polypropylene fibre

Color	White
Density	0.91 g/cm^3
Elongation break	10–45%
Melting point	170 C
Modulus of elasticity	pa

43.4 Experiment and Result

43.4.1 Mix Design and Sample Preparation

Bituminous concrete grade-II mix design used in this research for performance analysis of bituminous mix having various proportion of polypropylene fibre as per the Indian guideline of MoRTH-5th revision (Ministry for Road Transport and Highways 2013), Fig. 43.2 show the details of aggregate proportion ratio used in this study. A representative sample of the asphalt mixture is collected and heated to a specified temperature to make it workable. The sample is compacted into a cylindrical mould using a Marshall Compaction apparatus. Typically, the compaction is done with a set number of blows per layer and a specified number of layers.

Striping Value Test

Striping value test was performed as per IS code to identify the adhesive bond in between bitumen and aggregate in compacted mixture, for that taken 200 gm of aggregate as per specification mention in IS code and bitumen heated at 160 °C mixing to a temperature of 150 °C mixed with bitumen (IS:6241 1971). This mix taken into beaker and rest it to cool down at room temperature for 2 h, than add water

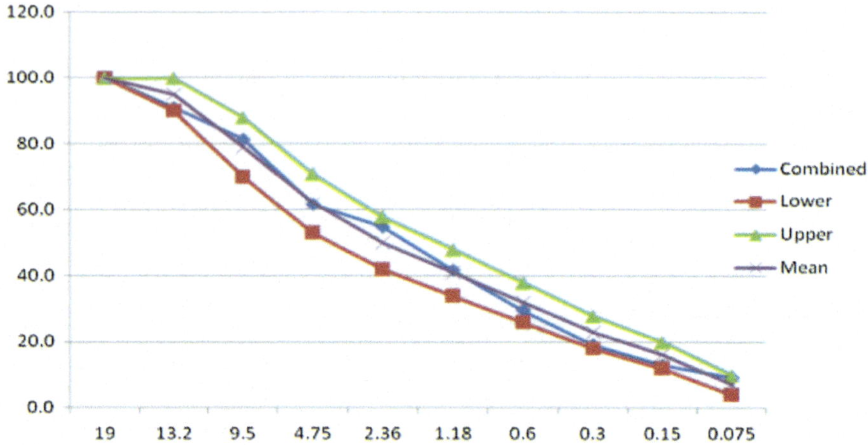

Fig. 43.2 Gradation chart

in beaker and maintain temperature 40 °C, and after that the beaker was remove and allowed to cool at room temperature and determined striping value as per IS code. Figure 43.3 shows the results aggregate stripping value.

Marshall Stability Test

Marshall Stability test was performed for different binder content like 4.9, 5.4 and 5.9% with varying doses of polypropylene fibre of 0, 0.1, 0.2, 0.3, and 0.4%. Three specimens have been prepared for each combination of different binder and fibre and considered average value of it. Further calculation has been carried out for volumetric parameter. Figure 43.4a, b showing the result of various marshal parameter versus binder content with different polypropylene fibre range.

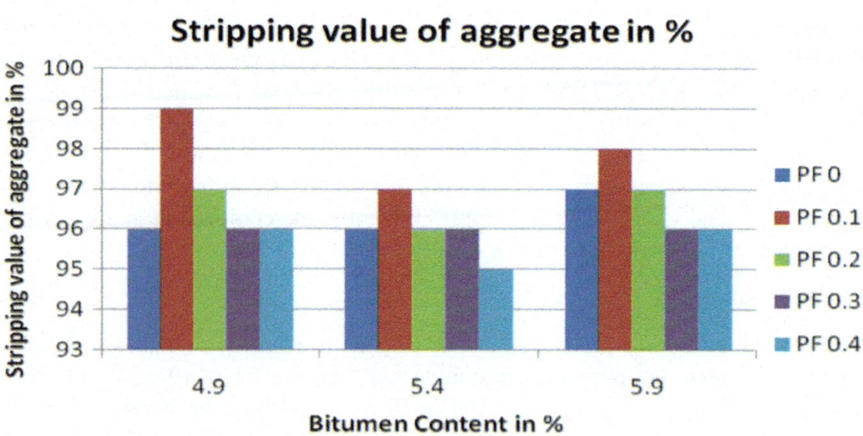

Fig. 43.3 Stripping value result

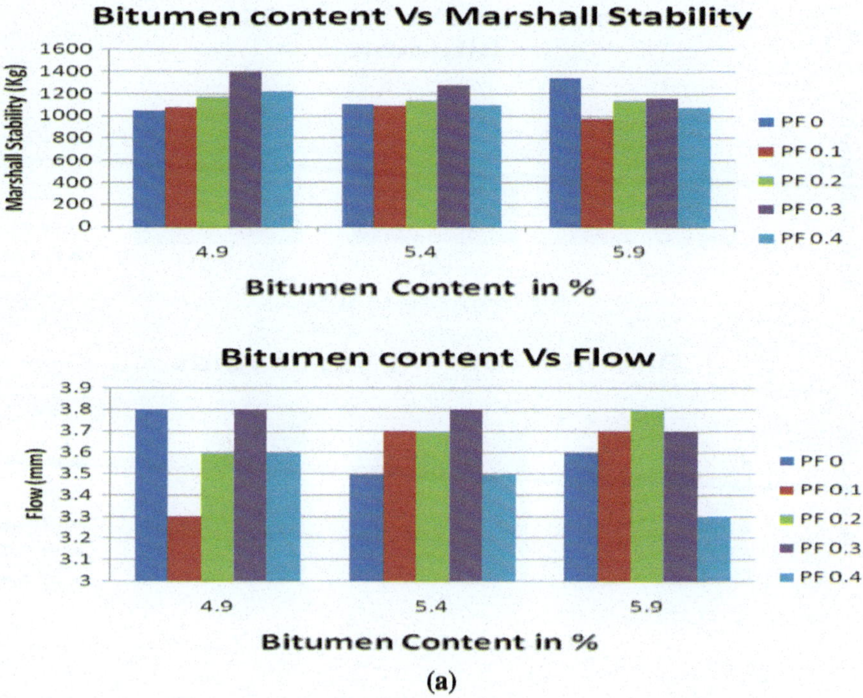

Fig. 43.4 **a** Bitumen content versus Marshal test result. **b** Bitumen content versus Marshal test result

(ITS)Indirect Tensile Strength Experiment

ITS testing is a common method used to assess strength of tensile or resistance to cracking of bituminous (asphalt) mixtures. It measures the ability of the mix of asphalt to withstand tensile forces, which is essential for evaluating its performance under traffic condition and weather condition. The ITS test involves subjecting to tensile stresses, rather than direct compressive forces.

A vertical load is applied to the top loading plate, causing the specimen to split or deform along the vertical axis. The applied force induces tensile stresses within the specimen. In these experiments both ITS unconditional and conditional evaluated and also determined ITR, this all results are shows in Fig. 43.5. Calculated tensile strength for each specimen by $St = (2000\,P)/(\pi td)$.

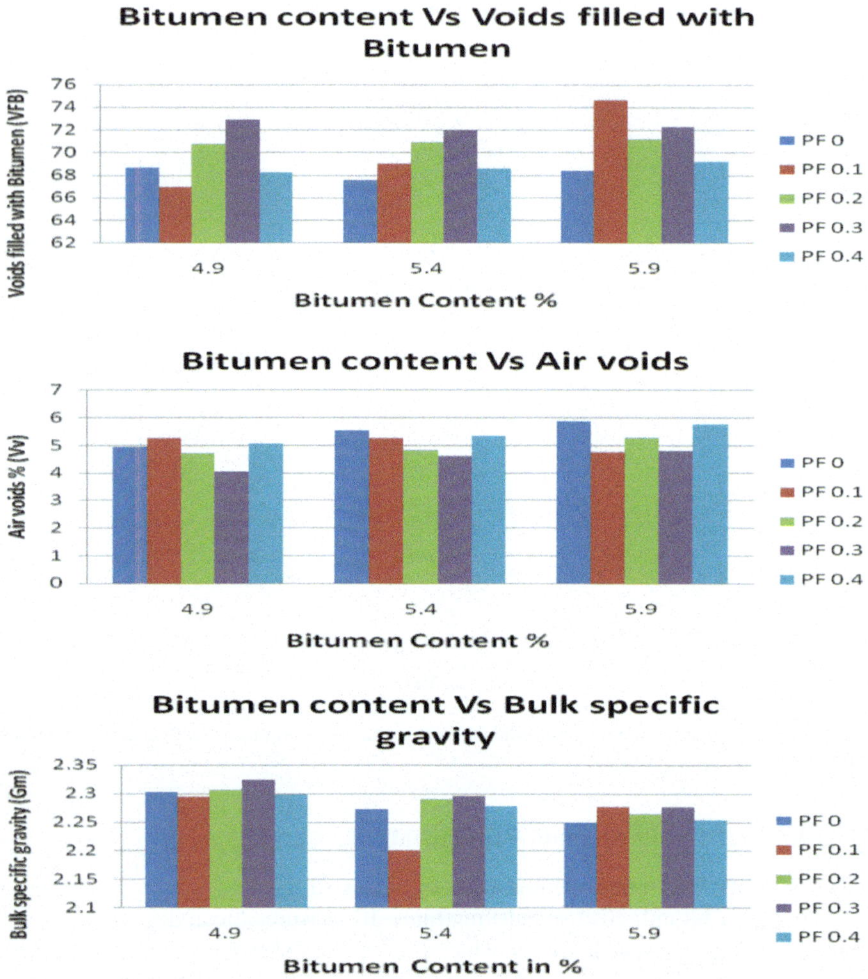

(b)

Fig. 43.4 (continued)

43.5 Conclusion

It is clearly observed from experiment that Marshall Stability has been improved by addition of 0.3% Polypropylene fibre with 4.9%binder content as compared to all other combination of polypropylene fibre and binder content. Stripping value test has been fulfilled minimum requirement by addition of 0.3% polypropylene fibre. Also, observed from experiment that indirect tensile strength has improved and fulfilled minimum requirement by adding 0.3% Polypropylene fibre with all binder content. It is concluded that polypropylene fibre and binder content are directly affecting in

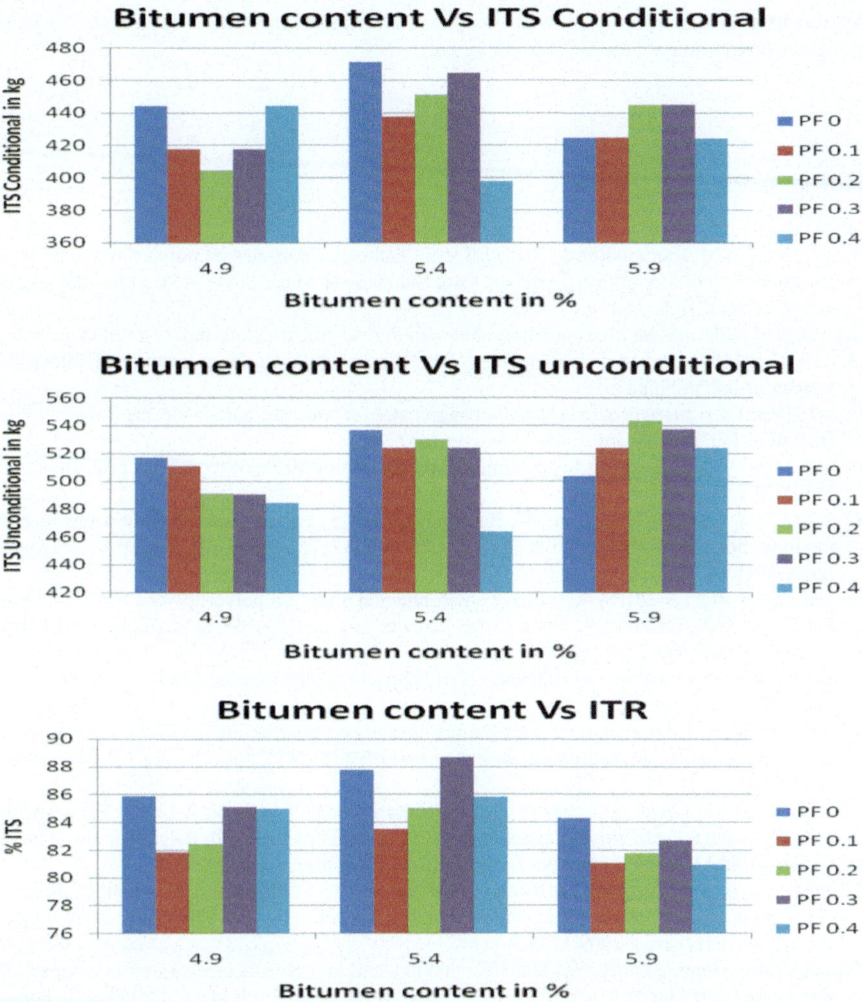

Fig. 43.5 ITS and ITR

strength of bituminous mix. Incorporating polypropylene fibers has the potential to improve the stability of the bituminous mix; this means that the mix is better able to withstand deformation and rutting caused by heavy traffic loads. Use polypropylene fiber can allow for the construction of thinner bituminous pavements while still maintaining the required structural integrity and performance. Depending on the specific project and mix design, the use of polypropylene fibers might contribute to sustainability by enabling the use of thinner pavement layers, thus reducing the overall material usage.

Acknowledgements This study was supported by Atmiya university seed money financial assistance programme (Phase-II, May 2022).

References

Bala N, Napiah M, Kamaruddin I (2018) Effect of nanosilica particles on polypropylene polymer modified asphalt mixture performance. Case Stud Constr Mater 8:447–454. https://doi.org/10.1016/j.cscm.2018.03.011

IRC (2010) Guidelines on use of modified bitumen in road construction. Indian Road Congress

IS:1201-1220 (1978) Indian standard methods for testing tar and bituminous materials, 1st Revision. Bureau of Indian Standards

IS:2386 (Part IV) (2016) Methods of test for aggregates for concrete, part 4: Mechanical properties. Bureau of Indian Standards, New Delhi, pp 1–37

IS:6241 (1971) Method of test for determination of stripping value of road aggregate. Bureau of Indian Standards

Klinsky LMG, Kaloush KE, Faria VC, Bardini VSS (2018) Performance characteristics of fiber modified hot mix asphalt. Constr Build Mater 176:747–752. https://doi.org/10.1016/j.conbuildmat.2018.04.221

Koçkal NU, Köfteci S (2016) Aggressive environmental effect on polypropylene fibre reinforced hot mix asphalt. In: Procedia engineering. Elsevier Ltd, pp 963–969. https://doi.org/10.1016/j.proeng.2016.08.834

Ministry for Road Transport and Highways (2013) Specifications for road and bridge work. Indian Roads Congress, pp 1–883

Pratul R, Satish K, Kumar SA (2021) Experimental study on fiber reinforced bituminous concrete. In: Materials today: proceedings. Elsevier Ltd, pp 11077–11083. https://doi.org/10.1016/j.matpr.2021.02.227

Slebi-Acevedo CJ, Lastra-González P, Castro-Fresno D, Vega-Zamanillo Á (2022) Experimental evaluation and recyclability potential of asphalt concrete mixtures with polyacrylonitrile fibers. Constr Build Mater 317. https://doi.org/10.1016/j.conbuildmat.2021.125829.

Tapkin S, Çevik A, Uşar Ü (2010) Prediction of marshall test results for polypropylene modified dense bituminous mixtures using neural networks. Expert Syst Appl 37(6):4660–4670. https://doi.org/10.1016/j.eswa.2009.12.042

Tapkin S (2013) Optimal polypropylene fiber amount determination by using gyratory compaction, static creep and Marshall stability and flow analyses. Constr Build Mater 44:399–410. https://doi.org/10.1016/j.conbuildmat.2013.02.060

Vishnuvardhan K, Rajkumar R, Ganesh VN, Santhosh JC (2020) An experimental performance evaluation of flex banner and polypropylene fibre reinforced bituminous mixes. In: Materials today: proceedings. Elsevier Ltd, pp 6224–6228. https://doi.org/10.1016/j.matpr.2020.10.583

Zachariah JP, Sarkar PP, Debnath B, Pal M (2018) Effect of polypropylene fibres on bituminous concrete with brick as aggregate. Constr Build Mater 168:867–876. https://doi.org/10.1016/j.conbuildmat.2018.02.016

Zachariah JP, Sarkar PP, Pal M (2021) A study on the moisture damage and rutting resistance of polypropylene modified bituminous mixes with crushed brick aggregate wastes. Constr Build Mater 269. https://doi.org/10.1016/j.conbuildmat.2020.121357

Chapter 44
Analysis of Geometric Parameters in Hill Road Sections to Improve Road Safety

Amar T. Shirke and Sujesh D. Ghodmare

Abstract Hill roads are having very complex alignments. It contains many horizontal curves and vertical curves also. Hence road user's safety is important in hilly sections. Because of its complexity in alignment accidents happening on the hill roads. To reduce rate of accident to negligible rate, the geometric design parameters of hill roads should be done carefully. Also, analyzing the road safety before designing and after construction of road is necessary. The hill roads have many design parameters. The geometric design of hill roads is varying from the plain and rolling terrain roads. While special standards for designing of hill roads are provided in special Indian Road Congress (IRC SP 48). The traffic road safety is mainly depending on major three factors i.e. vehicle, driver's behavior and the infrastructure. To improve traffic safety, it is necessary to take into consideration all this factors. But in the hill road sections infrastructure should be in proper manner to assist driver for riding their vehicles. The infrastructure includes road pavement, geometry of road, road markings and sign boards, slope protection structures etc. Geometric design of highway is designing of physical parameters of highway including elements in cross-section road, applicable sight distances, Horizontal & vertical alignments, elements of curves, superelevation, etc. India has surpassed China with 1.4286 Bn population as estimated by The United Nations Population Fund (UNFPA), and increases population progressively which causes increase of traffic volume too. So that, it is important to design and construct the geometric elements of the road in the initial stage itself by considering future traffic growth. In this research paper geometry of the road has been analyzed and modified according to an appropriate geometric standard to enhance road safety.

Keywords Hill roads · Hill road parameters · Geometric design · Road safety · Highway design

---

A. T. Shirke (✉)
M. Tech. (Transportation Engineering), G H Raisoni College of Engineering, Nagpur, India
e-mail: amarshirke23@gmail.com

S. D. Ghodmare
G H Raisoni College of Engineering, Nagpur, India

B. V. Khode et al. (eds.), *Innovations in Technologies: Pioneering Sustainable Infrastructure for a Resilient Future*, Sustainable Civil Infrastructures,
https://doi.org/10.1007/978-3-031-82133-2_44

44.1 Introduction

Hill road safety is a crucial aspect of ensuring safe and secure travel on mountainous terrains. Roads on hills are usually narrow and winding, with steep drops and dangerous curves. As such, it is important to adhere to strict safety protocols when driving on these roads, whether as a motorist or pedestrian. This involves observing traffic rules, maintaining appropriate speeds, and being aware of potential hazards such as rockfalls, landslides, and animal crossings. Hill road safety is not only essential for the protection of drivers, passengers, and pedestrians but also plays a critical role in preserving the natural environment and conserving the ecosystem. Hill roads can be a beautiful and scenic way to travel, but they can also be dangerous if proper safety precautions are not taken. Steep inclines, sharp curves, and narrow passages can make driving on hill roads challenging, especially during inclement weather. Hill road safety is essential to prevent accidents and ensure the safety of drivers and passengers. Understanding the risks and taking necessary safety measures can help prevent accidents and make the journey on hill roads more enjoyable. This topic is of utmost importance as hill roads are becoming more and more popular for adventure and tourism purposes, and it is important to ensure the safety of all travelers.

This project work will be focused on geometric design of road and underlines planning and designing of geometric features. Although there are number of factors influences on design of highway, but suitable geometric design having objective of giving optimum efficiency in traffic operation with contentment safety measures at reasonable cost. In the hill road one side hill and one side valley has very low scope to improve the geometry. Hence while constructing geometric parameters shall be taken in to consideration. And after constructing it shall be verified if those are serving the motive of design or not.

In India, out of 28 states and 8 U.T. 9 states are predominantly in hilly regions in north and North-East, whereas 9 other states have substantial hill areas (IRC Sp: 48-1998 1998). Also, India's North International border is having Himalaya mountain. Many residences are established in the valleys. J&K, Uttarakhand, Himachal Pradesh, Sikkim, Arunachal Pradesh, Nagaland, Manipur, Mizoram etc. states living its population in the valley. To connect these valleys government has engaged many agencies like NHIDCL, NHAI, BRO etc. But internal roads within the city and intercity roads and not as per standards. Hence research on safety on such hill roads should be done carefully to minimize the road accidents (Fig. 44.1).

44.1.1 Hill Road Design Parameters

While designing the hill road certain parameters should be taken into consideration. Some parameters are listed here: Survey and alignment of Hill roads, Formation

Fig. 44.1 Hill road photograph, Kullu, H.P. *Source* Project road site photograph

work, Drainage and cross drainage, Geometric design, Protective and soil stabilization works, Pavement design, Slope stability, erosion control and landslide correction, Maintenance of hill roads, Roadside Amenities, Safety on Hill roads, Traffic management, Environment and ecology.

44.1.2 Hill Road Safety

The hill road safety depends on many factors and parameters; mainly Geometric design of road, type of road, Road users, facilities on road, weather condition, etc. Also, road safety depends upon road condition. Among which geometric design play crucial role in hill road safety. On hill road the users may face sudden land sliding because of hilly terrain. Hence proper geometric design with all respects is very important in the hill road. The location of protective works like parapet walls, crash barriers, etc. should be assigned so that the hill road users get advantage of it at critical locations. Also, proper signages on hill road shall guide users for smooth riding on road.

Many accidents cause in hill roads due to following reasons; (1) Complex geometry of hill roads like sharp curves with inadequate sight distance, (2) Due many curves in Hill road the speed will be varying at entry and exit from every curve, (3) So many narrow bridges to cross streams, also alignment passes through major valley to one side valley, (4) The hill roads are at high altitude and vertical alignment is also having steep gradients and continuously changing gradients up and down.

44.1.3 Engineering Solutions

Geometric: Proper roadway width, curve radii, gradients and sight distance. **Engineering**: Design of drainage, protection works, retaining walls, breast walls, snow fences, snow shelters, netting for rolling bolder stone, etc. **Traffic management**: Proper sign boards, signals, road markings, warning systems, delineators, etc. **Maintenance**: Maintenance of road and its integrated parts **Monitoring**: Monitoring includes noting of the road conditions and helping road owner for upgradation of roads.

44.2 Literature Review

IRC SP:48-1998 The Hill road Manual gives the guidelines for design, construction and maintenance of the hill roads. IRC SP 48 is a comprehensive manual for the design, construction, and maintenance of hill roads in India. The manual was published by the Indian Roads Congress (IRC), a professional body that formulates and disseminates guidelines and standards for road construction in India. The IRC SP 48 manual is based on extensive research and practical experience in the design and construction of hill roads. It incorporates the latest developments in the field of hill road engineering and provides detailed guidance on the various aspects of hill road construction. The manual covers a wide range of topics related to hill road engineering, including road alignment, geometric design, pavement design, drainage, slope stabilization, and erosion control measures (IRC Sp:48-1998 1998). **IRC SP:44-1996 Highway Safety code** specifies the rules for all road users, discusses about safety of pedestrians, animal drawn vehicles, motorized vehicles, bicycles. It also gives additional requirements of truckers and bus drivers and discusses about emergencies for four wheeled vehicles (IRC Sp:44-1996 1996). **IRC SP:88–2019 Manual for road safety Audit** gives the standard procedure for conducting the road safety audit. Key steps, checklist for audit, management during the road safety, and case studies of road safety audits are discussed in this IRC code (IRC Sp:88-2019 2019). **Dinesh et al.** says that the root cause of an accidents intends to perception, intellection emotion and violation. His approach towards research is to get minimal setback/casualties of the road. In order to gain the best possible course of action, the stretch of 8 KM of National highway (NH-66) situated in a plain terrain in the district of Alappuzha, Kerala India. They collected, accident data from NHAI office and Police station with proper analysis by Accident Severity Index (ASI) method has been carried out. They worked out 10 black spot location along with the estimate after analyzing short term measures (Yadav et al. 2021). **Rishabh et al.** finds that the number of accidents is rising up every year due to increasing vehicles population. They analyzed accident data using accident frequency and severity index method. The safety deficiencies were detected to minimize accidents and save the road users and suggested further improvement for deficiencies along with the measures (Gupta and Kaur 2021). **Anoop et al**. finds that a large daily increase in the population in the last 10 years, then the most important traffic safety parameter for the audit process is calculated: sufficient drainage, street width, legal resource opening, traffic density in emergency. They audited the hilly roads and found that calculated point speed measurements, accident data acquisition and provide access to service lanes that are also required for emergencies (Kumar and Mishra 2021). **Payal et al**. prepared the highway geometry with the help of civil 3D and finds that a typical design of roadway with the support of AutoCAD Civil 3D. to achieve exact highway design standards AutoCAD Civil 3D design software helps in a comfortable and relaxed way. Civil 3D design is simple and easy to grasp to build alignment, AUTOCAD civil 3D uses the topography and survey data collected from LIDAR or similar technology. Once the alignment prepared, the geometry can be tested and evaluated using the IRC standard and requirements using the civil 3D built-up software. By completing design of horizontal alignment, the vertical alignments can be designed as per standards. For avoiding tired calculations Civil 3D uses integrated in built checks for transition length and sight distance to analyze geometry (Gaikwad and Ghodmare 2020). **Rishabh Chourasia et al**. was working on effect of the consistency of geometric design to reduce chances of accidents. His findings also focus on the road safety basis on the Potential Safety Improvement Index (PFI). His study also uses many road safety audit (RSA) principles in assessing the road design. This research would help researchers in finding accidental prone areas or the areas where accidents are likely to happen due to poor geometric design. mainly the study analyzes and give scope of the reduction of accidents and providing safe and forgiving roads by improving the inconsistency in the geometrics of the road (Chaurasiya et al. 2020). **Manasa et al**. discusses in their research paper about accidental analysis and road safety auditing with the help of case study of NH-75. Detailed analysis of national highway NH-75 carried out from the point of view of road safety and geometric design aspects. The research aims to identify deficiencies, improving design aspects, enhancing credibility of the roads (Mansa et al. 2020).

44.3 Methodology

44.3.1 Site Selection

To research on hilly roads and its safety analysis we choose the hilly area in Kullu valley of Himachal Pradesh state. The site comprises the road staring from Kullu and ends at Ramshila near Gammon bridge. The road having one side hill and valley at other side. The starting point and end point coordinates of the road are Lat.-31.938131, Long.-77.110867 and Lat.-31.976934, Long.-77.125025 respectively. Below figure shows the location of site.

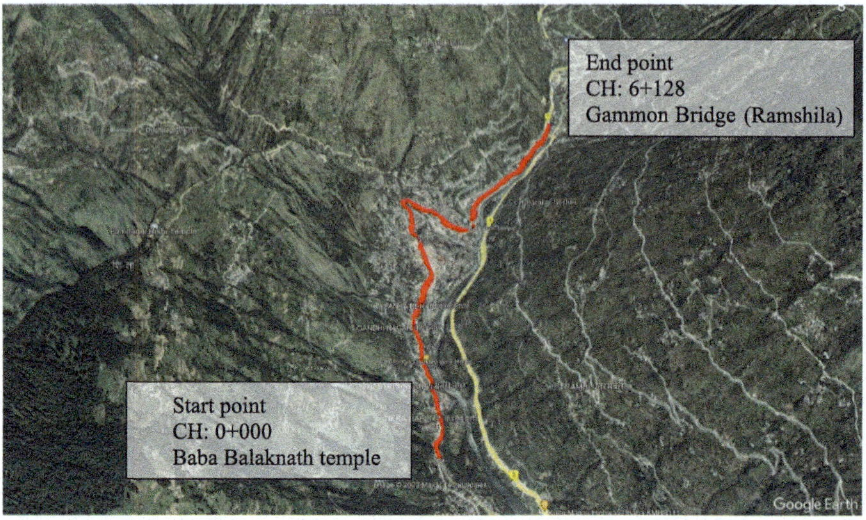

Fig. 44.2 Location map

44.3.2 Data Collection

To know about the road and its previous improvements it is required to collect old data from its owner department. The HPPWD is the owner department of this road. We collected map of the existing road and existing structures on the road and information about its past improvements. Also, it is was necessary to collect information about accident happened in last few years to analysis the road safety on the road hence we collected the accident data from the local police station. Also collected the traffic data and data about the Dashera festival since we heard about it from local people that around 4–5 lakhs people visit Kullu to during the festival.

44.3.3 Survey

Various surveys and investigations are required to analyze the geometry of road in hilly area. The survey includes traffic survey, Topographic survey, soil investigation survey, road condition survey, road inventory survey, structure survey, etc (Fig. 44.2).

44.3.4 Topographic Survey

Basic requirement of geometric design of road is topographic survey. It gives levels at every point on road with certain intervals. Topographic survey plays very important

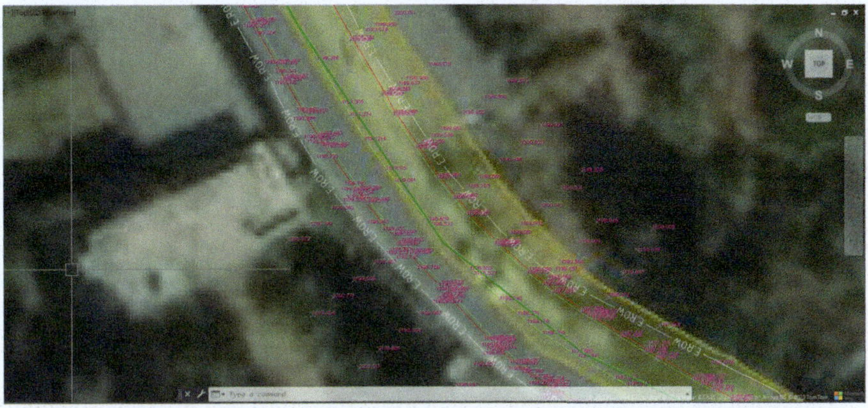

Fig. 44.3 Topographic survey for Kullu-Ramshila road google earth view

role in analysis of geometrid parameters of roads. The topographic survey was carried out on the road with the help of Total station. DGPS points were set and levels shifted along the roads and noted levels of center line of main carriageway, road edge, side shoulders, embankments, Junctions, drainage lines along the road and across the road, water pipe lines, cross drainage, minor bridge, etc. After preparation of survey drawing the drawing was checked thoroughly for mean sea levels and also checked for geolocation with help of AutoCAD civil 3D software. Screenshots of topographic survey is shown in below figure (Fig. 44.3).

44.3.5 Data Analysis

After collection of all data we started analyzing the data we analyzed it with respect to road safety. The road starts from the Baba Balaknath temple say chainage 0 + 000 and ends at Ramshila near Gammon bridge at chainage 6 + 128. This road falls within hilly terrain and having length of 6.128 km out of which 4.15 km length runs through built-up section while 1.978 km goes through open country. The road components are as follows; 1. Total road length- 6.128 km, 2. Existing width of the road-7.0 m, 3. Available ROW-12–15 m, 4. Side shoulders-1.0–2.0 m, 5. Major Junction-3 numbers (Dhalpur, Bhutti Chowk and Ramshila), 6. Drain-Open drain and Concrete drain (0.6 m wide), 7. Footpath-0.5–1.5m wide, 8. Retaining walls-200–300 m along the road, 9. Breast walls-100–150 m, 10. Structures-26 numbers, 1-minor bridge, 1-box culvert, 11-pipe culverts and 13-Slab culverts, 11. Bus shelters-7 numbers, 12. Toilet blocks-3 numbers, 12. City entry gates-1 number.

Identification of accident-prone area along the road:

The reviewed accident data marked on the Kullu ramshila road (SH-24); namely these locations are; Tikri Bawri, Sarvari Bridge at Shitla Mata temple, Gammon bridge at manali end, Gammon bridge at Ramshila road, Landslide area at meat market, these locations are shown in figure below marked on the google earth;

44.3.6 Analysis of Road Safety

Accident data shows the percentage of accidents happened in the past 5–6 years. The below table shows that accidents are likely increased. The increase of accidents due to many reasons but geometry of road is one the reason for it. At Tikri bawari and Sarvari Bridge there are sharp curves. Hence these sharp curves and other sharp curves also need to be improved (Fig. 44.4 and Table 44.1).

The geometric improvements suggested with some protection works like retaining walls, breast walls and parapet walls at hill and valley sections. Sight distances at horizontal and vertical curves shall be appropriate to reduce accidents on road. The sight distance has been improved.

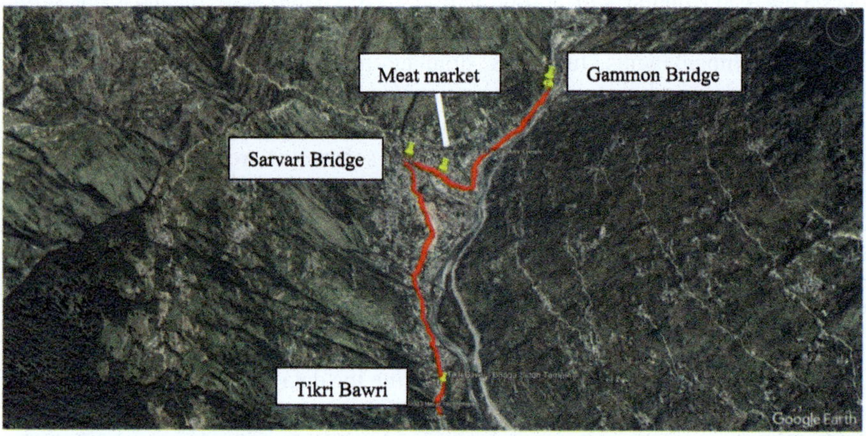

Fig. 44.4 Accident-prone area marked on map. *Source* Google earth

Table 44.1 Percentage of accidents in a year

Year	Accidents	Percentage % (%)
2016	21	24
2017	15	17
2018	11	13
2019	20	24
2020	19	22
Total	86	100

Source Accident data

44.3.7 Improvements

For the spotted accidents prone areas and other minor locations where, existing geometry seems inappropriate we designed it as per IRC codes. Total existing horizontal curves present in the road are 102 within just 6 km (approx. 17 curves per km). Also, there are many broken back curves present in the road which increases chances of accidents by misguiding drivers. The list of some existing curves is as following table (Table 44.2).

The total vertical curves are 108 within 6 km (18 curves per Km). We also listed some existing vertical curves, there are as following table (Table 44.3).

These locations where sharp curves were present in the road are designed so that the road user will get comfort while driving. The horizontal curves reduced from 102 to 37 and vertical curves from 108 to 48. These geometric parameters were designed

Table 44.2 List existing horizontal curves

No	Length	Radius	Start station	End station	Delta angle	Chord length	PI station	Degree of curvature by arc
1	31.624	65	195	227	27.88	31.313	211	26.87
2	10.231	65	298	308	9.02	10.221	303	26.87
3	29.317	55	592	621	30.54	28.971	607	31.75
4	15.611	30	628	644	29.81	15.435	636	58.21
5	16.336	35	651	667	26.74	16.188	659	49.90
6	34.689	55	689	724	36.14	34.117	707	31.75
7	24.475	50	1263	1287	28.05	24.231	1275	34.93
8	7.682	20	1468	1475	22.01	7.635	1472	87.32

Source AutoCAD Civil 3D Software

Table 44.3 List of existing vertical curves

No	PVI station	Grade in (%)	Grade out (%)	A (grade change) (%)	Profile curve type	Profile curve length	K value
1	38.7	3.82	6.66	2.83	Sag	30	10.5854
2	65.2	6.66	5.96	0.70	Crest	10	14.3842
3	84.9	5.96	7.74	1.77	Sag	15	8.4578
4	95.0	7.74	6.42	1.32	Crest	5	3.8008
5	113.7	6.42	0.54	5.88	Crest	30	5.1051
6	144.2	0.54	2.20	1.66	Sag	10	6.0428
7	170.8	2.20	7.53	5.33	Sag	30	5.6306
8	258.8	7.53	−3.21	10.74	Crest	80	7.4509

Source AutoCAD Civil 3D Software

and provided for the road. The plan and profile of the road can be design with the help of AutoCAD civil 3D.

44.4 Conclusion

This Kullu-Ramshila road is very busiest road in the Kullu city, and faces the traffic jam too. The geometry of the road has been modified which is leading to do some more works like protection works (Retaining walls, toe walls, crash barriers, etc.). As some locations small curves have been removed to ease the driveway for drivers. Some curved sections have been modified and tried to make its radii bigger. The modifications and improvements in the geometry of this road is very beneficial to the users. The findings of this research paper will definitely help the authorities of road to manage traffic on the road and maintenance of the road in future. Whereas more research is needed on the geometry of this road for better speed achievement and alternate alignments to ease the traffic movement.

References

Chaurasiya R, Agrawal HK, Nagar B (2020) Case study on the effect of geometric design consistency on level of safety for rural roads under Indian conditions. Int Res J Eng Technol (IRJET) 07(06). ISSN: 2395-0072

Gaikwad P, Ghodmare SD (2020) A review-geometric design of highway with the help of autocad civil 3D. Int J Res Appl Sci Eng Technol (IJRASET) 8(V). ISSN: 2321-9653

Gupta R, Kaur P (2021) Impacts of road condition, traffic and manmade features on road safety. Int J Eng Res Technol (IJERT) 10(07). ISSN: 2278-0181

IRC SP:48-1998 (1998) The Hill road manual. Indian Roads Congress, New Delhi

IRC SP:44-1996 (1996) Highway safety code. Indian Roads Congress, New Delhi

IRC SP:88-2019 (2019) Manual on road safety audit. Indian Roads Congress, New Delhi

Kumar A, Mishra AK (2021) Hill area road safety audit: case study on NH-34 Uttarakhand (INDIA). In: 1st international conference on emerging scientific applications in the field of engineering and technology, Vishwanathrao Deshpande Institute of Technology, Haliyal. ISBN: 978-93-91535-10-0

Mansa SR, Chandrashekhar HS, Vidyashree, Rachana, Shivakumar (2020) Accidental analysis and road safety auditing for NH-75. Int J Res Publ Rev 1(5):125–132

Yadav DK, Ghodmare S, Naveen Kumar N (2021) Mitigation of black-spot's on highways by the application of safe system approach. Int J Sci Res Sci Technol. Print ISSN: 2395-6011, Online ISSN: 2395-602X (www.ijsrst.com)

Chapter 45
A Critical Exploration of Machine Learning in Source Identification, Quantification Analysis, and Prediction Models for Residential Building Construction Waste

Akshay Gulghane, Nikhil Pitale, and R. L. Sharma

Abstract The construction industry is a significant contributor to global waste generation, with residential building construction being a prominent source. In an effort to address the environmental impact of construction waste, this study critically explores the application of machine learning (ML) techniques in source identification, quantification analysis, and prediction models specific to residential building construction waste. The research investigates existing methodologies and identifies gaps in traditional approaches, motivating the integration of ML for enhanced accuracy and efficiency. The study begins by providing a comprehensive overview of current waste management practices in the residential construction sector. Leveraging ML algorithms for source identification, the research evaluates the potential of supervised learning and unsupervised learning models in categorizing and attributing waste to specific sources within the construction process. Moreover, the research delves into the quantification analysis of construction waste, examining the efficacy of ML algorithms in estimating waste volumes based on diverse parameters such as project scale, construction materials, and building design. By examining historical data and incorporating real-time variables, the study aims to develop predictive models that can adapt to changing construction practices and environmental regulations. This aspect is crucial for sustainable waste management and proactive decision-making. Furthermore, the study assesses the challenges and ethical considerations associated with implementing ML in construction waste management, emphasizing the importance of transparency, accountability, and inclusivity in model development and deployment. In conclusion, this research provides a thorough examination of

A. Gulghane (✉) · N. Pitale
G H Raisoni College of Engineering, Nagpur, Maharashtra, India
e-mail: akshaygulghane@gmail.com

R. L. Sharma
Lovely Professional University, Phagwara, Punjab, India

B. V. Khode et al. (eds.), *Innovations in Technologies: Pioneering Sustainable Infrastructure for a Resilient Future*, Sustainable Civil Infrastructures,
https://doi.org/10.1007/978-3-031-82133-2_45

the role of machine learning in transforming the identification, quantification, and prediction of residential building construction waste.

Keywords Construction waste management · Quantification of waste and waste minimization · Machine learning · Prediction model · Source identification

45.1 Introduction

In contemporary urban landscapes, the burgeoning challenges associated with rapid population growth, urbanization, and sustainable resource management necessitate a paradigm shift in the construction industry. Residential building construction, a pivotal contributor to urban development, is concurrently a major source of waste generation. The efficient management of construction waste is imperative for sustainable urban development, resource conservation, and environmental stewardship. This study embarks on a critical exploration of machine learning applications in the realm of source identification, quantification analysis, and prediction models pertaining to residential building construction waste. As traditional waste management approaches prove inadequate to cope with the escalating magnitude of construction waste, integrating advanced technologies becomes paramount. Machine learning, a subset of artificial intelligence, offers a promising avenue for revolutionizing the construction waste management paradigm by providing data-driven insights, automation, and predictive capabilities. The primary focus of this exploration is threefold: firstly, to elucidate the sources of residential construction waste with a granular precision enabled by machine learning algorithms; secondly, to develop robust models for accurately quantifying the volume and composition of construction waste, thereby facilitating efficient resource allocation for waste management strategies; and finally, to delve into the potential of machine learning in predicting future waste generation patterns, aiding policymakers, construction industry stakeholders, and urban planners in devising proactive and sustainable waste management strategies (Coskuner et al. 2020; Markandeya Raju and Kameswari 2015).

The significance of this study lies in its potential to bridge the existing gaps in knowledge and practice within the construction industry, as well as its contribution to the broader discourse on sustainable urban development. By harnessing the power of machine learning, this research endeavours to provide actionable insights that empower stakeholders to optimize waste management processes, minimize environmental impact, and foster a more resilient and sustainable future for residential construction. As we navigate the intricate intersection of technology and sustainability, this critical exploration serves as a beacon guiding the construction industry toward a more informed, adaptive, and ecologically conscious approach to managing residential building construction waste. Machine Learning (ML) has emerged as a transformative force across various domains, offering unprecedented capabilities in data analysis, pattern recognition, and predictive modelling (Hu et al. 2021). The integration of ML into the construction waste management paradigm holds immense potential for enhancing source identification, quantification analysis, and prediction models. This research embarks on a critical exploration of the applications of ML in addressing the complexities of residential building construction waste. Source identification, the initial phase in waste management, poses a significant challenge due to the diverse origins and types of construction waste. Traditional methods rely on manual sorting and visual inspection, often proving time-consuming and prone to errors. ML algorithms, equipped with advanced image recognition and data analysis capabilities, present an opportunity to streamline the source identification process (Elshaboury 2022). By analysing historical data and learning from patterns, ML models can accurately categorize and identify the sources of construction waste, providing a foundation for targeted waste reduction strategies. Quantification analysis is crucial for assessing the volume and composition of construction waste, enabling informed decision-making in waste management practices. ML algorithms excel in handling large datasets and can process information from various sources, such as project plans, material specifications, and on-site observations. By leveraging ML, researchers and practitioners can develop robust quantification models capable of estimating the amount and types of waste generated during different construction phases, leading to optimized waste diversion strategies and resource allocation. Furthermore, prediction models powered by ML algorithms can play a pivotal role in anticipating future construction waste scenarios (Kolaventi et al. 2019). These models can take into account variables such as project scale, design complexity, and regional factors to forecast waste generation trends. Such predictive capabilities empower stakeholders to proactively implement waste reduction measures, promote circular economy principles, and contribute to the overall sustainability of the construction industry. This critical exploration aims to analyse the current state of ML applications in residential building construction waste management, identifying gaps, challenges, and opportunities for future research. By evaluating the effectiveness and limitations of existing ML models, this study seeks to contribute to the refinement and advancement of machine learning methodologies tailored to the unique intricacies of residential construction waste management. Ultimately, the integration of ML in this context has the potential to revolutionize traditional waste management practices, fostering a more sustainable and environmentally conscious approach to construction activities (Hassan et al. 2020; Quiñones and Llatas 2022).

45.2 Management of Construction Waste

Construction waste refers to all undesirable materials generated from construction or industrial operations, whether produced directly or indirectly. This category encompasses various types of waste, including concrete, steel, wood, and mixed site clearance materials. They are produced by projects like land excavation, building construction, demolition, roadwork, and building restoration (Coskuner et al. 2020; Markandeya Raju and Kameswari 2015). The issue of building waste management is a worldwide environmental hazard that impacts every country (Hu et al. 2021;

Elshaboury 2022). Following a review of the literature, 81 variables that affect building waste were identified. Seven categories were created to classify these factors, which encompassed design, workforce, administration, site circumstances, acquisition, and outside influences (Kolaventi et al. 2019). The building industry recognizes that material waste is a significant problem, this has a significant impact on the sector's efficiency as well as the environmental effects of building ventures. A waste management system that effectively controls and manages waste at the source throughout the construction project is required to solve this issue. The assessment of waste is crucial for evaluating the effectiveness of production processes. Within the realm of building project management, waste management holds significant importance due to the adverse impact the industry has on the environment, including resource depletion, waste generation, and pollution (Hassan et al. 2020; Quiñones and Llatas 2022). In order to minimize the impact caused by construction activities, construction professionals should initially identify the primary factors contributing to waste generation. Prior to commencing construction, they can employ various methods to manage building waste effectively, which fall into categories like waste classification, waste management strategies, and waste disposal technologies. By implementing these methods, the detrimental environmental consequences of construction can be curbed, and the overall efficiency of construction projects can be enhanced. Notably, the building sector in India generates substantial waste throughout all stages of development, encompassing bricks, demolished building materials, and mild steel. Prioritizing waste reduction and recycling will lower expenses and increase economic and environmental benefits (Wu et al. 2014). To avoid obstructing traffic and municipal governments, waste should be stored at the point of generation. To accomplish this objective, a hierarchical structure based on priority levels must be created. To manage construction waste effectively, construction companies should implement a waste management plan that includes waste reduction, reuse, and recycling strategies. The plan should begin during the design stage, where architects and engineers can specify materials that are recyclable, durable, and have minimal environmental impact (Foo et al. 2013; Li 2016). During the construction stage, workers should sort and segregate waste materials and place them in designated collection areas. Construction companies can also reduce waste by using prefabricated materials and modular construction techniques that produce less waste on-site. Unwanted materials should be generously donated to charitable organizations or repurposed for other construction endeavours. To ensure efficient handling of construction waste, a well-rounded waste management plan is essential, integrating strategies for waste reduction, reuse, and recycling. Construction firms can diminish the ecological consequences of their activities, conserve natural resources, and decrease expenses linked to waste management by implementing sustainable approaches. Figure 45.1 represents the life cycles process of building.

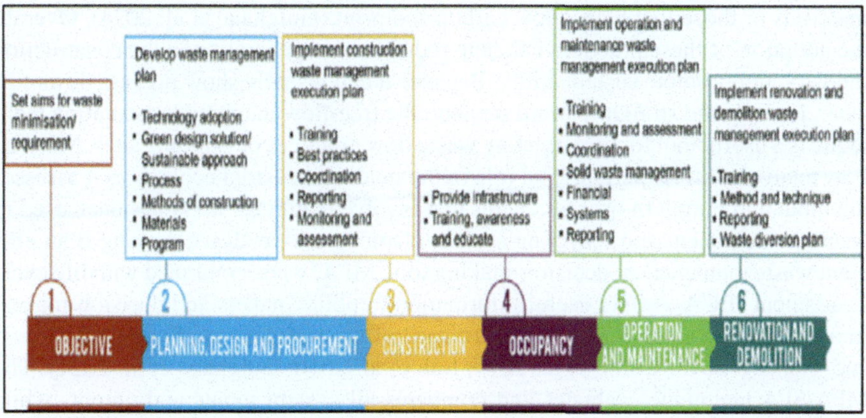

Fig. 45.1 Systematic process for construction waste management through life cycle of a building.

45.3 Practices in Construction Industry

There are several ways to reduce waste in building projects, including adopting a zero-waste mindset, making wise choices during the planning stage, effectively managing the site, standardizing materials, and implementing waste management systems (Ram and Kalidindi 2017). To reduce waste generation, all project employees should work together. Implementing the waste rate estimation technique can result in improved material handling, reduced waste rates, and higher productivity. The 3R and 4R ideas (reduce, reuse, recycle, and recover) can also be useful in reducing building material waste throughout the product and service lifecycle (Cha et al. 2021). Preventing and reducing waste, reusing materials, recycling, recovering energy, treating waste, and disposing of garbage are all part of the waste hierarchy, which emphasizes resource efficiency. However, there are several practices that can be implemented to manage the generation of construction waste (Gulghane et al. 2024, 2023b). Firstly, the use of Building Information Modelling (BIM) technology can help optimize the design process to reduce waste generation. This technology can simulate the construction process and identify areas where materials can be saved or reused (Ujong et al. 2022; Gulghane et al. 2023b). Furthermore, the incorporation of prefabrication and modular construction techniques presents a valuable opportunity to minimize on-site waste generation. By fabricating building components off-site and subsequently transporting them for assembly, a considerable reduction in waste can be achieved. Additionally, by establishing a well-designed waste management plan, we can further mitigate waste generation and ensure its appropriate disposal. This plan should include procedures for sorting, recycling, and disposing of waste (Ali et al. 2012). Additionally, training and educating construction workers on waste management practices can help improve their understanding of waste reduction and disposal. Lastly, using recyclable resources and behaviours can greatly reduce debris generation. This involves the use of materials that are eco-friendly, such as recycled materials or those that have a low carbon footprint (Gulghane et al. 2024). Overall, the adoption of these practices can help reduce waste generation in the construction industry and promote sustainability. Because it can visually show process flows in a clear, logical, and straightforward manner, the free-flow mapping presentation technique is a useful tool for investigating waste flow practices on building sites. Material flow analysis (MFA) plays a vital role in the field of industrial ecology as it assesses the input and output of process chains that involve materials. MFA has been used in waste management and recycling system planning and analysis, making it an efficient waste management decision-making tool. MFA, when combined with life cycle assessment (LCA), offers useful information to policymakers and decision-makers in assessing the environmental performance of waste management systems (Pawar and Thakare 2013; Coffie et al. 2019; Suk et al. 2016; Gulghane et al. 2024). The MFA/SFA technique evaluates and compares all system inputs and outputs while monitoring waste flows and substances within the waste management model. MFA is also useful for mapping total waste movements and management procedures, as it provides transfer factors for all residue processes across multiple organizations, which can be used in LCA calculations. Using MFA as a foundation, it is feasible to build an input-dependent waste management model using LCAs. Overall, these techniques provide a thorough knowledge of waste management processes and aid in the identification of potential system improvements (Coffie et al. 2019).

Waste flow prediction in construction can be modelled using building components during the construction phases. Reducing garbage generation, maximizing reuse and recycling, and minimizing mixed waste intake at landfills are all effective waste management methods. Additionally, it is possible to encourage the use of green building methods like large panel systems, prefabricated parts, and decreased wet trade applications (Foo et al. 2013). In addition to sorting mixed construction debris and encouraging reuse and recycling whenever practical, a well-managed public filling program with adequate facilities and access can be beneficial (Kaveh and Rahimi Bondarabady 2004). Figure 45.2 shows the Material Flow on the construction site for the projects. Effective tactics include training construction workers, improving supervision, coordinating with stores to avoid over-ordering, and correctly storing and handling materials on-site (Cha et al. 2021). Furthermore, strategies for reducing building and demolition waste can include standardizing designs, controlling stock to avoid over-ordering, and providing environmental education to employees. The government's introduction of the Construction Waste Disposal Charging Scheme (CWDCS) aims to incentivize construction and demolition companies to minimize their waste production and encourage the widespread adoption of recycling and reusing methods. This initiative aims to foster sustainable waste management approaches. To achieve a reduction in waste at building sites, the government can implement various policies, including a landfill tax, increased taxes on the use of new materials, and tax incentives for recycling efforts (Suk et al. 2016). Lean and Six Sigma methodologies can be used to identify and get rid of waste in the building industry. Construction waste illegally dumped poses a significant problem for the industry and has detrimental effects on the economy, ecology, and public health. Six Sigma aims to boost client satisfaction and profitability by lowering defects (Ali et al. 2012; Golbaz et al. 2019; Kaveh and Khalegi 2019). Lean thinking and Six Sigma tools and techniques can also be useful to public and semi-public agencies. Research data can be analyzed through various techniques, including the Mean Score method, Ranking method, and Mann–Whitney U Test (Gulghane et al. 2024, 2023a, 2023b). A cost–benefit assessment is used to evaluate the economic viability of waste management on building sites, with a focus on economics. Benefit–cost analysis (BCA), a useful instrument for determining the economic viability of construction waste management, evaluates potential cost savings. Construction waste can be quantified using either the global index method or the component index approach, or by using a common construction model (Bekr 2014). Often, the most efficient environmental solution is to reduce waste generation. If waste reduction is not an option, items and resources can be reused for the identical or another purpose. Waste can be recycled, composted, or used for energy recovery if reuse is not feasible. Disposal should only be used as a last measure, with the best available environmental option being used instead. According to research, increasing worker knowledge and education on construction waste management is critical for on-site waste minimization. Additional efficient strategies comprise the implementation of low-waste building technology, segregation of waste through designated waste skips based on material types, standardization of design and materials, and proper handling of construction materials (Azadi and Karimi-Jashni 2015; AbouRizk 2010). The Circular Economy (CE) offers a promising solution to mitigate the adverse consequences of construction and demolition activities. This innovative approach was devised as a response to the environmental repercussions caused by the waste generated from Construction and Demolition (C&D) processes. CE maximizes the use of raw materials, guarantees their value throughout their lifecycle, and reduces the production of excess waste. But there are many challenges involved in making the switch to CE in the construction and dismantling industries (Kaveh et al. 2008). To ensure successful execution of CE, it is crucial to have a thorough understanding of multiple aspects such as societal, governmental, economic, behavioural, technological, and environmental dimensions. The CE principles call for reforming design standards, cutting back on the purchase of excess raw materials, and cutting, reusing, and recycling trash. By promoting recycling and reducing waste, these practices effectively stop environmental deterioration (Gavali and Halder 2020; Venkatesh and Aarthy 2012; Pawar and Thakare 2013). The difference between two economies i.e., linear and a circular economy is shown in Fig. 45.3 A novel strategy for reducing harmful environmental effects and fostering economic growth for long-term development is the use of CE in the building sector.

Implementing sustainable waste management practices in the construction industry encounters various potential limitations and barriers, spanning economic, regulatory, technological, and cultural dimensions. A primary challenge is the perceived or actual increase in costs associated with sustainable practices. The need for investments in new technologies, training programs, and waste sorting facilities can deter stakeholders in a cost-sensitive industry. Additionally, a lack of awareness and education regarding the long-term benefits of sustainable waste management poses a hurdle, with stakeholders potentially resisting or remaining indifferent due

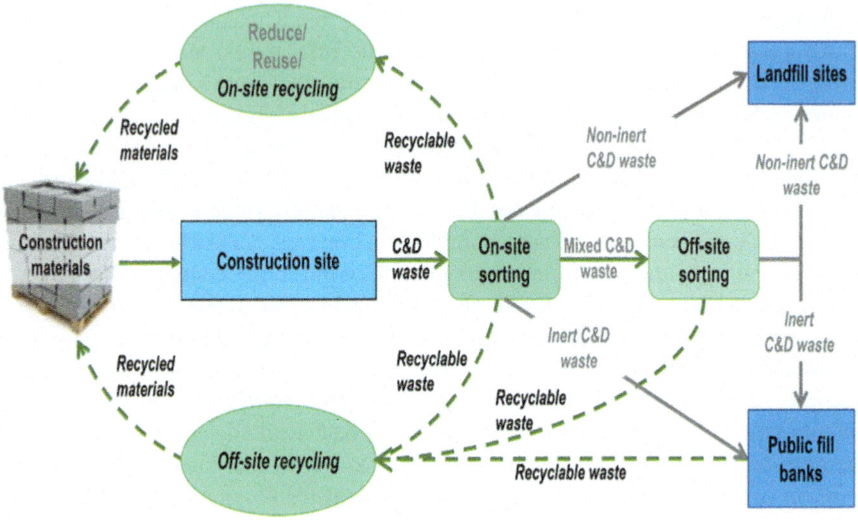

Fig. 45.2 Material Flow on the construction site

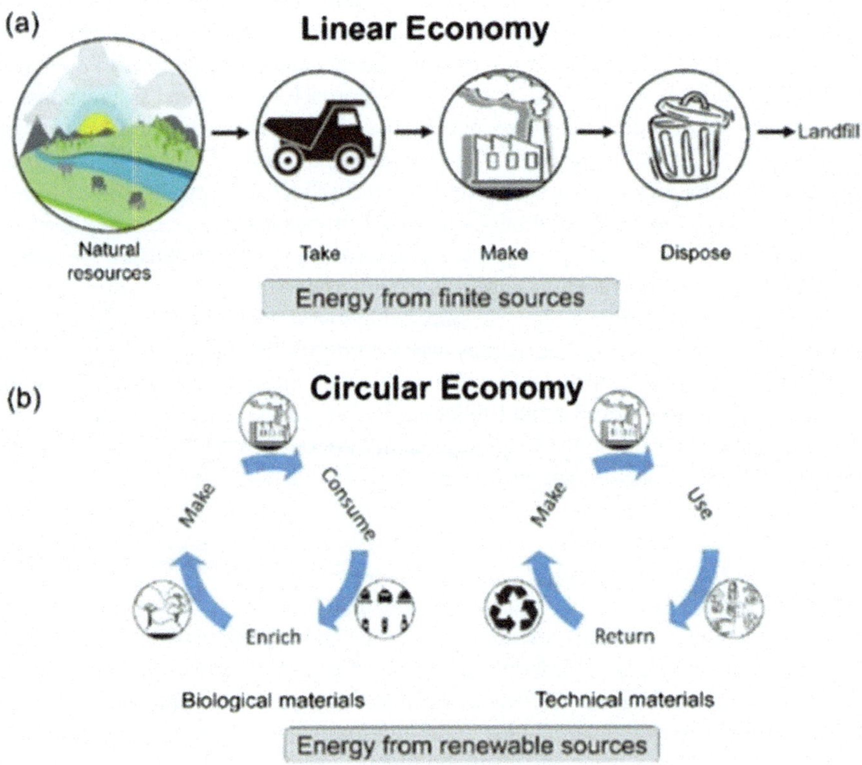

Fig. 45.3 Recycling economy (*Source*) Ram VG et al. 2017

to inadequate information. Regulatory compliance, though advancing, faces issues of inconsistent frameworks and insufficient monitoring, impacting the widespread adoption of sustainable practices. Limited infrastructure for sorting, recycling, and disposing of construction waste is a substantial barrier, hindering effective diversion from landfills. Technological limitations, especially in regions lacking access to advanced recycling technologies, and the industry's resistance to adopting new technologies contribute to the challenges. The fragmented supply chain in the construction industry, with diverse stakeholder priorities, impedes coordinated waste management efforts. Overcoming inertia and fostering a culture of sustainability within a traditionally resistant industry require dedicated time, effort, and effective communication. The urgency of project timelines often takes precedence over considerations for sustainable waste management, leading to neglect of environmentally friendly practices. Moreover, the perceived inconvenience of sorting waste and adhering to new procedures can discourage the adoption of sustainable practices in construction projects. Addressing these multifaceted challenges is crucial for achieving meaningful progress in sustainable waste management within the construction sector (Golbaz et al. 2019; Abu Hammad et al. 2008; Bekr 2014).

45.4 Sources of Construction Waste

The initial and crucial phase in waste management is recognizing the primary origins of waste generation. There are thirty-four potential factors that contribute to waste production, which can be classified into seven main groups. These categories encompass procurement, handling, storage, workers, site management and supervision, design and contract documentation aspects, as well as external factors. Furthermore, researchers found that eight waste sources are significant. The process of converting materials contributes significantly to production uncertainty, resulting in to an increase in non-value-added activities. Decision-making can be improved and appropriate strategies for mitigating construction waste created by determining the contribution rates of various waste sources (Coffie et al. 2019; Suk et al. 2016; Gulghane et al. 2024, 2023a). During the design, logistics, and physical building procedures, construction material waste can occur. In this research, construction waste refers to damaged materials that are destined for disposal, reuse, or recycling. These wastes are brought to the building site from a variety of sources, including design, procurement, material handling, operations, residuals, and others. Lack of current knowledge about on-site inventories is a common issue on big construction projects, which can result in the repeated purchase of the same materials, creating unnecessary waste. Waste can happen at any level of the structure and can come from both internal construction activities and external factors like theft and vandalism (Aravindh et al. 2022; Bekr 2014; Ali and Rahmat 2010; Gavali and Halder 2020; Venkatesh and Aarthy 2012; Pawar and Thakare 2013). Classified garbage sources into four categories, including operational waste. (mistakes by laborers or skilled workers, accidents caused by carelessness, damage to completed work due to subsequent trades, use of wrong materials, unclear planning of required quantities, malfunctioning equipment, bad weather), Waste in design (inadequate attention given to dimensional coordination, design changes made during construction, lack of experience in construction method and sequence, lack of attention to market-available standard sizes, lack of knowledge about substitute goods, intricacy, mistakes, and inadequate contractual paperwork, as well as opting for inferior quality items., errors in construction method and sequence, complexity, errors, and incomplete contract documents. Material handling issues (damage during transportation, improper storage, supply of loose materials, use of whatever materials are available nearby, unfriendly behaviour on the part of the project team and workers, theft), and procurement loss (ordering mistakes, the failure to request or purchase limited amounts and obtaining goods that do not meet the required standards.) (Ali and Rahmat 2010; Gavali and Halder 2020; Venkatesh and Aarthy 2012; Pawar and Thakare 2013).

Table 45.1 depicts a comprehensive mapping of the causative variables contributing to construction waste, as well as the frequency with which they appear, as cited in various study articles. These variables can be used as input parameters when creating pertinent questionnaires and models. Waste is produced throughout the entire construction process, starting from the initial planning phase and extending to the final stages. Understanding the various factors that contribute to this waste generation is essential for effectively managing and minimizing it. By addressing these fundamental causes, we can significantly reduce the overall amount of construction waste produced. Other major sources of building waste include incomplete information, poor material management, unskilled labour, and damage during transportation (Coffie et al. 2019; Suk et al. 2016). Construction waste creation has increased significantly as a result of the spike in construction activity, particularly in nations experiencing fast urbanisation and population growth. In response to this issue, a study was conducted to identify the key factors contributing to infrastructure waste in Thailand's construction sector. The article explains how construction waste is mainly generated by 28 important factors, which can be categorized into four groups: design, construction techniques, material sourcing, planning strategies, and human resources. They determined a significance level for each factor and designed a questionnaire to gather construction waste data. The findings demonstrate the primary classifications that are responsible for producing waste, and the elements within each group are arranged in order of their level of impact. Material administration, handling, transportation, storage, and site procedures were identified as common characteristics among the underlying causes. The identified underlying causes can help construction professionals reduce the waste generation in the early stages of the construction process (Parsamehr et al. 2022; Al Mamari et al. 2022; Aravindh et al. 2022; Bekr 2014; Ali and Rahmat 2010; Gavali and Halder 2020; Venkatesh and Aarthy 2012; Pawar and Thakare 2013; Coffie et al. 2019; Suk et al. 2016; Gulghane et al. 2024, 2023a). Construction waste generation is influenced by a myriad of specific factors and challenges, leading to significant environmental impacts and resource depletion. One primary factor is the inherent complexity of construction projects, which often involve diverse materials, technologies, and stakeholders. The lack of standardized processes for waste management in the construction industry exacerbates the problem, as there is often insufficient coordination and communication among project participants. Additionally, the prevailing culture of overestimation in material quantities during project planning contributes to excess material procurement and subsequent waste generation. Inadequate recycling infrastructure and limited awareness of sustainable construction practices further compound the challenge. The cumulative effect of these factors underscores the need for comprehensive strategies to minimize construction waste and promote a more sustainable approach to building projects (Suk et al. 2016; Gulghane et al. 2024, 2023a).

45.5 Sources of Construction Waste

Efficiently managing construction waste is essential for sustainable development and reducing its negative impact on the environment. A crucial aspect of this management involves accurately assessing the volume of waste generated throughout the construction process (Elshaboury 2022; Golbaz et al. 2019). There are different approaches to quantify construction waste, such as manual counting, weighbridge measurement, volume estimation, and modelling. Manual counting involves the physical counting of construction waste by workers or trained personnel. This technique is suitable for small-scale construction projects and can be used to determine the types and quantities of waste generated. However, it can be time-consuming and may not be accurate, especially when dealing with large volumes of waste. Weighbridge measurement involves weighing the construction waste using a weighbridge or scale. This technique is more accurate than manual counting and can be used to track the weight of waste generated over time. However, it may not be suitable for measuring bulky waste or waste with irregular shapes. Volume estimation involves estimating the volume of construction waste based on the size and shape of the waste. This technique can be used for waste with irregular shapes, and it is suitable for small-scale projects. However, it may not be accurate for bulky waste, and it may require trained personnel to estimate the volume correctly (Bekr 2014; Pawar and Thakare 2013).

Modelling involves using computer programs to model the waste generated during the construction process. This technique can provide accurate data on the types and quantities of waste generated and can be used for large-scale construction projects. However, it may require specialized software and skilled personnel to develop the model. Regenerate response determining how much is produced is the first step in efficiently managing construction and demolition (C&D) waste. A waste quantification model tailored to regional or national construction and demolition (C&D) waste generation is essential for achieving this goal. Accurate trash quantity estimation is critical in resolving building waste issues (Gavali and Halder 2020). Waste quantification entails keeping detailed records of site accounting and waste characterization, which is necessary for determining the composition of building waste. It is feasible to quantify waste and determine the potential for waste reduction by estimating the volume of construction waste produced. Making educated decisions for

Table. 45.1. Matrix of sources for generation of Construction waste (Abu Hammad et al. 2008; Aravindh et al. 2022; Gulghane et al. 2023a)

Clusters	Major causes	Frequency of occurrence
Designing related		
	Frequent change in construction design	18
	Major errors in drawings and Design	11
	Insufficiency in design information's	7
	Abysmal quality of design	5
	Incorrect layout for sites	4
	Complications in designs	2
	Inexperience designer	2
	Communication barriers	2
Material handling related	Incorrect material storage	14
	Improper handling of materials	13
	Transportation damage to material	8
	Inferior quality of material	8
	Failure of equipment's and tools	6
	Delay in allocation	5
	Use of non-suitable tools	3
Human factor related	Human Error	10
	Incompetent and low skilled worker	6
	Deplorable attitude of workers	5
	Damage caused by negligence of workers	4
	Insufficient training for workers	3
	Lack of experience and skills	3
	Shortage of skilled workers	3
	Use of inappropriate materials	3
	Poor workmanship quality and methods	2
	Improper intent towards work	2
	Inappropriate use of safety gears	1
	Fatigue problems	1
Planning and management	Substandard planning	11
	Lamentable management of site	9
	Poor controlling	9
	Execrable supervision on site	7

(continued)

Table. 45.1. (continued)

Clusters	Major causes	Frequency of occurrence
	Inadvisable use of construction methods	7
	Coordination barriers amongst departments	7
	Inadequate information quality	7
	Shortfall in Communication	6
	Scarcity of construction equipment and tools	6
	Ineffective Resources allocation	4
	Frequent Rework	4
	Inefficient waste management strategies	3
	Lack of awareness among stakeholders	1
Execution condition related	Materials residues on site	5
	Improper condition and layout of site	4
	Waste resulting from packaging	3
	Waste Congestion of the site	2
	Inadequate illumination on site	1
	Unwanted Crews interference	1
Material procurement related	Inaccurate procurement	11
	Careless Transportation	3
	Inappropriate quantity surveys	3
	Incorrect specifications and standards	2
Miscellionus and external	Unavoidable Weather effect	13
	Accidents on site	6
	Pilferage	4
	On site Vandalism	2
	Festivities	1
	Unpredictable local conditions	1
	Lack of legislative enforcement	1

minimizing and managing waste sustainably requires the use of quantification, which is a crucial technique for determining the actual magnitude of waste (Ali and Rahmat 2010). A construction waste assessment or audit is required to acquire this critical data. This technique advances our understanding of the reasons, sources, volume, and composition of construction refuse generation. Accurate waste assessment data should be collected, so a method for recording quantitative data should be developed. Construction officials are accountable for this because they can create rules and rewards to promote eco-friendly construction methods and waste management procedures (Aravindh et al. 2022; Venkatesh and Aarthy 2012). To reduce construction waste, standardized record-keeping methods and useful guidelines should be implemented. Enforcing requirements and norms are critical to ensuring compliance. For conventional buildings, a four-step waste quantification technique can be used, which entails identifying construction stages, selecting similar construction sites, sorting and weighing mixed waste, and thoroughly analysing the recorded amount (Aravindh et al. 2022; Pawar and Thakare 2013).

Construction officials can evaluate the viability of C&D refuse recycling programs and estimate the lifespan of depleted landfill areas. Waste measurement is a useful technique for pinpointing inefficiencies and potential development areas in manufacturing systems. However, precisely measuring waste in various stages of construction projects can be difficult, making waste estimation difficult and unreliable (Ujong et al. 2022; Bekr 2014). As a result, precise waste measurement is critical for evaluating the performance of manufacturing processes. Various methods for quantifying building waste have been suggested in previous research, including the percentage method. The third and final point of discussion is waste quantification, which can be accomplished using a variety of techniques. One such approach is the use of a conversion factor (Suk et al. 2016). A construction trash index can also be calculated using site audits, which include frequent visits, checklists, and the estimation of disposal records. Cutting and managing waste should be emphasized when it comes to addressing waste causes. As proposed design interventions can help reduce cutting waste, while proper organization and supervision of site activities can help avoid management waste. Quantifying waste is a crucial step in understanding its true size and taking appropriate actions to minimize and manage it sustainably (Abu Hammad et al. 2008).

Waste characterization is the first and most important step of data collection in waste management. This procedure entails determining the kinds of waste that are produced. Several studies on the characterization of different types of construction waste materials have been performed. However, most studies have concentrated on the main kinds of C&D waste that account for a significant portion, such as concrete, bricks, wood, and steel. They also introduced the concept of "wastage level," which measures the proportion of building materials that are likely to be wasted by contrasting the quantity of materials used or completed with the quantity of materials actually purchased (Gavali and Halder 2020; Pawar and Thakare 2013). This idea is based on earlier work source 1 assessment and quantification of C&D waste. It is an excellent way to cut waste and assess performance on-site, particularly in material processing, storage, and transportation. The sources of building waste production, such as design, procurement, handling, operation, and residual refuse, are identified using waste generation-rate models. Estimate that 1 to 10% of the materials bought for construction projects in the Dutch industry are wasted, with the majority of this waste coming from leftover cut-offs, plan changes, and subpar workmanship. As a result, waste reduction at the source is critical for efficient waste management and productivity in construction initiatives (Suk et al. 2016; Gulghane et al. 2024). Using waste indices to make calculations can help estimate the amount of waste that will be produced at different stages of a construction project. This can lead to increased awareness of waste reduction, better planning, and more efficient environmental management. In Hong Kong, this method has been applied to assess and standardize the environmental performance of construction projects, particularly in public housing. On average, construction waste generated by contractors in the building sector in Hong Kong ranges from 0.125 m^3 to 0.25 m^3 per gross floor area. (GFA) (Ujong et al. 2022). It is simpler to estimate waste in a unit amount/area of activity by using the "Global Index" as an indicator for a particular construction style and for projects of a similar nature in the future. This method can also be used to determine how much building waste is generated locally or nationally. In order to quantify the distribution of construction waste in a place, using a cutting-edge technique. They divided the layouts into four categories—stacked, gathered, scattered, and stockpiled—and then multiplied the expected volume by the estimated unit weight of each form to determine the weight of waste generated. A method for measuring the quantity of various waste products, such as 1 m^2 of a partition wall, 1 kg of reinforcing steel, and 1 m^3 of concrete (Al Mamari et al. 2022). Numerous surveys from various project sites were used to determine the fixed waste factors, along with soft measurement techniques like interviews, questionnaires, and statistical data estimation. Hard measurement techniques, on the other hand, included the Material Flow Analysis Approach (MFA) and sorting and weighing waste materials. Finally, waste quantification models are critical for identifying and reducing building waste production, with waste index calculations and the "Global Index" providing useful insights for efficient waste management. Furthermore, novel methods such as quantifying construction waste layout and developing detailed coded-classification systems can aid in accurately quantifying construction waste and informing future waste management practices (Ujong et al. 2022).

45.6 Sources of Construction Waste

Construction waste estimation is important in construction management because it allows project managers to establish effective waste control solutions. To achieve precise predictions regarding the waste produced during a construction project, a comprehensive grasp of the project's scope, materials employed, construction techniques, and potential waste sources becomes essential. This knowledge greatly aids in the effective planning and execution of waste reduction measures. Usually, project managers rely on historical data, such as waste records from prior projects or industry benchmarks, to anticipate the quantity of waste that will be generated (Ali and Rahmat 2010). Moreover, digital tools like Building Information Modelling (BIM) software offer project managers the ability to pinpoint potential waste generation hotspots and optimize construction techniques, leading to waste reduction. Accurate predictions of construction waste amounts empower project managers to implement effective strategies like recycling, repurposing, and landfill diversion. These measures not only decrease the environmental impact but also lead to cost savings associated with waste disposal (Gavali and Halder 2020). Understanding current waste levels and accurately estimating future waste generation are critical components of waste from construction and demolition management. Although C&D waste output is decreasing in many places, accurate forecasting is still required to plan for waste treatment and recycling capacity. Forecasting C&D waste generation can provide valuable information for refuse disposal choices. Quantifying and forecasting waste is critical for effective trash management and should be investigated further. Future waste generation is inextricably linked to past waste generation because waste output is time-series data (Venkatesh and Aarthy 2012). Using the ARIMA model and GRC method, a waste prediction system that relies on time-series can anticipate the generation of waste with a smaller amount of data. Based on the model's predictions, it appears that Sichuan is producing a greater amount of waste in general, but some specific sectors are generating an imbalanced proportion of it (Pawar and Thakare 2013).

Several techniques are employed to predict and anticipate the generation of construction and demolition waste. Estimating building area, doing material circulation analysis, and utilizing geospatial data management techniques are among these ways. Various methodologies, such as grey models, regression using linearity, and the integrative autoregressive moving average model, are used to forecast waste output (Coffie et al. 2019). These methods have been used in numerous studies to calculate and forecast the production of C&D garbage. Researchers have extensively used multiple regression analysis, BP neural network models, and Grey Model prediction methods to anticipate construction waste generation. Construction waste prediction methods are essential for efficient waste management in the construction industry (Suk et al. 2016). There are several methods for predicting construction waste, including qualitative and quantitative approaches. Qualitative methods rely on expert knowledge and experience to identify potential waste sources, such as identifying materials that are likely to generate waste, assessing construction site conditions, and evaluating project plans and specifications. The utilization of these techniques can furnish significant perceptions regarding the production of waste and can advise tactics for reducing waste. Conversely, quantitative approaches apply mathematical models to gauge the amount of waste produced while constructing (Coffie et al. 2019; Suk et al. 2016). These models can be based on data from previous construction projects or on statistical analysis of factors that affect waste generation. Other methods for predicting construction waste include waste audits, waste tracking systems, and construction waste management plans. By using these methods, construction companies can improve their waste management practices, reduce waste generation, and minimize the environmental impact of their operations (Gulghane et al. 2024, 2023a, 2023b).

In the past decade, there has been an increase in the use of machine learning and artificial intelligence (AI) approaches for construction waste prediction. These advanced methods utilize historical data from construction projects to create models facilitating precisely anticipating the amount of waste produced during specific endeavors. By analyzing various key factors, including project type, size, location, and materials employed, these models can deliver highly accurate waste predictions. With the aid of machine learning algorithms, construction firms can identify patterns and trends in waste production over time. This valuable insight enables them to pinpoint areas where waste reduction strategies can be most effectively implemented to achieve maximum impact. AI methods can also be used to optimize construction processes to minimize waste generation, such as identifying the most efficient use of materials and reducing over-ordering. Overall, machine learning and AI methods provide a powerful tool for construction companies to reduce waste and improve their sustainability efforts (Ali and Rahmat 2010; Gavali and Halder 2020). The objective of this study was to forecast the yearly increase of C&D waste using three methodologies: quadratic exponential smoothing prediction, the Mann–Kendall trend test, and the building area estimation method. A comprehensive case study was conducted to achieve this aim. The results showed an increase in C&D trash. To forecast building end-of-life values and evaluate recoverable materials during the demolition stage, a machine learning algorithm was also developed (Venkatesh and Aarthy 2012; Pawar and Thakare 2013). The random forest and decision tree methods were used in the model, which was appropriate for a smaller dataset with categorical data. The use of these machine learning algorithms can help to enhance waste management systems and make accurate predictions about the production of construction waste. In general, accurate measurement and forecasting of C&D garbage generation are essential for efficient waste management. A number of techniques, such as time-series waste prediction frameworks, estimation and prediction techniques, and Machine learning algorithms have the ability to make predictions and projections about the production of construction and demolition waste. Figure 45.4 shows the typical Structure of the decision tree (DT) algorithm (Gavali and Halder 2020; Venkatesh and Aarthy 2012; Suk et al. 2016).

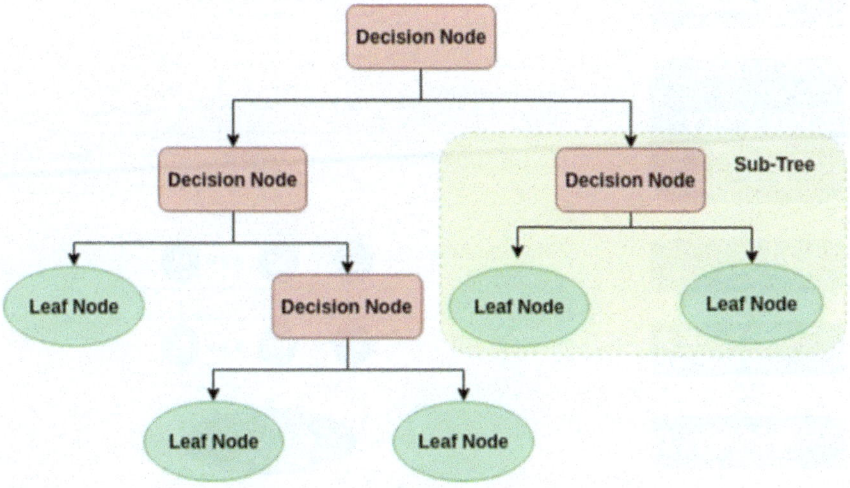

Fig. 45.4 Structure of the decision tree (DT) algorithm

Several models, such as the neural network model known as BP network and the grey model prediction technique, are utilized to forecast the amount of waste produced. The former needs a large amount of data, whereas the latter depends on known data to generate a regular time series for prediction. Depending on the amount of data used for prediction, waste prediction models can be classified as macro or micro models. A few models that are frequently used are linear and regression models, S-curve and Artificial Neural Network models, waste design, and Big Data framework. Research has indicated that for GIS-based prediction models, it is necessary to have comprehensive data collection, with a greater emphasis on mid- and long-term forecasts. Accurately and efficiently estimating the production of Construction waste is crucial. This can be achieved by harnessing open data from statistical departments and creating practical prediction models to promote the optimal utilization of waste resources. Furthermore, there have been suggestions to leverage artificial intelligence and machine learning technologies for forecasting waste production (Ali and Rahmat 2010; Coffie et al. 2019; Suk et al. 2016; Gulghane et al. 2023b) (Fig. 45.5).

The diagram 5 above illustrates the typical framework of the random forest method, commonly employed for waste forecasting. When predicting C&D waste production, numerous studies have utilized machine learning models optimized for continuous data inputs. These models encompass adaptive neuro-fuzzy reasoning systems, support vector machines, decision trees, linear regression analysis, and artificial neural networks. However, when dealing with categorical data, these algorithms may not yield optimal results. Therefore, most approaches to predict C&D waste involve employing machine learning and statistical analysis techniques such as adaptive neuro-fuzzy inference systems, ANNs, SVMs, DTs, LR analysis, and GAs. Incorporating historical data, Algorithms based on machine learning can be programmed

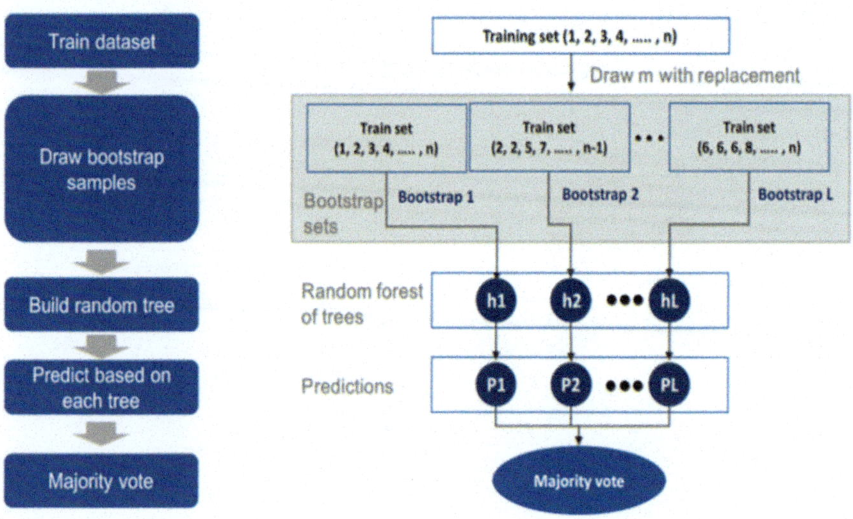

Fig. 45.5 Structure of the random forest (RF) algorithm

to recognize trends and anticipate future construction waste. For instance, decision trees, random forests, and neural networks can be used to forecast the quantity of waste produced based on factors such as project type, location, and construction materials used. Text data from construction documents and contracts can be analysed using Natural Language Processing approaches for identifying potential waste sources and estimating the quantity of debris that will be generated. This can help construction companies to optimize their waste management strategies (Gavali and Halder 2020; Venkatesh and Aarthy 2012; Gulghane et al. 2024). By utilizing computer vision methods, it is possible to examine pictures and videos captured at construction sites in order to recognize discarded materials and make an approximation of the amount of waste produced. This can help construction companies to identify areas where waste reduction measures can be implemented. Internet of Things devices such as sensors can be deployed at construction sites to collect data on waste generation and monitor waste disposal practices. This data can be analysed using ML algorithms to identify trends and patterns that can be used to optimize waste management strategies. Forecasting the quantity of waste produced can be accomplished through the application of predictive analytics methods that consider different factors like construction activities, project schedules, and weather patterns. This can help construction companies to plan their waste management activities more effectively.

The burgeoning field of sustainable construction has increasingly turned to machine learning (ML) techniques to address the challenges associated with the prediction and management of construction waste. This literature explores and consolidate the existing knowledge on the potential applications of ML in enhancing sustainability within the construction industry. Various studies have demonstrated the efficacy of ML algorithms in predicting construction waste generation by analysing historical data, project specifications, and other relevant variables. Employed machine learning models to forecast construction waste generation based on project characteristics, highlighting the ability of these models to provide accurate estimates (Aravindh et al. 2022). Additionally, Jones and Wang conducted a comparative analysis of different ML algorithms for predicting construction waste, emphasizing the versatility of these tools in adapting to various project scenarios (Kaveh and Servati 2001). Furthermore, the literature reveals the effectiveness of ML in optimizing waste management processes and resource allocation. Recent research by Chen and Li investigated the integration of ML algorithms into construction waste management systems, demonstrating improvements in efficiency and cost-effectiveness (Ali and Rahmat 2010). The potential of ML-based decision support systems in dynamically adjusting waste management strategies based on real-time data (Gavali and Halder 2020).

45.7 Conclusion

In conclusion, addressing the generation of construction waste necessitates the implementation of a comprehensive and strategic waste management plan. This plan must encompass multifaceted approaches, emphasizing the reduction, reuse, and recycling of waste. By committing to sustainable practices, construction firms not only contribute to environmental conservation but also stand to benefit from reduced expenses associated with waste management. The life cycle of construction projects, from planning to completion, witnesses the continuous production of waste. Understanding the various factors contributing to this waste generation is fundamental to effective management. Recognizing and addressing these underlying factors become imperative in curbing the quantity of waste produced at each stage. To minimize construction waste, a proactive approach involves targeting the root causes. Quantifying the amount of waste generated depends on factors such as project scale, complexity, waste characteristics, and the availability of resources. Employing a diverse set of techniques can yield accurate data on waste generation, thereby facilitating the formulation of tailored waste management strategies. In this context, the integration of machine learning and AI techniques emerges as a powerful tool for construction companies. These technologies enable precise data analysis, predictive modelling, and decision support, enhancing the ability to curb waste effectively. By harnessing the potential of artificial intelligence, construction firms can not only reduce their ecological footprint but also make significant strides in improving overall sustainability practices. In essence, a holistic waste management plan, coupled with advanced technologies, is essential for the construction industry to navigate the challenges posed by construction waste. Embracing these strategies not only aligns with environmental stewardship but also positions construction companies as responsible entities contributing to a more sustainable and eco-friendly future.

References

AbouRizk S (2010) Role of simulation in construction engineering and management. J Constr Eng Manag 136:1140–1153

Abu Hammad A, Alhaj Ali S, Sweis G, Bashir A (2008) Prediction model for construction cost and duration in Jordan. Jordan J Civil Eng 2(3):250–266

Al Mamari AHS, Al Ghafri RSHH, Aravind N et al (2022) Experimental study and development of machine learning model using random forest classifier on shear strength prediction of RC beam with externally bonded GFRP composites. Asian J Civ Eng. https://doi.org/10.1007/s42107-022-00502-3

Ali J, Khan R, Ahmad N, Maqsood I (2012) Random forests and decision trees. Int J Comput Sci 9(5):272–278

Ali AS, Rahmat I (2010) The performance measurement of construction projects managed by ISO-certified contractors in Malaysia. J Retail Leis Prop 9(1):25–35. https://doi.org/10.1057/RLP.2009.20/TABLES/3

Aravindh MD, Nakkeeran G, Krishnaraj L et al (2022) Evaluation and optimization of lean waste in construction industry. Asian J Civ Eng 23:741–752. https://doi.org/10.1007/s42107-022-00453-9

Azadi S, Karimi-Jashni A (2015) Verifying the performance of artificial neural network and multiple linear regression in predicting the mean seasonal municipal solid waste generation rate: a case study of Fars province, Iran. Waste Manag 48:14–23. https://doi.org/10.1016/j.wasman.2015.09.034

Bekr GA (2014) Study of the causes and magnitude of wastage of materials on construction sites in Jordan. J Constr Eng 2014:1–6. https://doi.org/10.1155/2014/283298

Cha G-W, Moon HJ, Kim Y-M, Hong W-H, Hwang J-H, Park W-J, Kim Y-C (2021) Development of a prediction model for demolition waste generation using a random forest algorithm based on small data sets. Int J Environ Res Public Health 17(19):6997

Coffie GH, Aigbavboa CO, Thwala WD (2019) Modelling construction completion cost in Ghana public sector building projects. Asian J Civ Eng 20:1063–1070. https://doi.org/10.1007/s42107-019-00165-7

Coskuner G, Jassim MS, Zontul M, Karateke S (2020) Application of artificial intelligence neural network modeling to predict the generation of domestic, commercial and construction wastes. Waste Manag Res: J Sustain Circ Econ 39(3):499–507

Elshaboury N, Al-Sakkaf A, Abdelkader EM, Alfalah G (2022) Construction and demolition waste management research: a science mapping analysis. Int J Environ Res Public Health 19(8):4496

Foo LC, Rahman IA, Asmi A, Nagapan S, Khalid KI (2013) Classification and quantification of construction waste at housing project site. Int J Zero Waste Gener 1(1):1–7

Gavali A, Halder S (2020) Identifying critical success factors of ERP in the construction industry. Asian J Civ Eng 21:311–329. https://doi.org/10.1007/s42107-019-00192-4

Golbaz S, Nabizadeh R, Sadat Sajadi H (2019) Comparative study of predicting hospital solid waste generation using multiple linear regression and artificial intelligence. J Environ Health Sci Eng 17(1):41–51. https://doi.org/10.1007/s40201-018-00324-z

Gulghane A, Sharma RL, Borkar P (2023a) Quantification analysis and prediction model for residential building construction waste using machine learning technique. Asian J Civ Eng 24:1459–1473. https://doi.org/10.1007/s42107-023-00580-x

Gulghane A, Sharma RL, Borkar P (2023b) Performance analysis of machine learning-based prediction models for residential building construction waste. Asian J Civ Eng 24:3265–3276. https://doi.org/10.1007/s42107-023-00708-z

Gulghane A, Sharma RL, Borkar P (2024) A formal evaluation of KNN and decision tree algorithms for waste generation prediction in residential projects: a comparative approach. Asian J Civ Eng 25:265–280. https://doi.org/10.1007/s42107-023-00772-5

Hassan SH, Aziz HA, Daud NM, Keria 1, Noor SM, Johari I, Shah SMR (2020). The methods of waste quantification in the construction sites (a review). In: Advances in civil engineering and science technology, AIP conference proceedings, 2020, pp 020056-1–020056-6

Hu R, Chen K, Chen W, Wang Q, Luo H (2021) Estimation of construction waste generation based on an improved on-site measurement and SVM-based prediction model: a case of commercial buildings in China. Waste Manag 126:791–799

Kaveh A, Khalegi A (2019) Prediction of strength for concrete specimens using Artificial Neural Networks. In: 1st international conference on engineering computational technology/4th international conference on computational structures technology, pp 165–171. WOS:000077305500020

Kaveh A, Rahimi Bondarabady HA (2004) Wavefront reduction using graphs, neural networks and genetic algorithm. Int J Numer Methods Eng 60:1803–1815. https://doi.org/10.1002/nme.1023

Kaveh A, Gholipour Y, Rahami H (2008) Optimal design of transmission towers using genetic algorithm and neural networks. Int J Space Struct 1(23):1–19. https://doi.org/10.1260/026635108785342073

Kaveh A, Servati H (2001) Design of double layer grids using back-propagation neural networks. Comput Struct 79:1561–1568. https://doi.org/10.1016/S0045-7949(01)00034-7

Kolaventi SS, Tezeswi TP, Siva Kumar MVN (2019) An assessment of construction waste management in India: a statistical approach. Waste Manag Res 38(4):444–459. https://doi.org/10.1177/0734242X19867754

Li Y, Zhang X, Ding G, Feng Z (2016) Developing a quantitative construction waste estimation model for building construction projects. Resour, Conserv Recycl 106:9–20. https://doi.org/10.1016/j.resconrec.2015.11.001

Markandeya Raju P, Kameswari L (2015) Construction and demolition waste management—a review. Int J Adv Sci Technol 84:19–46

Parsamehr M, Perera US, Dodanwala TC et al (2022) A review of construction management challenges and BIM-based solutions: perspectives from the schedule, cost, quality, and safety management. Asian J Civ Eng. https://doi.org/10.1007/s42107-022-00501-4

Pawar SN, Thakare A (2013) Cost analysis and project planning using ERP in construction industry. Int J Latest Trends Eng Technol 2(4):16–29

Quiñones R, Llatas C, Montes MV, Cortés I (2022) Quantification of construction waste in early design stages using bim-based tool. Recycling 7(5):63. https://doi.org/10.3390/recycling7050063

Ram VG, Kalidindi SN (2017) Estimation of construction and demolition waste using waste generation rates in Chennai, India. Waste Manag & Res: J Sustain Circ Econ 35(6):610–617. https://doi.org/10.1177/0734242X17693297

Suk SJ, Chi S, Mulva SP, Caldas CH, An SH (2016) Quantifying combination effects of project management practices on cost performance. KSCE J Civil Eng. https://doi.org/10.1007/s12205-016-0499-0

Ujong JA, Mbadike EM, Alaneme GU (2022) Prediction of cost and duration of building construction using artificial neural network. Asian J Civ Eng 23:1117–1139. https://doi.org/10.1007/s42107-022-00474-4

Venkatesh J, Aarthy C (2012) Emerging erp systems—exemplary for the construction industry. EXCEL Int J Multidiscip Manag Stud 2(6):24–31

Wu Z, Yu ATW, Shen L, Liu G (2014) Quantifying construction and demolition waste: an analytical review. Waste Manag 34(9):1683–1692

Chapter 46
The Behavior of Reinforced Concrete Under Impact Loading: Review

Aman A. Satfale and N. H. Pitale

Abstract The behavior of impact-loaded reinforced concrete structures has piqued the interest of many people because to its relevance to various sectors, including structural design, infrastructure protection, and civil engineering. This research seeks to offer a thorough understanding of the dynamic behavior of reinforced concrete under impact loading circumstances. The study looks at the fundamental mechanics of how reinforced concrete reacts to unexpected, high-energy loads. It investigates the qualities of the material, structural designs, and other elements. This paper covers the issues of impact loading on reinforced concrete structures by combining data from theoretical models, empirical studies, and numerical simulations. Important subjects discussed include strain rate effects, fragmentation, and spalling, three dynamic failure mechanisms in concrete materials subjected to rapid stress. Study is also done on the impact structure's overall performance as a function of material properties, types of reinforcement, and system organization. The advances in modeling strategies and analytical approaches for analyzing and forecasting reinforced concrete building responses to impact loads are also highlighted in this review. By using these techniques, designers may create designs that are stronger and more resilient to impact. Future research paths are based on the synthesis of previously published information and the identification of knowledge gaps in the field of reinforced concrete structure behavior under impact loading. In the conclusion, this research advances our knowledge of reinforced concrete's dynamic behavior and provides suggestions for bolstering the security and resilience of buildings exposed to impact events.

Keywords Reinforced concrete · Impact loading · Dynamic response · Structural behavior · Material properties

A. A. Satfale (✉) · N. H. Pitale
G.H. Raisoni College of Engineering, Nagpur, India
e-mail: aman.satfale.mtechstr@ghrce.raisoni.net

N. H. Pitale
e-mail: nikhil.pitale@raisoni.net

B. V. Khode et al. (eds.), *Innovations in Technologies: Pioneering Sustainable Infrastructure for a Resilient Future*, Sustainable Civil Infrastructures,
https://doi.org/10.1007/978-3-031-82133-2_46

46.1 Introduction

Because they offer the structural support needed to keep a structure stable, columns are an essential component of public infrastructure, including buildings and bridges. The substance (wood, metal, composite materials, stone, etc.), size, direction, and section geometry (square, angle, etc.) of the lines used in this model may vary.

When vehicles collide with the system, rear impact is usually the result. These mishaps may be intentional, like a horrific crime (Anderson 2018) or suicidal thoughts (Hogan 2020), or they could be unintentional, like an out-of-control car or a negligent driver (Gutschke 2020). This activity may result in fatalities as well as the collapse of lines and structures. At 2002, the I-40 Bridge, a concrete construction located at Webbs Falls, Oklahoma, that across the Arkansas River, collapsed. Fourteen individuals lost their life in the disaster as a consequence of the boat and support colliding and a portion of the bridge collapsing (Burton 2017). Commercial complexes and large structures are vulnerable to catastrophic failures that result in significant damage. If we wish to stop this rate of fatalities, we must find strategies to lessen the explosion risk. The use of support materials like steel pipe or fiber-reinforced polymer (FRP) and design decisions like enhancing system impact resistance are some ways to lessen failures brought on by outside factors. A variety of studies on materials that lessen the impact of loading will be looked at in this review of the literature, including FRP tubular concrete, spiral steel bars (SR), concrete fiber reinforced concrete, and FRP steel double-layer tubular concrete. (Fibre package UHP). The purpose of this research is to assess the effects of fiber-glass reinforced concrete (GFRP) pipes and determine whether or not they can enhance the performance of steel bars. Consequences associated with carrying since dynamic qualities are hard to compute precisely, approximations are frequently employed. Additional numerical equations based on hypotheses. Errors in estimating the static attribute balance are frequently very high. Because of this, it is difficult to strengthen columns' resistance to dynamic loads without a complete grasp of how this information enhances column performance. To gain a better knowledge of how impact loads affect the structural integrity of each column and the structure as a whole, the impacts of impact on the outside of the columns are investigated. This enhanced comprehension will enable the development of more precise estimates for external energy systems and apparatus or the use of tactics to raise the efficiency of the impacted lines.

46.2 Background

46.2.1 Testing Methods

Impact testing should ideally imitate the failures and reactions that could happen at that precise moment and match the conditions that arise under load action. In the study under evaluation, three primary criteria were employed to evaluate the adverse effects of the model. A weight loss model, a horizontal truck or tram collision, and a finite element (FE) model that mimics the lines of a vehicle contact are a few of these techniques. Qasrawi et al. (2015) also evaluated the FRP sheathing's ability to support a static load using the monotonic test. A range of weights, shocks, and impact velocities were used in the studies. Every experimental technique used to quantify unfavorable effects has pros and downsides.

While weight loss devices are the most widely used method of evaluating impacts because they more accurately replicate impact damage in real-world settings, certain methods are still unable to accurately capture the force of the weight. When calculating the impact, that additional weight must be taken into consideration if it travels with the sample after it is struck rather than immediately leaving it (Chen et al. 2020). The height of the shock must be increased or its ends must be strengthened to withstand the weight if a greater impact or quicker speed is required.

The horizontal impact testing system (HITS) was created by Aghdamy et al. (2016) and applies the impact using a pneumatic instrumented impactor. Due to their limited ability to transmit impact force, HITS and weight loss have the drawback of making it difficult to simulate the effects of an automobile collision. Chen et al. employed a trolley machine as a horizontal intervention. Consequently, the system can produce thorough crash testing (Chen et al. 2020). Several researchers have also examined the impact of collisions with columns using similar devices. The impacts of complete collisions, which occur at speeds often observed in the surroundings of systems, are also examined using finite element models. This method of assessing impact could be effective, but only if standards are applied correctly, which might necessitate a procedure for testing and standard comparison. As a result, the data needed to validate the findings determine the validity of the finite model, and gathering experimental data for research can be a laborious process. This approach helps explain how different materials, which are commonly employed to support structural parts, enhance their effects.

46.2.2 Effects of Strengthening Materials on Lateral Impact Resistance

46.2.2.1 Effects of Spiral Reinforcement on Lateral Impact Resistance

A mathematical analysis of reinforced concrete's dynamic reaction was conducted by Cai et al. (2017). The three stages involved in determining the effect on time history are the oscillation phase, stability phase, and attenuation phase. The initial contact causes the oscillation phase, which is when the impact force reaches its greatest amplitude and the weight and speed increase. At the plateau level, the curve then rapidly deteriorates before progressively rising once more. The impact process ends when the impact energy is depleted and the impact force drops to zero during the damping period. The deflection-time history curve displays stages of growing, declining, and damped vibration, just like the force–time history curve. Demartino et al.'s mathematical study (2017) yielded a force–time history that was comparable to what Cai et al. (2017). The two energy-time histories differ in a few ways, most notably in Demartino et al. (2017). The date does not correspond to the stable phase that Cai et al. (2017) discovered. The major cause of this disparity is Cai et al. (2017) The colon has a square cross-section, and the afflicted area is in the top portion of the colon, as noted by Demartino et al. (2017) The impact point is located close to the round shaft's base.

A research by Demartino et al. (2017) found that impacts result in damage more quickly and that the severity of the damage increased with impact speed. Moreover, Cai et al. (2017) came to the conclusion that a higher impact velocity led to a higher maximum displacement. By reducing the distance between concrete columns, strengthening the shear resistance, and offering adequate protection in case the concrete falls, the extent of damage can be lessened. Chen et al. (2020) recommended adding spiral steel bars. Enhance the load-impacted lines' behavior.

46.2.2.2 Impact Resistance of Lateral Impacts Due to FRP

Chen et al.'s experiment (2020), which included force–time history, demonstrated that the inclusion of GFRP tubes improved the column's overall impact characteristics and shear performance. GFRP becomes increasingly significant when impact resistance rises. Because thicker GFRP pipes have more local strength and damping, they may be able to sustain more impacts. A concrete-filled glass fiber reinforced plastic tube (CFFT) column's force–time history is comparable to that of concrete: The force increases sharply at first, then quickly decreases and stabilizes, and then gradually decreases until it reaches zero. Tests using thick GFRP tubes shown smaller decline than those using thin GFRP tubes. As the implant velocity increases, the shallow water supply in the narrow tube model stays constant. This suggests that thickening the GFRP pipe will enhance the column's damping, lengthening its total period and reducing the impact force. The impact of concrete can be increased by using FRP materials, as per the findings of Pham and Hao's (2016) study. The inclusion of FRP improves the beam's ability to sustain loads while simultaneously increasing its ductility and energy-absorbing capability. The results of Chen et al. support this (2020) came to the conclusion that energy absorption rises and the maximum displacement of the CFFT tube array decreases as GFRP tube thickness increases. Moreover, they discovered that although the FRP tubes' addition would have contributed to the rise in compressive strength, the FRP's tensile strength increased as well. According to Qasrawi et al.'s experimental results (2015), adding FRP coating greatly raised the required tensile strength of concrete. Additionally, the FRP covering lessens the possibility of the concrete in the column crushing or spalling and experiencing permanent deformation. The ductility of the column is increased by the FRP coating. For each column with 1.2% and 2.4% reinforcement, the energy needed to fail increased by 1223% and 467%, respectively; this suggests a better value for columns with less contrast. According to Wang et al. (2015), the thickness of FRP tubes lowers impact loads and residual deformation. Additionally, it is stated that FRP fibers are orientated parallel to the hoop's direction; however, more research is required to ascertain how this orientation affects the final result.

46.2.2.3 Effects of FRP and Spiral Reinforcement on Lateral Impact Resistance

The use of fiber-reinforced polymer tubes or spiral reinforcement increases the resistance of concrete columns to lateral impacts. According to Qasrawi et al. (2015), adding FRP tubes to a spirally reinforced column significantly increased the impact resistance of the column. Chen et al.'s study (2020) found that concrete columns reinforced exclusively with fiber-reinforced polymer (FRP) tubes had a smaller maximum displacement and absorbed more energy, but also showed a brittle failure mechanism. This work suggests that employing steel fasteners can reduce the brittleness of fiber-reinforced plastic (FRP) pipes. As a result, employing spiral reinforcements and FRP pipes improves the product's ductility while also raising the pipe string's resistance. It was found that the installation of spiral reinforcement and FRP pipe efficiently decreased external impacts for every line with greater axial loads. According to Huang et al.'s significant axial investigation (2017), installing spiral and FRP materials together provides higher resistance than using a single material.

46.2.2.4 Effects of Other Strengthening Materials on Lateral Impact Loads

One viable material for reinforcing concrete columns to increase their resistance to lateral impacts is steel tube. The residual lateral deformations of concrete columns reinforced with only a fiber-reinforced polymer tube (CFFT) were approximately 40% more than those of columns with an internal hollow steel tube (DSTC), according a study by Chen et al. (2020). This suggested that the steel tube members had greater global stiffness and reduced local stiffness. At the same impact velocity, the DSTC members may also absorb more energy than the CFFT members.

Ultra-high-performance fiber-reinforced concrete (UHPFRC) is thought to hold great potential for enhancing resistance to impact loads. In Fan et al.'s work (2018), the ramifications of UHPFRC were assessed by numerical analysis. When compared to conventional remote control systems, the UHPFRC sleeve minimizes displacement by at least 50% and exhibits less damage in comparable circumstances. The effect of the line below the lowest impact is influenced by the UHPFRC sheath thickness; at higher impact rates, the sheath thickness has the lowest capacity following the impact.

46.2.2.5 An Overview of the Damage to the Column

The different materials that are utilized to reinforce a column can have an impact on how quickly the column deteriorates. It should go without saying that increasing material would enhance impact resistance and decrease damage. FRP double-layer pipes are more durable than reinforced concrete systems, which use concrete along with other materials. They can withstand the greatest forces while sustaining the least amount of damage.

46.2.2.6 Steel Tubes' Impact on Axial Impact Loads

Zhu et al. (2020) examined the effectiveness of concrete steel tube (CFST) columns in axial resistance, despite the fact that steel pipes have been demonstrated to have excellent resistance to external impacts on concrete slabs. According to studies, during their short existence, cylinders experience the least amount of change and can absorb most hits. Because of its characteristics that allowed the force to be distributed more widely throughout the concrete structural body, parallel bars were determined to be the best option for withstanding axial impact loads. The axial impact of CFST can be increased by reinforcing steel, however this effect is more noticeable in non-circular beam types since they are more prone to local buckling.

46.2.2.7 Impact of Spiral Reinforcement and Fiber Reinforcement on Axial Impact Loads

It has been demonstrated that installing spiral steel rod and FRP in a continuous line is a reliable way to guard against outside influences. Examining the axial impacts of GFRP and SR rollers was Huang et al. (Huang et al. 2017). The findings indicate that the pipe array's resistance to damage is increased when GFRP and SR are combined, as opposed to when those two extra materials are used alone. In this design, the increasing voltage responsiveness and higher voltage level following the peak demonstrated how insulation enhances the line's ability to dissipate energy. The dynamic peak stress and static peak stress may converge more closely with higher GFRP layer counts and SR volume percentages. These results align with the study conducted by Huang et al. (2018), which demonstrated that columns that combined the use of SR and CFRP had a higher resistance to impact loading than columns that only used these two materials. Similar to Huang and colleagues [11th], the rope behaves less brittlely and becomes more ductile when CFRP and SR are combined. The findings of both investigations indicated an increase.

46.2.3 *Prospective Aspects to Take into Account*

Future research projects must address the discrepancies in data found throughout these investigations. Although the FRP fibers in this experiment were oriented along the edges of the components, Wang et al. (2015) note that it is unclear how this orientation impacts impact resistance. Owing to this uncertainty, it is necessary to ascertain the fiber orientation in the FRP for every test piece in order to guarantee test specimen uniformity. Numerous research on the FRP tunnel strength offered by concrete structures were examined by Pham and Hao (2016). The investigation found that there are still unanswered concerns regarding the relationship between stress and damage as well as the impact loading of fiber-reinforced polymers (FRP). More investigation is required to find out how loading affects these problems.

46.3 Methods of Research

The interaction between composite systems and GFRP tube-confined CFFT columns is being systematically studied, and this inquiry is the initial step in that process. The effects of various material-based concrete admixtures have, in fact, been the subject of much investigation. Nevertheless, there aren't many research out there right now that particularly assess the enhancements in external effect of CFFT columns with composite systems. To increase belief in the benefits and comprehension of fairness, more research on the effects of the CFFT line is required. At Carleton University, research is now being conducted to evaluate the effectiveness of CFFT with reinforced columns. In order to simulate the impacts of traffic accidents with piers, this work aims to examine the performance of columns placed with CFFT materials in comparison to reinforced concrete. Even though simulating traffic accidents is the most widely used method for assessing how system components would behave under impact, it is challenging since waste materials may weaken the model. Pendulum effects have been applied to concrete features in a number of studies (Buth et al. 2011; Silva et al. 2009; Zhang et al. 2016), and they have shown to be an effective method for simulating the characteristics of a car. As a result, specific loadings were applied to the sample in this experiment using a 221 kg pendulum machine. The graphic displays the testing. 1. Using a winch-connected pulley system, the pendulum was lifted at certain angles and then released to provide impact force on the tops of the columns. Pendulum weight removal to the extent specified by the impact angle (Fig. 46.1).

The impact angle using a winch crank. A steel reaction frame held the winch in place. A high tensile strength rope that functioned as the pulley system's link connected the winch, a manufactured rolling rod above it, and a hook on the pendulum weight system. A synthetic rod with a rope fastened to it served as the releasing mechanism. The rod was fastened to its hook in such a way that when the rope connected to it was tugged, the rod would separate from the hook.

Fig. 46.1 Test set-up of lateral impact test using pendulum machine

Every specimen has two different interior diameters of 152 and 210 mm and stands 1800 mm tall. Every pipeline is supported by six to ten meter-long steel bars. To improve performance, the column's height and the link between the 10 M and 100 mm volume have been raised. A hollow steel profile steel base with dimensions of 152 mm in diameter and 450 mm in height has been built to offer stability at the base of the column. The steel base is fastened to the laboratory floor using four anchor bolts. Standard strength concrete is used to pour the structure. The maximum frame size is 20 mm. The average compressive strength of concrete is around 35 MPa. The longitudinal steel bars and connecting rods possess a yield strength of 447 MPa and an elastic modulus of 200 GPa. Glass fiber reinforced plastic (FRP) tubes with different wall thicknesses, measuring 2.9 mm and 6.4 mm, were used to test the impact of different tube properties. The walls of the pipe are surrounded by a continuous glass surface that is encased in an epoxy resin matrix. Table 46.1 displays the GFRP tubes' dimensions and characteristics. A 50 kip compression balance weight sensor, a Celesco wire potentiometer (SP1-25), and an oscilloscope (Yokogawa DL750P) were some of the instruments utilized in this experiment. In the laboratory, the functionality, efficacy, and capacity of load cells and chord potentiometers are examined. To record the stresses in the steel bars and GFRP pipes upon impact, test strips were also positioned on each column.

Table 46.1 Dimensions and mechanical properties of fiber-reinforced polymer tubes

Tube type	Internal diameter (mm)	Thickness (mm)	Number of layers	Stacking sequence	E_1 (MPa)	E_2 (MPa)
I	152	2.9	8	[±60]4	20,860	8785
II	152	6.4	16	[±612, ± 432, ± 614]	23,630	9270
III	210	6.4	26	[±74, ± 604, ± 74, ± 604, ± 744]	24,720	10,250

46.4 Conclusion

Many excellent articles exist that make the case for enhancing the benefits of concrete stones. Spiral rebar, FRP, FRP-SR combinations, steel pipe, and UHPFRC coating are a few of these materials.

1. It has been demonstrated that impact loading can lessen the extent of damage by making each spiral reinforcement's shear strength stronger. This study also demonstrates how decreasing the brittle fracture mode using spiral reinforcing improves the beam's ductility. The impact force of circular reinforced concrete columns rapidly drops after the first peak value, according to a force time history examination of the columns.
2. Application of fiber reinforced polymer (FRP) restraint can enhance concrete's energy absorption, load bearing capacity, and ductility. It was shown that the results sent to the tower were enhanced by thickening the FRP pipe.
3. The combination of spiral and FRP material greatly improves impact resistance when compared to either material alone. If the line is required to carry a sufficiently high axial load, this configuration is also believed to be the most effective way to prevent interference.
4. It has been shown that hollow steel tubes may reduce residual forest loss by 40%, suggesting that their application might enhance concrete performance.
5. Under high-speed impact loading, UHPFRC coating significantly boosts concrete strength by fortifying columns' resistance to impact.
6. The steel pipe retainer increases the affected capacity relative to the concrete's axial impact resistance, shortens the impact length and maximum displacement, and significantly improves the column's axial impact. Because of its unique qualities, the steel pipe may increase the axial impact while having little effect on the column. It was demonstrated that using the FRP-SR combination to increase the axial impact resistance of columns is a smart idea. It has been demonstrated that raising the FRP thickness or SR volume % may also improve the resistance of the member and raise the system's fracture strength.

7. The lateral impact resistance of concrete panels has been considerably enhanced by the FRP-SR configuration. Future research will compare this GFRP construction with continuous concrete to examine how these beams react to impact loads.

References

Aghdamy S, Thambiratnam DP, Dhanasekar M (2016) Experimental investigation on lateral impact response of concrete-filled double skin tube columns using horizontal-impact-testing system. Exp Mech 56:1133–1153

Anderson C (2018) Uzbek asylum seeker gets life in prison for deadly truck attack in Sweden. New York Times. https://www.nytimes.com/2018/06/07/world/europe/sweden-truckattack-rakhmat-akilov.html

Burton W (2017) Ceremony set to honor I-40 bridge collapse victims. Muskogee Phoenix. https://www.muskogeephoenix.com/news/ceremony-set-to-honor-i-40-bridge-collapse

Buth CE, Brackin MS, Williams WF, Fry GT (2011) Collision loads on bridge piers: phase 2. Report of guidelines for designing bridge piers and abutments for vehicle collisions. Rep FHWA/TX-11/9-4973-2, Texas Transportation Inst, College Station, TX; 2011.victims/article_374575c6-27e0-582c-b632-a3fab0b6b319.html

Cai J, Ye J, Wang Y, Chen Q (2017) Numerical study on dynamic response of reinforced concrete columns under low-speed horizontal impact loading. Proc Eng 210:334–340

Chen Z, Wang J, Chen J, GangaRao H, Liang R, Liu W (2020) Responses of concrete-filled FRP tubular and concrete-filled FRP-steel double skin tubular columns under horizontal impact. Thin-Walled Struct 155

Demartino C, Wu J, Xiao Y (2017) Experimental and numerical study on the behavior of circular RC columns under impact loading. Proc Eng 199:2457–2462

Fan W, Xu X, Zhang Z, Shao X (2018) Performance and sensitivity analysis of UHPFRC-strengthened bridge columns subjected to vehicle collisions. Eng Struct 173:251–268

Gutschke L (2020) Rule woman dies after vehicle strikes bridge column near I-20 in north Abilene. Abilene Reporter News. https://www.reporternews.com/story/news/2020/06/26/rulewoman-traci-lynn-farwell-dies-after-vehicle-crash-north-abilene-near-i-20-interestate-20/3262790001/

Hogan J (2020) Woman crashes vehicle into bridge at M-24/I-69 in suicide attempt. The County Press. https://thecountypress.mihomepaper.com/articles/woman-crashes-vehicle-intobridge-at-m-24-i-69-in-suicide-attempt/

Huang L, Kasal B, Yan L, Sun X (2017) Impact behavior of concrete columns confined by both GFRP tube and steel spiral reinforcement. Constr Build Mater 131:438–448

Huang L, Kasal B, Yu T, Yan L, Gao C (2018) Experimental and numerical studies of CFRP tube and steel spiral dual-confined concrete composite columns under axial impact loading. Comp Part B 152:193–208

Pham TM, Hao H (2016) Review of concrete structures strengthened with FRP against impact loading. Structures 7:59–70

Qasrawi Y, Heffernan PJ, Fam A (2015) Dynamic behaviour of concrete filled FRP tubes subjected to impact loading. Eng Struct 100:212–225

Silva PF, Mesia WD, Marzougui D, Badie SS (2009) Performance evaluation of flexure impact resistance capacity of reinforced concrete members. ACI Struct J 106(5):726

Wang R, Han LH, Zhong T (2015) Behavior of FRP-concrete-steel double skin tubular members under lateral impact: experimental study. Thin-Walled Struct 95:363–373

Zhang X, Hao H, Li C (2016) Experimental investigation of the response of precast segmental columns subjected to impact loading. Int J Impact Eng 95:105–124; Omran M, Karani A (2022)

Cable loss performance investigation of cable-stayed bridge equipped with Roll-N-Cage isolator. Structures 41:1329–1344
Zhu Y, Yang H, Yang X, Sun F (2020) Behavior of concrete-filled steel tubes subjected to axial impact loading. J Construct Steel Res 173

The manufacturer's authorised representative in the EU is Springer Nature Customer Service Centre GmbH, Europaplatz 3, 69115 Heidelberg, Germany. If you have any concerns regarding our products, please contact ProductSafety@springernature.com

Printed and bound by CPI Group (UK) Ltd, Croydon, CR0 4YY
23/09/2026
02221536-0003